D0433996

Humic Substances in Soils, Peats and Waters

Health and Environmental Aspects

Humic Substances in Soils, Peats and Waters

Health and Environmental Aspects

Edited by

M.H.B. Hayes

The University of Birmingham, UK

W.S. Wilson

University of Essex, UK

Invitations for contributions were based on presentations from the international conference on Humic Substances in Soils, Peats and Water: Implications for Plant Growth, Industry and Sustainable Environment held in Dublin on 18 and 19 September 1995 and the meeting on the Role of Organic Matter in Sustainable Agriculture held in Silsoe on 14 and 15 April 1994.

Special Publication No. 172

ISBN 0-85404-699-2

A catalogue record for this book is available from the British Library

Published by The Royal Society of Chemistry,
Thomas Graham House, Science Park, Milton Road,
Cambridge CB4 4WF, UK

Printed by Athenaeum Press Ltd, Gateshead, Tyne and Wear, UK

Preface

The interest that is focused on soil organic matter as a carbon pool provided one of the stimuli for this Volume. The others arose from the renewed interest in the utilisation on (or the disposal to) land of sewage sludges and composts from municipal and industrial organic refuse. These matters were 'aired' at a special symposium sponsored by the Royal Society of Chemistry and the British and Irish Chapter of the International Humic Substances Society at the Spring meeting of the British Soil Science Society which celebrated the launch of the European Journal of Soil Science in 1994, and at a symposium supported by the same sponsors convened in University College, Dublin in September, 1995. Some who presented papers at these conferences were invited to contribute to this Volume. Others were invited to do so because their work was relevant to the theme.

Much is written and said about the contributions of fossil fuels to increased emissions of carbon dioxide, and about the depletion of a valuable carbon sink as tropical forests are felled, and yet it is recognised only by those close to the field that the input of carbon to the atmosphere from fossil fuels is only 8 - 9% of that from soil respiration. Each soil type has an organic carbon balance with a steady state related to the management applied. Pressures for increased soil productivity arise from an expanding population, and from market forces in the more affluent societies which seek inexpensive supplies of food. The practices needed to respond to the demands lead to the depletion of soil organic carbon, and thus to a decline in the innate fertility of soils. Civilisations have vanished in antiquity when over-use of soil led to the depletion of fertility, and the practises of modern agriculture would suggest that essential lessons have not been learned from history. A major objective should be to maintain soil organic matter levels close to those for the steady state characteristic of a soil under its natural climax vegetation. Such aims cannot realistically be met where soils are in long term cultivation, but consideration should be given to returning to soil all of the organic waste products that came from its use. The directives which prevent disposal of sludges at sea after 1998, and the land fill taxes in Britain are re-awakening interest in composting and in the utilisation of sewage sludges on land. Because modern agriculture is based on the extensive uses of inorganic fertilizers, and on 'efficient' industrial type practices which often have little consideration for the future, there will need to be incentives or legislation to convince producers of the benefits of supplying some of the needs of crops with organic manures.

A comprehensive treatise on the subject matter in the title is outside the scope of this Volume, and so we have tried to blend aspects of review with accounts of ongoing research. We feel that considerations of the uses and the behaviour of organic matter in soil and water environments should, in so far as possible, be based on an awareness of the compositions and structures of the components. Account is taken of these in Sections 1 and 2 with regard to different humic components. Many who have regarded organic matter as a "black box" to which some generalized properties of the gross mixtures can be assigned, and who have not been, by emotion or by physical efforts, involved in studies of fundamental aspects of their compositions and structures, are dismissive of the field. They should realise that awareness of the nature of the interactions of humic substances in the environment has come from an improved knowledge of composition and structure. Overall, the advances made in this generation have been encouraging, and those being made in this decade are striking. We have tried, therefore, to give to the readership an up-to-date account of aspects of humic compositions and structures, and to introduce some of the modern experimentation and reasoning that is being directed in the subject area.

Section 2, dealing with interactions, has concentrated on aspects of the experimental approach. Though this is the age of the modeller, we have reservations about applications of models when the modeller lacks awareness of structures of the major reactants. The structural/compositional properties for the substrate core in many models are naive, and in others these are irrelevant to the theme. The suspected role of humic substances in the transport of anthropogenic chemicals to waters is an example. Considerations in this Volume of that subject matter has used humic substances from drainage waters, and the results obtained run counter to concepts that are in vogue in the literature.

The environmental impacts of humic substances and organic matter (Section 3) are far reaching. We have not addressed the vital role of organic matter in the formation and stabilization of soil aggregates, but emphasis is given to humic substances of podzols because many soils of the cool temperate climates would, without management, tend towards podzols in the long term. Upland peats are major carbon sinks, and management practices which utilise peat for fuel, agriculture, or forestry will release carbon from the sink. Planting of blanket peats to forests has not taken account of the possible detrimental effects from the practice. Responses of peats to atmospheric acid depositions has also been ignored to a degree. Such important but shelved considerations are highlighted.

There has always been anecdotal evidence to link desirable health benefits to peats and to humic substances. Attempts are now being made to replace the anecdotal with scientific evidence. Epidemiological evidence links health damaging effects to the chlorination of waters rich in humic substances, and these effects are put in perspective in Section 4. The same applies for curative effects, but this science is in the early stages of development.

We would like to promote systems that amend soils with organic matter, and are not injurious to soil and to plant and animal health. The directives which will ban disposal of sludge at sea, and the embryonic interest in the utilization of composted municipal wastes may well boost this approach. Utilization of composts and sludges on land will have to compete with disposal by incineration and the winning of energy from the combustion. Dangers of contamination from anthropogenic chemicals and heavy metals in sludges and composts will always be present unless there is careful control of inputs to these. It is important to know the behaviour in soil of such substances. Thus we have included a review in Section 5, as well as research reports on the influences of heavy metals in sludges on plant growth and the soil accumulation of these. Composts and organic manures have growth promotional and plant disease suppression properties that cannot be predicted from their compositions, and aspects of these subjects are reviewed in Section 5.

We are indebted to our symposium sponsors, and also to the Perry Foundation, the Sigma BioSciences, Poole, Dorset, the Sheppy Fertilizers, and to others for support given in various ways. We appreciate the work of Dr. Jim Collins and his colleagues of the Department of Environmental Resource Management of University College, Dublin who managed the Dublin conference efficiently, and the help of the secretarial staff of the Department of Biological Sciences, the University of Essex. We especially thank Janet Freshwater of the RSC for her helpful advice and consideration during the preparation of this Volume, and also Dr. Barbara Watt of the School of Chemistry of the University of Birmingham for her involvement in and advice on editorial matters.

M.H.B. Hayes (The University of Birmingham)
W.S. Wilson (The University of Essex).

Contents

1 Compositions and Aspects of the Structures of Humic Substances 1

Emerging Concepts of the Compositions and Structures of Humic Substances 3
M.H.B. Hayes

Extractability, Chemical Composition, and Reactivities of Soil Organic Matter of Irish Grassland Soils 31
D. McGrath

Investigations of Some Structural Properties of Humic Substances by Fluorescence Quenching 39
I.P. Kenworthy and M.H.B. Hayes

Applications of NMR Spectroscopy for Studies of the Molecular Compositions of Humic Substances 46
A.J. Simpson, R.E. Boersma, W.L Kingery , R.P. Hicks, and M.H.B. Hayes

Effect of Ascorbate Reduction on the Electron Spin Resonance Spectra of Humic Acid Radical Components 63
D.B. McPhail and M.V. Cheshire

Humic Substances from Podzols under Oak Forest and a Cleared Forest Site I. Isolation and Characterization 73
A. J. Simpson, B.E. Watt, C.L. Graham, and M.H.B. Hayes

Humic Substances from Podzols under Oak Forest and a Cleared Forest Site II. Spectroscopic Studies 83
A. J. Simpson, J. Burdon, C.L. Graham, and M.H.B. Hayes

Studies of Humic Substances at Different pH Values Using Scanning Electron Microscopy, Scanning Tunnelling Electron Microscopy, and Electron Probe X-Ray Micro Analysis 93
A.J. Simpson, C.L. Graham, M.H.B. Hayes, K.A. Stagg, and P. Stanley

Investigations Into the Nature of Phosphorus in Soil Humic Acids Using ^{31}P NMR Spectroscopy 100
E.C. Norman, C.L. Graham, and M.H.B. Hayes

Dissolved Humic Substances in Waters from Drained and Undrained Grazed Grassland in SW England 107
T.M. Hayes, B.E. Watt, M.H.B. Hayes, D. Scholefield, C.E. Clapp, R.S. Swift, and J.O. Skjemstad

Soil Organic Matter: Does Physical or Chemical Stabilization Predominate? 121
J.A. Meredith

Humic Substances from Interment Sites I. Isolation and Characterization 136
M.H.B. Hayes, Lorraine J. Stewart, and P. Bethel

Humic Substances from Interment Sites II. Digest Products of Sequential Degradation Reactions 148
Lorraine J. Stewart, J. Burdon, and M.H.B. Hayes

Natural Abundances of ^{13}C in Soils and Waters 158
C.E. Clapp, M.F. Layese, M.H.B. Hayes, D.R. Huggins, and R.R. Alimaras

Humic Substances from a Tropical Soil 176
Thomas B. Rick Yormah and Michael H.B. Hayes

2 Interactions of Humic Substances 187

An Automated Procedure for Studying Sorption by Humic Substances Using a Combined Continuous-Flow and Flow-Injection Analysis Process 189
I.A. Law, J.J. Tuck, C.L. Graham, and M.H.B. Hayes

Sorption of Imidazolinone Herbicides by Humic Acid Preparations 199
Michael J. Häusler and Michael H.B. Hayes

The Influence of Humic Substances from Drainage Waters on the Transportation of Anthropogenic Organic Chemicals 208
T.M. Hayes, M.H.B. Hayes, and L.V. Vaidyanathan

Preparation of Iron Pillared Clays and Their Applications for Sorption of Humic Substances 219
A.L. Lacey, M.H.B. Hayes, and L.V. Vaidyanathan

Uses of Schists- and Clay-Derived Materials to Remove Humic Substances from Waters 226
J.J. Farnworth, L.V. Vaidyanathan, and M.H B. Hayes

Investigations of the Impacts of Humic-Type Substances on the Bayer Process 237
E.C. Norman, I.R. Dixon, C.L. Graham, M.H.B. Hayes, and S.C. Grocott

3 Environmental Impacts of Humic Substances and Organic Matter 247

A Review of the Characteristics and Distributions of Organic Matter-Related Properties in Irish Podzols 249
Declan J. Little

The Influence of Forestry on Blanket Peatland 262
Kenneth A. Byrne and Edward P. Farrell

Is Peat Chemistry Useful in Assessing Acid Deposition Effects on Mineral Soils? 278
Catherine C. White, Abdulkadir M. Dawod, and Malcolm S. Cresser

The Direct Relationship between Base Cation Ratios in Precipitation and River Water Draining from Peats 288
Abdulkadir M. Dawod, and Malcolm S. Cresser

The Effects of Acidification on the Carbohydrate Contents of Humic Substances from the Watershed Soils and Waters of Lake Skjervatjern 299
M.H.B. Hayes, I.P. Kenworthy, M.T. Quane, and B.E. Watt,

Decomposition in Soil of C4 and C3 Plant Material Grown at Ambient and at Elevated Atmospheric CO_2 Concentrations 311
A.S. Ball and J. Bullimore

Humic Substances Research in Polar Waters 319
R.J. Lara, U. Hubberten, D.N. Thomas, and G. Kattner.

The Effects of Humic Substances on Enhanced Chemiluminescence Produced by the HRP-Catalyzed Oxidation of Luminol 326
B.E. Watt, E.E. Robinson, K.E. Sawcer, G.H.G. Thorpe, and M.H.B. Hayes

4 Biological Impacts of Humic Substances 335

Does the Genetic Toxicity of Water Chlorination Products Pose a Risk to Public Health? 337
D. L. MacDonald and J.K. Chipman

Aspects of the Biochemistry of the Healing Effects of Humic substances from Peat 346
W. Flaig

5 Composts, Peats, and Sludges 357

Composts and the Control of Plant Diseases 359
Harry A.J. Hoitink and Marcella E. Grebus

Peat - A Valuable Resource 367
A.E. Johnston, M.V. Hewitt, P.R. Poulton, and P.W. Lane

Influence of Lime Stabilized Sewage Sludge Cake on Heavy Metals and Dissolved Organic Substances in the Soil Solution 410
Yongming Luo and Peter Christie

The Effect of Long-Term Annual Sewage Sludge Applications on the Heavy Metal Content of Soils and Plants 425
M.N. Aitken and D.I. Cummins

Effect of Repeated Application of Sewage Sludge to Pasture on Metal Levels in Soil and Herbage 438
E.G. O'Riordan and D. McGrath

Comparative Effects of Organic Manures on the Spring Herbage Production of a Grass/White Clover (c.v. Grasslands Huia) Sward 446
J. Humphreys, T. Jansen, N. Culleton, and F. MacNaeidhe

Cheese Whey Applications to Two Contrasting Soil Types: Effects on Grass Growth, Soil Properties, and Drainage Water Composition 462
M.V. Ross and G. J Mullen

The Use of Peat in Treating Landfill Leachate 475
H.J. Lyons and T.J. Reidy

Subject Index 487

Section 1

Compositions and Aspects of the Structures of Humic Substances

Humic substances (HS) are in all soils and waters, and their abundance in soils is of the order of two to three times greater than the living mass of organic matter (OM) on and above the surface of earth. HS, though probably the most resistant of the transformation products of living organic matter, will eventually transform to carbon dioxide and water. Their best protection from degradative influences is probably provided by the soil mineralogy (see the contribution by Meredith, p. 121 - 135, and by Yormah and Hayes, p. 176 - 186. This may be inferred also from the contribution by Hayes *et al.*, p. 136 - 147, and especially from the humified body outline in p. 137. Some of the influences which HS have in soil and water are referenced in page 5.

The genesis of HS is complex. A variety of chemical reactions and biological transformation processes are involved in their formation. There is not genetic control of their biological synthesis, and the chemical synthesis is likely to be random. Applications of $\delta^{13}C$ techniques (see Clapp *et al.*, p. 158 - 175) can help follow the transformations of plant tissues when mixtures of C3 and C4 plants are humified in incubation experiments, but the problem of 'sorting' out the mixture of humic products formed still remains.

For meaningful studies of composition and structure, it is desirable, as far as possible, to work with substances that are relatively homogeneous. It is unlikely, however, that it will be possible to isolate in amounts necessary for detailed structural studies sufficient quantities of HS that are pure compounds. At best, it might be possible to isolate substances that have a degree of homogeneity with respect to molecular size and charge density. There is discussion (p. 7 - 13) about procedures for the isolation and fractionation of HS from soils and waters, and aspects of this theme are dealt with further in p. 32 - 34, p. 74, p. 109, p. 125-127, p. 137 - 138, and for tropical soil samples in p. 178 - 179. None of the approaches described can claim to have isolated humic fractions that are homogeneous, even from the viewpoints of molecular sizes and charge density values.

There is genetic/biological control of the synthesis of the peptide and carbohydrate moieties associated with HS. These moieties can contribute 20% and more to the compositions of soil HS, as defined classically. In contrast, the contributions of these to aquatic HS are of the order of 10% of those for soils. There are several possibilities for associations between such biological molecules and HS. Van der Wals attractions and hydrogen bonding provide plausible physisorption processes that can be disrupted by uses of reagents such as urea, guanidine, and dimethyl sulphoxide (see p. 8, p. 138) and the sugar and amino acid contents of the classical humic fractions are lessened accordingly. There are possibilities too for covalent linkages giving, for example, phenolic glycosides, and imine structures. It is likely that peptide and carbohydrate moieties associated with, or forming part of the humic structures are protected sterically from microbial/enzymatic attack. Changes in the conformations of the HS molecules could allow access of enzymes to the hydrolyzable peptide and glycosidic linkages. The cleavage of such structures, should they link humic 'cores', would release smaller, water soluble humic molecules.

Studies of composition generally begin with isolation and fractionation, and this topic is covered by Hayes (p. 7 - 13). McGrath (p. 31 - 38) has dealt with the extractability, chemical composition, and reactivities of OM of Irish grassland soils. Emphasis was placed (p. 9) on the mechanism by which acids such as ethanoic acid can be used to break up the 'assemblies' of humic molecules, and this approach would suggest that the classical concepts of high molecular weight (HMW) HS may be wrong. Sodium pyrophosphate complexes metals which insolubilize HS, and this reagent can be used at different pH values for the isolation of HS. The study of Norman *et al.* (p. 100 - 106), using ^{31}P NMR, draws attention to the fact that P can be retained by the samples in some circumstances. Kenworthy and Hayes (p. 39 - 45) have interpreted the loss of protection provided by HS against the quenching of pyrene by bromide as an indication that the 'assemblies' of humic molecules arising from hydrophobic bonding are dispersed when the samples are treated with ethanoic acid followed by base. The concept that a pseudo HMW property arises from a 'self-assembly' of humic components is discussed in p. 11 - 12, and in p. 24..

We are in a decade in which rapid advances are being made in the humic sciences. That is not a reflection on the qualities of scientists who have worked with HS in previous decades. We live in an age of rapid advances in instrumentation and in computational capabilities, and developments, especially in nuclear magnetic resonance (NMR) instrumentation have had major impacts on the advances. Cross polarization magic angle spinning (CPMAS) ^{13}C NMR is now extensively used in studies of compositions of HS, and although one dimensional (1-D) NMR (^{1}H and ^{13}C) can give valuable information about functionalities in complex molecules such as HS, significantly more information is provided by 2-D techniques. The contribution by Simpson *et al.* (p. 46 -.62) outlines applications of 1-D and of 2-D NMR which will have increasing impacts in studies of humic compositions/structures. However, structural information about the more complex components will require better fractionation than has been achieved so far. The classical degradation techniques, and the uses of sequential degradation reactions, identifications of digest products, and interpretations based on degradative mechanisms (see p. 19 -22, and the contribution by Stewart *et al.*, p. 148 - 157) still have a role to play.

McPhail and Cheshire (p. 63 - 72) have described an elegant use of electron spin resonance (ESR) to differentiate between radical species that give rise to structured components and those representative of unstructured components.

Podzols provide HS where a degree of fractionation of the substances has taken place during the genesis of the soil profiles. It is likely that the humic components in the B_h horizons were formed, and partially 'matured' in the A_h horizons. Those which migrated to the B_h horizon were either solubilized, or formed a colloidal dispersion which could be transported within the profile. The sequence of papers by Simpson *et al.* (p. 73 -99) describes the isolation and partial characterization (p. 73 - 82) of humic fractions isolated from the A_h and B_h horizons of a podzol under oak forest, and from the B_h horizon of a similar soil where the forest was cleared about 400 y BP. The study has looked at the spectroscopic characteristics, and especially the fluorescence and proton NMR properties (p. 83 - 92). Samples for scanning electron microscopy, scanning tunnelling electron microscopy, and for electron probe X-ray micro analyses (p. 93 - 99) were prepared using an improved sample preparation procedure, and the results obtained may be of samples that represent the humic configurations/conformations that exist in nature.

The study by Hayes *et al.* (p. 107 - 120) has shown definite compositional differences between HS isolated in drainage waters at different times of year, and between those from waters collected as runoff and those which had passed through the soil.

Emerging Concepts of the Compositions and Structures of Humic Substances

M.H.B. Hayes

THE UNIVERSITY OF BIRMINGHAM, SCHOOL OF CHEMISTRY, EDGBASTON, BIRMINGHAM, B15 2TT, ENGLAND

Abstract

The innate compositional and structural complexities of humic substances have delayed our awareness of a detailed understanding of the mechanisms by which these operate in soil and water environments. The development of advanced scientific instrumentation, especially in the area of spectroscopy, has allowed very significant advances to be made in this generation in our understanding of composition. The information which is now emerging might indicate that HS are largely composed of a self-assembly of molecules which give a pseudo high molecular weight property. Gradually, because of the better awareness that is emerging with regard to composition, it might in this generation be possible to obtain a good working knowledge of the types of humic structures which react and interact with such significant effects in soil and water environments.

1 Introduction

Soil Organic Matter, Humic Substances, and the Carbon Cycle

Soil organic matter (SOM) is a very important pool of carbon in the global carbon cycle. Estimates for global soil organic carbon (SOC) range from 11 x 10^{17} to 30 x 10^{17} g (Schlesinger, 1984), and although the estimates are based on a large number of measurements it is very difficult to obtain accurate data because of the spatial variability of soils. The most widely accepted estimates would suggest that globally SOM is composed of 14 - 15 x 10^{17} g of C (Eswaran *et al.*, 1993; Schlesinger, 1984, 1995).

Schlesinger (1995) and others have provided useful diagrammatic representations of the global carbon cycle. These show that the largest exchange takes place between the atmosphere and land plants. About 120 x 10^{15} g of carbon y^{-1} is fixed by land plants (the gross primary production) and about 60 x 10^{15} g is respired. Thus the net primary production (NPP) is ca 60 x 10^{15} g, and from estimates of the terrestrial biomass (5.6 x 10^{17} g) the mean residence of carbon in live biomass is of the order of 9 y. Schlesinger has summarised how the annual oscillations in the concentrations of atmospheric carbon dioxide reflect the seasonal storage of carbon in vegetation.

Should the pool of organic carbon in soils be considered to be ca 15 x 10^{17} g, then this pool would be 2.5 - 3 times greater than the terrestrial biomass. In broad terms, the pool represents a steady state between the input of plant and animal debris from the NPP and losses from biological transformations (soil respiration) of the order of 68 x 10^{15} g. The excess (8 x 10^{15} g) over the input (NPP) from dead vegetation is attributed by Raich and Schlesinger (1992) to the contribution from plant respiration below ground. The flux of CO_2 from soil would suggest a mean residence time of 32 y for the carbon in SOM. This mean value will include the relatively short turnover time of readily bio-transformed organic substances and the long residence time of the more stable organic materials that are relatively intractable or/and are protected from enzymatic attack through their associations with mineral colloids. It is of interest to note that the input to the atmosphere of carbon from fossil fuels amounts to 6 x 10^{15} g y^{-1}, or 8-9% of that from soil respiration.

Labile plant materials are decomposed rapidly on entering the soil environment, whereas some more resistant components transform very slowly. It is impossible, because of the diversities in composition, and because of the differences in the states of transformation of the components, to give precise definitions for the gross mixture that composes SOM. Hayes and Swift (1978) have outlined a useful delineation between recognizable remains of plant/animal debris and the highly transformed materials which do not contain recognizable plant, animal, or microbial structures. All of the recognizable plant debris and identifiable classes of organic macromolecules, such as carbohydrates, peptides, and nucleic acids are classified as non-humic substances, and the highly transformed, amorphous, dark coloured materials are classified as humic substances (HS).

The recognizable plant remains constitute a small percentage of the SOM of mineral soils. In mineral soils the transformed materials, often referred to as humus, can constitute more than 90% of the SOM, and HS can compose 70-80% of the total. Carbohydrate, peptide, and lipophilic materials from plant and from microbial remains and processes are other products present. HS have a degree of resistance to further microbial degradation. That resistance can be attributed to the self associations of the molecules, to the protection provided by associations with the soil mineral colloids, and to entrapment in soil aggregates. In these situations availability to transforming microorganisms and to their degradative enzymes is impeded. Thus some of the labile molecules can survive. However, all are eventually biodegradable. If that were not so, the surface of the earth would be deeply covered in HS (Jenkinson, 1981), and waters would be dark coloured 'soups' of these. From the foregoing considerations it is clear that HS are by far the most abundant of the organic macromolecules of nature, and as indicated, their carbon contents can be expected to be two to three times greater than that in all living matter.

HS are present in all waters, and the compositions and amounts of these in terrestrial waters are related to the HS of the soils of the watersheds (Watt *et al.*, 1996). HS in estuarine environments are related to the terrestrial aquatic HS, but those in the oceans have different origins and different compositions (Harvey and Boran, 1985). Schlesinger (1995) estimated that the upper limit for the global formation of soil HS is of the order of 4 x 10^{14} g C y^{-1}, or ca 0.7% of the terrestrial NPP. That value is of the same order as the amount transported to the sea in riverine systems and would suggest a steady state system for SOM globally. The accumulation of HS on land is balanced by their loss to sea where they may ultimately be stored in ocean sediments (Lugo and Brown, 1986).

Cultivation of soils gives rise to a rapid depletion of the SOM, and between 20 and 40% of the SOM is lost when virgin soils are cultivated (Schlesinger, 1986; Mann, 1986). However, the rate slows after 20 years, and after 30 - 40 years of cultivation a steady state is approached (Johnston, 1991). The part that is decomposed is considered to be labile OM and the resistant material is referred to as the refractory fraction. Schlesinger (1995) estimated that mechanized agriculture gave rise to a loss of 36 x 10^{15} g of C between 1860 and 1960, and that the current rate of loss is of the order of 0.8 x 10^{15} g C y^{-1}. That, of course, represents a serious depletion of the reserves of SOM. However, according to Jenkinson *et al.* (1991), even more (up to 61 x 10^{15} g) may be lost from the global pool in soils as the result of global warming in the next 60 years.

The Roles of Humic Substances in Soil and Water Environments

Even though the HS contents in mineral soils are generally in the range of one to five per cent, these have significant influences on soil properties and performance. The reader is referred to Hayes and Swift (1978, 1990) and to Stevenson (1994) for detailed listings of the properties. SOM is vital to the formation and stabilization of soil aggregates, and Swift (1991) has outlined how polysaccharide and HS have important but different roles in the formation and in the stabilization of the aggregates. Polysaccharides are involved in the aggregate formation process, and HS are involved in their stabilization. HS are important as cation exchangers, in the release of plant nutrients (especially, N, P, and S) when mineralized, and in the binding of anthropogenic organic chemicals (AOCs). The biological activity of most aromatic AOCs is decreased or lost when these are contacted by HS. HS in waters do not present a hazard to health. In fact there are schools of research (see Flaig, p. 346 - 354, this Volume) which indicate that HS can benefit health and have healing effects. However, when waters are chlorinated, dissolved HS can give rise to products which can have deleterious effects on health (Watt *et al.*, 1996; MacDonald and Chipman, p. 337 - 345, this Volume).

Objectives

Humic substances are derived from a variety of precursors and processes and are therefore heterogeneous mixtures. For meaningful studies of structure it is desirable to deal with pure substances. The best that can be hoped for in so far as the homogeneity of HS is concerned is to isolate substances that are homogeneous with respect to size and to charge density, and that goal has not yet been fully attained. It is proposed to review briefly the classical attempts (pre 1990) at the isolation, fractionation, and approaches to studies of composition and structure, and then to discuss the advances which are currently being made that should lead to a better awareness of composition and of aspects of structure.

2 Definitions

The classical definitions of HS are operational, and may be considered to be the contents of SOM that are solubilized in base. Aiken *et al.* (1985) defined ***humic substances*** as *"a general category of naturally occurring heterogeneous organic substances that can generally be characterized as being yellow to black in color, of high molecular weight, and refractory"*. That broad definition is still valid but, as will be evident from what follows in Sections 4 and 5 of this manuscript, the use of the term high molecular weight (HMW) may not be as applicable now as it was even a few years ago. The terms humic acids (HAs), fulvic acids (FAs), and humins are still used to describe fractions of HS.

There is not a clear cut off point between these fractions. HAs and FAs, especially, can be considered to be a continuum, and the cut off point (pH 1 for soils) is arbitrarily chosen.

Humic Acids (HAs) are defined by Aiken *et al.* (1985) as *"the fraction of HS that is not soluble in water under acid conditions but becomes soluble at greater pH"*. A pH value of 2 is the level of acidification used by water scientists, and soil scientists generally use pH 1 as their standard. In the definition by Aiken *et al.*, specific mention is made to the 'fraction of humic substances', and as discussed in Section 1, HS do not include biomolecules such as peptides, sugars, nucleic acid residues, fats, etc. Some such molecules can be sorbed to or co-precipitated (at pH 1 or 2) with the HAs. Häusler and Hayes (1996) have shown that by dissolving HAs in dimethyl sulfoxide (DMSO) containing 1% HCl, and passing the solution onto an XAD-8 [(poly)methylmethacrylate] resin column, the amino acid and neutral sugar contents of the HAs can be decreased significantly. Hence, the precipitate at pH 1 is better described as the *"humic acid fraction"*. The procedure of Ping *et al.* (1985), using XAD-8 resins, may be employed to remove from the HAs non-humic molecules that are associated (but not by covalent links) with these. They dissolved the HA fraction precipitated at pH 1 in 0.1M NaOH, then diluted the solution to < 100 ppm, and passed it onto an XAD-8 resin column. The polar non-humic components of the HA fraction passed through the resin, and the HAs which sorbed to the resin were recovered by back elution in base, H^+-exchanged by passing through a cation exchange resin (in the H^+ form), and recovered by freeze drying. A modification of this approach is described by Hayes *et al.* (1996), and is refered to in Section 3 of this manuscript.

Fulvic Acids (FAs) are defined by Aiken *et al.* as the *"fraction of humic substances that is soluble under all pH conditions"*. Soil scientists have considered FAs to be the fraction which stays in solution when soil extracts in aqueous base are adjusted to pH 1. Materials from the acidified base will contain non-humic materials. Thus the extracts are best defined as the *"fulvic acid fraction"*. In the IHSS (International Humic Substances Society) procedure for isolating FAs (p. 9) the acidic FA fraction is passed on to XAD-8 resin, then treated as described for the HAs.

Humin, in the definition of Aiken *et al.* (1985) is *"that fraction of humic substances that is not soluble in water at any pH value"*. On the basis of that definition, humin can include any humic-type material that is dissolved in non-aqueous solvents after the soil has been exhaustively extracted with basic aqueous solvents. It is generally considered that the humin substances are intimately associated with the soil inorganic colloids. Clapp and Hayes (1996) have carried out a sequential extraction in which the soil was exhaustively extracted at one pH value before passing to an aqueous solvent at a higher pH. After application of the final aqueous solution (0.1M NaOH at pH 12.6), a DMSO/HCl medium was applied. Although the solvent-soluble substances could normally be classified as humin, the properties of the materials isolated were similar to those of HAs/FAs. It was concluded that the materials did not dissolve in the aqueous media because the polar 'faces' of the molecules had intimate associations with the inorganic colloids and exposed to the exterior the hydrophobic moities of the macromolecules. It could also mean that strong associations between the polar moieties exposed to the outside hydrophobic faces which were not solvated in the polar aqueous media.

There is a need to provide more precise classifications and definitions of HS and of the various fractions of HS. Water scientists have used XAD-8 and XAD-4 (styrenedivinylbenzene) resins to isolate the HAs, FAs, and the so-called XAD-4 acids from waters, and Malcolm and MacCarthy (1992) have described the uses of the XAD-8 and XAD-4 resins in tandem for the isolation of HS from water. The principle involves applying the water to the resins at pH 2 (at which the acidic functionalities are not dissociated). Under these conditions hydrophobic interactions take place between the solute materials and the resins. The less polar fractions (which include the HAs and the FAs) are sorbed to the XAD-8, and some materials which are more hydrophilic are sorbed to the XAD-4. Fractionation on the basis of charge density differences can be achieved by back eluting with aqueous solvents of different pH values. The material that is not eluted in aqueous base can be removed in ethanol and in acetonitrile, and that organic soluble fraction is called the ***"Hydrophobic Neutrals"***.

The works of Ping *et al.* (1995), and of Hayes *et al.* (1996) have shown that HAs can be fractionated using the XAD-8 resin approach. Hayes *et al.* diluted their solution (in base) to ca 20 ppm, adjusted the pH to 2, and applied the soluble materials to the resin. Some precipitation can take place in the pH range 2.5 to 2.0, and the precipitated materials are very different from those that are soluble at pH 2 (see Tables 1 and 2).

XAD-4 Acids are not included in the classical definitions of the HS fractions. Some of the hydrophilic fractions which are not sorbed by XAD-8 are retained by the highly non-polar XAD-4 resin, and the most polar fractions are washed through. Back elution in dilute base desorbs the so-called XAD-4 acids from the resin. The substances which do not elute in base but are eluted in ethanol/acetonitrile are called the ***"Hydrophilic Neutral"*** fraction.

The author is not convinced that the XAD-4 acids represent a legitimate fraction of HS. It is certain that they do contain some brown materials that can be classified as HS, but the high neutral sugar (NS) contents (Table 2) suggest that carbohydrate-related materials are major components of the fraction.

3 Isolation of Humic Substances from Soils and Waters

Isolation of HS from Soils

Hayes (1985) has presented a detailed account of the principles and procedures which apply for the isolation of HS from soil, and recently Swift (1996) has reviewed methods for the isolation, fractionation, and characterization of soil HS. The reader is referred to this recent publication by Swift, and it is intended here to refer only to the principles involved in the isolation processes and to outline new isolation procedures that have emerged since the middle of the last decade.

Criteria for good solvents for the isolation of HS from soils are reviewed in the articles by Hayes (1985) and by Swift (1996). Solubility is best achieved through solvation of anionic functionalities in the HS, and the abundances of these increase as the pH is raised. However, the nature and the charge of the cations balancing the negative charges on the HS are important. Aspects of the behaviour and properties of H^+-exchanged HS (achieved by use of mineral acids) will be similar to those of neutral polar macromolecules that are hydrogen bonded to each other. The combination of hydrogen bonding and associations of the non-polar moieties through hydrophobic bonding/van der Waals forces, and charge transfer processes provides the molecules with a pseudo high molecular weight (HMW) property, and only the most polar and the least associated (with other humic molecules) of these will dissolve in water. As the pH of the medium is raised, the most strongly acidic functionalities are first to dissociate, and the conjugate bases (generally carboxylates) then solvate. As the pH is raised further, less strongly acidic functionalities dissociate and solvate, and eventually the most weakly acidic functionalities (some phenols, enols) dissociate and solvate. Use can be made of these principles to fractionate HS on the basis of charge density differences (vide infra).

The dominant exchangeable cations in agricultural soils are di- and trivalent (especially Ca^{2+}, Mg^{2+}, Fe^{3+} and Al^{3+}), and these cations are strongly held by the humic anionic functionalities. Such cations form inter- and intramolecular bridges between anionic sites, suppressing repulsion, and inhibiting solvation. Cation bridging does not take place when charge balancing metal cations are monovalent. Thus repulsion between the charged species takes place, the macromolecules (or the molecular associations) assume expanded conformations, and solvation can occur readily (provided the ratio of charged to neutral moieties is adequate). The cationic bridging (as well as the hydrogen bonding) and the hydrophobic interactions/van der Waals association effects (which apply also for the H^+-exchanged species) cause the molecules to assume shrunken (or condensed) conformations, and water is partially excluded from the matrix. These effects can be overcome by replacing the divalent/polyvalent cations by monovalent species (other than H^+). Thus, the addition of sodium pyrophosphate ($Na_4P_2O_7$) to the system allows the pyrophosphate to complex the divalent/polyvalent cations and the sodium to neutralise the negative charges. The charged species then solvate in water. These principles were behind the approach used by Clapp and Hayes (1996) and by Hayes (1996) for the isolation of HS from soils. Exhaustive extraction with water can be expected to isolate the highly charged, and predominantly FA and XAD-4-type acids. The use of sodium pyrophosphosphate (Pyro, 0.1M) adjusted with mineral acid (often phosphoric) to pH 7 complexes the divalent/polyvalent metals and the sodium ions balance the charges on the organic acids that are dissociated at pH 7. The result is that the humic fractions with relatively high charge densities (contributed from dissociations of carboxyl groups) are solvated. The next convenient solvent in the sequence is 0.1M Pyro (pH of 10.6), and in theory that solvent system would cause phenols to dissociate and to be solvated. The final aqueous solvent in the sequence (0.1M Pyro + 0.1M NaOH) would be expected to cause very weak acids (including enols) to dissociate and to solvate. Clapp and Hayes (1996) carried out a final single extraction with dimethyl sulphoxide (DMSO, 94%) and 12M HCl (6%). DMSO, a dipolar aprotic solvent, is a poor solvent for anions but a good solvent for cations (Martin and Hauthal, 1975, p. 131). Although the metal cations which neutralise the charges on humates would be solvated, the conjugate bases (carboxylates and phenates) would not. Hence DMSO is a poor solvent for ionized humates. When the humates are H^+-exchanged, however, they behave essentially as polar molecules. These become associated through hydrogen bonding (as well as van der Waals forces). The strong acid in the medium H^+-exchanges the conjugate base sites. There are strong interactions between the polar face of DMSO and water, carboxyl, and phenolic hydroxyl groups (Martin and Hauthal, 1975, p. 77), and DMSO-water interactions are stronger than the associations between water molecules. Thus DMSO will associate with the phenolic and carboxyl groups to break the intra- and intermolecular hydrogen bonds. The non-polar face of the solvent can associate with the hydrophobic moieties in the macromolecule, and the combination of properties of hydrogen bond breaking and disruption of hydrophobic associations makes DMSO a very effective solvent for HS. The removal of the solvent does not present a problem because the HS will sorb to XAD-8 resin and the DMSO and acid are washed through. (The resin procedure for treatments of DMSO extracts is described by Häusler and Hayes, 1996, and by Clapp and Hayes 1996).

Hayes (1985) concluded that good organic solvents for HS have electrostatic factor (the product of the relative permittivity and dipole moment) values greater than 140, and pK_{HB} (a measure of the strength of the solvent as an acceptor in hydrogen bonding) values greater than 2. Dimethylformamide (DMF) and DMSO meet these requirements and both were shown to be good solvents for H^+-exchanged HAs, with DMSO the better of the two. Hayes has also discussed applications of solubility parameter data in order to explain differences in the abilities of organic solvents to solvate HAs. The best of the organic solvents employed for dissolving H^+-exchanged HAs had d_p (dispersion force), d_h (hydrogen bonding), and d_b (proton acceptor) parameters greater than 6, 5, and 5, respectively. Solvation is greatest when the product of d_a (solvent) x d_b (solute), or *vice versa* are maximum. Water has very large values for d_h, d_a, and d_b which reflect the extents of self association of water molecules as the result of hydrogen bonding. Thus, the d_a or d_b values of H^+-exchanged HAs are not sufficient to disrupt these attractive forces, and although water also satisfies all of these criteria, it is a poor solvent for H^+- and divalent/polyvalent cation exchanged HAs. An alternative to the exhaustive sequential extraction procedure using a series of solvents would be to extract with the NaOH/Pyro system, and subsequently to fractionate by eluting at different pH values from XAD resins (*vide infra*). Extraction with DMSO/HCl would follow the exhaustive extraction in base.

Swift (1996) has described the procedure used to isolate the soil HA and FA standards of the IHSS. The procedure involves H^+-exchanging the soil (using HCl), centrifuging, and separating the supernatant (FA Fraction 1), adjusting the soil residue to pH 7 (with 1M NaOH), then extracting with 0.1M NaOH (under dinitrogen gas) using a soil to solution ratio of 1:10, centrifuging and adjusting the pH of the supernatant to 1 (using 6M HCl). The HAs are precipitated and separated from the FA fraction (Fraction 2, in solution) by centrifugation. The HA fraction is redissolved in 0.1M KOH (under N_2) and KCl is added to a concentration of 0.3M $[K^+]$, and then centrifuged to remove solids. The HA is reprecipitated (6M HCl) at pH 1, and the HA precipitate is suspended overnight in 0.1M HCl/0.3M HF (plastic container). After centrifugation the HCl/HF treatment is repeated as often as necessary to lower the ash content to $< 1\%$, the precipitate is dialyzed, then freeze dried. The HCl/HF treatment followed by dialysis leads to considerable losses of the humic fractions. Clapp and Hayes (1996) and Hayes *et al.* (1996) found that filtration through partially clogged 0.45 μm or 0.2 μm filters was very effective in lowering the ash content (mainly finely divided inorganic soil colloids) of the humic samples.

The FA (supernatant) fraction is passed onto a column of XAD-8 , and the FAs are sorbed to the resin and the polar associated components in the FA fraction are washed through. The FAs are then recovered as described for water samples (*vide infra*).

Ping *et al.* (1995), Hayes (1996), and Hayes *et al.* (1996) have combined the classical procedures for the isolation of HS from soils with those for the isolation of HS from water. The procedures used by Hayes involved sequential and exhaustive extractions in water, Pyro, and Pyro + NaOH, neutralizing to pH 7, filtering, diluting the filtrates to < 2 mg L^{-1}, and passing each fraction through XAD-8 and XAD-4 resins in tandem (Malcolm and MacCarthy, 1992), and processing in the manner used to isolate HS from waters.

Isolation of HS from Waters

Aiken (1985) has described filtration procedures and applications of co-precipitation, ultrafiltration, reverse osmosis, solvent extraction, ion exchange, and sorption, including uses of alumina, carbon, and non-ionic macroporous and weak anion exchange resins (with secondary amines and their salts providing the active functionalities).

The XAD Resin Procedure. Leenheer (1981) has referred to six fractions of dissolved organic carbon (DOC). These are the hydrophobic acids, bases, and neutrals, and the hydrophilic acids, bases, and neutrals. The hydrophobic fractions are retained by XAD-8 resins, and that principle was used to isolate the IHSS Standard aquatic HAs and FAs from the Suwannee River. The procedure is described by Thurman and Malcolm (1981), by Thurman (1985), and is appropriately dealt with by Aiken (1985). It involves pre-filtering the samples through 0.45 μm or 0.2 μm filters (the effectiveness of the filtration is enhanced as the filter pores become partially clogged), adjusting the the filtrate to pH 2 in order to suppress ionization of the acidic groups, and introducing the water to the resin. The binding mechanism involves some (weak) hydrogen bonding between the HS and the ester functionalities of the resin, and hydrophobic/van der Waals forces. The sorbed HS are recovered by back elution from the XAD-8 resin using 0.1M NaOH, and lowering the pH immediately to avoid oxidation of the materials. The HS are reconcentrated on a smaller XAD-8 column until the DOC is greater than 500 mg L^{-1}, the pH is adjusted to 1.0 with HCl to precipitate the HAs, and the HAs are washed to remove Cl^-. The HAs are dissolved in 0.1M NaOH, and H^+-exchanged by passing the solution through a cation exchange resin (in the H^+-exchanged form). The FAs are reapplied at pH 2 to the XAD-8 resin and desalted by rinsing the column with a void volume of distilled water, then back eluting with 0.1M NaOH. The eluate is H^+-exchanged by passing through the cation exchange resin (H^+-form), and the HAs and FAs are freeze dried. The hydrophobic neutrals are not back eluted in the aqueous base but can be recovered using solvents such as ethanol or ethanolic NaOH. The procedure for cleaning the resin (soxhlet extraction using acetonitrile and ethanol) desorbs tightly bound materials. However, the heating processes involved in soxhlet extraction gives artefacts.

Malcolm and MacCarthy (1992) have described an XAD-8 and XAD-4 resin in tandem process to isolate the hydrophilic as well as the hydrophobic humic fractions. The so-called XAD-4 acids are recovered in the back eluate (with 0.1M NaOH), and processed as outlined for the eluates from XAD-8, and the hydrophilic neutrals (which are not removed in the base) are recovered using the organic solvent system.

The Reverse Osmosis (RO) Procedure. The RO process concentrates OM readily from large volumes of water under ambient conditions (Deinzer *et al.*, 1975). Serkiz and Perdue (1990), and Sun *et al.* (1995) have described instrumentation which processes 150 - 200 L of water h^{-1} with 90% recovery of organic carbon without exposing the DOM to harsh chemical reagents. Because all the solutes, including inorganic salts and non-HS, are concentrated, the concentrates must be subjected to further processing. The resin treatments can be employed for further processing of the concentrates.

4 Fractionation of Humic Substances

Hayes and Swift (1978) have reviewed the techniques that were in operation 20 years ago for the fractionation of soil HS, and Swift (1985, 1996) and Stevenson (1994) have referred to the techniques that are more generally used nowadays. Leenheer (1985) and Thurman (1985) have reviewed the principles and the procedures used to fractionate HS from water. The techniques that apply for the fractionation of aquatic HS apply equally well to the fractionation of those from soil, and no attempt is made here to describe procedures specific for soil or for water HS.

Considerable emphasis has been placed on the uses of gel chromatography, or gel permeation chromatography (GPC), for the fractionation of HS on the basis of molecular size differences (MSD), and extensive reference is made to applications of this technique by Swift and Posner (1971), Cameron *et al.* (1972a), Hayes and Swift (1978), Swift (1985). Swift (1996) has provided a diagrammatic representation of the tedious procedure required to produce fractions that have a degree of molecular size homogeneity. In principle the procedure involves the reprocessing (through the gel column) of substances eluted between specific volume boundaries from a standardized gel column until all of the material is contained within these volume boundaries. In theory it should be possible to isolate several homogeneous fractions in that way. However, as the fractions are reprocessed, smaller sized components are released. At best the reprocessing gives concentrations of components that would appear to have similar molecular sizes. Recent work by Piccolo *et al.* (1996) suggests that it might be appropriate to re-evaluate the effectiveness of GPC for the fractionation of HS on the basis of MSD.

Piccolo *et al.* (1996) subjected two HAs from different sources to GPC and collected, dialyzed, and freeze dried the components eluted in 0.02M borate ($Na_2B_4O_7$) in the void volume of a Sephadex G-100 gel. [This material was considered to have a nominal molecular weight (MW) value > 100 000 D]. From this material a stock solution was made by suspending the HAs in water and raising the pH to 11.8 (using 0.5M KOH), and storing under N_2 gas. When these solutions were eluted in a 0.02M solution of $Na_2B_4O_7$ through an LKB K 16-70 analytical GPC column packed with Biorad P100 Biogel (molecular range 5-100 KD) in the same buffer, a large peak was obtained at an elution volume corresponding to the void volume (V_o) of the column, and a smaller peak (eluted at a larger volume) indicated that materials of smaller molecular sizes had separated from the macromolecules in the stock solution. When methanoic, ethanoic, propanoic or butanoic acids were added to adjust the pH of the stock solution to 2.1, and when this material (at pH 2.1) was applied to the column, all of the HA materials were eluted (in the borate solution) in a volume that suggested that the MW of the materials was < 25 KD. Benzoic acid did not affect the elution volume. When the pH was readjusted from 2.1 to 3.5, 4.0, 6.0, and 8.5, and the materials were transferred to GPC columns and eluted with the borate buffer, it was evident that reassociations of the humic molecules had taken place at pH 4.5 and above, and most of the substances were eluted in the column void volume. Mineral acids did not have any influence in the GPC performance of the HAs.

Wershaw (1986) considered that hydrophobic bonding causes humic molecules to associate in micelle-like aggregates. Burchill *et al.* (1981) have used the concept of micelles to explain associations between hydrophobic organic chemicals and surfactants. As the concentration of surfactant in solution is increased, a break occurs in the continuity of the plot of the specific conductance per g-equivalent of solute (equivalent activity) versus the square root of the normality of the solution. The break takes place at the *critical micelle concentration* (CMC). The free energy of the system is increased when the structure of the solvent is disrupted during additions of the surface active material. This increase is minimized when the molecules concentrate at the surface and orientate so that their hydrophobic groups are directed away from the liquid. Alternatively, if the surface active molecules aggregate into clusters or micelles in which the hydrophobic groups are oriented towards each other and towards the interior, and the hydrophilic functionalities are oriented towards the exterior, clusters or micelles are formed and the free energy increase is minimized. The hydrophilic groups in the micelle structures can interact through hydrogen bonding, and through dipole-dipole interactions with the water molecules. Entropy is increased as water structure is disrupted, and is decreased with micelle formation. There is, however, a favourable increase in energy when roughly spherical micelles are formed (with radius roughly equivalent to the length of the hydrophobic group in linear surfactants) when the solution concentration is less than ten times the CMC. At higher concentrations non-spherical micelles can form.

It can be assumed that at the lower pH values inter-micelle associations might be expected to take place also between the non-dissociated acid groups of the outer core of the micelles. This would give rise to large molecular associations which would exceed the capacity of the associations to remain soluble in water. In the concept of micelles, the reasoning of Piccolo *et al* (1996) would suggest that the organic acids would penetrate into the inner (hydrophobic) core of the micelle structures, and thereby give a potentially pseudo high charge density molecular structure. This association between the organic acids and the humic macromolecules can be considered to be feasible because the acids have a hydrophobic as well as a hydrophilic face, and the hydrophobic face would associate with the hydrophobic functionalities in the inner core. As the pH is raised the sorbed acids in the inner core, and the acidic functionalities in the outer core of the HAs would dissociate, and the aggregate and the micelle structures would be "blown apart" because of repulsion between the negatively charged groups.

It might be argued that the compositions of humic molecules are too heterogeneous to form regular micelle structures, and so it may be more appropriate to consider self association phenomena which could also give rise to pseudo high MW molecules.

Other fractionation techniques that are based on molecular size/shape differences include dialysis, and ultrafiltration (uf). Membranes are available with pores from the nm to the mm sizes. In theory, applications of a range of such membrances should allow fractions of relatively discrete molecular sizes to be isolated. However, the question of self associations would also apply to applications of uf techniques.

Fractionation based on charge density differences includes applications of electrophoresis, which involves the movement of charged species in solution in response to an applied electrical potential. Duxbury (1989) has reviewed applications of the different electrophoretic separation methods, including zone electrophoresis, moving boundary electrophoresis, isotachophoresis, and isoelectric focusing (IEF). Preparative column electrophoresis (Clapp, 1957), and continuous flow paper electrophoresis techniques (Hayes *et al.*, 1985) have been used in studies of HS. Although separations of polysaccharide from coloured humic components were obtained, fractionation of the coloured humic materials was not achieved.

Ciavatta *et al.* (1996) have looked into the influences of the carrier ampholite (CA) on the applicability of IEF for the preparative fractionation of SOM. They concluded that the possible formation of complexes between the CAs and the HS makes unreliable applications of the technique to preparative fractionations.

The data of Trubetskoj *et al.* (1996) suggest that polyacrylamide gel electrophoresis (PAGE) can provide a fractionation of HS, and that the fractionation is based on MW differences. HAs were fractionated into three zones when subjected to PAGE. When the HAs were eluted from Sephadex G-75 in 7M urea three fractions were collected, and each of these fractions when subjected to PAGE gave a pattern that corresponded to one of the zones in the electrophoretogram of the unfractionated material.

Klavins *et al.* (1996) have investigated the uses of several different sorbent and solvent systems for the fractionation of HS, and have concluded that the best separation efficiencies are based on the uses of sorbents which could engage in hydrophobic interactions. Good separation was obtained using a sorbent with a polymer matrix and a weakly basic (1,2,4-aminotriazole) functionality. This provided a facility to interact with the hydrophobic and the anionic functionalities. The eluent was a gradient of 0.1 to 10% dioxane in water. Five fractions were isolated, each with different elemental compositions, different capabilities to complex metals, and different spectral characteristics.

Reference was made (p. 7 *et seq*) to applications of the XAD resins which provide useful fractionations based on charge density differences. In the cases of XAD-8, some hydrogen bonding is expected between the carboxyl (on the HS) and the ester functionality (on the resin), and hydrophobic bonding will also take place. As the pH is raised, the strongly acidic functionalities dissociate first. Where such functionalities are in close proximities, charge repulsion effects will change the conformations of molecules in contact with the resin, and as the molecules solvate they are removed in the effluent. As the pH is raised further, acidic functionalities of higher pK_a values will dissociate, solvate, and the same principles will apply for their removal. Finally, at the higher pH values, acidic functionalities rich in weakly dissociable acidic groups will be removed.

The classical fractionation of HS is based on solubility differences in aqueous media at different pH values. In principle, the HA fractions which have weakly dissociable acids should be first to precipitate as the pH is lowered, and the strongest acids would be last.

Additions of salt decrease the intra- and intermolecular charge repulsion giving rise to a shrinking of the molecules and exclusion of water from the matrix. Also the suppression of the electrical double layer allows the molecules to approach more closely, and to promote coagulation (which may be expressed as the formation of self association or pseudo micelle-type structures). These are considered to be salting out effects. Selective precipitation can be achieved using heavy metals.

Mention was made of the properties of good organic solvents for HS, and of the fact that the best solvents are in the dipolar aprotic class. However, HS have low solubilities in most of the conventional solvents (see Hayes, 1985, p. 355), and some of those solvents such as ethanol, methanol, and acetone can be used to fractionally precipitate HAs from alkaline solutions. Attempts to partition HS between alkaline solutions and non-(or sparingly) miscible organic solvents can give a concentration of HS at the water-solvent interface. The hydrophobic components of the HS are attracted to the organic solvent at the interface, but the polar moieties are not solvated and hence the HS do not cross the interface. The development by Eberle and Schweer (1974) of a hydrophobic extractable ion pair by dissolving long chain tertiary or quaternary amines in chloroform and extracting HS and lignosulphonic acids from water at pH 5, then recovering the HS at pH 10, gave promise of applying counter current distribution techniques for the fractionation of HS. Such techniques are tedious and more emphasis has been given to fractionations based on molecular size and charge density differences.

5 Compositions of Humic Substances

The compositions of HS refer to the elemental composition, to the functional groups, and to the 'building blocks' or the molecules which compose the humic structures. The results from analyses of data relating to these compositional characteristics are most meaningful when the samples analysed and compared have been subjected to the same isolation and fractionation procedures. It is for that reason that the IHSS have isolated (p. 9), using specified procedures, HA and FA samples from water (from the Suwannee River), from soil (from a Mollisol), from a sapric histosol (from the Everglades, Florida) and from lignite. The approach allows researchers to obtain standard samples and source materials from the IHSS, and to compare the compositional data obtained using their own instrumentation and approaches with those from the analyses provided by the IHSS. Also, using the same procedures, it allows comparisons of the standard samples with those obtained from different sources. Reference humic materials are also available, as well as information about the procedures used to isolate, fractionate, and to characterize these.

The procedures for ***elemental analyses*** are well documented. Refer to Huffman and Stuber (1985) for the procedures used for determinations of moisture, ash, and the elements of the Standard and Reference IHSS samples.

Hayes and Swift (1978) have reviewed and referenced the *wet chemical* and some of the spectroscopy procedures that were used 20 years ago for determinations of functional groups in HS, and Swift (1996) has updated that information. Comprehensive reviews have been presented about acidic functionalities (Perdue, 1985), and about applications of various spectroscopy procedures, such as infrared (MacCarthy and Rice, 1985), both proton and ^{13}C nuclear magnetic resonance (NMR) (Steelink *et al.*, 1989; Wilson, 1989; Malcolm, 1989), electron spin resonance (ESR) (Senesi and Steelink, 1989), and Bloom and Leenheer (1989) have reviewed vibrational, electronic, and high energy spectroscopic methods for characterizing humic substances.

Acidic, aromatic, and aliphatic hydrocarbon functionalities provide the major components in HS. The charge, and hence the cation exchange capacities (CEC) of HS are pH dependent, and the CEC increases as the pH of the medium is raised. Leenheer *et al.* (1995) have shown that there are some very strong acidic functionalities (provided by certain activating substituents alpha to carboxyl groups), and Perdue (1985) has stressed how identical functional groups can have different pK_a values because of the influences of the chemical environments on the functionalities.

IR spectra show broad bands characteristic of functional groups in a variety of environments. Diffuse reflectance Fourier-transform IR instrumentation (DRIFT) improves resolution and allows analysis of whole soil samples. Neimeyer *et al.* (1992) have provided DRIFT spectra of IHSS Reference HAs, and these spectra emphasise that that IR provides only limited information about the compositions of HS.

Raman spectroscopy is used for studying vibrational modes and is complimentary to IR spectroscopy. Applications in the past of Raman spectroscopy to the study of HS were inhibited because of the fluorescence properties of the HS. Very recently Yang and Wang (1997) sorbed humic samples on the wall of a truncated NMR tube and spectra were recorded in air and at room temperature using a Brucker IFS-66 spectrometer connected with a Raman module. Radiation of 1064 nm from an Nd:YAG laser was used for the excitation with a laser power of 200 mW, and a back scattering geometry was employed. The FT-Raman spectra gave two broad bands of varying intensities at ca 1300 cm^{-1} and at ca 1600 cm^{-1}. The authors interpreted their results as indicative of graphite-like quality for the HS. However, the FT-Raman spectra of the samples in their acidic forms were different. Downshifts were observed that could be attributed to amorphism of the samples. On neutralizing the samples, the spectra were typical of graphite-like carbons.

When organic species are sorbed on noble metal surfaces their Raman signals are enhanced 10^3- 10^7-fold, and the effect is called Surface-Enhanced Raman Spectroscopy (SERS). Current work by Liang *et al.* (1997) is using SERS for studies of binding sites and of the chemical characterizations of HS. They have obtained excellent signal to noise ratios, and have observed several distinct bands. For example, they have seen bands at 1594 and 1586 cm^{-1}, and at 738 and 1190 cm^{-1} corresponding to in-plane ring vibrations, and bands at 1038, 1026, and 1010 cm^{-1} assignable to ring breathing modes. They deduced that monoaromatic rings and/or pyridine-like rings sorbed in perpendicular orientation on the Cu electrode surface, and assigned bands at 1390 and 1378 cm^{-1} to COO^- stretching vibrations, bands at 2066 and 2026 to ethyne stretching vibrations, and those at 2926 and 2850 cm^{-1} to the symmetric and asymmetric stretching vibrations of aliphatic C-H. Their data would suggest that the chemical heterogeneity of HS is far less than this author suspects. Certainly the new developments in Raman spectroscopy can have significant applications in HS studies, and it will be appropriate to apply the SERS technique to a range of well characterized humic fractions.

Ultra violet-visible (UV-Vis) spectroscopy has limited applications in the humic sciences. The peaks of the UV-Vis spectra are also broad, and provide little information that is interpretable. Use is made of E_4/E_6 (absorbance at 465/665 nm) ratio values for comparing humic samples from different sources and environments (see Simpson *et al.*, p 73 - 82, this Volume). The ratio values can show distinct differences, but there are no definite interpretations of the meanings of these.

ESR provides information about free radicals, and although radicals can significantly influence interactions of HS, should these radicals be accessible, they are relatively few in terms of the total compositions of HS.

NMR gives the most useful compositional information, and the development of ***cross polarization magic angle spinning (CPMAS) ^{13}C NMR*** has provided an excellent finger print for humic fractions from different sources. The CPMAS ^{13}C NMR data of Watt *et al.* (1996) illustrate compositional similarities and differences between humic fractions from waters from different pristine sources. Their data would suggest close similarities in the humic fractions from waters from watersheds of similar soil types. ^{13}C NMR data also provide an abundance of information about functionalities.

Malcolm (1989) has listed chemical shift data that are useful for assignments of functionalities to humic structures. The chemical shift data in Figure 1 were obtained using the Chem Windows Program with the attached C-13 Module Version 1 (from SoftShell International, 1600 Ute Avenue, Grand Junction, Colorado 81501-4614), and it gives chemical shift data for each carbon of a lignin-type segment. The data show resonances for aromatic carbon ranging from 112 to 148. Data for HS showing such a broad spread would suggest that lignin-/tannin-type residues persist.

Figure 2 presents CPMAS ^{13}C NMR spectra (from Hayes, 1996) of HAs, FAs and XAD-4 acids isolated from the 0 - 15 cm layer (4.1% organic carbon) of a long term grassland Hallsworth series soil from the BBSRC IGER Station, North Wyke, Devon. Spectra 1, 4, and 7 are for the HAs, FAs, and XAD-4 acids, respectively, isolated in sodium pyrophosphate (Pyro) at pH 7, spectra 2 and 8 are for the HAs and XAD-4 acids isolated in Pyro at pH 10.6, and spectra 3, 6, and 9 are for the HAs, FAs, and XAD-4 acids isolated in Pyro and NaOH at pH 12.6.

Table 1 gives the integrated areas of seven resonance bands of the spectra. The spectra (Figure 2) show distinct differences between the same operationally defined fractions isolated at the different pH values, and the data for the integrated areas of the selected

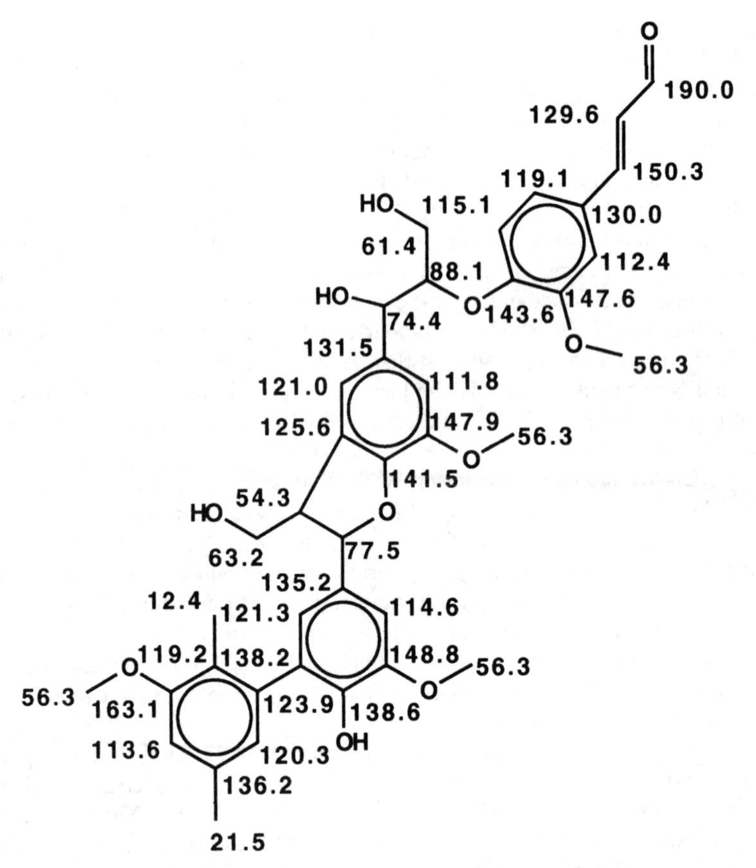

Figure 1 *^{13}C NMR chemical shift assignments for each carbon in a segment of a lignin-like structure, as determined using a computer programme (p. 15)*

resonance bands reflect these differences. The HAs isolated at pH 7 was the most highly aromatic (spectrum 1), and the aromaticity of this fraction was almost double that for the HAs isolated at pH 10.6 and at 12.6. The integrated areas for the resonances at 160 - 190 ppm (spectra 2 and 3), which represent the carbonyl functionalities of carboxylic acids, esters, and amides, indicate that such functionalities contribute about equally to each of the three fractions. The contribution of amide structures to this resonance is often overlooked, but it is clear from Table 2 that amino acids (probably in peptide structures) are significant contributors to the compositions of the HAs, and the contribution of amino acids to the overall composition of the HAs isolated at pH 12 is high (16.3%). Although the integrated areas suggest that the HAs isolated at pH 7 are richest in O-aromatic functionality (hydroxy,-methoxy-, alkoxy-), the spectra would indicate more clear cut evidence for such functionalities in the cases of the HAs isolated at pH 10.6 and 12.6. The 65 - 110 ppm resonance includes O-alkyl and C-OH resonances, and it is sometimes considered to be the saccharide region of the spectrum. That interpretation is valid only when the anomeric carbon resonance, centered at 105 ppm, has an integrated area of the order of one-fourth to one- fifth of that for the 65 to 110 ppm resonance. The neutral sugar contents of the HAs (Table 2) are of the order of 5% to 7% for the three HAs, and

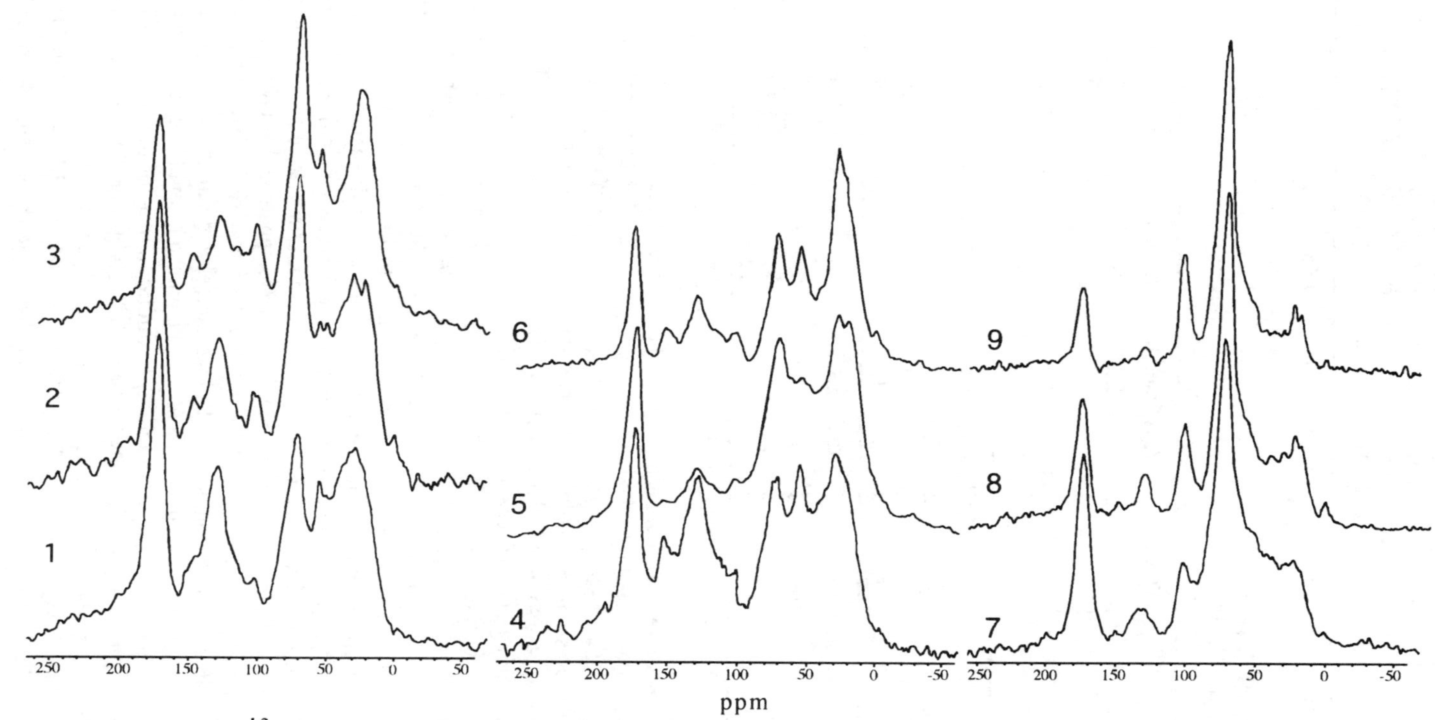

Figure 2 CPMAS ^{13}C-NMR spectra for the soil humic (1, 2, and 3), fulvic (4 and 6) and XAD-4 (7,8, and 9) acids isolated at pH 7 (1, 4, 7), pH 10.6 (2, 8), and at pH 12.6 (3, 6, 9) from the Hallsworth series soil (p. 108) and processed following the XAD-8 and XAD-4 resin in tandem procedure. Spectrum (5) is for the HA isolated at pH 7, subjected to XAD-8 resin treatment, and precipitated at pH 2.0-2.5 (see p. 17)

Table 1 *Integrated areas of seven resonance bands of the CPMAS ^{13}C NMR spectra of the humic (HA), fulvic (FA), and XAD-4 acids isolated from the Hallsworth series surface soil, using aqueous solvents at pH 7 (7), pH 10.6 (10), pH 12.6 (12), and processed using the XAD-8 and XAD-4 resin in tandem procedure*

Chemical shift (ppm)	220-190	190-160	160-140	140-110	110-65	65-45	45-10	
Sample	Ketonic/ aldehyde	Carboxyl/ Ester/ Amide	O-Aryl	Aromatic	O-Alkyl/ Sugars		Alkyl C	*fa
	Devon Soil HA isolated using the XAD-8 resin							
R 7	4	16	7	22	18	12	21	29
R 7pt‡	2	17	3	8	36	-	35	11
R 10	4	14	5	10	34	-	34	15
R 12	2	12	4	13	17	13	39	17
R 12pt	0	13	3	10	18	13	43	13
	Devon Soil FA isolated using the XAD-8 resin							
R 7	5	21	4	21	17	11	22	25
R 10	3	17	5	14	26	8	28	19
R 12	3	15	4	15	25	12	27	19
	Devon Soil XAD-4 acids HS isolated using the XAD4 resin							
R 7	1	19	1	6	53	-	21	7
R 10	3	12	3	6	58	-	18	9
R 12	2	9	1	4	72	-	13	5

*fa = Total aromaticity R = recovered from XAD-8 resin ‡see text for isolation procedure

yet the anomeric carbon resonance is seen to increase as the pH of the extractant is raised. It is likely, therefore, that ether functionalities, or sugar derived/related structures are major contributors to the 65 - 90 ppm resonance, at least in the case of the HAs isolated at pH 7. Peptide structures can also contribute to the resonance in this region.

The resonance at 65 - 45 ppm includes methoxyl (a sharp peak at 56 ppm). This peak (especially prominent in lignins) is characteristic of HS in the process of humification.

The compositions of the HAs which gave spectrum 5 are different from those which provided spectra 1, 2, and 3 (Figure 1). In the preparation process dilute soil extracts were adjusted to pH 2 and passed in tandem onto XAD-8 and XAD-4 resin columns. The materials which sorbed to the resins were back eluted in base, and the HAs were precipitated at pH 1 from the eluate from XAD-8. The HAs were redissolved in base, and as the pH was being adjusted to 2 (in order to desalt using the resin column), a precipitate formed in the pH range 2.5 to 2. That precipitate was found, as spectrum 5 shows, to be different in composition from the material which did not precipitate at that pH (spectrum 1). The data in Table 1 show that aromaticity was low (for the HAs in spectrum 5), and that the major contributors to the composition had resonances in the aliphatic hydrocarbon functionality region (10 - 45 ppm) and in the 65 - 110 ppm band. Evidence for anomeric carbon is weak, and the sugar content (Table 2) is similar to that for the HAs at pH 7. However, the amino acid content is high (14%), and it is likely that peptide functionalities contribute significantly to the resonances at 160 to 190, and at 65 to 90 ppm.

The spectra for the FAs isolated at pH 7 (spectrum 4) and pH 12.6 (spectrum 6) are also significantly different from each other, and both have features in common with the

Table 2 *Amino acid (AA) contents (%) and neutral sugars (NS) contents (%) of the humic (HA), fulvic (FA), and XAD-4 acids isolated from the Hallsworth soil at pH 7 (7HA, 7FA, and 7XAD), pH 10.6 (10HA, 10FA, and 10XAD) and at pH 12.6 (12HA, 12FA, and 12XAD) and processed using XAD-8 and XAD-4 resins (R). HA7pt and HA12pt refer to the HAs precipitated at pH 2.5 as the pH of the HA solutions were lowered to 2. Values are calculated on a dry ash-free basis.*

Sample	%AA	% NS
R 7HA	7.9	5.0
R 7HApt	14.0	6.7
R 7XAD	3.9	15.2
R 7FA	5.0	4.5
R 10HA	9.4	7.3
R 10XAD	7.4	22.4
R 10 FA	5.6	11.3
R 12HA	16.3	6.3
R 12HApt	13.6	6.2
R 12XAD	5.8	17.0
R 12FA	6.5	11.1

HAs isolated at the same pH values. Both have significant aromaticity, and the anomeric carbon signal is strong in the case of the FAs isolated at pH 12.6, and the sugar and amino

acid contents (Table 2) are likely to be the major contributors to the resonance at 65 to 110 ppm. There are clear indications for methoxyl in both samples; that is unusual for FAs.

The spectra for the XAD-4 acids are different from those for the FAs and HAs, and there are distinct similarities between the spectra for the samples isolated at the different pH values. Aromaticity was low for each sample, and especially in the case of that isolated at pH 12.6. In all cases the 65 - 110 ppm resonance was strongest, and there is clear evidence for anomeric carbon. There was not an especial enrichment in amino acids.

Applications of one-dimensional (1-D) proton and of 2-D NMR procedures for studies of humic substances. As for ^{13}C NMR, assignments of functional groups can be made from chemical shift data for proton (^{1}H) NMR, and humic scientists will be familiar with the treatise of Wershaw (1985) in this area. Excellent descriptions of the fundamentals and applications of 1-D NMR are in texts such as Derome (1987) and Williams and Fleming (1989), and the reader is referred to Simpson *et al.* (p. 46 - 62, this Volume) and to Boersma *et al.* (1997) who discuss applications of 1-D and of 2-D NMR for studies of structures of HS. Applications of 2-D NMR to the HS field are new, although there are a number of reports of applications of the techniques in structural studies of lignins (e.g. Garver *et al.*, 1996).

The development of powerful high field 500 MHz and 600 MHz spectrometers has refocused interest in ^{1}H NMR, and the combination of proton with ^{13}C NMR in 2-D ^{1}H-^{13}C Heteronuclear Multiple Quantum Coherence (HMQC) spectroscopy provides information on the specificity of proton-carbon bonding, and will improve confidence in functional group assignments.

The contribution by Simpson *et al.* (p. 46 - 62) has summarised (p. 48) the kinds of compositional information which 1-D ^{1}H and 1-D ^{13}C procedures provide, and it gives an outline of that which can be obtained from modern 2-D homonuclear (^{1}H - ^{1}H) procedures such as COrrelation SpectroscopY (COSY), 2-D HOmonuclear HArtmann HAhn (HOHAHA), TOtal Correlation SpectroscopY (TOCSY), Nuclear Overhauser Effect SpectroscopY (NOESY), and from the 2-D Heteronuclear (^1H - ^{13}C) Multiple Quantum Coherence (HMQC) and Heteronuclear Multiple Bond Coherence (HMBC) correlation techniques. They have provided 1-D ^1H, and 2-D COSY, 2-D TOCSY, and 2-D HMQC (^1H - ^{13}C) NMR spectra for a FA fraction isolated at pH 12.6, from the humifying roots of a moss sample growing on a rock face that had been exhaustively extracted under less alkaline conditions (pH 7 and at pH 10.6). The spectra for the FAs isolated at pH 12.6 were the best resolved of those for all of the different fractions isolated and the heterogeneity was less for this sample than for those isolated at lower pH values. Interpretations of the spectra from the HMQC experiments of Boersma *et al.* (1997), carried out on the IHSS soil HA standard (isolated from a Mollisol), could suggest evidence for quinones, unsaturated lactones, ortho-coupling of para-substituted phenols, primary amides, and several carbohydrate-related substances, and amino acids (the hydrolyzates of this standard are rich in sugars and amino acids). There is much still to be done before the full potential of applications of 2-D NMR spectroscopy to studies of composition and aspects of structures of HS will be realised.

Chemical degradation procedures are now less favoured than in the past for studies of the component molecules of HS. Arguably these still provide the best methods of providing awareness of the molecules that compose the macromolecules.

It is relatively easy to cleave the components of proteins, polysaccharides, and nucleic acids because the links between the molecules are hydrolyzable. However, although HS are made up of a variety of relatively simple *'building blocks'*, the units are linked together by bonds which are, for the most part, difficult to cleave. The major structural units consist of a *'backbone'* whose component molecules are linked by carbon to carbon bonds, ether linkages, and by other intractable units. Hydrolysis in some instances (see Parsons, 1989) can cause up to 50% of the masses of soil HAs to be lost as CO_2, and as soluble molecules which include sugars, amino acids, small amounts of purine and pyrimidine bases, and phenolic substances. The amounts of sugars and amino acids detected in the hydrolysates of aquatic humic substances are relatively small percentages of those detected in the hydrolysates of the similar nominal fractions from soils. However, such comparisons will not be truly valid until analyses are carried out on the hydrolysates of soil HAs and FAs subjected to treatment with XAD-8 resin, as outlined. Although losses of CO_2 during hydrolysis (through decarboxylation of activated carboxyl groups, such as β-keto acids, and of hydroxybenzenecarboxylic acids) are significant, the total acidities of soil HAs are not decreased by the hydrolysis. That suggests that new acid groups are formed (e.g. from esters and lactones).

A boron trifluoride-methanol transesterification procedure has been adapted by Almendros and Sans (1991, 1992) for studies of the compositions of HAs and humins. This mild procedure allowed the removal of humic components linked by ester functionalities, and the labile (derivatized) structures could then be identified. Yields of identifiable products were of the order of 30-35%, and were similar to those for oxidative degradation procedures. A variety of monobasic straight chain and branched fatty acids, long chain dicarboxylic acids, di-hydroxymonobasic acids, tri-hydroxymonobasic acids, methoxy benzenecarboxylic acids, di-, tri-, and tetra-methoxy benzenecarboxylic acids, and a variety of other miscellaneous acids were detected in the digests of humins. It seems likely that these acids were present as esters in the humins, and are typical of depolymerization components of plants, such as cutins and suberins.

There have been more than 100 products identified in digests of a variety of oxidative degradations of HS. This cannot be considered as evidence for the presence of more than 100 'building blocks' because in the course of a degradation several different products can form from a single precursor. The products depend on the type and concentration of reagent used, on the temperature, on the time of reaction, etc. The majority of the compounds identified are acids, and the aromatic acids range from benzenedi- to benzehexacarboxylic acids, and there is abundant evidence also for mono to tri-hydroxy (or methoxy), and of hydrocarbon substituents on the aromatic nuclei.

The abundance of benzenecarboxylic acids in the digests of alkaline permanganate oxidative degradations of HS might be interpreted in terms of contributions to the macromolecular structures of fused aromatic components (see Hayes and Swift, 1978; Griffith and Schnitzer, 1989; Christman *et al.*, 1989). However, the same acids are found in the digests of alkaline cupric oxide degradation processes, and such reagents would not degrade fused aromatic structures to benzene polycarboxylic acids. Therefore, it is reasonable to assume that the carboxylic acid substituents on the aromatic nuclei were formed from oxidizable substituents, such as aldehydes, alcohols, olefinic and other structures that are cleaved under the reaction conditions. There are possibilities for carbonylation reactions to give rise to benzenecarboxylic acids, and this possibility has not been resolved satisfactorily for degradations of HS in alkaline digests. Phenols are degraded in alkaline oxidative media, but their ether derivatives resist degradation. Hence it is necessary to methylate the substrates prior to oxidative degradation processes, and the phenolic hydroxyls persist as methoxy substituents. Because methoxyl substituents on aromatic nuclei, as well as phenolic hydroxyls are indigenous to humic substances (these could, for example, have their origins in lignins), we cannot say whether or not the methoxyl groups detected in methylated digests were present in indigenous structures, or were formed from phenolic hydroxyls.

Aliphatic dicarboxylic acids, ranging from ethanedioic to decanedioic, are abundant in the digests of oxidative degradation reactions, and there is evidence also for aliphatic tricarboxylic acids. The carboxyls could arise from cleavages of unsaturated groups along aliphatic chains, and could be formed from oxidizable functionalities, such as alcohols and aldehydes, and inevitably some are present as carboxyl groups on the aliphatic structures. Hayes and Swift (1978) have discussed mechanisms of release of long chain acids and hydrocarbon compounds from structures in the macromolecules.

In general, it is unlikely that more than one carboxyl substituent resides on a single aromatic unit in humic molecules. This is evident from the products identified in digests of reductive degradation reactions using sodium amalgam (see reviews by Hayes and Swift, 1978, and by Stevenson, 1989). None of the aromatic structures identified from such digests have more than one carboxyl substituent. There is evidence for the presence of one to three hydroxyl or methoxyl substituents on the aromatic nuclei, and such activating substituents could promote the decarboxylation of di- and poly-hydroxybenzenecarboxylic acids. A variety of non-carboxyl bearing hydroxy/methoxy benzene structures have been identified in the digests, and these might have resulted from decarboxylation of hydroxy/methoxy substituted benzenecarboxylic acids.

In many ways, the products identified in the digests of degradations with sodium sulphide at temperatures of 250 °C compliment the information from degradations with sodium amalgam. Hayes and O'Callaghan (1989) have outlined the mechanisms of degradation in saturated (10%) Na_2S solutions, and provide evidence to suggest that the formation of quinone methide intermediates are important contributors to the degradation mechanisms. They were able to explain possible origins in the humic structures of products identified in the degradative digests. Most of the aromatic compounds identified had one or two methoxy substituents (hydroxyl would have been converted to methoxyl in the methylation process needed for analysis by GCMS), one or two methyl or other aliphatic substituents and, with the exception of phthalic acid, did not have more than one carboxyl substituent attached to the benzene nucleus. (This contrasts with the benzenepolycarboxylic acids found in the digests where permanganate and alkaline cupric oxide were used.) The methyl substituents would form through the quinone methide mechanism when hydroxyl or ether functionalities are present on the carbon attached to the ring. Also, methyl substituents can arise as artefacts during the methylation of phenols, and could, of course, be present as substituents on some of the aromatic groups which compose HS. Ethyl and alkyl substituents would not be artefacts, but could arise from appropriate secondary alcohol and ether structures attached to a carbon substituent on the aromatic ring, and *ortho* or *para* to a phenolic hydroxyl (or to an ether substituent which would give a hydroxyl group during reaction in the Na_2S medium). This arrangement gives quinone methide intermediates arising from the breaking of relevant C-O bonds. Hayes and O'Callaghan (1989) pointed out that many of the products identified in their digests could have origins in phenylpropane structures of the types associated with lignins, and in which two or three hydroxy or ether substituents were present in the propyl side chain.

Hayes and O'Callaghan (1989) have outlined some mechanisms of degradation of HAs in phenol under reflux conditions. The process gives extensive degradations of HS. Examination of the digest products, and evaluations of the mechanisms which are considered to operate would suggest that there are hydrocarbon linkages between aromatic nuclei, and some of the aromatic chains may well be long.

Some concepts of the structures of HS were influenced for a time by results from zinc dust distillation and fusion experiments. The majority of the digest products identified were fused aromatic structures, and some of these had heterocyclic nitrogen (the digest products are listed by Hayes and Swift, 1978, and by Stevenson, 1989). Yields of polycyclic aromatic hydrocarbons (PAHs) from zinc dust distillation and fusion of HAs and FAs are invariably very low, and are often less than 1% of the starting material. Such low yields are not unusual even for alkaloids subjected to zinc dust distillation and fusion. On the basis of the results obtained, Cheshire *et al.* (1967) considered that humic structures are built around a polynuclear aromatic 'core' to which polysaccharides, simple phenols, proteins or peptides, and metals are attached (see Hayes and Swift, 1978). However, as pointed out by Hansen and Schnitzer (1969a, 1969b), the drastic reaction conditions associated with zinc dust distillation can result in excess bond breaking, and dehydrogenation and recombinations of the fragments can give rise to condensed ring structures. Cheshire *et al.* (1968) have shown that PAHs are generated from the zinc dust distillation of furfural (formed from the dehydration of pentose sugars), from quinone polymers, and even from dihydroxybenzoic acid.

Because fused aromatic structures have been detected only in high temperature degradation processes (whether in zinc dust distillations/fusions, or in pyrolysis) it is tempting to infer that these structures are not components of naturally occurring HS (in soils and waters), but are artefacts of the highly energetic degradative processes.

The yields of identifiable products in the different chemical degradations of HS are rarely more than 40% of the masses of the starting materials. It is likely that some of the degradation processes, and especially those carried out in alkaline media, give new (synthetic) macromolecules from components released in the digests from the parent humic macromolecules. Thus, some at least of the residues which do not dissolve in the organic extracting solvents might well be very different from the naturally occurring HS used in the degradation reactions. There is a need to learn more about these 'recalcitrant' fractions, and it would be appropriate to use sequential degradation procedures (as suggested by Hayes and Swift, 1978) to see to what extents the macromolecular residues of previous degradations would be depolymerized (or degraded to identifiable products) when a succession of chemicals and procedures are used which provide different mechanisms of degradation. Appropriate sequences might include boron trifluoride-methanol, followed by sodium amalgam, followed by phenol under reflux conditions, followed by sodium sulphide, and finally by alkaline permanganate or alkaline Cu(II)O.

Pyrolysis offers a rapid procedure for the degradation of HS. Pyrolyzates of soil HS give compounds indicative of polypeptide, lignin, or polyphenol origins, and to a lesser extent some of the compounds would appear to have origins in polysaccharide, or in so-called 'pseudopolysaccharide' structures (Bracewell *et al.*, 1989). The evidence obtained by these authors suggested that 'pseudopolysaccharide' (a term used to describe saccharide-type or -derived/related materials which give, on pyrolysis, products such as furan, methylfuran, dimethylfuran, furfural, and methylfurfural structures) are especially prevalent in the fulvic acid fractions. However, because the pyrolysis was carried out on FA samples that had not been subjected to a 'cleansing' treatment (e.g. using XAD-8), it is likely that the substrates examined contained considerable amounts of polysaccharides. This view is substantiated by the fact that evidence for peptide- and for carbohydrate-derived structures was diminished when the samples were first hydrolysed in 6M HCl. It was seen that the abundance of phenol pyrolysis products, with likely origins in lignins and in microbially synthesized polyphenol substances, was increased by hydrolysis.

Schulten and Schnitzer (1995) pyrolyzed (at 500 °C) in a Fisher Curie-point pyrolyzer attached to a GCMS system HAs having a 46 to 400 KD mass range. They reported the major thermal products to be benzene and alkyl benzenes, with C_1 to C_{13} *n*-alkyl benzenes especially predominant. There was evidence also for branched members of the same series. Some trimethylbenzenes, alkylnaphthalenes, and alkylphenanthrenes were also detected, and that might be considered as evidence for two and three ring fused aromatic structures. However, the only evidence for fused aromatic structures in the degradation products of HS has been obtained when high temperatues were used, and it is likely that these were artifacts.

Fast-atom bombardment mass spectrometry (FABS), laser desorption mass specrtrometry (LDMS), and ***matrix assisted laser desorption mass spectrometry*** (MALDI) are new techniques which would suggest that MW values of HS anre in the range of several hundreds to a few thousand (Brown *et al.*, 1998).

6 Emerging Concepts of Structures of Humic Substances

To resolve the structure of a macromolecule it is essential to know the molecular 'building blocks' and their exact sequences in the molecule, as well as the shapes, sizes, configurations, and conformations of the macromolecules. This information is vital in order to be able to predict accurately the reactivities of the macromolecules. Chemically synthesized polymers can have regularized structures, and the biological macromolecules of nature, whose syntheses are genetically or biologically controlled, will have regular, and reproducable compositions, structures, and properties. Because there is no evidence of biological control of the synthesis of HS, it is unlikely that the component molecules will be arranged in any specific sequences, as applies for the biological polymers.

About eight years ago Hayes *et al.* (1989), in summarising the contributions of several authors to *Humic Substances II,* regarded HS as highly polydisperse materials. They drew on the work of Cameron *et al.* (1972a, 1972b) who isolated HAs from a sapric histosol and carefully isolated 11 reasonably discrete fractions from these using gel filtration and pressure filtration techniques, and by means of the ultracentrifuge determined that the MW values of these ranged from 2.6×10^3 to 1.36×10^6 (see Hayes and Swift, 1978; Swift, 1989). By use of frictional coefficient data (obtained using the ultracentrifuge) the HAs in solution were considered likely to assume random coil conformations, and less likely discoid structures. The negative charges distributed along the strands would give rise to repulsion effects, and the strands would coil randomly with respect to time and space to give roughly spherical shapes. There would be a Gaussian distribution of molecular mass, and with the mass densities greatest at the centre and decreasing to zero at the outer edges. Whether or not this coil would be tightly or loosely 'wound' would depend on the extent of solvent penetration, the charge densities, the nature of the cross linking, and the nature of the counter ions. The evidence suggested that the extents of cross linking are small. Hayes and Swift (1978) extrapolated the concept of the random coil solution conformation to the solid/gel state that is characteristic of the soil. When H^+, or divalent and polyvalent cations are the exchangeable species introduced to the system, the molecules shrink because of cation bridging and hydrogen bonding effects, the solvent in the macromolecular matrix is decreased, and solidification occurs.

The information which is emerging is questioning these concepts. All who have attempted to fractionate HS on the basis of molecular size differences have been frustrated to find that, in the course of sequential fractionation processes, materials of smaller sizes will continually be set free. The studies of Piccolo *et al.* (1996) and the MALDI experiments (referred to in p. 22) may indicate that associations of molecules, or microaggregates, or molecular assemblies can remain in association (whether or not these are in true solution) during gel filtration. Kenworthy and Hayes (p. 39 - 45, this Volume) lend strength to the concept of arrangements of HS in molecular assemblies which provide hydrophobic sites or cages which can be disrupted as the molecular conformations change, or as the molecular associations are broken up. The new information would suggest that the high MW values for HS might reflect molecular associations in self assembly-type structures, or in micelle-like arrangements of the molecules.

It is evident from Section 5 (this chapter) that the component molecules of HS are still not known unambiguously, and so far the best indications of the natures of these can be deduced from identifications of products of degradation reactions, and from an awareness of the mechanisms involved in the degradation processes. Awareness of likely component molecules, and of functionality [provided by techniques such as wet chemical methods, NMR, FTIR, titration data, and developments in Raman spectroscopy (which may provide valuable information in the near future)] has not changed dramatically in the last eight years. We knew then, as now, that 25 to 45% of the components of soil HAs are aromatic (Malcolm, 1989), and that value may be higher in some instances, e.g. for samples from the B_h horizons of some podzols. HAs from aqueous systems are less aromatic and more highly oxidized than those from soils. The consensus 8 years ago regarded the aromaticity as being composed of single ring compounds (Hayes *et al.*, 1989), and with 3 to 5 ring substituents. That view has not changed.

Much data have emerged about sugar and amino acid contents of humic fractions. Aquatic HS have significantly less sugars and amino acids than those in soils, and the contents in the aquatic FAs are especially low. That might suggest that the sugars and amino acids are sterically protected in their humic associations (or physically protected through associations with mineral colloids, and in soil aggregates) from enzymatic attack. Should the associations involve covalent links between the biological molecules and the humic core, then the MW values of the systems would be increased. It is plausible to consider that conformational changes to the HS in the soil environment, or the disruption of aggregates, could expose glycosidic and peptide linkages to microbial attack releasing core humic structures, some of which dissolve in the soil solution. The lower MW values of aquatic HS (compared with soil HS) might be explained by the hydrolysis thesis.

The hydrogens in three to five of the (aromatic) ring positions may be replaced by substituent groups, and these consist of hydroxyl and methoxyl, aliphatic hydrocarbon structures, and some of these may be involved in linking aromatic structures. There is also evidence for aldehyde and keto functional groups attached to some of the aromatic nuclei, for phenylpropane (3-carbon chains attached to the aromatic rings) units, and for hydroxyl and methoxyl substituents. The phenyl propane structures, and hydroxy/methoxy substituents in the 3- and 4-, and in the 3- , 4-, and 5- ring positions would suggest origins in lignins, whereas the presence of these substituents in the 3- and 5- ring positions suggests origins in the skeletal structures, or the metabolic products of microorganisms.

There is evidence that ether functionalities link aromatic structures, and it is logical to assume that aromatic-aliphatic ethers are also present. It is not certain that all of the aliphatic units are linked to aromatic structures, and long chain hydrocarbons, with origins in plant cuticles and in algal metabolites, could possibly be present as 'impurities' held to the humic structures by van der Waals forces. Some of the hydrocarbon structures are olefinic, as evidenced by the presence of mono- and di- (and sometimes tri-) carboxylic acids in the digests of oxidative degradation reactions (Section 5, this chapter).

Fatty acids in degradation digests may be be released from esters of phenols and other hydroxyl groups in the 'backbone' structures. Also these could arise from waxes and suberins associated with HS, and could contribute to the hydrophobic properties which HS display in some instances, and have a role in the self association concept.

Titration data indicate that the acid groups in HS provide a continuum of dissociable protons representing a range of acid strengths, from the strong to the very weak. The strongest acids are carboxylic, and some are activated by appropriate adjacent groups. Phenolic hydroxyls also contribute to the acidity, and these are most abundant in the HA fraction. It would seem that their contribution to the total acidity is greatest in newly formed humic substances, and especially in those with origins in the lignified components of plants. As oxidation takes place, the phenols are oxidized, and eventually carboxylic acids are formed. Some of the phenolic substituents can have enhanced acidity because of the influences of other substituents on the aromatic structures. Enols and other weakly dissociable groups also contribute to the charge characteristics under alkaline conditions

FAs have many compositional characterisatics similar to those of HAs, but they have differences as well. FAs tend to be less aromatic than than HAs (and can contain as little as 25 per cent aromatic components in some instances), and they appear to be smaller, more polar, and more highly charged. These effects would cause the FAs to be less self associated than HAs, and in terms of the random coil concept, to be more linear than coiled. Fulvic acids are not precipitated under acid conditions, or by low concentrations of divalent metals. They might be expected to be readily removed from soils in drainage waters. That this does not happen is an indication of interactions between fulvic acids and insoluble components of soils. Fulvic acids in mineral soils are held by the inorganic colloids, and by entrapment within (associations with) the more hydrophobic HAs.

The XAD-4 acids would seem to be an association of HS with carbohydrate and peptide type materials. It will be possible to fractionationate these substances further.

Because of their polydispersity, it would be pointless at this stage to attempt to provide exact structures for HS fractions. It is not, however, necessary to know the exact structures in order to have a good understanding of composition, and to be able to predict how the macromolecules will react with other species. The awareness of structure which is emerging allows us a degree of understanding of the processes involved in their binding of metals and of anthropogenic organic chemicals, in their adsorption to clays and to (hydr)oxides, and of the ways in which they respond to chemical treatments of waters.

References

Aiken, G.R. 1985. Isolation and concentration techniques for aquatic humic substances. p. 363-385. *In* G.R. Aiken, P. MacCarthy, R.L. Malcolm, and R.S. Swift (eds), *Humic Substances in Soil, Sediment, and Water*. Wiley, New York.

Aiken, G.R., D.M. McKnight, R.L. Wershaw, and P. MacCarthy. 1985. An introduction to humic substances in soil, sediment, and water. p. 1-9. *In* G.R. Aiken, P. MacCarthy, R.L. Malcolm, and R.S. Swift (eds), *Humic Substances in Soil, Sediment, and Water*. Wiley, New York.

Almendros, G. and J. Sans. 1991. Structural study on the soil humin fraction-Boron trifluoride-methanol transesterification of soil humin preparations. *Soil Biol. Biochem.* **23**:1147-1154.

Almendros, G. and J. Sans. 1992. A structural study of alkyl polymers in soil after perborate degradation of humin. *Geoderma* **53**:79-95.

Bloom, P.R., and J.A. Leenheer. 1989. Vibrational, electronic, and high-energy spectroscopic methods for characterizing humic substances. p. 409-446. *In* M.H.B. Hayes, P. MacCarthy, R.L. Malcolm, and R.S. Swift (eds), *Humic Substances II: In Search of Structure*. Wiley, Chichester.

Boersma, R.E., W.L. Kingery, M.H.B. Hayes, M.A. Locke, and R.P. Hicks. 1997. Applications of homonuclear and heteronuclear two-dimensional NMR in structural studies of humic substances. *Soil Sci.* Submitted.

Bracewell, J.M., K. Haider, S.R. Larter and H.-R. Schulten. 1989. Thermal degradation relevant to structural studies of humic studies. p. 181-222. *In* M.H.B. Hayes, P. MacCarthy, R.L. Malcolm, and R.S. Swift (eds), *Humic Substances II: In Search of Structure*. Wiley, Chichester.

Brown, T.L., F.J. Novotony, and J.A. Rice. 1998. Comparison of desorption mass spectrometry techniques for the characterization of fulvic acid. *In* C.E. Clapp *et al.* (eds), *Humic Substances IV*. For publication by IHSS.

Burchill, S., M.H.B. Hayes, and D.J. Greenland. 1981. Adsorption. p. 221-400. *In* D.J. Greenland and M.H.B. Hayes (eds), *The Chemistry of Soil Processes*. Wiley, Chichester.

Cameron, R.S., R.S. Swift, B.K. Thornton, and A.M. Posner. 1972a. Calibration of gel permeation chromatography procedures for use with humic acid. *J. Soil Sci.* **23**:342-349.

Cameron, R.S., B.K. Thornton, R.S. Swift, and A.M. Posner. 1972b. Molecular weight and shape of humic acid from sedimentation and diffusion measurements on fractionated extracts. *J. Soil Sci.* **23**:394-408.

Cheshire, M.V., P.A. Cranwell, C.P. Falshaw, A.J. Floyd, and R.D. Haworth. 1967. Humic acid.- II. Structure of humic acids. *Tetrahedron* **23**:1669-1682.

Cheshire, M.V., P.A. Cranwell, and R.D. Haworth. 1968. Humic acid -III. *Tetrahedron* **24**:5155-5167.
Christman, R.F., D.L. Norwood, Y. Seo, and F.H. Frimmel. 1989. Oxidative degradation of humic substances from freshwater environments. p. 33-67. *In* M.H.B. Hayes, P. MacCarthy, R.L. Malcolm, and R.S. Swift (eds), *Humic Substances II: In Search of Structure.* Wiley, Chichester.
Ciavatta, C., M. Govi, G. Bonoretti, D. Montecchio, and C. Gessa. 1996. Isoelectric focusing of soil organic matter: The role of the carrier ampholites. p. 53-56. *In* C.E. Clapp, M.H.B. Hayes, N. Senesi, and S.M. Griffith (eds), *Humic Substances and Organic Matter in Soil and Water Environments.* IHSS, University of Minnesota, St. Paul.
Clapp, C.E. 1957. High molecular weight water soluble muck: Isolation and determination of constituent sugars of a borate complex-forming polysaccharide employing electrophoretic techniques. Ph.D Thesis, Cornell University.
Clapp, C.E. and M.H.B. Hayes. 1996. Isolation of humic substances from an agricultural soil using a sequential and exhaustive extraction process. p. 3-11. *In* C.E. Clapp, M.H.B. Hayes, N. Senesi, and S.M. Griffith (eds), *Humic Substances and Organic Matter in Soil and Water Environments.* IHSS, University of Minnesota, St. Paul.
Deinzer, M., R. Melton, and D. Mitchell. 1975. Trace organic contaminants in drinking water; their concentration by reverse osmosis. *Water Res.* **9**:799-805.
Derome, A.E. 1987. Modern NMR techniques for chemistry research. p. 183-234. *In* J.E. Baldwin (ed.), *Organic Chemistry Series, Vol 6.* Pergamon, Oxford.
Duxbury, J.M. 1989. Studies of the molecular size and charge of humic substances by electrophoresis. p. 593-620. *In* M.H.B. Hayes, P. MacCarthy, R.L. Malcolm, and R.S. Swift (eds), *Humic Substances II: In Search of Structure.* Wiley, Chichester.
Eberle, S.H. and K.H. Schweer. 1974. Bestimmung von Huminsaure und Ligninsulfonsaure im Wasser durch Flussig-Flussingextraktion. *Vom Wasser* **41**:27-44.
Eswaran, H., E. Van den Berg, and P. Reich. 1993. Organic carbon in soils of the world. *Soil Sci. Soc. Am. J.* **57**:192-194.
Gadian, D.G. 1982. *Nuclear Magnetic Resonance and Its Application to Living Systems.* Oxford University Press.
Garver, T.M., K.J. Maa, and K. Marat. 1996. Conformational analysis of 2D NMR assignment strategies for lignin model compounds. The structure of acetoguaiacyl-dehydro-diisoeugenol methyl ether. *Can. J. Chem.* **74**:173-184.
Griffith, S.M., and M. Schnitzer. 1989. Oxidative degradation of soil humic substances. p. 69-98. *In* M.H.B. Hayes, P. MacCarthy, R.L. Malcolm, and R.S. Swift (eds), *Humic Substances II: In Search of Structure.* Wiley, Chichester.
Hansen, E.H. and M. Schnitzer 1969a. Zinc dust distillation of soil humic compounds. *Fuel* **48**:41-46.
Hansen, E.H. and M. Schnitzer. 1969b. Zinc dust distillation and fusion of a soil humic acid. *Soil Sci. Soc. Amer. Proc.* **33**:29-36.
Harvey, G.R. and D.A. Boran. 1985. Geochemistry of humic substances in seawater. p. 233-247. *In* G.R. Aiken, P. MacCarthy, R.L. Malcolm, and R.S. Swift (eds), *Humic Substances in Soil, Sediment, and Water.* Wiley, New York.
Häusler, M.J. and M.H.B. Hayes. 1996. Uses of XAD-8 resin and acidified dimethylsulfoxide in studies of humic acids. p. 25-32. *In* C.E. Clapp, M.H.B. Hayes, N. Senesi, and S.M. Griffith (eds), *Humic Substances and Organic Matter in Soil and Water Environments.* IHSS, University of Minnesota, St. Paul.
Hayes, M.H.B. 1985. Extraction of humic substances from soil. p. 329-362. *In* G.R. Aiken, P. MacCarthy, R.L. Malcolm, and R.S. Swift (eds), *Humic Substances in Soil, Sediment, and Water.* Wiley, New York.
Hayes, M.H.B., J.E. Dawson, J.L. Mortensen, and C.E. Clapp. 1985. Electrophoretic characteristics of extracts from sapric histosol soils. p. 31-41. *In* M.H.B. Hayes and R.S. Swift (eds), *Volunteered Papers, 2nd Intern Conf., International Humic Substances Society.* School of Chemistry, The University of Birmingham.

Hayes, M.H.B., P. MacCarthy, R.L. Malcolm and R.S. Swift. 1989. The search for structure: setting the scene. p. 3-31. *In* M.H.B. Hayes, P. MacCarthy, R.L. Malcolm, and R.S. Swift (eds), *Humic Substances II: In Search of Structure.* Wiley, Chichester.

Hayes, M.H.B., and M.R. O'Callaghan. 1989. Degradations with sodium sulfide and with phenol. p. 143-180. *In* M.H.B. Hayes, P. MacCarthy, R.L. Malcolm, and R.S. Swift (eds), *Humic Substances II: In Search of Structure.* Wiley, Chichester.

Hayes, M.H.B. and R.S. Swift. 1978. The chemistry of soil organic colloids. p. 179-320. *In* D.J. Gredenland and M.H.B. Hayes (eds), *The Chemistry of Soil Constituents.* Wiley, Chichester.

Hayes, M.H.B. and R.S. Swift. 1990. Genesis, isolation, composition and structures of soil humic substances. p. 245-305. *In* M.F. DeBoodt, M.H.B. Hayes, and A. Herbillon (eds), *Soil Colloids and their Associations in Aggregates.* Plenum, New York.

Hayes T.M. 1996. Study of the humic substances from soils and waters and their interactions with anthropogenic organic chemicals. PhD thesis, Univ. of Birmingham.

Hayes, T.M., M.H.B. Hayes, J.O. Skjemstad, R.S. Swift, and R.L. Malcolm. 1996. Isolation of humic substances from soil using aqueous extractants of different pH and XAD resins, and their characterization by 13C-NMR. p. 13-24. *In* C.E. Clapp, M.H.B. Hayes, N. Senesi, and S.M. Griffith (eds), *Humic Substances and Organic Matter in Soil and Water Environments.* Proc. 7th Intern. Conf. IHSS. IHSS, The University of Minnesota, St. Paul.

Huffman, E.W.D. and H.A. Stuber. 1985. Analytical methodology for elemental analysis of humic substances. p. 433-455. *In* G.R. Aiken, P. MacCarthy, R.L. Malcolm, and R.S. Swift (eds), *Humic Substances in Soil, Sediment, and Water.* Wiley, New York.

Jenkinson, D.S. 1981. The fate of plant and animal residues in soil. p. 505-561. *In* D.J. Greenland and M.H.B. Hayes (eds), *The Chemistry of Soil Processes.* Wiley, Chichester.

Jenkinson, D.S., D.E. Adams, and A. Wild. 1991. Model estimates of CO_2 emissions from soil in response to global warming. *Nature* **351**:304-306.

Johnston, A. E. 1991. Soil fertility and soil organic matter. p. 299-313. *In* W.S. Wilson (ed), *Advances in Soil Organic Matter Research: the Impact on Agriculture and the Environment.* Royal Society of Chemistry, Cambridge.

Klavins, M., M. Purite, and E. Apsite. 1996. Fractionation of soil aquatic and soil humic substances. p.41-46. *In* C.E. Clapp, M.H.B. Hayes, N. Senesi, and S.M. Griffith (eds), Humic Substances and Organic matter in Soil and Water Environments. *Proceedings of the Seventh International Conference, IHSS, Trinidad and Tobago.* IHSS, University of Minnesota, St. Paul.

Leenheer, J.A. 1981. Comprehensive approach to preparative isolation and fractionation of dissolved organic carbon from natural waters and wastewaters. *Environ. Sci. Technol.* **15**:578-587.

Leenheer, J.A. 1985. Fractionation techniques for aquatic humic substances. p. 409-429. *In* G.R. Aiken, P. MacCarthy, R.L. Malcolm, and R.S. Swift (eds), *Humic Substances in Soil, Sediment, and Water.* Wiley, New York.

Leenheer, J.A., R.L. Wershaw, and M.M. Reddy. 1995. Strong-acid, carboxyl-group structures in fulvic acid from the Suwannee River, Georgia. 1. Minor structures. *Environmental Sci. and Technol.* **29**:393-398.

Liang, E.J., Y-h.Yang, and W. Kiefer. 1997. Surface-enhanced raman spectra of fulvic and humic acids adsorbed on copper electrode. *Vibrational Spectroscopy.* Submitted.

Lugo, A.E. and S. Brown. 1986. Steady state terrestrial ecosystems and the global carbon cycle. *Vegetation* **68**:83-90.

MacCarthy, P. and J.A. Rice. 1985. Spectroscopic methods other than NMR for determining functionality in humic substances. p. 527-559. *In* G.R. Aiken, D.M. McKnight, R.L. Wershaw, and P. MacCarthy (eds), *Humic Substances in Soil, Sediment and Water.* Wiley-Interscience, New York.

Malcolm, R.L. 1989. Applicaions of solid-state ^{13}C NMR spectroscopy to geochemical studies of humic substances. p. 340-372. *In* M.H.B. Hayes, P. MacCarthy, R.L.

Malcolm, and R.S. Swift (eds), *Humic Substances II: In Search of Structure*. Wiley, Chichester.

Malcolm, R.L. 1991. Factors to be considered in the isolation and characterisation of aquatic humic substances. p. 369-391. *In* H. Boren and B. Allard (eds). *Humic Substances in the Aquatic and Terrestrial Environment*. Wiley, Chichester.

Malcolm, R.L. and P. MacCarthy. 1992. Quantitative evaluation of XAD-8 and XAD-4 resins used in tandem for removing organic solutes from water. *Environ. Int.* **18**:597-607.

Mann, L.K. 1986. Changes in soil carbon storage after cultivation. *Soil Sci.* **142**:279-288.

Martin, D. and H.G. Hauthal. 1975. *Dimethyl Sulphoxide* (translated by E.S. Halberstadt). Van Nostrand-Reinhold, New York.

Niemeyer, J., Y. Chen, and J.-M. Bollag. 1992. Characterization of humic acids, composts, and peat by diffuse reflectance Fourier Transform Infrared Spectroscopy. *Soil Sci. Soc. Am. J.* **56**:135-140.

Parsons, J.W. 1989. Hydrolytic degradation of humic substances. p. 99-120. *In* M.H.B. Hayes, P. MacCarthy, R.L. Malcolm, and R.S. Swift (eds), *Humic Substances II: In Search of Structure*. Wiley, Chichester.

Perdue, E.M. 1985. Acidic functional groups of humic substances. p. 493-526. *In* G.R. Aiken, D.M. McKnight, R.L. Wershaw, and P. MacCarthy (eds), *Humic Substances in Soil, Sediment and Water*. Wiley-Interscience, New York.

Piccolo, A., S. Nardi, and G. Concheri. 1996. Macromolecular changes of humic substances induced by interaction with organic acids. *European J. Soil Sci.* **47**:319-328.

Ping, C.L., G.J. Michaelson, and R.L. Malcolm. 1995. Fractionation and carbon balance of soil organic matter in selected cryic soils in Alaska. p. 307-314. *In* R. Lal, J. Kimble, E. Levine, and B.A. Stewart (eds), *Soils and Global Change*. CRC Lewis Publishers, Boca Raton.

Raich, J.W. and W.H. Schlesinger. 1992. The global carbon dioxide flux in soil respiration and its relationship to vegetation and climate. *Tellus* **44B**:81-99.

Schlesinger, W.H. 1984. Soil organic matter: A source of atmospheric CO_2. p. 111-127. *In* G.M. Woodwell (ed.), The role of terrestrial vegetation in the global carbon cycle: Measurement by remote sensing. SCOPE, Wiley, Chichester.

Schlesinger, W.H. 1986. Changes in soil carbon storage and associated properties with disturbance and recovery. p. 194 - 220. *In* J.R. Tabralkaand D.E. Reichle (eds), *The Changing Carbon Cycle: A Global Analysis*. Springer, New York.

Schlesinger, W.H. 1995. An overview of the carbon cycle. p. 9-25. *In* R. Lal, J. Kimble, E. Levine, and B.A. Stewart (eds), *Soils and Global Change*. CRC Lewis Publishers, Boca Raton.

Schulten, H.-R. and M. Schnitzer. 1995. Three-dimensional models for humic acids and soil organic matter. *Naturwissenschaften* **82**:487-498.

Senesi, N. and C. Steelink. 1989. Application of ESR spectroscopy to the study of humic substances. p. 373-408. *In* M.H.B. Hayes, P. MacCarthy, R.L. Malcolm, and R.S. Swift (eds), *Humic Substances II: In Search of Structure*. Wiley, Chichester.

Serkiz, S.M. and E.M. Perdue. 1990. Isolation of dissolved organic-matter from the Suwannee River using reverse osmosis. *Water Res.* **24**:911-916.

Steelink, C., R.L. Wershaw, K.A. Thorn, and M.A. Wilson. 1989. Application of liquid-state NMR spectroscopy to humic substances. p. 282-308. *In* M.H.B. Hayes, P. MacCarthy, R.L. Malcolm, and R.S. Swift (eds), *Humic Substances II: In Search of Structure*. Wiley, Chichester.

Stevenson, F.J. 1989. Reductive cleavage of humic substances. p. 122-142. *In* M.H.B. Hayes, P. MacCarthy, R.L. Malcolm, and R.S. Swift (eds), *Humic Substances II: In Search of Structure*. Wiley, Chichester.

Stevenson, F.J. 1994. *Humus Chemistry: Genesis, Composition, Reactions*. Second edition. Wiley, New York.

Sun, L., E.M. Perdue, and J.F. MacCarthy. 1995. Using reverse-osmosis to obtain organic-matter from surface and ground waters. *Water Res.* **29**:1471-1477.

Swift, R.S. 1985. Fractionation of soil humic substances. p. 387-408. *In* G.R. Aiken, P. MacCarthy, R.L. Malcolm, and R.S. Swift (eds), *Humic Substances in Soil, Sediment, and Water.* Wiley, New York.

Swift, R.S. 1989. Molecular weight, size, shape, and charge characteristics of humic substances: some basic considerations. p. 449-466. *In* M.H.B. Hayes, P. MacCarthy, R.L. Malcolm, and R.S. Swift (eds), *Humic Substances II: In Search of Structure.* Wiley, Chichester.

Swift, R.S. 1991. Effects of humic substances and polysaccharides on soil aggregation. p. 153-162. *In* W.S. Wilson (ed.), *Advances in Soil Organic Matter Research: the Impact on Agriculture and the Environment.* Royal Society of Chemistry, Cambridge.

Swift, R.S. 1996. Organic matter characterization. p. 1011-1069. *In* D.L. Sparks *et al.* (eds), *Methods of Soil Analysis. Part 3. Chemical Methods.* SSSA Book Series no. 5. Soil Science Soc. of America and Am. Soc. Agronomy, Madison, Wisconsin.

Swift, R.S. and A.M. Posner. 1971. Gel chromatography of humic acid. *J. Soil Sci.* **22**:237-249.

Thurman, E.M. and R.L. Malcolm. 1981. Preparative isolation of aquatic humic substances. *Environ. Sci. Technol.* **15**:463-466.

Thurman, E.M. 1985. *Organic Geochemistry of Natural Waters.* Martinus Nijhoff/Dr W. Junk Publishers, Dordrecht, The Netherlands.

Trubetskoj, O.A., O.E. Trubetskaya, G.V. Afanasieva, and O.I. Reznikova. 1996. Characterization of humic acids by polyacrylamide gel electrophoresis following preparative gel filtration. p. -47-51. *In* C.E. Clapp, M.H.B. Hayes, N. Senesi, and S.M. Griffith (eds), Humic Substances and Organic matter in Soil and Water Environments. *Proceedings of the Seventh International Conference, IHSS, Trinidad and Tobago.* IHSS, University of Minnesota, St. Paul.

Watt, B.E., R.L. Malcolm, M.H.B. Hayes, N.W.E. Clark, and J.K Chipman. 1996. The chemistry and potential mutagenicity of humic substances in waters from different watersheds in Britain and Ireland. *Water Res.* **6**:1502-1516.

Wershaw, R.L. 1985. Application of nuclear magnetic resonance spectroscopy for determining functionality in humic substances. p. 561-582. *In* G.R. Aiken, P. MacCarthy, R.L. Malcolm, and R.S. Swift (eds), *Humic Substances in Soil, Sediment, and Water.* Wiley, NY.

Wershaw, R.L. 1986. A new model for humic materials and their interactions with hydrophobic organic chemicals in soil-water or sediment-water systems. *J. Contaminant Hydrology* **1**:29-45.

Williams D.H, and I. Fleming, 1989. *Spectroscopic Methods in Organic Chemistry.* McGraw Hill, Maidenhead, Berkshire, England.

Wilson, M.A. 1989. Solid-state nuclear magnetic resonance spectroscopy of humic substances: Basic concepts and techniques. p. 309-338. *In* M.H.B. Hayes, P. MacCarthy, R.L. Malcolm, and R.S. Swift (eds), *Humic Substances II: In Search of Structure.* Wiley, Chichester.

Yang, Y., and T. Wang. 1997. Fourier transform Raman spectroscopic characterization of humic substances. *Vibrational Spectroscopy* **14**:105-112. Also *Vibrational Spectroscopy,.* Submitted for publication.

Extractability, Chemical Composition, and Reactivities of Soil Organic Matter of Irish Grassland Soils

D. McGrath

TEAGASC, JOHNSTOWN CASTLE, WEXFORD, IRELAND

Abstract

Regression relationships between soil parameters and soil organic carbon (SOC) are reviewed using data that have not been reported previously, as well as data that have been reported. For SOC determination, the Walkley and Black method was the method of choice for soil organic carbon (SOC) determinations, and loss-on-ignition served as a very good substitute ($r^2 = 0.90$). The colour of extracts could be a useful indicator of SOC ($r^2 = 0.56$), but soil colour (Munsell) was not of use ($r^2 = 0.10$).

Humic (humic and fulvic acids) and carbohydrate constituents (uronic acids, sugars released by mild hydrolysis, and total sugars) of soil polysaccharides were linearly related to SOC. The proportion of humic acid extracted with sodium pyrophosphate/sodium hydroxide solution increased with increasing SOC, whereas that of fulvic acid decreased. There was evidence for enrichment of carbohydrate in some soil types.

Extractability by chelating agents of Zn, but not of other trace metals, was strongly influenced by SOC. Iodine levels in sandy soils were weakly influenced by SOC.

Adsorption of organic micropollutants, polar and non-polar, showed a strong positive correlation with SOC. Phytotoxicity of soil-applied herbicides and of diesel oil was strongly and negatively correlated with SOC.

1 Introduction

Numerous soil characteristics are known, or are reputed to be quantitatively related to the soil organic carbon (SOC) content. These include the ability to provide for the nutrient requirements of crops and animals, and the capacity to withstand various natural and human-induced pressures. SOC levels are known to be elevated in Irish soils (Brogan, 1966; McGrath, 1973a; McGrath, 1980), with a mean of 53 mg kg^{-1} for soils under permanent pasture. This arises partly from the fact that tillage accounts for only a small proportion of land usage in Ireland (at the present time for only 6%). In addition, climatic conditions, moderate temperatures allied to high rainfall, are such as to favour SOC accumulation over degradation. It would seem that a similar situation would obtain in many other areas with comparable climatic conditions and land use; similarly high levels of SOC are probably widely distributed in many areas, including New Zealand (Burney *et al.*, 1975), parts of Great Britain (McGrath and Loveland, 1991), and parts of Canada (Dormaar, 1972). Because of the quantitative significance of soil organic matter in Irish soils, it was considered desirable to assemble, to review, and to generalise on proven relationships.

2 Materials and Methods

Soils

Soils used have been described in different publications relating to carbohydrates and humic substances (McGrath, 1973b)), to iodine (McGrath, 1988a), heavy metals and organic micropollutants (McGrath, 1995a; McGrath, 1995b)), mineral oil (McGrath, 1992), herbicide toxicity (McGrath and McCormack, 1979) and adsorption (McGrath, 1996).

Estimation of Organic Matter

SOC was estimated by the Walkley and Black method as described by Byrne, (1979). Loss on ignition was determined by heating soil (which had previously been dried at 105 °C) to 450 °C overnight. The colour of the dried soil was determined by visual matching to Munsell colour charts.

Other Analyses

Methods have been described for the estimation of humic substances (HS) content (McGrath, 1988b), carbohydrates (McGrath, 1973b), iodine (McGrath, 1988b), heavy metals and organic micropollutants (McGrath, 1995a; McGrath, 1995b), phytotoxicity (McGrath and McCormack, 1979), and adsorption (McGrath, 1996).

3 Results and Discussion

Measurement of Soil Organic Matter

Perhaps the most widely used method of quantifying total organic matter in soil is through measurements of SOC, using one of the many modifications of the Walkley and Black method. This method has the advantage of simplicity and robustness, but unfortunately it utilises strong H_2SO_4 and, more regrettably, hexavalent Cr. At various times we have investigated, for different reasons, surrogate methods, including loss on ignition, extractability using NaOH, and soil colour.

Loss-on-ignition was found to be closely related to SOC (Table 1), and can well serve as a measure of SOC for Irish soils. Measurement of NaOH-extractable organic matter (OM), or light absorbance (at 465 nm), which have been suggested as

Table 1 *Relationships between alternative assays, loss on ignition and absorbance of extractable (0.5M NaOH) organic matter, and SOC**

	Regression equation		r^2	P	n	Reference
Loss on ignition	= 2.000 SOC	+ 16.1	0.90	0.001	678	Brogan (1966)
	= 1.94 SOC	+ 16.0	0.96	0.001	30	McGrath (1996)
Extract Absorbance	= 0.0116 SOC	+ 0.09	0.56	0.001	30	McGrath (1996)
at 465 nm)	= 0.0157 SOC	- 0.37	0.62	0.001	38	McGrath (1988)

*Units, mg g^{-1}

indicators of herbicide effectivities in soils (McGrath and McCormack, 1979; Strek *et al.*, 1990) would appear to be less useful (Table 1). Sodium hydroxide (NaOH, 0.5M) removed about one third of SOC in a single extraction (McGrath, 1988b). Absorbance of the NaOH extract was significantly correlated ($r^2 = 0.62$) with SOC. Absorbance of extracts correlated well ($r^2 = 0.94$) with carbon in solution (McGrath, 1988b). Regression of extracted carbon on SOC ($r^2 = 0.76$) was not a much better measure (McGrath, 1988b). Consideration of soil hue and value and chroma (for tillage soils) showed that, whereas a weak significant correlation existed between the latter two parameters (at $P < 0.05$) and SOC, they were much too poor (r2 < 0.10) to have any predictive value (McGrath, 1973c).

Humic (HA) and Fulvic (FA) Acids

In one investigation (McGrath, 1988b), OM was extracted with 0.1M sodium pyrophosphate in 0.1M sodium hydroxide (Table 2). Again, about one third of the soil carbon was removed in a single extraction. Proportionately more HA was removed with increasing SOC and proportionately less FA. For that reason the HA/FA ratio increased significantly with increasing SOC. Soil pH influenced FA, but not HA. Whereas the weaker relationship with soil pH had previously been noted (Tan *et al.*, 1972), the relationships to soil carbon appear to be more exceptional.

Table 2 *Relationships* between extractable (0.1M sodium pyrophosphate:0.1M sodium hydroxide) humic substances and ‡SOC (n = 38)*

	Regression equation		r^2	P
Extractable carbon	= 0.343 SOC	- 0.48	0.93	0.001
(Humic acid C/SOC) 100	= 0.618 SOC	+ 9.83	0.36	0.001
(Fulvic acid C/SOC) 100	= - 0.576 SOC	+ 25.20	0.16	0.05
Humic acid C/Fulvic acid C	= 0.054 SOC	+ 0.334	0.55	0.001
	= - 0.149 pH	+ 1.494	0.11	0.05

*Units mg $^{-1}$; ‡SOC = soil organic carbon

Table 3 *Regression*of carbohydrate components on soil organic carbon (SOC)*

	Regression equation	r^2	P
Uronic acid	= 0.63 - 0.0188 SOC	0.80	0.001
Hemicellulose sugars	= 4 00 + 0.177 SOC	0.83	0.001
Total sugar	= 8.77 + 0.367 SOC	0.86	0.001

*Units, mg g^{-1}; n = 38

Soil Carbohydrates

Regression analyses were carried out on SOC for various carbohydrate fractions (McGrath, 1973b). Not surprisingly, relationships were all highly significant, viz., carbohydrate levels in soils were dependant on the amount of organic carbon (Table 3). Uronic acid comprised approximately 1% of organic matter, hemicellulose sugars 15%, and total sugars 25%. It would seem likely that much carbohydrate derived from fibre and, since all soils were from permanent pasture, this would not be surprising.

However, when "soil categories", soil series and soils of low and high SOC, were examined by analysis of variance another picture emerged (Table 4). There is no apparent reason why two soil "categories" should be different from the others. Soils from the Clonroche series, would on the basis of SOC, age of pasture and drainage characteristics, be expected to resemble low rather than high carbon soils as was found. Perusal of SOC levels and of 'fibre' content (Brogan, 1966) offered no explanation.

Inorganic Components of Soils

In a recent study, (McGrath, 1995a; McGrath, 1995b), total and EDTA - extractable Cd, Cr, Cu, Hg, Ni, Pb and Zn were measured in a range of Irish soils. Remarkably strong relationships (at $P \leq 0.001$) with SOC were found for the amount of Zn extracted ($r^2 = 0.81$) and for the proportion of Zn in soil extracted ($r^2 = 0.85$). Total zinc in soil was not related (at $P \leq 0.05$) to SOC, nor was the amount of zinc extracted

Table 4 *Analysis of variance of the relative proportions of uronic acid and neutral sugars in different soil categories*

Soil category	n	SOC* (mg g^{-1})	Uronic acid/SOC	Hemicellulose sugar/SOC	Total sugar/SOC
Clonroche	9	47.4a†	0.028ab	0.203a	0.471a
Elton	5	64.4a	0.036bc	0.309b	0.634b
Drumlin	5	56.3a	0.032b	0.304b	0.621b
Low carbon	9	24.3b	0.044a	0.309b	0.609b
High carbon	10	137.2c	0.023a	0.195a	0.414a

†Values followed by different letters were significantly different from each other using analysis of variance. *SOC = soil organic carbon

Table 5 *Regression of K_d for different pesticides on SOC and on other soil parameters (mg g^{-1})*

		Regression equation	n	r^2
K_d Lindane	=	1.75 SOC - 6.7	40	0.96
	=	0.862 LOI† - 16.8		0.91
	=	5.59 Extractable carbon* + 0.95		0.77
	=	8.50 Absorbance + 29.0		0.54
	=	0.276 Clay + 13.3		0.29
K_d Atrazine	=	0.087 SOC + 0.21	40	0.91
	=	0.043 LOI - 0.34		0.88
	=	0.276 Extractable carbon + 0.59		0.74
	=	0.420 Absorbance + 1.98		0.51
	=	0.0136 Clay + 1.22		0.27
K_d 2,4-D	=	0.027 SOC + 0.21	36	0.65
	=	0.013 LOI + 0.19		0.54
	=	0.108 Extractable carbon + 0.10		0.79
	=	0.167 Absorbance + 0.67		0.52
	=	0.0052 Clay + 0.37		0.28

†LOI, Loss on ignition: *Extracted with 0.5M NaOH

related to the amount in soil. Comparable relationships to that for Zn were not found for other metals. This behaviour of zinc needs further investigation. A similar instance involving Cu but not Zn has been reported (Filipek and Powlowski, 1990). However, it has been found for Scottish soils that the amount of Zn removed by a number of extractants, including one incorporating EDTA, was positively correlated with soil cation exchange capacity for soils of high organic OM contents (Jahiruddin *et al.*, 1992)

Iodine retention in soil is dependent on SOC levels as well as on extractable Al (Whitehead, 1978; McGrath, 1988a). For most soils the influence of the former (SOC) is much less than the latter and would appear to contribute little to overall retention. However, for loamy sands and sandy loams, the influence of SOC on iodine retention was most apparent (McGrath, 1988a). Regression analyses of iodine on SOC, and combined extractable Al for these soils ($n = 18$) gave r^2 values of 0.34 and 0.47, respectively, and a combined r^2 value in multiple regression of 0.77.

Adsorption of Lipophilic Organics

In an investigation with 30 soils, selected for regional and soil type representation, the adsorption coefficient (K_d) was measured for three representative chemical types, lindane (1α, 2α, 3β, 4α, 5α, 6β hexachlorocyclohexane), atrazine (2-chloro-4-ethylamino-6-isopropylamino-1,3,5-triazine) and 2, 4 D [(2,4-dichlorophenoxy) acetic acid] (McGrath, 1996). Regression analyses were carried out between K_d and SOC and related parameters, loss on ignition, absorbance of NaOH extracts, clay, sesquioxides

Table 6 *Regression of ED50[1] on SOC*

SOC	Atrazine	C IPC	Linuron	Pyrazone
Low* (n=5)	0.87*[3]	0.90**	0.85**	0.67*
High* (n=9)	0.91**	0.86**	0.95**	0.57*
Combined	0.82**	0.75**	0.86**	0.34*

[1] ED50 = concentration of test substance in soil (µg g^{-1}) which gives a 50% depression in growth of oat or turnip seedlings
+SOC ranged from 18.0 to 35.0 mg g^{-1} (low) and 36.1 to 95.5 mg g^{-1} (high); [3]*P < 0.05; **P < 0.01

and pH. Relationships were weak or non-significant except with SOC and related parameters. These relationships involving lindane, atrazine, and 2,4-D with SOC (Table 5), are perhaps more compelling than most derived elsewhere, indicating the strong influence of organic matter in Irish soils. In particular, atrazine adsorption indicated no large departure from normality as is occasionally found in other European soils (Brouwer *et al.*, 1990; Hermann, 1992).. Adsorption of 2,4-D by SOC is rarely reported as being significant without additional consideration of pH (Moreale and van Bladel, 1980; Barriuso and Calvet, 1992). Significance did increase slightly ($r^2 = 0.73$ compared to 0.65) when pH was considered in addition to SOC. It was anticipated that PCB (polychlorinated biphenyl) levels in areas of low contamination would be influenced by SOC, but this was not found (McGrath, 1995) to be the case..

Herbicide Phytoxicity

Tests have been conducted at field, plot, and pot level to evaluate the influence of SOC on the effectivity of soil applied herbicides. The concentration of substance (µg) in soil (g) which gave a 50% depression in yield of test plant, normally oat or turnip, was used as an index of phytotoxicity. Phytotoxicity was strongly and negatively correlated with SOC for the four herbicides studied. Evidence was obtained (Table 6) for a quantitative difference in the effect of OM in low carbon as distinct from high carbon soils. This effect was supported by a statistically significant difference in oat yield when OM extracted from different soils, using sodium hydroxide, was subsequently returned to a control soil of low OC content (McGrath and McCormack, 1979). Depression of phytotoxicity was not confined to micropollutants. At percentage levels in soil, the depressive effect of diesel oil on ryegrass (*Lolium perenne)* plant growth was also found to be inversely correlated with SOC (McGrath, 1992).

4 Conclusions

These may be summarised as follows:

1. Organic components of soil organic matter, humic substances, and carbohydrates increase with increasing SOC in grassland soils. However, for carbohydrates there is superimposed a soil type effect, and for humic substances a pH effect.

2. Extractability (EDTA) of zinc from soil and retention of iodine by soil are influenced by SOC.

3. Adsorption of lipophilic organics (and polar organics such as 2, 4-D) are strongly and positively influenced by SOC; similarly, herbicide and mineral oil phytotoxicity is negatively influenced by SOC. Evidence was obtained for a qualitative as distinct from a quantitative effect by organic matter in reducing herbicide phytotoxicity.

References

Barriuso, E. and R. Calvet 1992. Soil type and herbicide adsorption. *Intern.J. of Envir. and Anal. Chem.* **46**:117-128.

Brogan, J.C. 1966. Organic carbon in Irish pasture soil. *Irish J. agr. Res.* **5**:169-176.

Brouwer, W.W.M., J.J.T.I. Boestens, and W.G. Siegers. 1990. Adsorption of transformation products of atrazine by soil. *Weed Res.* **30**:123-128

Burney, B., A. Rahman, G.A.C.. Oomen, and J.M. Whittam. 1975. The organic matter, status of some mineral soils in New Zealand, *28th N.Z., Weed and Pest Control Conf.* pp. 101-103.

Byrne, E. 1979. *Chemical Analyses of Agricultural Materials.* An Foras Taluntais, Dublin.

Dormaar, J.F. 1972. Chemical properties of organic matter extracted from a number of horizons by a number of methods. *Can. J. Soil, Sci.* **52**:67-77

Filipek, T. and L. Powlowski. 1990. Total and heavy metal content of some soils of the Lublin mining region. *The Sci. of the Total Environ,* **96**:131-137.

Herrmann, M. 1992. Evaluation of the EEC laboratory ring test. Adsorption/desorption of chemicals in soils. p. 78-142.. In G. Kuhnt and H. Muntau (eds), *Eurosoils. Identification, Collection, Treatment, Characterisation.* Ispra.

Jahiruddin, M., B.J Chambers, .M.S. Cresser, . and N.T. Livesey. 1992. Effects of soil properties on the extraction of zinc. *Geoderma* **52**:199-208.

.McGrath, D. 1973a. Frequency distribution of the percentage of organic carbon in Irish tillage soils. *Irish J. agric. Res.* **12**:109-111.

McGrath, D. 1973b. Sugars and uronic acids in Irish soils. *Geoderma* **10**: 227-235.

McGrath, D. 1973c. Organic carbon in tillage soils. ' Soils: Research. p 64. Report. Soils. An Foras Taluntais, Dublin.

McGrath, D. 1980. Organic carbon levels in Irish soils. p. 259-268. In D. Boels, D.B. Davies and A.E. Johnston (eds), *Soil Degradation.* Wageningen

McGrath, D. 1988a. Iodine levels in Irish soils and grasses. *Irish J. agric. Res.* **27**:75-81.

McGrath, D. 1988b. Extraction of organic matter from Irish soils. *Irish J. agric. Res.* **27**:187-192.

McGrath, D. 1992. A note on the effects of diesel oil spillage on grass growth. *Irish J. agric. Res.* **31**:77-80.

McGrath, D. 1995a. Organic micropollutant and trace element pollution of Irish soils. *The Sci. of the Total Environ.* **164**:125-133

McGrath, D. 1995b. Application of single and sequential extraction procedures to polluted and unpolluted Irish soils. *The Sci. of the Total Environ.* **178**:37-44.

McGrath, D. 1996. A note on the adsorption characteristics of organic pollutants in Irish soils. *Irish J. agric. Res.* **35**:55-61.

McGrath, S.P and Loveland, P.J. 1991. *The Soil Geochemical Atlas of England and Wales.* Blackie.

McGrath, D. and McCormack, R.J. 1979. Herbicide activities in Irish soils. *Irish J. agric. Res.* **18**:89-96.

Moreale, A. and van Bladel, R. 1980. Behaviour of 2, 4-D in Belgian soils. *J. Environ. Qual.* **9**:627-633.

Strek, H.A., Dulka, J.J. and Parsells, A.J. 1990. Humic acid content vs organic matter content for making herbicide recommendations. Comm. In *Soil Science and Plant Anal.* **21**:1985-1995.

Tan, K.H., Beaty, E.R., McCreery, R.A. and Powell, J.D. 1972. Humic-fulvic acid content in soils as related to ley clipping management and fertilisation. *Soil Sci. Soc.Amer. Proc.* **36**:565-567

Whitehead, 1978. Iodine in soil profiles in relation to iron and aluminium oxides and organic matter. *J. Soil Sci.* **29**:88-94.

Investigations of Some Structural Properties of Humic Substances by Fluorescence Quenching

I.P. Kenworthy and M.H.B. Hayes

THE UNIVERSITY OF BIRMINGHAM, SCHOOL OF CHEMISTRY, EDGBASTON, BIRMINGHAM B15 2TT, ENGLAND

Abstract

Humic substances (HS) were sequentially and exhaustively isolated from a podzolic A_o horizon. The sequence of solvents used was distilled water, sodium pyrophosphate (Pyro, 0.1M, pH 7), Pyro (0.1M, pH 10.6), and a sodium hydroxide (0.1M)/Pyro (0.1M) mixture (pH 12.6). Isolates were fractionated using XAD-8 and XAD-4 resins in tandem. Binding constants (K_{oc}) for pyrene were calculated for the various fractions using fluorescence spectroscopy. A general trend was seen in which the less highly charged molecules exhibited larger K_{oc} values. The suggested cage-like properties of HS were investigated by monitoring the fluorescence properties of a pyrene probe placed in solutions of the fractionated HS. Quenching of the pyrene fluorescence by bromide ions in the presence of the HS was consistent with previous work which suggests that the HS provide a hydrophobic environment shielding the probe from the quenching ion. Treatment of the HS with ethanoic acid and base removed this protection, which may indicate that the HS in solution could be associations of lower molecular masses held together by hydrophobic bonding.

1 Introduction

The quenching of pyrene fluorescence by bromide ions has been shown to be a dynamic process (Engebretson and Von Wandruska 1994, Grätzel and Thomas 1973, Abuin *et al.*, 1984), and is, therefore, directly dependent on the availability of the fluorophore to the quencher. Previous experiments using micellar systems have shown that pyrene enters the hydrophobic environment within the micelle, and the 'cage like' structure of the inner core protects the quenching anion from access by the pyrene fluorophore.

Several authors have proposed that humic substances (HS) aggregate to form micellar structures at critical micelle concentrations (CMC) of ca 10 g L^{-1} (Engebretson and Von Wandruska 1994, Gueltzoff and Rice 1994). Piccolo *et al.* (1996), however, suggests that HS are found naturally as micelle-like structures, and that the molecular weight (MW) values of soil and aquatic HS given in current literature are not true values. Using gel permeation chromatography (gpc) they demonstrated the ability of organic acids to alter the molecular weight distribution of a humic sample from above 100 000 Daltons (excluded in the void volume of the gel) to below 20 000 Daltons (the molecular weight cut-off of the gel). They postulated that this change in mass distribution is due to the break up of micelle-like aggregates of the HS. In their hypothesis the small organic acid molecules enter the hydrophobic cage-like structure of the HS, and these are ionized upon addition of base. The repulsive forces between the newly deprotonated carboxylic functionalities are sufficient to 'blow apart' the humic associations, giving rise to a range of smaller humic associations and molecules (see also Hayes, p. 11, this Volume). Mineral acids produced no effect on the molecular mass distribution of the HS. If this hypothesis is correct, the resulting smaller disaggregated humic structures should provide much less protection to pyrene molecules against the quenching effect of the bromide ion.

2 Materials and Methods

Isolation and Fractionation of Humic Substances

Soil (1 kg) from the A_o horizon of a podzol soil from an oak forest near Killarney, Co. Kerry, Ireland, was sequentially extracted using distilled water, sodium pyrophosphate (Pyro, 0.1M, pH 7), Pyro (0.1M, pH 10.6), and a sodium hydroxide (0.1M)/Pyro (0.1M) mixture (pH 12.6). Extraction was repeated with each solvent until the optical density values of the extracts were less than 1.

HS were isolated [as humic acids (HAs), fulvic acids (FAs), and XAD-4 acids] using XAD-8 [(poly)methyl methacrylate] and XAD-4 (styrene divinylbenzene) resins in tandem, as described by Malcolm and MacCarthy (1992) and by Hayes *et al.* (1996). The fractions are referred to as Killarney HAs or FAs.

Quenching of Pyrene Fluorescence by Humic Substances

Aqueous solutions of pyrene were prepared by stirring an excess of the solid in a solution of 0.01M sodium perchlorate adjusted to pH 5 using HCl. The solution was stirred for 24 h and then passed through a 0.2 μm polycarbonate filter to remove excess solid pyrene, then stored in the dark at 5 °C. Increasing amounts of HA and FA in solution were added to the pyrene solution (5 mL) to give final DOC concentrations ranging from 10^{-6} to 10^{-5}M. Samples and blanks were allowed to equilibrate overnight. Corrections for background scattered light were made by recording spectra for 0.01M sodium perchlorate at pH 5. Emission spectra were recorded at 373 nm, with the excitation wavelength set to 335 nm. All results were corrected for the inner filter effect using the equation of Gauthier *et al.* (1986), and Stern-Volmer plots were generated using the equation of Gauthier *et al.* (1986).

Bromide Quenching of Pyrene Fluorescence in the Presence of Humic Substances

Aqueous solutions of pyrene used in salt (KBr) quenching experiments were prepared by stirring an excess of the solid in distilled water for 24 h and then filtering through 0.2 μm pore size polycarbonate filter to remove excess solid, then stored in the dark at 5°C. Quenching of pyrene fluorescence by bromide in the presence of HS was measured by adding increasing amounts of 2M potassium bromide solution to an aqueous solution of pyrene amended with 10 ppm HS. The salt solution contained the same concentration of pyrene and HS to prevent any dilution effects. Spectra were recorded with the excitation

Table 1 *Interaction of pyrene with humic (HA) and fulvic (FA) acids extracted in water, in 0.1M pyrophosphate at pH 7 (7), in pyrophosphate at pH 10.6 (10), and in 0.1M pyrophosphate plus 0.1M NaOH at pH 12.6 (12)*

Sample	Slope, L mg^{-1}	Intercept	r^2	K_{oc} mL g^{-1}
FAWATER	0.089	0.90	0.97	1.65×10^5
FA7	0.045	1.00	0.97	0.83×10^5
FA10	0.074	1.09	0.99	1.36×10^5
FA12	0.090	0.92	0.97	1.57×10^5
FABULK*	0.087	1.01	0.97	1.57×10^5
HAWATER	0.101	0.97	0.98	1.79×10^5
HA7	0.100	1.01	0.99	1.79×10^5
HA10	0.115	0.94	0.98	2.06×10^5
HA12	0.119	0.93	0.99	2.07×10^5
HABULK	0.096	1.11	0.97	1.68×10^5

*BULK refers to HAs and FAs isolated from soil using 0.1M sodium hydroxide

wavelength set to 335 nm and the emission was measured at 373 nm. Absorbance measurements of the HA and FA solutions were made with a Perkin Elmer Lambda 20 uv/vis spectrometer. All results were corrected for the inner filter effect using the equation of Gauthier *et al.* (1986).

3 Results and Discussion

The data in Table 1 give the variation of slopes for the Stern-Volmer plots and the resulting binding constants for the interaction of the fractions of HA and FA with pyrene.

As in previous experiments of this nature, which include one that uses HS isolated from a podzolic soil, the pyrene sorbed more strongly to the HA than to the FA. No single mechanism is accepted at this time for the interaction of natural organic matter (NOM) and polyaromatic hydrocarbons. Gauthier *et al.* (1987) found strong relationships between a number of chemical properties of OM and the K_{OC} values. The degree of unsaturation, and in particular the aromaticity as determined by solid state ^{13}C NMR, produced a correlation factor of $r = 0.997$ with K_{OC}. The absorptivities at 272 nm and H/C ratios of the OM also gave very good correlations with binding constants. Gauthier *et al.* concluded from their study that increasing aromatic character of NOM will increase its reactivity with pyrene. However, for more nonplanar hydrocarbon pollutants, the aliphatic nature of the organic matter may play a much more significant role. The mechanism proposed for the interaction of pyrene with NOM is "hydrophobic bonding" due to van der Waals forces and a thermodynamic gradient forcing the poly aromatic hydrocarbons (PAHs) out of solution. An alternative hypothesis for the interaction of PAHs and NOM would suggest that the PAHs are contained within micelle-like structures formed either intra or intermolecularly by the NOM (Engebretson *et al.*, 1996, Chiou *et al.*, 1986, Gschwend and Wu 1985). In this 'partition-like' mechanism polarity (C/O ratio), size, and flexibility of the macromolecule are the most important factors. Smaller inflexible humic molecules are unable to form intramolecular 'cage-like' structures, but may form intermolecular aggregates (eg. through self-association processes) which will provide hydrophobic environments to which the PAHs can enter. In complete contradiction with the earlier mechanism, aromaticity is seen as explanation for low association constants because aromatic structures increase rigidity within molecules preventing intramolecular interactions from taking place (Engebretson *et al.*, 1996).

The binding experiments with the Killarney HS, and the bromide quenching experiments described in the next section would suggest that both mechanisms may be in operation. The work of several authors supports the view of 'cage-like' structures which provide hydrophobic environments into which pyrene molecules can enter (Engebretson *et al.*, 1996, Chiou *et al.*, 1986, Gschwend and Wu 1985, Gratzel and Thomas 1973, Gueltzoff and Rice 1994), and would suggest that HAs produce larger K_{oc} values than FAs from the same source. That might suggest that molecular mass is an important factor. However, the K_{oc} values for the Killarney HS would suggest that the charge on the macromolecules is also of great significance. The general trend suggests that the less highly charged molecules exhibit larger K_{oc} values, and that is the reverse of what would be expected if the partition-like mechanism were the only factor in HS-PAH interactions because the C/O ratios are highest for the molecules with greater charge. If C/O ratios or polarity were the main factor in these associations, FAs would be expected to interact most strongly with PAHs. It is likely, therefore, that the increased aromatic character of HAs are responsible for their larger K_{oc} values. It is possible that the mass, polarity, and flexibility of the macromolecule (or assembly of macromolecules) controls the nature of the 'cage-like' structures formed by the HS. However, once the PAH molecule has entered this hydrophobic environment, interactions are governed by the degree of unsaturation within the micelle-like structure. If the exterior of the micelle is highly charged, diffusion of the PAH into the hydrophobic interior may become more difficult.

Effects of Bromide on Pyrene Fluorescence in an Aqueous Solution of Humic Substances

Engebretso *et al.* (1996), and Engebretso and Von Wandruszka (1994) showed that HS inhibit the quenching effect of bromide ions on fluorescence. They also showed that pyrene fluorescence actually increased on addition of small concentrations of KBr in the presence of some HA samples, and this observation was attributed to intramolecular interactions within the humic molecule.

Figure 1 shows the quenching effect of bromide ions on pyrene in the presence and in the absence of Killarney HAs and FAs. A decrease in bromide quenching is seen in the presence of both HAs and FAs. The salt can be expected to give a suppression of the electrical double layer effects allowing the molecules to come closer together, and the intramolecular repulsions to be lessened. This effect could be expected to be greater as the salt additions are increased, and the molecular associations could have 'cage-like' properties. The intramolecular associations are more tightly bound than the intermolecular associations, and are more active in preventing diffusion of bromide ions to the pyrene. This observation is reinforced by the work of Schlautman and Morgan (1993) who showed that binding constants increased at higher pH values in the presence of Ca^{2+} (relative to those in the presence of Na^{+}). The divalent Ca^{2+} will give rise to intramolecular cation bridging especially, which will cause the molecules to shrink and to provide 'cage-like' structures. This effect will be accentuated at higher pH values because dissociations of the weakly acidic groups will provide additional ionized groups that are able to interact with the Ca^{2+}. None of the Killarney samples produced fluorescence enhancements on

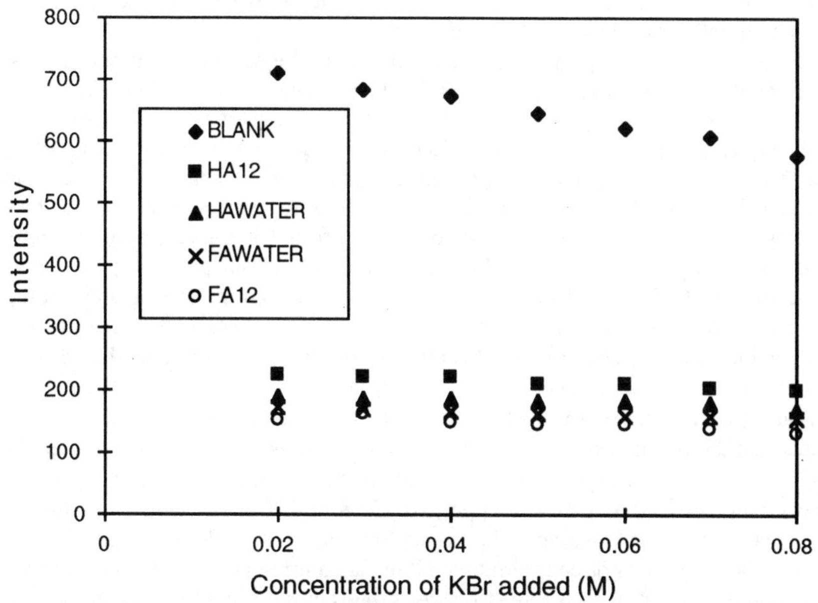

Figure 1 *Fluorescence intensity of pyrene as a function of KBr concentration in the presence of 10 mg L^{-1} Killarney humic (HA) and fulvic (FA) acids isolated in water (HA and FAWATER), and at pH 12 (HA12, FA12)*

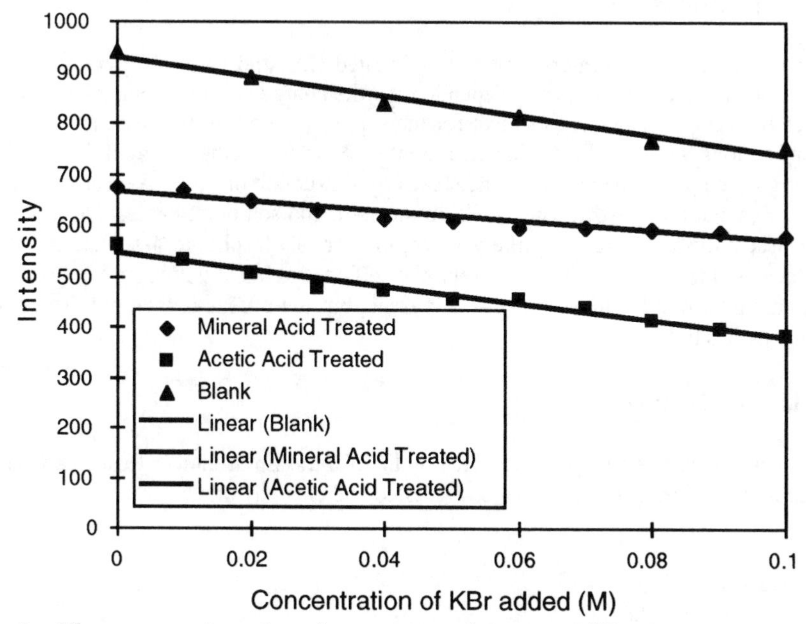

Figure 2 *Fluorescence intensity of pyrene as a function of KBr concentration in the presence of Killarney humic acids treated with mineral acid or ethanoic acid. Linear denotes results of linear regression analyses*

addition of small concentrations of KBr. This could be the result of the molecules being too small or inflexible (Engebretson *et al*., 1996).

Figure 2 shows the quenching effects on pyrene fluorescence of bromide ions in the presence of Killarney HAs treated with mineral acid and then adjusted to pH 9, and in the presence of HA previously treated with ethanoic acid and then adjusted to pH 9.

The quenching of pyrene by bromide in the presence of the HA treated with mineral acid shows a lessening in quenching (relative to that for the blank) as the concentration of KBr added was increased. It is considered that the decrease in quenching is attributable to the presence of intermolecular associations which form 'cage-like' structures that protect the pyrene from the bromide ions. In contrast, the slope of the plot of intensity versus the concentration of KBr added shows that quenching followed the same pattern in the cases of the ethanoic acid treated HA and the blank. The increased fluorecence inhibition displayed by the HA treated with ethanoic acid (in comparison to that shown by the HA treated with mineral acid) could be the result of interactions of pyrene with newly released hydrophobic sites on the HA molecules released from HA aggregates. However, it would appear that the pyrene associated with these humic molecules was not protected from the fluorescence inhibiting properties of the bromide ions. In all experiments, the concentrations of the HS were 10 ppm, which is ca 100 times lower than the CMC as predicted by previous research. This result, and the work of Piccolo *et al.* (1996) would suggest that, regardless of concentration, HS isolated using current procedures can exist as micelle-like aggregates, or as self associated groups of molecules which can have a pseudo micelle type property in solution.

4 Conclusions

The K_{OC} values for the interaction of fractionated HAs and FAs with pyrene suggest that the interactions are, to a degree, dependent on the charge and functionalities of the humic macromolecules. The trends shown for the suppression of the fluorescence of pyrene by bromide ions would indicate that treatment of HAs with ethanoic acid followed by the raising of the pH to the alkaline range leads to destruction of 'cage-like' (or self assembly) structures that are characteristic of HS in the solid and solution phases. These 'cage-like'/ self assembly structures are likely to result from hydrophobic associations of humic molecules (Piccolo *et al.*, 1996). When the self-assembled structures are broken up more sites are provided for interactions with pyrene, but the pyrene is then not protected from attack by bromide.

Acknowledgements

This work is part of a project supported by the Natural Environment Research Council for studies of humic substances in paired (forested and cleared) sites.

References

Abuin, E.A.L., L. Seplveda, and F.H. Quina. 1984. Ion exchange between monovalent and divalent counterions in cationic micellar solution, *J. Phys. Chem.* **88**:81-85.

Chiou, C.T., R.L. Malcolm, T.I. Brinton, and D.E. Kite. 1986. Water solubility enhancement of some organic pollutants and pesticides by dissolved humic and fulvic acids. *Environ. Sci. Technol.* **20**:502-508.

Engebretson, R.R., T. Amos, and R. Von Wandruszka. 1996. Quantitative approach to humic acid associations. *Environ. Sci. Technol.* **30**:990-997.

Engebretson, R.R. and R. Von Wandruszka. 1994. Microorganization in dissolved humic acids. *Environ. Sci. Technol.* **28**:1934-1941.

Gauthier, T.D., E.C. Shane, W.F. Guerin, W.R. Seitz, and C.L. Grant. 1986. Fluorescence quenching method for polycyclic aromatic hydrocarbons binding to dissolved humic materials. *Environ. Sci. Technol.* **20**:1162-1166.

Gauthier, T.D., W.R. Seitz, and C.L. Grant. 1987. Effects of structural and compositional variations of dissolved humic materials on pyrene K_{oc} values. *Environ. Sci. Technol.* **21**:243-248.

Grätzel, M., and J.K.Thomas. 1973. On the dynamics of pyrene fluorescence quenching in aqueous ionic micellar systems. Factors affecting the permeability of micelles. *J. Am. Chem. Soc.* **95**:6885-6889.

Gschwend, P.M., and S.Wu. 1985. On the constancy of sediment water partition coefficients of hydrophobic organic pollutants. *Environ. Sci. Technol.* **19**:90-96.

Gueltzoff, T.F. and J.A. Rice 994. Does humic acid form a micelle? *Sci. Total Environ.* **152**:31-35.

Piccolo, A., S. Nardi, and G. Concheri. (1996). Macromolecular changes of humic substances induced by interaction with organic acids. *European Journal of Soil Science* **47**:319-328.

Schlautman, M.A., and J.J. Morgan. 1993. Effects of aqueous chemistry on the binding of polycyclic aromatic hydrocarbons by dissolved humic materials. *Environ. Sci. Technol.* **27**:961-969.

Applications of NMR Spectroscopy for Studies of the Molecular Compositions of Humic Substances

A.J. Simpson, R.E. Boersma[1], W.L. Kingery[2], R.P. Hicks[1], and M.H.B. Hayes

THE UNIVERSITY OF BIRMINGHAM, SCHOOL OF CHEMISTRY, EDGBASTON, BIRMINGHAM B15 2TT, ENGLAND
[1]DEPARTMENT OF CHEMISTRY, MISSISSIPPI STATE UNIVERSITY, MS 39762, USA
[2]DEPARTMENT OF SOIL AND PLANT SCIENCES, MISSISSIPPI STATE UNIVERSITY

Abstract

Considerable effort has focused in recent years on applications of cross polarization magic angle spinning (CPMAS) ^{13}C nuclear magnetic resonance (NMR) spectroscopy to studies of the compositions of humic substances (HS). This technique has provided valuable information on some functionalities, and about the relative abundances of these in the macromolecules. The spectra provide an excellent 'fingerprint' technique for comparing fractions from the same source, and similar fractions from different sources. However, in its present state of development, CPMAS ^{13}C NMR spectroscopy cannot provide detailed structural information about the heterogeneous mixture of components that compose HS.

The development of powerful NMR spectrometers has caused interest to focus on applications of proton (^{1}H) NMR spectroscopy to studies of humic structures. The present contribution highlights the uses of ^{1}H NMR and combinations of ^{1}H and ^{13}C NMR for studies of aspects of the compositions of humic fractions. A fulvic acid isolated at pH 12.6 from the humified portions of a moss culture growing on a rock surface was studied using one dimensional (1-D) ^{1}H NMR, 2-D ^{1}H-^{1}H COrrelation SpectroscopY (COSY), 2-D ^{1}H-^{1}H TOtal Correlation SpectroscopY (TOCSY), and 2-D ^{1}H-^{13}C Heteronuclear Multiple Quantum Coherence (HMQC) spectroscopy using high field 500 MHz and 600 MHz spectrometers. The spectra illustrate the potential of the techniques for determinations of structures. However, it will be necessary to develop improved fractionation procedures in order to be able to obtain an advanced awareness of the structures in the more highly humified HS.

1 Introduction

Macromolecules of terrestrial and aquatic origins help form essential linkages between the atmosphere, lithosphere, hydrosphere, and biosphere (Kingery and Clapp, 1997). Humic substances (HS), the most abundant of the organic macromolecules of nature, have an essential role in these linkages. It is important to have an awareness of the chemical structures of naturally occurring substances in order to understand properly the reactivities of these in the physical, chemical, biological, and environmental processes that are requisite to the development of viable ecosystem restoration and/or management planning. However, because the genesis of HS involves combinations of several reaction pathways and a wide variety of chemical bonding systems (Hayes *et al*., 1989a; Stevenson, 1994), it has been very difficult to evolve any clear concepts of their structures.

Hayes *et al.* (1989a) have outlined the difficulties in applying the classical concepts of structural chemistry (i.e., empirical, molecular, and structural formulae, configuration, conformation, and primary through to quaternary structure, to complex humic macromolecules. Nevertheless, as Hayes *et al.* (1989b) and Stevenson (1994) have pointed out, considerable progress has been made this generation in providing an awareness of some of the gross features of humic structures. Many of the classical methods were based on elemental compositions, and manipulations of such data have been made for more than 170 years of studies of HS (Orlov, 1985). Steelink (1985) has emphasized the significant contributions that elemental analyses and atomic ratio data can make to compositional information. It is, however, essential to emphasize that elemental analyses data represent averages for agglomerations of molecules, and it is impossible to derive precise empirical formulae from these (Hayes *et al.*, 1989).

A second logical line of enquiry is based on the notion that a knowledge of structures could evolve from identifications of the products of chemical degradations of HS. In the first contribution in this Volume, Hayes (p. 20 - 23) has outlined the principles involved in that approach. Chemical degradation techniques such as acid/base catalyzed hydrolysis (Parsons, 1989), oxidative (Christman *et al.*, 1989; Griffith and Schnitzer, 1989; Hayes and O'Callaghan, 1989) and reductive (Stevenson, 1989; Hayes and O'Callaghan, 1989) processes, and thermal procedures (Bracewell *et al.*, 1989; Schulten and Schnitzer, 1995) have yielded valuable information on possible chemical constituents and building blocks of HS. However, because the major linkages in the 'core' structures of HS are not hydrolyzable, the energy inputs needed to cleave the links between the component molecules give rise to digest products that can be vastly different from the molecules that compose the macromolecular structures. Hayes (p.20, this Volume) has referred to the uses of degradative mechanisms in order to link the compounds identified in the degradation digests with possible structures in the macromolecules. The same kind of approach would apply for thermal degradation procedures.

More recently, various sprectroscopic procedures have been employed for investigations of aspects of the compositions and structures of HS (MacCarthy and Rice, 1985; Wershaw, 1985, Bloom and Leenheer, 1989; Steelink *et al.*, 1989; Wilson 1989, 1990; Malcolm, 1989; Preston, 1996). Infrared has been the spectroscopy technique most widely used for studies of HS (MacCarthy and Rice, 1985; Bloom and Leenheer, 1989), although indications of the natures of only a few of the functionalities in the substances can be obtained by this method. Other spectroscopic approaches, which include electron spin resonance (Senesi and Steelink, 1989), Raman, ultraviolet-visible, fluorescence, X-ray photoelectron (Bloom and Leenheer, 1989) and Mössbauer (Goodman and Cheshire, 1985), have made contributions to the study of compositions and aspects of the structures of HS.

This contribution will focus on applications of nuclear magnetic resonance (NMR) spectroscopy for studies of aspects of the compositions and structures of HS, and with special emphasis on contributions from proton (^1H) NMR spectroscopy.

2 Nuclear Magnetic Resonance Spectroscopy (NMR)

Skoog and Leary (1992) have commented that few developments in analytical chemistry have ever experienced such a short delay between initial discovery and widespread acceptance and application as has NMR. (Bloch at Stanford and Purcell at Harvard shared the Nobel Prize in 1952 for their work in 1946 which demonstrated the proposal first made by Pauli in 1924 that certain atomic nuclei, upon exposure to a magnetic field, would experience splitting of their energy levels. Varian Associates marketed the first high-resolution NMR spectrometer in 1953.) NMR is amongst the most powerful tools available for determinations of structure of both organic and inorganic species. It is likely that the first recorded use of NMR for the examination of soil humics was made in 1963 (Preston, 1996). The earlier studies employed ^1H and ^{13}C liquid-state NMR. The early spectra did not, however, display sufficient resolution to provide useful information (Malcolm, 1989). Substantial progress was made after the introduction of techniques such as multiple pulse Fourier transform NMR, dipolar decoupling, cross-polarization nuclear induction spectroscopy, and cross polarization magic-angle spinning (CPMAS) NMR. These allowed for improvements in functional group assignments and provided potential for their quantification (Malcolm, 1989). Despite these gains, one dimensional (1-D) proton and carbon spectra of HS are still broad and relatively ill-defined, and yield only limited structural information. The complex, heterogeneous nature of these macromolecules results in overlapping of individual signals which prevents unambiguous mapping of the hydrogen-carbon-oxygen framework.

Table 1 *Summary of NMR spectroscopy experiments applicable to structural studies of humic substances*

NMR Experiment	Structural Information
1-D ^1H	Chemical shift provides information on the functional groups present
1-D ^{13}C	Chemical shift-functional groups
2-D COSY	^1H-^1H two and three bond coupling
2-D TOCSY or HOHAHA	^1H-^1H coupling throughout a complete spin system
2-D ROESY or NOESY	^1H-^1H dipolar (through space) coupling information which provides inter-proton distances. Helps to refine structure of the molecule
2-D ^1H-^{13}C HMQC	Via one bond, provides ^1H-^{13}C connectivity information on the specificity of proton-carbon bonding. Supports functional group assignments.
2-D ^1H-^{13}C HMBC	Via two and three bonds, provides ^1H-^{13}C connectivity information to confirm functional group assignments and substitution patterns

Over the past ten to fifteen years, a large number of two-dimensional NMR methods have been developed to determine the structures of organic compounds and biomolecules. Only recently, though, have these powerful techniques been applied to the study of the compositions and aspects of the structures of HS. The liquid state one- (1-D) and two-dimensional (2-D) NMR experiments are examples of these procedures, and the information provided by each is listed in Table 1.

One-dimensional Experiments

In ^1H and ^{13}C NMR, assignments of functional groups can be made according to the relative chemical shifts. These are the kinds of experiments that were first used in applications of NMR to studies of compositions of HS.

Homonuclear Experiments

Two-dimensional (2-D) COSY (COrrelation SpectroscopY) can be obtained using any one of the many different variations of the basic COSY pulse sequence (Derome, 1987). The COSY experiment provides information on ^1H-^1H two and three bond coupling. As with any homonuclear 2-D experiment, the COSY spectrum consists of a diagonal (pseudo 1-D spectrum) running from the lower left-hand corner to the upper right-hand corner. Coupling between protons are indicated by a pair of off-diagonal cross peaks.

Two-dimensional, phase-sensitive HOmonuclear HArtmann HAhn (HOHAHA) spectra, or TOtal Correlation SpectroscopY (TOCSY) provides information additional to that obtained in the COSY experiment. The TOCSY experiments provide, in addition to two and three bond ^1H-^1H coupling, long-range coupling information throughout the ^1H-^1H coupling network.

Two-dimensional NOESY (Nuclear Overhauser Effect SpectroscopY) or ROESY (Rotating frame Overhauser Enhancement SpectroscopY) spectra will provide inter-proton distance information which can be very useful for the determination of three dimensional (3-D) structure.

Heteronuclear Experiments

The ^1H-^{13}C heteronuclear correlation techniques are as important as the homonuclear 2-D experiments. Inverse detection techniques should be used to assign the natural abundance ^{13}C spectrum of the soil samples by way of Heteronuclear Multiple Quantum Coherence (HMQC) and Heteronuclear Multiple Bond Coherence (HMBC) experiments. In inverse detection experiments, the chemical shift and coupling information of the insensitive nuclei, in this case ^{13}C, is transferred to the more sensitive ^1H nuclei. This technique gives a significant increase in sensitivity, due to the larger natural abundance of ^1H compared to ^{13}C. Thus, fewer scans are required to obtain an acceptable signal-to-noise ratio. Heteronuclear 2-D experiments, unlike homonuclear experiments, do not exhibit a spectrum diagonal. The HMQC experiment is used to determine one bond ^1H-^{13}C coupling. This information is used to determine which protons are directly bonded to which carbons.

The HMBC experiment is used to characterise two and three bond ^1H-^{13}C coupling. The information obtained indicates which protons are two and three bonds away from a particular carbon atom. This information is very useful in the assignment of unprotonated carbons. Generally, three bond ^{1}H-^{13}C coupling constants are larger than two bond ^{1}H-^{13}C coupling constants. That should be kept in mind when interpreting the HMBC spectra. It must be remembered that the HMBC experiment is much less sensitive than the HMQC experiment and requires either a much higher concentration of sample or a large number of scans. HMBC spectra often contain artefacts of strong one bond ^{1}H-^{13}C couplings. Such artefacts yield doublets, triplets, or quartets on the proton axis, depending on the number of equivalent protons that are strongly coupled to the carbon atom. These artefacts will not normally line up with the observed proton chemical shifts. Care must be taken not to assign mistakenly strong coupling artefacts to two and three bond ^{1}H-^{13}C couplings.

Sample Preparation

The quality of the humic sample presented to the spectrometer sets an obvious limit to the merit of the information obtained from NMR experiments. HS are associated with each other and with other components, such as polysaccharides and peptides, charge-neutralizing cations, and mineral colloids, and for meaningful structural studies it is critical to be able to separate HS from non-humic materials (Hayes and Swift, 1990), and to provide samples that are relatively homogeneous. Criteria for the properties of solvents (Hayes, 1985) for the extraction and for the fractionation (Swift, 1985) of HS have been presented elsewhere. Fractionation of HS has typically depended on molecular size and charge density differences, and the variability of charge with pH provides a facile method for fractionation on the basis of charge density differences. Adsorption of HS to resins such as XAD-8 [poly(methylmethacrylate)] (Thurman and Malcolm, 1981) and XAD-4 (styrene divinylbenzene) can provide fractionation on the basis of charge density and hydrophobicity/hydrophilicity, and Malcolm and MacCarthy (1992) and Hayes *et al.* (1996) have used these resins in tandem for the fractionation of HS. Using CPMAS ^{13}C NMR, Hayes *et al.* (1996) have observed compositional differences between humic fractions eluted at different pH values from an XAD-8 resin. Fractionation and isolation of HS in this way can enhance the degree of homogeneity, which can allow for more highly resolved NMR spectra. Greater spectral resolution has the potential for confirmation of linkages, and possibly even confirmation of structural units.

3 Materials and Methods

HS were isolated from the B_h horizon of a cleared forest podzol site (the oak trees were felled ca 400 y BP) adjacent to Urah Wood, Lough Inchiquin, Kenmare, Co. Kerry, Ireland (Grid reference: V.83.62), and from the humified root region of a moss monoculture growing on the surface of a large rock within the forest area. Detailed descriptions of the soil profiles in the forested and cleared sites are given by Little (1994), and some relevant details of the sampling and methods of fractionation are contained in the contribution by Simpson *et al.*, (1997a). The sequence of extracting solvents used was water, 0.1M sodium pyrophosphate (Pyro, pH 7), Pyro (0.1M) pH 10.6, and 0.1M Pyro plus 0.1M NaOH at pH 12.6. The humic (HA) and fulvic (FA) acids were isolated as described in the contribution by Simpson *et al.* (1997a), and fractionated using the XAD-8 and XAD-4 resins in tandem technique (Hayes, 1996; Hayes *et al*, 1996). This contribution is primarily concerned with the FA fraction isolated at pH 12.6 from the humified moss.

Table 2 *NMR experimental parameters*

NMR Experiment	Sample Amount	Spectrometer	#Data Points	#Process Points	#Scans
1D ^1H	10 mg	Bruker AMX-600	32768	16384	64
2D ^1H-^1H COSY	10 mg	Bruker DRX-500	512(F1) 2048(F2)	1024(F1) 1024(F2)	32
2D ^1H-^1H TOCSY (F1)	10 mg	Bruker AMX-600	256(F1) 2048(F2)	512(F1) 1024(F2)	16
2D ^1H-^{13}C HMQC (F1)	50 mg	Bruker AMX-600	256(F1) 2048(F2)	512(F1) 1024(F2)	128

= number of

The data presented for the FAs isolated from the B_h horizon at pH 12.6, and for the IHSS Standard Soil (a Mollisol) HA are used to illustrate specific points. The humic materials were dissolved in DMSO and transferred to 5 mm NMR tubes for the experiments. The parameters for each experiment are given in Table 2.

4 Results and Discussion

Extensive NMR data have been obtained for the various humic fractions isolated. The complexities of the data will require extensive processing, and interpretations will take several months to complete. Results are given only for the FA fractions isolated at pH 12.6 from the humified parts of the moss cover close to the rock surface, and spectra for the FAs isolated at pH 12.6 from the deforested B_h horizon and for the HAs of the IHSS Standard Soil HA are given for comparison. The discussion will outline general findings, and more comprehensive interpretations will follow at a later date.

The 1-D ^1H NMR spectrum for the moss FA (Figure 1) is well resolved when compared with non-derivatized spectra for FAs presented in the literature (e.g. see the spectra in Leenheer and Noyes, 1989). This can be attributed to the improved homogeneity of the sample attributable to the extensive fractionation (Hayes *et al.*, 1996) and to the high field strength of the 600 MHz spectrometer used. The 1-D proton spectrum demonstrates in excess of 40 different aromatic/amide resonances, which is far in excess of any detail provided by CP-MAS ^{13}C NMR.

The 1-D proton spectrum can be split into three main regions: 0.8 - 3 ppm caused by aliphatic protons (Oka *et al.*, 1969; Simpson *et al.*, 1997b, this issue; Fleming and Williams, 1989); 3 - 5.5 ppm caused by protons associated with oxygen-containing functionalities [Oka *et al.*, 1969 assigned the region to lactone protons; Wershaw, 1985 assigned it to exchangable protons; more recently the region is thought to represent a wide range of protons associated with oxygen-containing functional groups, and these (protons) can be exchangeable and non-exchangeable (Simpson *et al.*, 1997b)]; and the 6-8.5 ppm region, caused by aromatic/amide protons. Table 3 gives the major functional group assignments that can be made from the 1-D proton spectrum of the moss FA.

It is important to note that the relative strengths (or even the absence) of peaks can provide important information. For example, the resonance at 1.5 - 2.0 ppm is relatively weak in comparison with those in the rest of the spectrum. Such could indicate that the

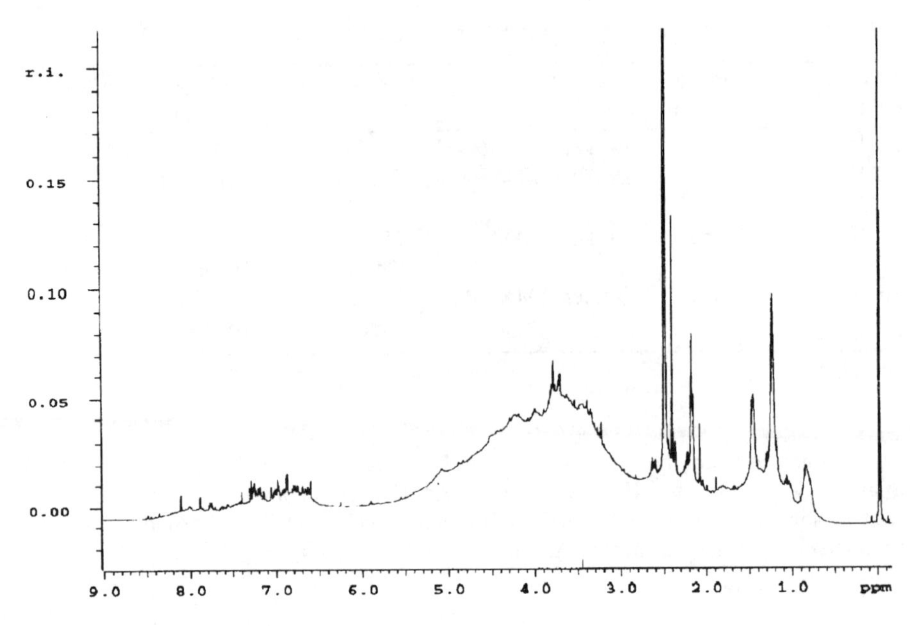

Figure 1 *One dimensional ^1H NMR (600 MHz) spectrum of the moss fulvic acid in DMSO d_6*

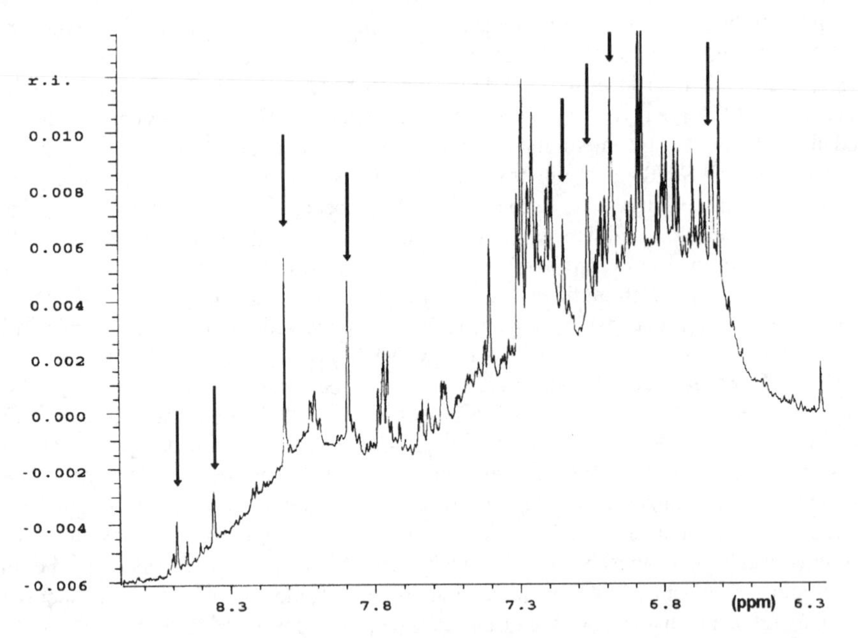

Figure 2 *One dimensional ^1H NMR (600 MHz) spectrum of the moss fulvic acid aromatic/amide resonance taken in DMSO d_6. Arrows indicate resonances from amide protons. These are seen to disappear on addition of D_2O*

Table 3. *Assignments of chemical shifts in the 1-D ^1H NMR spectrum of the FA fraction isolated at pH 12.6 from humified moss*

Chemical Shift (ppm)	Assignment
0.8	Methyl Protons
1.2	Methyl Protons (likely to be CH_3-C-O protons)
1.4	Methine Protons
2.4	CH_3-CO-OAr or R-CH_2-CO-R
2.5	Solvent, DMSOd_6
2.6-2.7	CH_3-CO-Ar or R-CH_2-Ar
3-5.5	Protons associated with oxygen containing functionalities
6-8.5	Aromatic/amide protons

protons of functionalities such as R-CH_2-C-O-R' (long chain ethers), R_1R_2CH-R_3 (the ethine protons of branched aliphatics), and CH_3-C=C/R-CH_2-C-C=C protons (associated with certain unsaturated systems) do not contribute to a significant extent to the FA structure. The relatively weak resonances at 2.3 and at 2.7 ppm, which are representative of CH_3-Ar and R-CH_2-Ar units, respectively, may indicate that alkyl groups may not be major substituents in the aromatic functionalities. The intensities of the peaks may also provide information. For example, in a compound containing long chain aliphatic hydrocarbons, the peak at 1.4 ppm (CH_2 protons) would predominate over those at 1.2 and 0.8 ppm (mainly methyl protons). However, that is not the case for the FAs under study. The data would suggest that long chain aliphatic components are not major contributors to the structures, and aliphatic substituents and bridges are more likely to be short, one or two unit structures.

Addition of D_2O to the solvent provides information about the exchangeable protons that might be present as indicated by the 1-D ^1H experiment. This technique gives rise to the exchange of amide and hydroxyl protons for deuterium, and thereby to the disappearance of the signals for the protons of these functionalities. Addition of D_2O to the moss FA sample decreased the intensity of the 3 - 5.5 ppm resonances, indicating the exchange, under acidic conditions, of amino/amido/hydroxyl protons. However, the D_2O treatment can give more vital information about the amide/aromatic resonances.

Figure 2 shows the expanded 1-D proton spectrum for the 6.3 to 8.5 ppm aromatic/amide region for the moss HA. More than 40 aromatic/amide proton resonances can be seen, and protons attributable to amide structures are marked by arrows. Addition of D_2O causes peaks to disappear (see Figure 3), and these peaks can be considered to represent amide protons. The decreased resolution in the aromatic resonance after D_2O exchange is attributable to H_2O contamination in the added D_2O. The high intensity of the water signal at δ = 3.3 ppm dominates the spectrum and dwarfs resonances in other regions, and the spectra can then appear to be less well defined on expansion. A watergate pulse sequence was used to partially quench the water signal (Gadian, 1982).

The COSY spectrum for the moss FA sample shows good resolution and connectivities, and couplings are observed through two and three bonds (Figure 4). The broad band from δ = 3 - 5.5 ppm, caused by a large number of overlapping signals, cannot be interpreted from the 1-D spectrum. However, the COSY spectrum indicates a number of coupled hydrogens in the form of cross peaks. Not only can these chemical shifts be

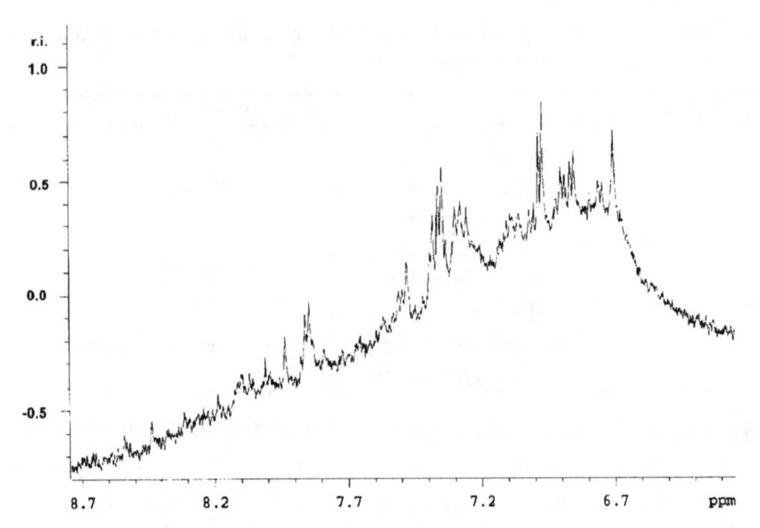

Figure 3 *One dimensional ^1H NMR (600 MHz) spectrum of the aromatic/amide resonance of the moss fulvic acid in DMSO d_6 after the addition of D_2O*

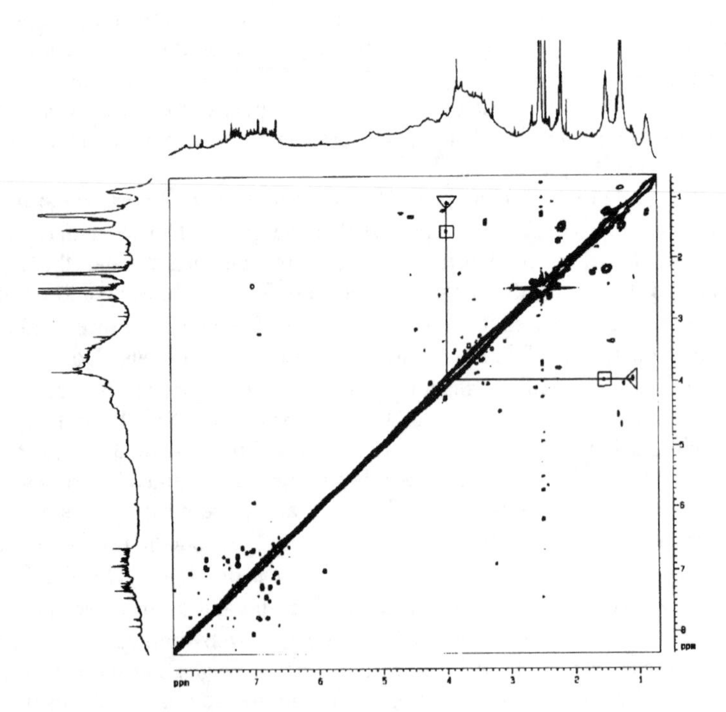

Figure 4 *Two dimensional COSY (^1H-^1H; 500 MHz) NMR spectrum of the moss fulvic acid in DMSO d_6. The cross peaks at $\delta = 4.1$ ppm represent R-CH_2-O-CO-R protons and are seen to couple to both the methylene (square marker) and the methyl (triangular marker) protons*

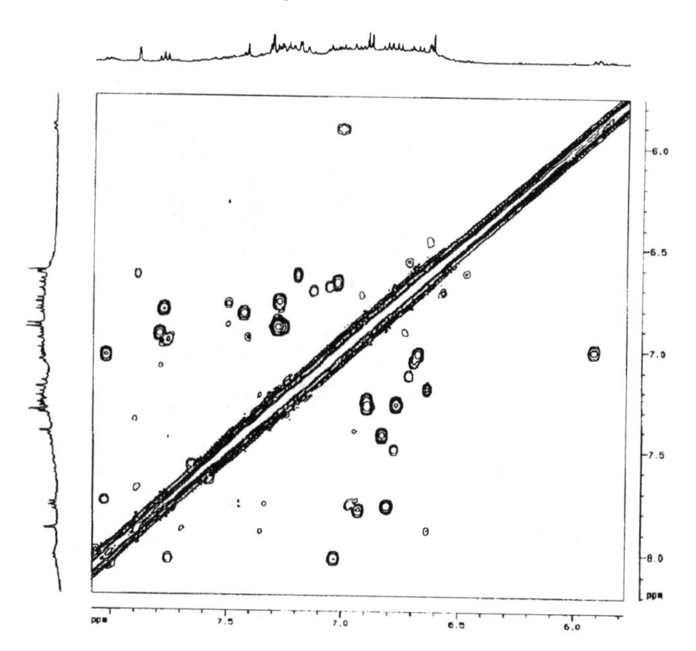

Figure 5 *Two dimensional COSY (^1H-^1H; 500 MHz) NMR spectrum of the aromatic/amide resonance of the moss fulvic acid in DMSO d_6*

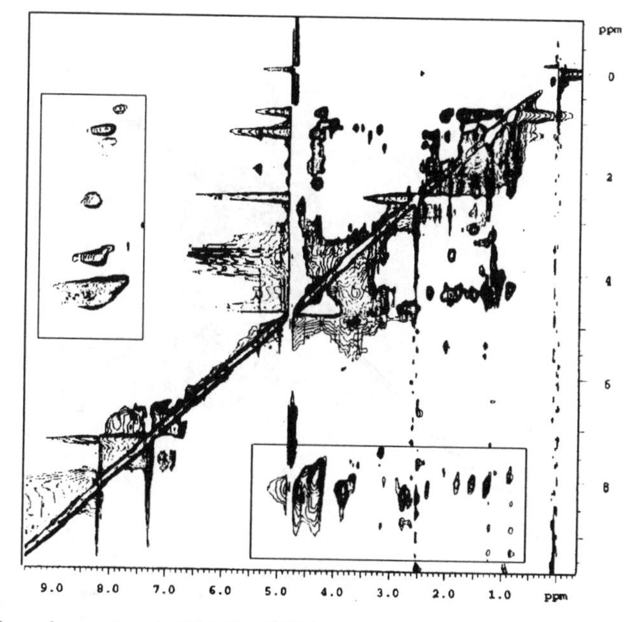

Figure 6 *Two dimensional TOCSY (^1H-^1H; 600 MHz) NMR spectrum of the IHSS Standard soil humic acid in DMSO d_6. The boxes highlight couplings from amino acid moieties*

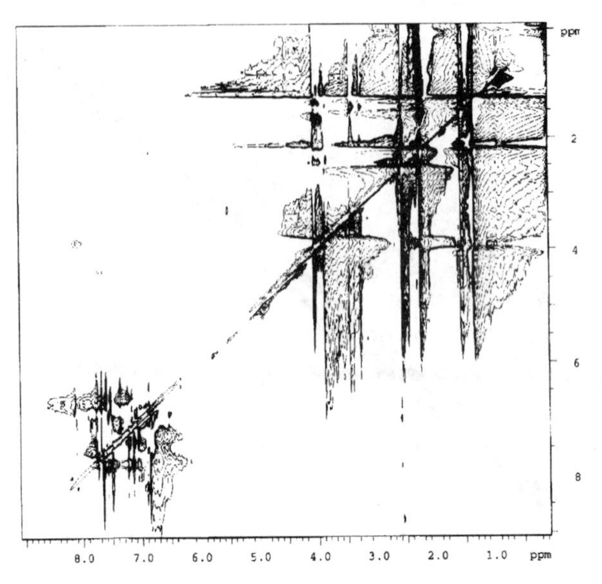

Figure 7 *Two dimensional TOCSY (^1H-^1H; 600 MHz) NMR spectrum of the fulvic acid from a podzol B_h horizon in DMSO d_6*

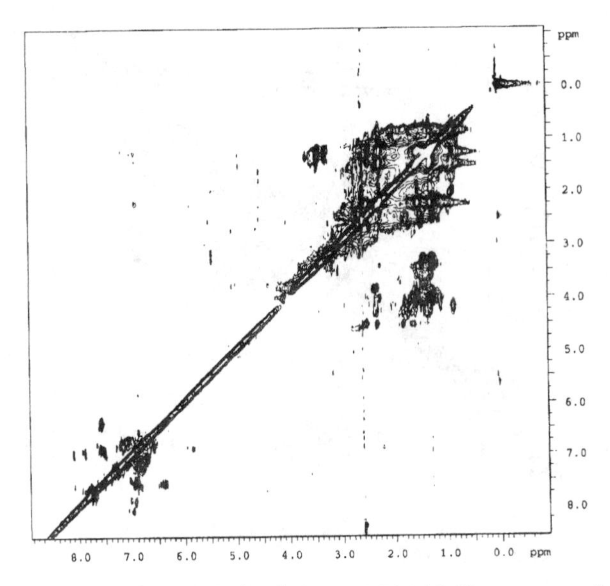

Figure 8 *Two dimensional TOCSY (^1H-^1H; 600 MHz) NMR spectrum of the fulvic acid from a B_h horizon in DMSO d_6 after the addition of D_2O*

used to identify possible moieties, but they can indicate which units may be connected to each other. To illustrate this point we select as an example the two cross peaks at 4.1 ppm. The chemical shift of the peaks would indicate protons on R-CH_2-O-CO-R units, and their connectivity to both the methyl and methylene regions infers the presence of both CH_3-CH_2-O-CO-R and R-CH_2-CH_2-O-CO-R units.

The COSY spectrum provides considerable information about the aromatic/amide region. To be coupled, aromatic and amide protons must be adjacent to each other, and the coupling of amide protons can be eliminated by D_2O exchange. From careful examination of the aromatic/amide resonance of the moss FA (Figure 5), 24 cross peaks can be seen representing aromatic protons, whereas the amide protons appear not to be coupled. In aromatic systems, the presence of neighbouring hydrogens gives information about the level and the pattern of substitution. When there are two adjacent hydrogens on a ring there cannot be substituents on the 1,3,5 ring positions, and rings with more than four substituents can be ruled out.

TOCSY experiments allow the short and long range coupling of hydrogens to be observed. Such data help compliment information found in COSY spectra and can give clues about how units will fit together. The TOCSY experiment is especially useful for detecting the couplings between the amido protons and protons found on the side chains of amino acid residues. Although COSY spectra allow adjacent couplings to be seen which indicate the presence of linked amino acids (as in peptide-type structures), TOCSY spectra allow long range couplings to be seen. Such information is useful in identifying the amino acids present. The TOCSY spectrum of the moss sample gives little information additional to that obtained from the COSY spectrum. However, the TOCSY spectrum for the IHSS Standard soil HA clearly demonstrated amino acid coupling (Figure 6). (This Standard contains ca 8% amino acids.) Amido couplings can be confirmed by taking advantage of the exchangeable properties of the amido group. The couplings disappeared upon the addition of D_2O.

The addition of D_2O was noted to have an enhancing effect on the resolution obtained in the TOCSY spectra of some samples. In the case of the FA extracted from the B_h horizon, the TOCSY spectrum was complex, and with large areas of blurring (see Figure 7). However, the resolution of the spectrum became greatly enhanced after the addition D_2O (see Figure 8). This can be attributed to a decrease in rolling in the baseline attributable to large, broad, exchangeable bands. This makes possible a more refined phasing of the data that allows signals to become apparent that previously were hidden by blurring.

The heteronuclear HMQC experiment produced a large range of data for all the H-C bonds present in the moss sample (Figure 9). We are focusing on the the $\delta = 3 - 5.5$ ppm resonance because little information could be obtained for this resonance from the 1-D and 2-D homonuclear experiments. The region becomes highly resolved when both the carbon and hydrogen chemical shifts are considered, and it can be seen that 20-30 different C-H bonds contribute to the resonance (Figure 10). The amino and hydroxyl groups do not show up on the HMQC spectrum, but are likely to cause the overlap and broadening on the 1-D spectrum.

The intense cross peak at $\delta = 3.9$ ppm (^1H) and at 55 ppm (^{13}C) is indicative of methoxyl bound to an aromatic ring, as is characteristic of structures derived from lignin. The resonance between 60 and 105 ppm in ^{13}C NMR spectra is often considered to be the 'carbohydrate region'. However, the HMQC data suggest that the cross peaks at

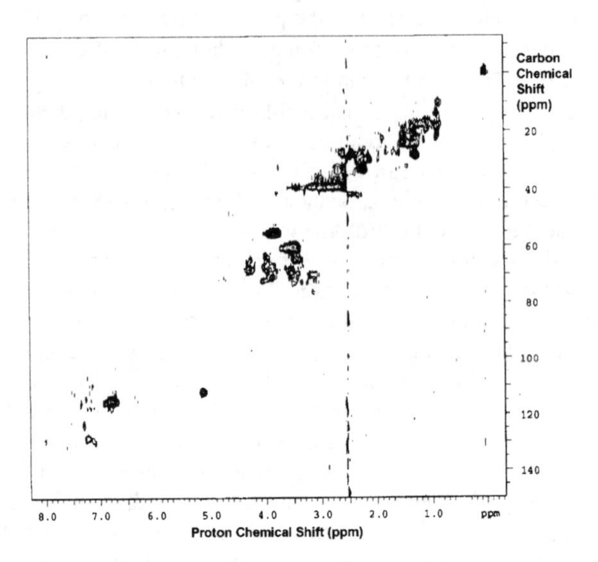

Figure 9 *Two dimensional HMQC (^{1}H-^{13}C; 600 MHz) NMR spectrum of the moss fulvic acid in DMSO d_6*

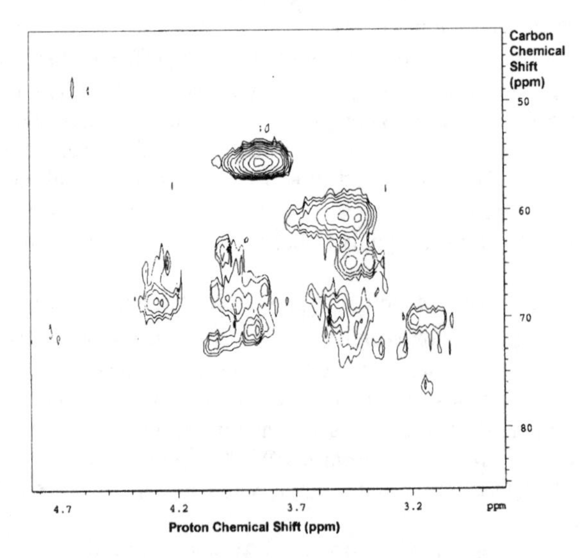

Figure 10 *Two dimensional HMQC (^{1}H-^{13}C; 600 MHz) NMR spectrum of the moss fulvic acid for the δ = 2.8 - 4.8 ppm resonance (assigned to protons associated with oxygen containing functionalities) in DMSO d_6*

between δ = 3.1 and 3.7 ppm (^1H) and 60 - 75 ppm (^{13}C) are consistent with protons bound to the carbon atoms of carbohydrate/carbohydrate-related, and of ether-type structures. (The neutral sugar content of the sample was 6.6%).

The cross peaks in the resonance between δ = 3.8 to 4.1 ppm (^1H) and 65 to 75 ppm (^{13}C) are consistent with ester functionalities, and those at δ = 4.3 ppm (^1H) and 70 ppm (^{13}C) are indicative of the CH_2 protons in RCH_2OAr -type functionalities.

5 Conclusions

Hayes *et al.* (1989a) have emphasized that the heterogeneity of the components of humic substances presents the biggest barrier to structural determinations. The exhaustive sequential extraction procedure applied by us achieved significant fractionation based on charge density differences. The material extracted at pH 12.6 was not as extensively humified as the fractions isolated under less basic conditions, and it is considered that the heterogeneity was less than for the samples dissolved at the lower pH values. It is clear from the data we have reported that considerable information can be obtained for the samples studied from 2-D NMR spectroscopy experiments. However, the spectra obtained for the humic fractions isolated at the lower pH values were less well resolved, although couplings could be observed.

The heteronuclear experiments help highlight structural components, and the homonuclear experiments indicate connectivities. When considered together the homonuclear and heteronuclear approaches provide very powerful tools for determinations of aspects of structure. Identification of products in chemical degradation digests can give leads with regard to the nature of the structural units (see Hayes, 1997, this issue). This can help in the identification of such units in the NMR spectra, and considerations of couplings can provide information about connectivities.

The interpretation offered in this paper is intended only to highlight the potential of the various NMR procedures employed. We have additional experiments underway, such as NOESY and ROESY which give conformational information. The ROESY experiment is providing information that is more useful than that from the NOESY. HMBC heteronuclear experiments that allow C-H coupling to be seen over 2 and 3 bonds are being investigated, and with some limited success. The major problem with the HMBC is likely to arise from the large molecular weight of the humic fractions which would cause relaxation before the long range couplings can be detected. Modified pulse sequences may lead to the acquisition of more refined structural data. Such could make calculation of functional group substituents detectable three bonds removed, and allow predictions of aromatic substitution patterns and the mapping of large areas of structure. Achievement of this type of objective will be made easier if more refined fractionations of the humic mixtures can be achieved.

References

Bloom, P.R., and J.A. Leenheer. 1989. Vibrational, electronic, and high-energy spectroscopic methods for characterizing humic substances. p. 409-446. *In* M.H.B. Hayes, P. MacCarthy, R.L. Malcolm, and R.S. Swift (eds), *Humic Substances II: In*

Search of Structure. Wiley, Chichester.

Bracewell, J.M., K. Haider, S.R. Larter and H.-R. Schulten. 1989. Thermal degradation relevant to structural studies of humic studies. p. 181-222. *In* M.H.B. Hayes, P. MacCarthy, R.L. Malcolm, and R.S. Swift (eds), *Humic Substances II: In Search of Structure*. Wiley, Chichester.

Christman, R.F., D.L. Norwood, Y. Seo and F.H. Frimmel. 1989. Oxidative degradation of humic substances from freshwater environments. p. 33-67. *In* M.H.B. Hayes, P. MacCarthy, R.L. Malcolm, and R.S. Swift (eds), *Humic Substances II: In Search of Structure*. Wiley, Chichester.

Derome, A.E. 1987. Modern NMR techniques for chemistry research. p. 183-234. *In* J.E. Baldwin (ed.), *Organic Chemistry Series, Vol 6*. Pergamon, Oxford.

Gadian, D.G. 1982. *Nuclear Magnetic Resonance and Its Application to Living Systems*. Oxford University Press.

Goodman, B.A. and M.V. Cheshire. 1985. A Mössbauer-effect study of the reduction of iron by fulvic acid. p. 180-182. In M.H.B. hayes and R.S. Swift (eds), *Volunteered papers, 2nd International Conference (Birmingham, 1984), International Humic Substances Society*. IHSS, University of Birmingham, School of Chemistry.

Griffith, S.M., and M. Schnitzer. 1989. Oxidative degradation of soil humic substances. p. 69-98. *In* M.H.B. Hayes, P. MacCarthy, R.L. Malcolm, and R.S. Swift (eds), *Humic Substances II: In Search of Structure*. Wiley, Chichester.

Hayes, M.H.B. 1985. Extraction of humic substances from soil. p. 329-362. *In* G.R. Aiken, P. MacCarthy, R.L. Malcolm, and R.S. Swift (eds), *Humic Substances in Soil, Sediment, and Water*. Wiley, New York.

Hayes, M.H.B., P. MacCarthy, R.L. Malcolm and R.S. Swift. 1989a. The search for structure: setting the scene. p. 3-31. *In* M.H.B. Hayes, P. MacCarthy, R.L. Malcolm, and R.S. Swift (eds), *Humic Substances II: In Search of Structure*. Wiley, Chichester.

Hayes, M.H.B., P. MacCarthy, R.L. Malcolm, and R.S. Swift. 1989b. Structures of humic substances; the emergence of forms. p. 689-733. *In* M.H.B. Hayes, P. MacCarthy, R.L. Malcolm, and R.S. Swift (eds), *Humic Substances II: In Search of Structure*. Wiley, Chichester.

Hayes, M.H.B., and M.R. O'Callaghan. 1989. Degradations with sodium sulfide and with phenol. p. 143-180. *In* M.H.B. Hayes, P. MacCarthy, R.L. Malcolm, and R.S. Swift (eds), *Humic Substances II: In Search of Structure*. Wiley, Chichester.

Hayes, M.H.B. and R.S. Swift. 1990. Genesis, isolation, composition and structures of soil humic substances. p. 245-305. *In* M.F. DeBoodt, M.H.B. Hayes, and A. Herbillon (eds), *Soil Colloids and their Associations in Aggregates*. Plenum, New York.

Hayes, T.M. 1996. Isolation and characterization of humic substances from soil, and the soil solution, and their interactions with anthropogenic organic chemicals. Ph.D. Thesis. The University of Birmingham.

Hayes, T.M., M.H.B. Hayes, J.O. Skjemstad, R.S. Swift, and R.L. Malcolm. 1996. Isolation of humic substances from soil using aqueous extractants of different pH and XAD resins, and their characterization by 13C-NMR. p. 13-24. In C.E. Clapp, M.H.B. Hayes, N. Senesi, and S.M. Griffith (eds), *Humic Substances and Organic Matter in Soil and Water Environments*. Proc. 7th Intern. Conf. IHSS. IHSS, The University of Minnesota, St. Paul.

Kingery, W.L. and C.E. Clapp. 1997. Soil organic matter: Chemistry and relation to environmental quality. *In* J.B. Dixon *et al.* (eds), *Environmental Soil Mineralogy*. Soil Science Society of America, Madison, WI (in review).

Leenheer, J.A. and T.I. Noyes. Derivatization of humic substances for structural studies. p. 257-280. In *In* M.H.B. Hayes, P. MacCarthy, R.L. Malcolm, and R.S. Swift (eds), *Humic Substances II: In Search of Structure*. Wiley, Chichester.

Little, D.J. 1994. Occurrence and characteristics of podzols under oak woodland in Ireland. PhD thesis, The National University of Ireland.

MacCarthy, P. and J.A. Rice. 1985. Spectroscopic methods other than NMR for determining functionality in humic substances. p.527-559. *In* G.R. Aiken, D.M. McKnight, R.L. Wershaw, and P. MacCarthy (eds), *Humic Substances in Soil, Sediment and Water*. Wiley-Interscience, New York.

Malcolm, R.L. 1989. Applicaions of solid-state ^{13}C NMR spectroscopy to geochemical studies of humic substances. p. 340-372. *In* M.H.B. Hayes, P. MacCarthy, R.L. Malcolm, and R.S. Swift (eds), *Humic Substances II: In Search of Structure*. Wiley, Chichester.

Malcolm, R.L. and P. MacCarthy. 1992. Quantitative evaluation of XAD-8 and XAD-4 resins used in tandem for removing organic solutes from water. *Environ. Int.* **18**:597-607.

Oka H, M. Sasaki , and Suzuku A, 1969. Study on the chemical structure of peat humic acids by high resolution nuclear magnetic resonance spectroscopy. *Nenryo Kyokai Shi* (J.Japanese Fuel Soc.) **48**:295-302

Orlov, D.S. 1985. Humus acids of soils (translated from Russian). Amerind Publishing, New Delhi.

Parsons, J.W. 1989. Hydrolytic degradation of humic substances. p. 99-120. *In* M.H.B. Hayes, P. MacCarthy, R.L. Malcolm, and R.S. Swift (eds), *Humic Substances II: In Search of Structure*. Wiley, Chichester.

Preston, C.M. 1996. Applications of NMR to soil organic matter analysis: History and prospects. *Soil Sci.* **161**:144-166.

Schulten, H.-R. and M. Schnitzer. 1995. Three-dimensional models for humic acids and soil organic matter. *Naturwissenschaften* **82**:487-498.

Senesi, N., and C. Steelink. 1989. Application of ESR spectroscopy to the study of humic substances. p. 373-408. *In* M.H.B. Hayes, P. MacCarthy, R.L. Malcolm, and R.S. Swift (eds), *Humic Substances II: In Search of Structure*. Wiley, Chichester.

Simpson, A.J., B.E. Watt, C.L. Graham, and M.H.B. Hayes. 1997a. Humic substances from podzols under oak forest and a cleared forest site I. Isolation and characterization. This volume.

Simpson A.J, J. Burdon, C.L. Graham, and M.H.B. Hayes. 1997b. Humic substances from podzols under oak forest and a cleared forest site II. Spectroscopic studies. This volume.

Skoog, D.A., and J.J. Leary. 1992. *Principles of Instrumental Analysis*. Fourth edition. Harcout Brace College Publishers, Fort Worth, TX.

Steelink, C. 1985. Implications of elemental characteristics of humic substances. P. 457-476. In G.R. Aiken, D.M. McKnight, R.L. Wershaw, and P. MacCarthy (ed.). Humic substances in soil, sediment and water. Wiley-Interscience, New York.

Steelink, C., R.L. Wershaw, K.A. Thorn and M.A. Wilson. 1989. Application of liquid-state NMR spectroscopy to humic substances. p. 282-308. *In* M.H.B. Hayes, P. MacCarthy, R.L. Malcolm, and R.S. Swift (eds), *Humic Substances II: In Search of Structure*. Wiley, Chichester.

Stevenson, F.J. 1989. Reductive cleavage of humic substances. p. 122-142. *In* M.H.B. Hayes, P. MacCarthy, R.L. Malcolm, and R.S. Swift (eds), *Humic Substances II: In*

Search of Structure. Wiley, Chichester.

Stevenson, F.J. 1994. *Humus Chemistry: Genesis, Composition, Reactions.* Second edition. Wiley, New York.

Swift, R.S. 1985. Fractionation of soil humic substances. p. 387-408. *In* G.R. Aiken, P. MacCarthy, R.L. Malcolm, and R.S. Swift (eds), *Humic Substances in Soil, Sediment, and Water*. Wiley, New York.

Thurman, E.M. and R.L. Malcolm. 1981. Preparative isolation of aquatic humic substances. *Environ. Sci. Technol.* **15**:463-466.

Wershaw, R.L. 1985. Application of nuclear magnetic resonance spectroscopy for determining functionality in humic substances. p. 561-582. *In* G.R. Aiken, P. MacCarthy, R.L. Malcolm, and R.S. Swift (eds), *Humic Substances in Soil, Sediment, and Water*. Wiley, NY.

Williams D.H, and Fleming I, 1989. *Spectroscopic Methods in Organic Chemistry.* McGraw Hill, Maidenhead, Berkshire, England.

Wilson, M.A. 1989. Solid-state nuclear magnetic resonance spectroscopy of humic substances: Basic concepts and techniques. p. 309-338. *In* M.H.B. Hayes, P. MacCarthy, R.L. Malcolm, and R.S. Swift (eds), *Humic Substances II: In Search of Structure*. Wiley, Chichester.

Wilson, M.A. 1990. Applications of nuclear magnetic spectroscopy to organic matter in whole soils. p. 221-260. *In* P. MacCarthy, C.E. Clapp, R.L. Malcolm, and P.R. Bloom (eds), *Humic Substances in Soil and Crop Sciences: Selected Readings.* American Society of Agronomy, Inc. and Soil Science Society of America, Inc., Madison, WI.

Effect of Ascorbate Reduction on the Electron Spin Resonance Spectra of Humic Acid Radical Components

D.B. McPhail and M.V. Cheshire

MACAULAY LAND USE RESEARCH INSTITUTE, CRAIGIEBUCKLER, ABERDEEN AB15 8QH, SCOTLAND

Abstract

Electron spin resonance (ESR) solution spectra of soil humic acids (HAs) in alkali show the presence of free radicals. Those giving rise to an unstructured spectrum appear to be very stable, whereas those displaying hyperfine structure are transient. ESR methods have been used to examine the interaction of ascorbate with HA from a peat and a mineral soil. The results demonstrate the ability of HAs to oxidize ascorbate to the ascorbyl radical rapidly as reduction of the HA proceeds. The HA radical species that give rise to structured components are suppressed, while a significant increase occurs in the radical species responsible for the broad, unstructured component, but the two events are not directly related.

1 Introduction

Electron spin resonance (ESR) spectroscopy can detect and characterise chemical species containing unpaired electrons, such as free radicals and paramagnetic transition metal ions (Knowles *et al.*, 1976). Such species are of considerable interest in soil systems because of the effects of these on redox chemistry and their involvement in polymerization reactions of low molecular weight organic constituents (Senesi, 1990). Examination of the ESR solution spectra of humic acids (HAs) from a range of soil types, including peat, has shown that these all possess a composite free radical signal with varying contributions from species displaying, or not displaying, hyperfine structure (Cheshire and McPhail, 1996). The observation of hyperfine structure, which results from the magnetic interaction of the unpaired electron with neighbouring nuclei of non-zero spin, such as ^1H, has not been consistently reported. Haworth and co-workers observed four-peak spectra with HA extracted from soils with pH < 4.2 (Atherton *et al.*, 1967), but not with HA samples when the soil pH was > 4.2. In contrast, Riffaldi and Schnitzer (1972) failed to detect hyperfine structure in HAs from 50 different sources, and Senesi and Steelink (1989) considered that the absence of structure was a characteristic of humic substances (HS), although Senesi had earlier observed hyperfine splitting in fulvic acid (Senesi *et al.*, 1977). This inconsistency is thought to arise from an inappropriate choice of instrument settings which result in overmodulation and power saturation of the signal, thereby obliterating the presence of structure through line broadening (Cheshire and McPhail, 1996). Furthermore, when the unstructured component of the signal is dominant, species giving rise to hyperfine structure may only be observed in the second-derivative mode (Cheshire and McPhail, 1996). Free radical species giving rise to hyperfine structure are most strongly expressed in soils with pH < 4.2, but the reasons for this are not understood. A few soils of higher pH show anomalous results which could be explained by historical soil pH change and the longevity of the HA (Cheshire and Cranwell, 1972). The ESR spectrum is thought to arise from the presence of semiquinone radicals that are in equilibrium with the corresponding hydroquinones or quinones (Senesi and Steelink, 1989; Senesi and Schnitzer, 1978; Steelink and Tollin, 1967). Spectral properties of the radical consistent with a semiquinone include linewidths, g-values, pH-dependency, temperature and visible light effects, radiation effects, and electrochemical behaviour (Senesi and Steelink, 1989).

In the present work the effect of the reducing potential of ascorbate on the ESR signal from HAs has been investigated in order to elucidate the redox chemistry of the free radical components.

2 Materials and Methods

Protocatechuic acid (3,4–dihydroxybenzoic acid), quercetin (3,3',4',5,7-pentahydroxyflavone), and sodium ascorbate were purchased from the Sigma-Aldrich Corporation. Sodium and potassium hydroxide (Analar grade) were purchased from BDH. An arable, mineral soil (pH 6.2) of the Insch Association and series, developed on a till derived from basic igneous rocks from Muirton of Barra (National Grid Reference: NJ 782 259), was sampled to a depth of 200 mm (A-horizon). A peat soil (pH 3.2) of the Countesswells Association, developed on a till derived from granitic rocks from Cairn O' Mount (National Grid Reference: NO 649 807), was sampled to a depth of 300 mm. Samples (1 kg) of air-dried soil, or partially-dried peat (15% moisture) were rolled in a glass jar on a ball mill with HCl (0.1 mol L^{-1}, 10 L) for 1 h. After standing for 15 h, the supernatant was decanted and the residue rolled for 1 h with NaOH (0.2 mol L^{-1}, 10 L^{-1}) and then stood for a further 15 h. The mixture was centrifuged at 10 000 g through a centrifugal separator (Alpha Laval). The supernatant was filtered through an X3 porosity sintered glass filter to remove floating debris. The solution was acidified to pH 2 with redistilled HCl (6 mol L^{-1}) and the precipitated HA was collected by centrifuging at 2000 g. The precipitate was frozen to -18 ^{o}C to obtain a granular material (Forsyth and Fraser, 1947), thawed, and recovered on a sintered glass filter. The solid was washed with water until free of chloride.

HA solutions (10 mg mL^{-1}) were prepared in KOH (0.1 mol L^{-1}). The solutions were stirred for 15 min whilst open to the atmosphere . A solution of sodium ascorbate (0.02 mol L^{-1}) was prepared in KOH (0.1 mol L^{-1}). The HA solution was let stand undisturbed. The ascorbate solution, and a solution of KOH (0.1 mol L^{-1}) were de-oxygenated under a stream of N_2 gas. Solutions for ESR analysis were prepared as follows:

1 HA solution (2 mL) + KOH (2 mL);
2 HA solution (2 mL) + ascorbate solution (2 mL); and
3 Ascorbate solution (2 mL) + KOH (2 mL).

An aliquot (1 mL) of sample was placed in a quartz solution cell and ESR spectra were recorded 2 min after mixing. Spectra were obtained at room temperature on a Bruker ECS 106 spectrometer operating at 9.5 GHz (X-band) frequency and equipped with a cylindrical TM110 mode cavity. Instrument settings are given in the Figure captions. Computer simulation of spectra was carried out using SimFonia v. 1.2 software (Bruker, Billerica, USA).

Semiquinone radicals formed by autoxidation of protocatechuic acid or quercetin were produced by dissolution in alkali (0.02 mol L^{-1} in 0.1 mol L^{-1} KOH) followed by agitation of the solution in air, then storage under N_2 gas. Aliquots were removed and added to an equal volume of de-oxygenated KOH (0.1 mol L^{-1}) as control, or de-oxygenated sodium ascorbate solution (0.02 mol L^{-1} ascorbate in 0.1 mol L^{-1} KOH). Spectra were recorded 2 min after mixing according to the HA protocol above. All spectroscopic procedures were conducted in duplicate.

3 Results and Discussion

The second derivative spectrum of the alkaline solution of the peat soil HA (Figure 1a) shows characteristic free radical hyperfine features (Cheshire and McPhail, 1996). In the presence of ascorbate (Figure 1b) the intensities of the hyperfine components have been decreased ca 10-fold, although the linewidths remained unchanged. A triplet of intensity ratio 1:2:1 is present downfield of the HA resonances, originating from a radical species in which the unpaired electron interacts with two magnetically-equivalent protons. This is not present in the ascorbate/KOH control (Figure 1c). Furthermore, a second triplet, overlapping part of the weakened HA signal, can be discerned. The hyperfine splittings (A-values) of the triplets are similar (ca 0.019 mT) suggesting that these are part of a larger doublet with a hyperfine splitting of 0.179 mT. This doublet splitting is almost identical to that of the ascorbyl radical observed in a microsomal system, (0.178 mT, Figure 2) and this can be observed in other biological systems as a result of ascorbate oxidation (Yamazaki and Piette, 1961). Observation of the triplet may be attributable to a narrowing of linewidths due to the very low (0.02 mT) modulation amplitude used allowing the weak hyperfine interaction with the methyl protons of the ascorbate molecule to be resolved. The doublet hyperfine coupling from the single ring proton is similar to that reported previously at physiological pH (Kihari *et al.*, 1995), and suggests that no structural change to the ring environment has occurred because of the high pH of the system being investigated here. The spectral data suggest that ascorbate reduces radical species in the HA which exhibit hyperfine structure, and in the process is oxidized to the ascorbyl radical. The first-derivative spectra (Figure 3 a,b,c) also show the destruction of hyperfine components in the ascorbate-treated sample. Double integration of the spectra, however, shows that the overall radical concentration increases by 33% upon ascorbate treatment. This increase cannot be accounted for in terms of the small contribution made to the signal by the ascorbyl radical.

HA from the mineral soil is representative of that in which the contribution to the ESR spectrum by radical species displaying hyperfine structure is low (Cheshire and McPhail, 1996). Treatment with ascorbate increases the concentration of the non-structured component almost 6-fold, as determined by double integration of the spectra (Figure 4 a,b). The results suggest that ascorbate has two effects on the ESR signal of humic acid solutions. Firstly, it appears to reduce radical species which display hyperfine structure to non-radical components. Secondly, it increases the concentration of radical

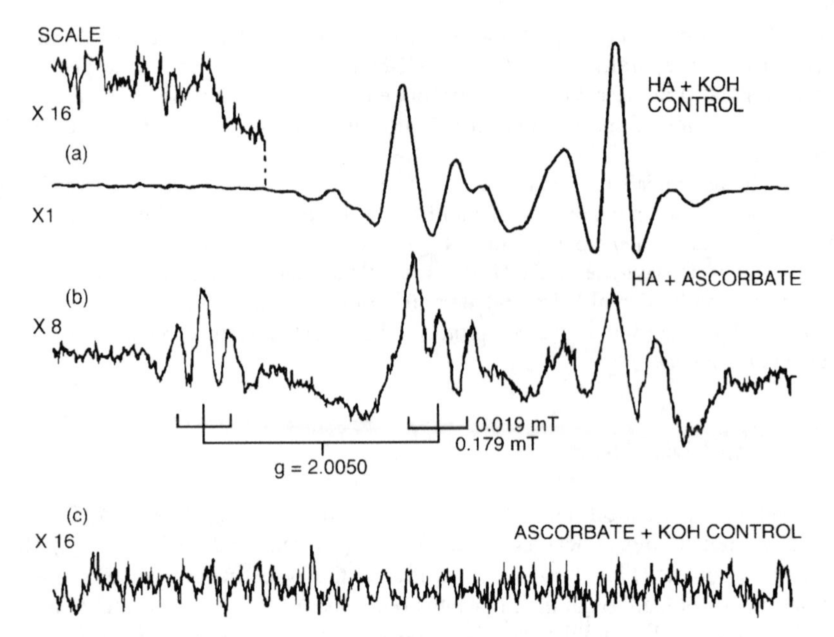

Figure 1 *Effect of ascorbate on the second-derivative ESR spectrum of the peat soil humic acid. Spectrometer settings: microwave power 1 mW, modulation amplitude 0.02 mT. A-values (indicated on stick diagram) ± 0.005 mT, g-value ± 0.0003*

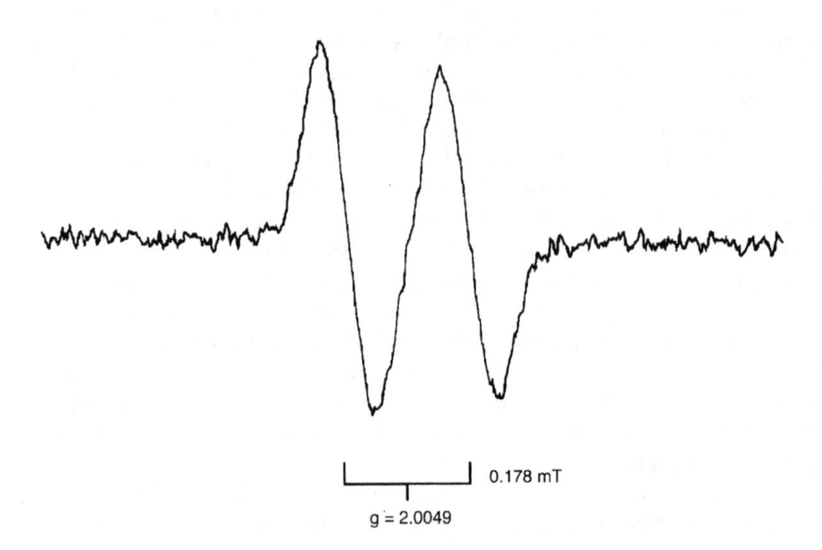

Figure 2 *First-derivative ESR spectrum of ascorbyl radical observed in a biological system (rat liver microsomes). Spectrometer settings: microwave power 2 mW, modulation amplitude 0.2 mT. A-value (indicated on stick diagram) ± 0.010 mT, g-value ± 0.0005*

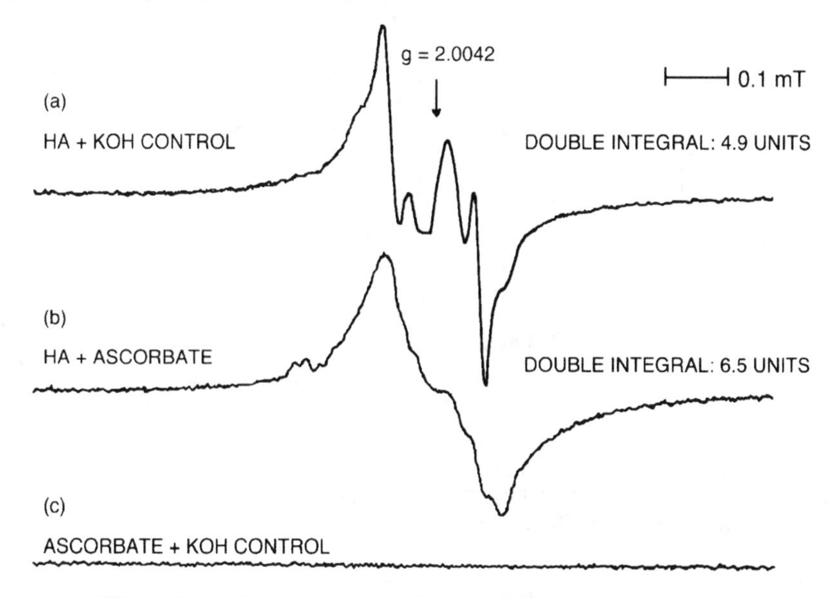

Figure 3 *Effect of ascorbate on the first-derivative ESR spectrum of the peat soil humic acid. Spectra on same scale. Spectrometer settings: microwave power 1 mW, modulation amplitude 0.02 mT. g-value ± 0.0002*

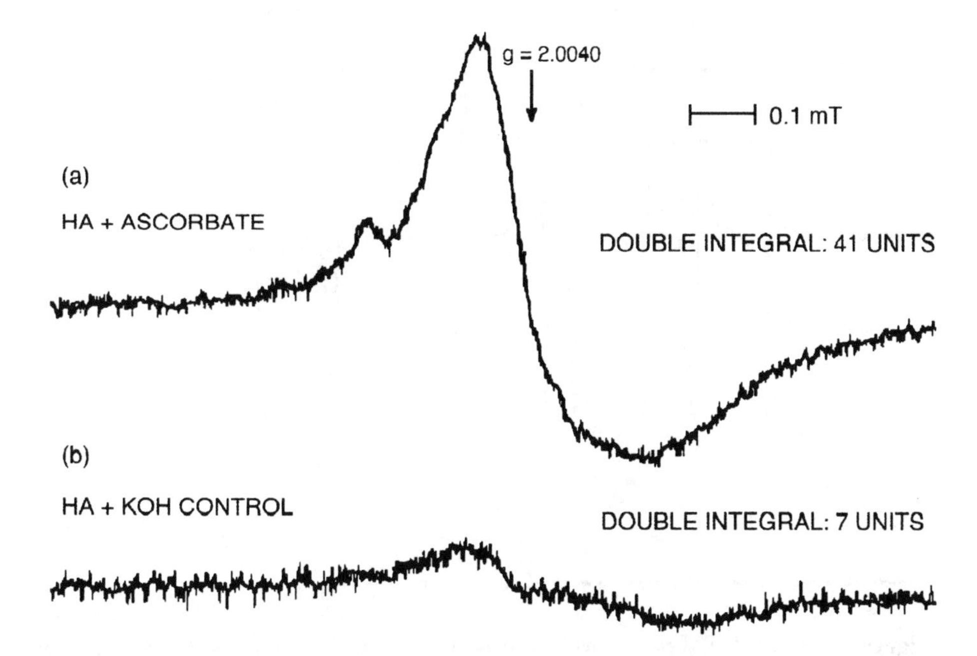

Figure 4 *Effect of ascorbate on the first-derivative ESR spectrum of the mineral soil humic acid. Spectra on same scale. Spectrometer settings: microwave power 1 mW, modulation amplitude 0.02 mT. g-value ± 0.0005*

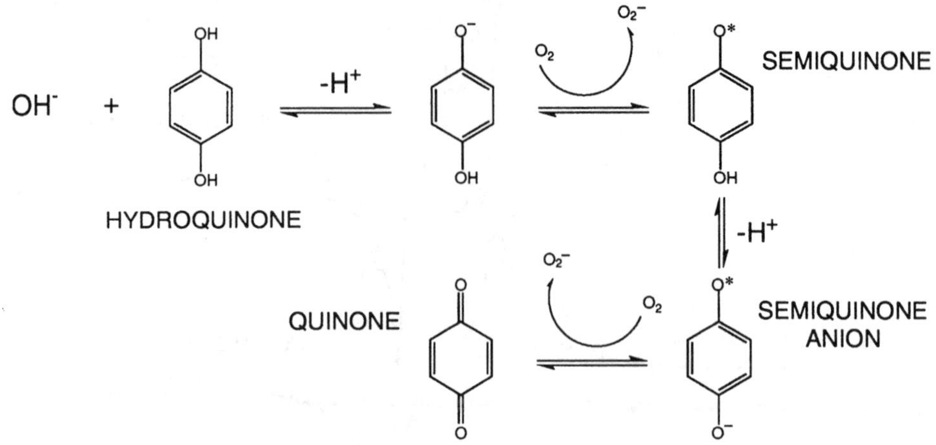

OH⁻ + HYDROQUINONE $\underset{}{\overset{-H^+}{\rightleftharpoons}}$ $\underset{}{\overset{O_2 \rightarrow O_2^-}{\rightleftharpoons}}$ SEMIQUINONE

$-H^+$

QUINONE $\underset{}{\overset{O_2^- \leftarrow O_2}{\rightleftharpoons}}$ SEMIQUINONE ANION

Figure 5 *Generation of semiquinone radical and quinone by autoxidation of hydroquinone in alkali.*

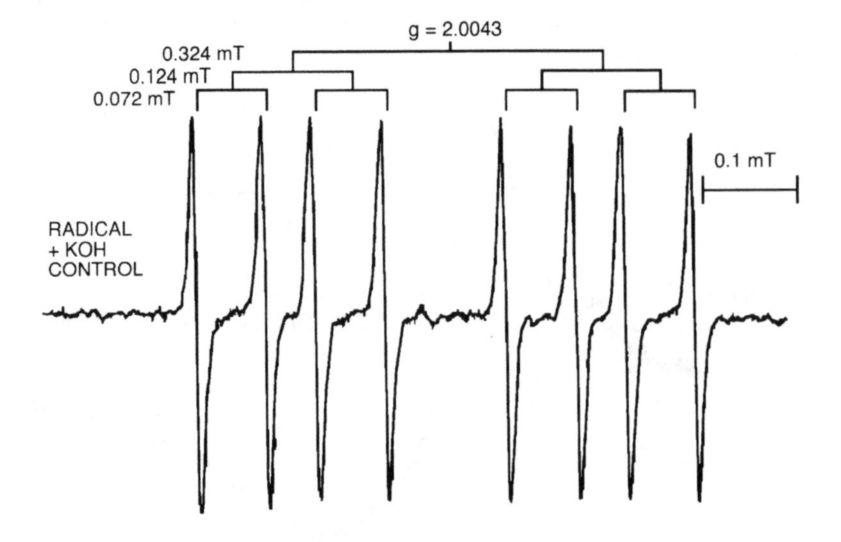

RADICAL + ASCORBATE

Figure 6 *Reduction of protocatechuic acid semiquinone radical by ascorbate. Spectra on same scale. Spectrometer settings: microwave power 0.2 mW, modulation amplitude 0.02 mT. A-values (indicated on stick diagram) ± 0.002 mT, g-value ± 0.0003.*

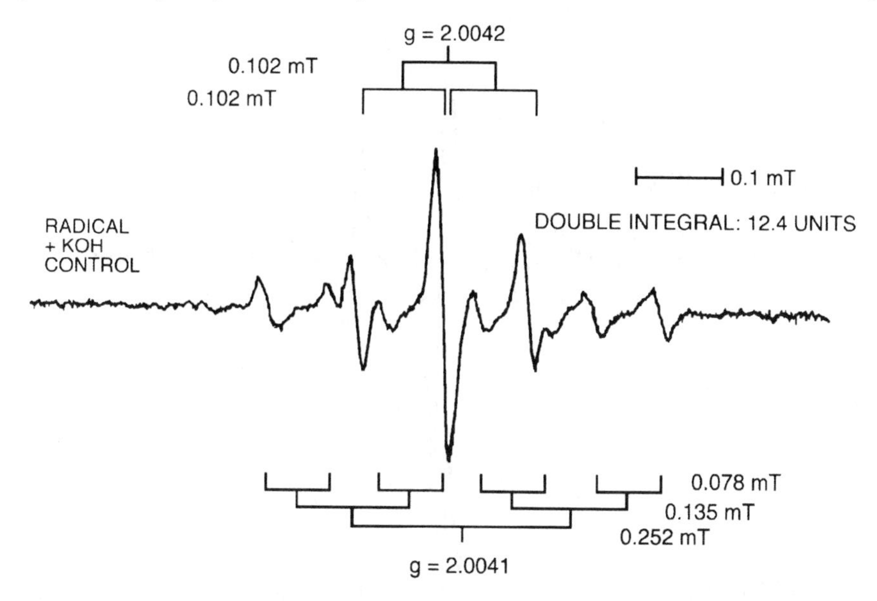

Figure 7 *Reduction of quercetin (flavonoid) semiquinone radical by ascorbate. Spectra on same scale. Spectrometer settings: microwave power 0.2 mW, modulation amplitude 0.02 mT. A-values (indicated on stick diagrams) ± 0.005 mT, g-values ± 0.0004*

species which contribute to a broad, unstructured feature from non-radical precursors. That a dramatic increase occurs in the signal from the mineral soil HA solution, where the contribution to the overall spin concentration by radicals displaying hyperfine structure is small, suggests that the increase in intensity is not a consequence of the interconversion of structured species into non-structured species. Whilst we have no direct experimental evidence to confirm this supposition, it is thought to be the most likely explanation.

Polymerization of simple phenols by oxidative processes is thought to be one of the mechanisms for the formation of HS (Stevenson, 1992), and more recently modifications of lignin structures are considered to provide major pathways in their formation (Haider, 1994). Similarities to radicals generated by base treatment of phenols containing catechol structures have been claimed by Steelink and Mikita (personal communication in Senesi and Steelink, 1989). Consequently, we have examined the ability of ascorbate to reduce semiquinone radical intermediates formed by the autoxidation in alkali (Figure 5) of two

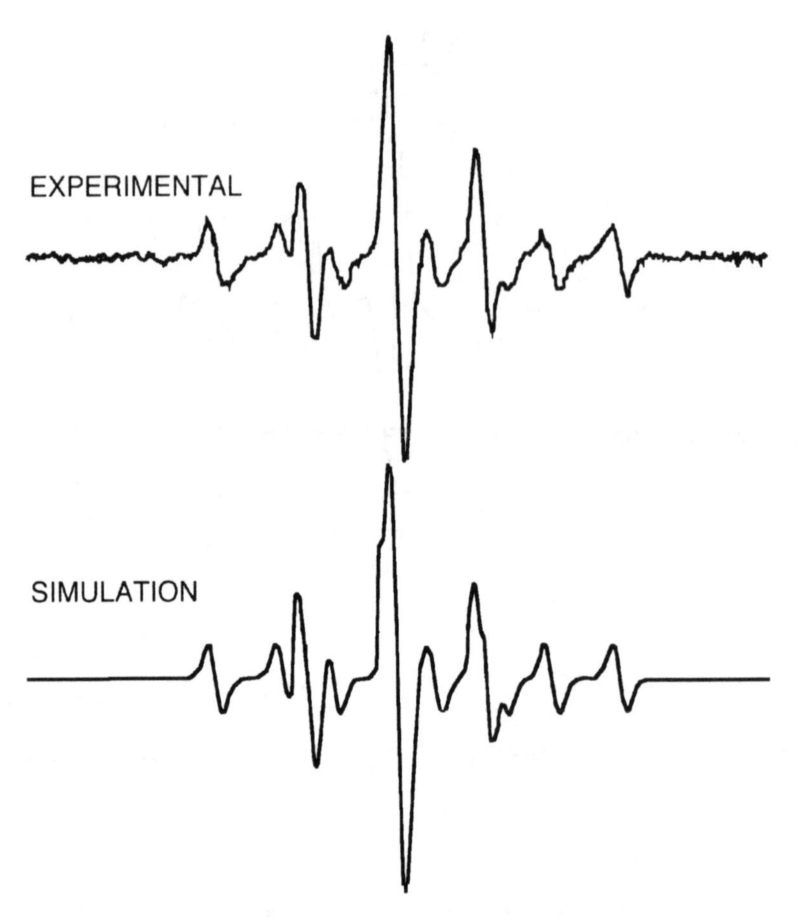

Figure 8 *Comparison of experimental and computer simulated spectra of the composite radical signal obtained by quercetin autoxidation*

naturally-occurring phenolics, protocatechuic acid and the flavonoid quercetin. The ESR spectrum of the protocatechuic acid solution (Figure 6) comprises a doublet of doublet of doublets which can be interpreted as arising from the interaction of the unpaired electron with the nuclear spin of the three inequivalent protons on the ring system. On treatment with ascorbate the radical is fully quenched. The spectrum from the quercetin solution (Figure 7) consists of two distinct radical species. One species comprises a triplet of intensity ratio 1:2:1 arising from the interaction of the unpaired electron with two magnetically-equivalent protons. This has been assigned to a radical intermediate formed by oxidation of the A-ring of the flavonoid. The second species consists of a doublet of doublet of doublets resulting from an interaction of the unpaired electron with three magnetically-inequivalent protons, this being indicative of oxidation of the catechol moiety on the flavonoid B-ring. Computer simulation of the composite spectrum (Figure 8) was successfully achieved using the hyperfine (A-value) parameters reported in Figure 7, thus validating the spectral interpretation. Treatment with ascorbate reduces both radical species and diminishes the overall spin concentration 7-fold (Figure 7).

Solution spectra of HAs in alkali appear to comprise varying proportions of radicals giving rise to resolved hyperfine structure and radicals producing a broad, slightly-asymmetric envelope in which no structure is observed. The radical species are thought to be semiquinone moieties. In this work we have demonstrated the ability of ascorbate to reduce the semiquinone radicals of two simple phenolics produced in alkali. In the peat soil HA, a similar effect is observed with respect to radical species producing hyperfine structure. Whether such species contribute to the unresolved spectrum observed in the solid state, or are formed by autoxidation of phenolic moieties present in the HA by dissolution in alkali, is not known. If the latter process occurs, however, this may be proved either by ESR spin-trapping studies aimed at detecting superoxide produced in the autoxidation process or by monitoring oxygen uptake that would also occur. The broad, unstructured component increases considerably on treatment with ascorbate. That this occurs in the mineral soil HA, where the structured components are of low concentration, would seem to preclude their interconversion by ascorbate into the unresolved features. The ascorbate must therefore be reducing a non-radical precursor to the radical species giving rise to the unstructured component unless both structured components and non-radical precursors contribute to the measured increase of the unstructured component. A possible explanation is that the ascorbate is reducing a quinone moiety, thereby shunting the quinone - semiquinone equilibrium in favour of the radical (Figure 5). The presence of two radical species in humic substances has been suggested previously (Steelink and Tollin, 1962; Alberts *et al.*, 1974; Senesi and Schnitzer, 1977). However, hyperfine structural differences have not been noted.

4 Conclusions

The ability of ascorbate to reduce semiquinone radicals formed by autoxidation of protocatechuic acid and quercetin in alkali has been demonstrated. In alkaline solutions of HA, ascorbate also reduces radical species responsible for the hyperfine structure observed in the ESR spectrum. However, ascorbate treatment increases the concentration of radical species giving rise to the non-structured component of the HA solution spectrum. The interconversion of structured to non-structured components by ascorbate can probably be discounted. These results may be explained by the presence of two distinct categories of semiquinone contributing to the composite ESR spectrum. The first category comprises semiquinones that are readily reduced to the corresponding phenols and which are responsible for the structured features observed in the spectrum. The second category comprises semiquinones which are in equilibrium with a larger pool of quinones. The effect of ascorbate reduction is to shunt the equilibrium towards the semiquinone. This latter category of radical species is responsible for the unstructured component observed in the ESR spectrum.

Treatment of HA solution by ascorbate, therefore, distinguishes two classes of free radicals that contribute to the ESR spectrum and which have distinct redox properties. Characterization of radical centres in humic HS, and an understanding of their chemical attributes may give insight into potential modes of interaction with anthropogenic agents such as herbicides.

References

Alberts, J.J., J.E. Schindler, R.W. Miller, and D.E. Nutter. 1974. Elemental mercury evolution mediated by humic acid. *Science* **184**:895-897.

Atherton, N.M., P.A. Cranwell, A.J. Floyd, and R.D. Haworth. 1967. ESR spectra of humic acids - I. *Tetrahedron* **23**:1653-1667.

Cheshire, M.V. and P.A. Cranwell. 1972. Electron spin resonance of humic acids from cultivated soils. *J. Soil Sci.* **23**:424-430.

Cheshire, M.V. and D.B. McPhail. 1996. Hyperfine splitting in the electron spin resonance solution spectra of humic substances. *European J. Soil Sci.* **47**:205-213..

Forsyth, W.G.C. and G.K. Fraser. 1947. Freezing as an aid in the drying and purification of humus and allied materials. *Nature* **160**:607.

Haider, K. 1994. Advances in the basic research of the biochemistry of Humic substances. p. 91-107. *In* N. Senesi and T.M. Miano (eds), *Humic Substances in the Global Environment and Implications on Human Health*. Elsevier, Amsterdam.

Kihari, T., S. Sakata, and M. Ikeda. 1995. Direct-detection of ascorbyl radical in experimental brain injury-microdialysis and an electron spin resonance spectroscopic study. *J. Neurochemistry* **65**:282-286.

Knowles, P.F., D. Marsh, and H.W.E. Rattle. 1976. *Magnetic Resonance of Biomolecules.* John Wiley and Sons, London.

Riffaldi, R. and M. Schnitzer. 1972. Electron spin resonance spectrometry of humic substances. *Soil Sci. Soc. Amer. Proc.* **36:**301-305.

Senesi, N. 1990. Application of electron spin resonance spectroscopy in soil chemistry. *In:* B.A. Stewart (ed.), *Advances in Soil Science* **14:**77-130. Springer-Verlag New York.

Senesi, N. and Schnitzer, M. 1978. Free radicals in humic substances. p. 467-481. *In* E.N. Krumbein (ed.), *Environmental Biogeochemistry and Geomicrobiology*. Ann Arbor Science.

Senesi, N., Y. Chen, and M. Schnitzer. 1977. Hyperfine splitting in electron spin resonance spectra of fulvic acid. *Soil Biol. Biochem.* **9**:371-372.

Senesi, N. and M. Schnitzer. 1977. Effects of pH, reaction time, chemical reduction and irradiation on ESR spectra of fulvic acid. *Soil Science* **123**:224-234.

Senesi, N. and C. Steelink. 1989. Application of ESR spectroscopy to the study of humic substances. p. 373-408. *In* M.H.B. Hayes, P. MacCarthy, R.L. Malcolm, and R.S. Swift (eds), *Humic Substances II: In Search of Structure.* Wiley, Chichester.

Steelink, C. and G. Tollin. 1967. Free radicals in soil. p. 147-169. *In* A.D. McLaren and G.M. Peterson (eds), . *Soil Biochemistry.* Marcel Dekker, New York.

Steelink, C. and G. Tollin. 1962. Stable free radicals in soil humic acid. *Biochimica Biophysica Acta* **59**:25-34.

Stevenson, F.J. 1982. *Humus Chemistry: Genesis, Composition, Reactions.* Wiley, New York.

Yamazaki, I. and L.H. Piette. 1961. Mechanisms of free radical formation and dissappearance during the ascorbic oxidase and peroxidase reactions. *Biochimica Biophysica Acta* **50**:62-69.

Humic Substances from Podzols Under Oak Forest and a Cleared Forest Site I. Isolation and Characterization

A.J. Simpson, B.E. Watt, C.L. Graham, and M.H.B. Hayes

THE UNIVERSITY OF BIRMINGHAM, SCHOOL OF CHEMISTRY, EDGBASTON, BIRMINGHAM B15 2TT, ENGLAND

Abstract

Humic substances (HS) were sequentially and exhaustively extracted from a podzolic A_h horizon under oak forest, from the corresponding B_h horizon, and from the B_h horizon from a neighbouring site where the forest was cleared ca 400 years BP. The sequence of solvents used was distilled water, sodium pyrophosphate (Pyro, 0.1M, pH 7), Pyro (0.1M, pH 10.6), and a sodium hydroxide (0.1M)/Pyro (0.1M) mixture (pH 12.6). Isolates were fractionated using XAD-8 and XAD-4 resins in tandem. Elemental data demonstrate that the humic acids had the highest carbon and nitrogen contents followed by the fulvic acids, and the fraction from the XAD-4 resin had the least. The E_4/E_6 ratio values were highest for the fractions isolated at pH 7, and that might reflect higher aromaticity and unsaturation in these samples. The lowest values were obtained for the XAD-4 acids, and these were the least aromatic. A decrease in the total sugar contents (although an increase in sugars from microbial sources) was observed in the HS of the B_h (compared with the A_h) horizon. The sugars in the less polar HS extracted at the higher pH values were more representative of materials from plant sources than were those of the more highly transformed substances.

1 Introduction

Podzols are leached profiles which demonstrate dark, highly organic A_o and A_h surface horizons overlaying an inorganic, leached, yellow-grey A_e horizon, and this in turn overlays either a yellow B_s horizon (where the accumulation of sesquioxides has occurred) or a black B_h horizon (where humic substances have accumulated).

There have been few studies which relate the vegetation cover to the chemical compositions and the physical properties of the humic substances (HS) in the B_h horizons of podzols. HS in the B_h horizon have been solubilized, transported, and redeposited, and can be considered to have undergone a 'maturation' process. It is of interest to know whether or not the compositions of the HS in the B_h horizon will change when the soil management is altered and the A_h horizon is lost.

This study set out to compare aspects of the compositions of the HS from two paired podzol profiles. One site was located in an ancient oak forest, known to be at least 2000 years old, and the second was located in a neighbouring site where the forest was cleared ca 400 years ago. It was considered appropriate to compare the HS isolated from the A_h and B_h horizons within the forest with those from the B_h of the cleared site. Differences between the compositions of the HS in the A_h and B_h horizons can give information about the genesis of the HS, and about changes which could be expected in the HS in the B_h horizon when the climax vegetation of a podzol is changed.

Part 1 of this study focuses on results from elemental and sugar analyses, and on ultra violet/visible spectroscopy data for the HS fractions isolated. In part 2, fluorescence, FT-IR, and ^{1}H NMR spectroscopy were used to study and to compare the same fractions.

2 Materials and Methods

Sources of Samples

Soil samples were collected from two sites:

1, A Forested Site, located at Uragh Wood, Lough Inchiquin, Kenmare, Co. Kerry, Ireland, (Grid reference: V.83.62). The soil type is a humo-ferric podzol with Devonian sandstone as the parent material. The vegetation consists of *Quercus petraea* and *Betula pubescens* as the dominant species, and with *Ilex aquifolium* and *Sorbus aucuparia* also present. The ground cover is dominated by *Calluna vulgaris, Vaccinium myrtillus, Luzula sylvatica,* and *Blechnum spicant* (Little, 1994). The A_h horizon, at 6-15 cm depth, is a humified dark brown/black layer containing root material. The B_h horizon, at 50-55 cm, is highly organic, black, and with no visible roots or fibrous plant material.

2, A Cleared Site, on a nearby ridge close to the forested site at Uragh, with similar topographical features. The site has been without trees for not more than 400 years (Little, 1994) and is now being grazed. The soil type is the same as for the forested site. There was no A_h horizon and samples were taken at 45 - 50 cm from the black, highly organic B_h horizon. There was no visible root or plant material in the soil sampled.

Isolation and Fractionation of Humic Substances

Soil (1 kg) from each B_h horizon, and from the forested A_h horizon was sequentially extracted using distilled water, sodium pyrophosphate (Pyro, 0.1M, pH 7), Pyro (0.1M, pH 10.6) and a sodium hydroxide (0.1M)/Pyro (0.1M) mixture (pH 12.6). Extraction was repeated with each solvent until the optical density values of the extracts were less than 1.

HS were isolated [as humic acids (HAs), fulvic acids (FAs), and XAD-4 acids] using XAD-8 [poly(methyl methacrylate)] and XAD-4 (styrene divinylbenzene) resins in tandem, as described by Malcolm and MacCarthy (1992) and by Hayes (1996).

Elemental Analyses

Elemental analyses were carried out by Dr. S Boyer at the SACS laboritories at the University of North London. Results are given on a dry, ash free basis.

E_4/E_6 Ratios

E_4/E_6 ratio values were measured as described by Chen *et al.* (1977). Samples (2 mg) were dissolved in 0.05M $NaHCO_3$ (10 mL), and absorbance values at 465 and at 665 nm were recorded against a 0.05M $NaHCO_3$ reference.

Determinations of Neutral Sugar Contents

The procedure was based on that of Blakeney *et al.* (1983). HAs, FAs, and XAD-4 acids were hydrolyzed for 2h in 2M trifluoroacetic acid (TFA) at 120 °C. The sugars were determined as the alditol acetate derivatives using a GC-FID (Perkin Elmer AutoSystem XL), and a BPX70 [25 m x 0.25 mm i.d., SGE (UK) Ltd.] column. Results are given on a dry, ash free basis.

3 Results and Discussion

Yields

The total yields of HAs and FAs obtained from the soil samples are given in Table 1. HS from the forested A_h, forested B_h, and cleared B_h horizons amounted to 11.94 g, 21.15 g, and 8.45 g, respectively. The lower recovery from the cleared B_h horizon could reflect the absence of an A_h horizon, and hence the failure to replenish (or add to) the precipitated or sorbed B_h horizon HS with soluble components from the A_h horizon.

Elemental Analyses

The elemental analyses data in Table 2 show that the HA fractions have the highest carbon contents, followed in order by the FAs and XAD-4 acids. The carbohydrate contents (Table 4) are in the reverse order, and that would partially explain the differences seen. The trends for N are less clear cut, and there are some striking differences in the N contents of the different samples. It is clear that the FAs isolated at the same pH values are more enriched in N in the A_h than in the B_h horizon of this soil. However, the N contents of the FAs isolated at pH 7 and pH 10.6 from the B_h horizon of the unforested soil are more

Table 1 *Yields of humic (HA), fulvic (FA), and XAD-4 acids isolated from the paired soils*

Soil and extraction solvent	FA (g)	HA (g)	XAD-4 (g)
Forested A_h pH 5.5	0.12	0.04	0.12
Forested A_h pH 7	3.79	6.73	Trace
Forested A_h pH 10.6	0.14	0.51	Trace
Forested A_h pH 12.6	0.26	0.46	0.05
Forested B_h pH 5.5	Trace	Trace	Trace
Forested B_h pH 7	3.33	16.17	1.06
Forested B_h pH 10.6	0.04	0.32	0.17
Forested B_h pH 12.6	HA/FA mix 0.03	HA/FA mix 0.03	Trace
Unforested B_h pH 5.5	Trace	Trace	Trace
Unforested B_h pH 7	0.69	4.7	0.06
Unforested B_h pH 10.6	0.11	1.6	0.29
Unforested B_h pH 12.6	0.26	0.74	Trace

enriched in N than the corresponding samples from the forested soil. With the exception of the sample isolated at pH 10.6, the N contents of the HAs are significantly higher in the B_h horizon of the unforested (compared to the forested) soil.

The higher the H:C ratio values in hydrocarbons, the lower is the aromaticity and/or unsaturation. The oxygen contents, and especially the extents of carboxyl functionality in a structure will also influence these ratio values. The data in Table 2 show that, in general, the lowest H:C ratios are associated with the isolates at pH 7. These are the most highly humified fractions, and have been shown in studies with non-podzol soil fractions to be the most aromatic of the fractions isolated at pH 7, 10.6, and 12.6 (Hayes, 1996). Titration data will reveal whether or not the low ratio value for the B_h XAD-4 acids isolated from the B_h horizon of the unforested soil can be attributed to carboxyl functionalities.

It is appropriate to compare the data for the N contents of the humic fractions isolated at the same pH values from the different horizons of the forested and unforested soils. With the exception of the fraction isolated at pH 12.6, the N contents of the FAs from the A_h horizon were greater than those of the FAs from the B_h horizons of the forested and

Table 2 *Elemental analyses data for the humic acids (HA), fulvic acids (FA), and XAD-4 acids isolated in water (W), in 0.1M sodium pyrophosphate, pH 7 (Pyro, 7), in 0.1M sodium pyrophosphate, pH 10.6 (Pyro, 10.6), and 0.1M Pyro + 0.1M sodium hydroxide (P + Sh, pH 12.6) from the paired podzol soils*

Sample	C (%)	N (%)	H (%)	N:C Ratio	H:C Ratio
Kerry forested A_h FA *(W)*	46.2	2.2	3.7	0.06	1.45
Kerry forested A_h FA *(Pyro 7)*	52.1	2.5	3.4	0.06	1.17
Kerry forested A_h FA *(Pyro 10.6)*	56.6	3.9	4.2	0.09	1.33
Kerry forested A_h FA *(P+Sh 12.6)*	51.2	0.6	3.3	0.01	1.17
Kerry forested A_h HA *(W)*	43.9	1.5	4.5	0.04	1.85
Kerry forested A_h HA *(Pyro 7)*	56.3	2.3	2.7	0.05	0.86
Kerry forested A_h HA *(Pyro 10.6)*	57.7	4.9	4.3	0.11	1.35
Kerry forested A_h HA *(P+Sh 12.6)*	54.8	1.0	2.6	0.02	0.87
Kerry forested A_h XAD-4 *(W)*	55.6	1.3	4.5	0.03	1.45
Kerry forested A_h XAD-4 *(P+Sh 12.6)*	55.2	6.2	4.5	0.15	1.46
Kerry forested B_h FA *(Pyro 7)*	49.4	1.6	2.7	0.04	0.97
Kerry forested B_h FA *(Pyro 10.6)*	53.4	1.7	3.6	0.04	1.22
Kerry forested B_h HA *(Pyro 7)*	57.4	1.7	2.5	0.04	0.77
Kerry forested B_h HA *(Pyro 10.6)*	63.4	3.4	5.8	0.07	1.63
Kerry forested B_h FA/HA *(P+Sh 12.6)*	53.8	2.3	3.4	0.06	1.15
Kerry forested B_h XAD-4 *(Pyro 7)*	44.5	3.0	3.6	0.09	1.47
Kerry forested B_h XAD-4 *(Pyro 10.6)*	44.7	2.5	3.6	0.07	1.47
Kerry unforested B_h FA *(Pyro 7)*	54.0	0.9	2.1	0.02	0.69
Kerry unforested B_h FA *(Pyro 10.6)*	57.7	3.7	5.7	0.08	1.77
Kerry unforested B_h FA *(P+Sh 12.6)*	59.4	2.0	4.9	0.04	1.48
Kerry unforested B_h HA *(Pyro 7)*	60.2	3.8	2.7	0.08	0.81
Kerry unforested B_h HA *(Pyro 10.6)*	61.2	3.4	4.5	0.07	1.33
Kerry unforested B_h HA *(P+Sh 12.6)*	64.5	3.5	5.7	0.07	1.58
Kerry unforested B_h XAD-4 *(Pyro 7)*	43.4	4.1	4.3	0.12	1.80
Kerry unforested B_h XAD-4 *(Pyro 10.6)*	33.8	0.7	0.7	0.03	0.36

unforested soils. In the case of the HAs, this trend applied for samples from the forested soils, but not for the contents in the HAs of the unforested soil.

The highest N content was in the XAD-4 fraction of the A_h of the forested soil isolated at pH 12.6. A relatively high content was found also in the XAD-4 sample isolated from the unforested B_h horizon at pH 7. There were no obvious trends in the N contents in the XAD-4 acids.

We await data for the amino acid (AA) contents of the humic fractions, but in general the highest contents of these are contained in the XAD-4 acids, and often in the humic fractions isolated from soils at the higher pH values (Hayes, 1996). The less well humified fractions are isolated at the higher pH values and these can contain the AAs of the partially humified substrates. The largest N:C ratios tended to be associated with the humic fractions isolated at the higher pH values.

UV/Visible Spectroscopy

Absorbance data at 465 and at 665 nm for the various humic fractions, and the ratios of these absorbances (the E_4/E_6 ratios) are given in Table 3. There are no reliable or agreed scientific interpretations of the meanings of these ratio values. Kononova (1966) considered that the magnitude of the E_4/E_6 ratios is related to the degree of condensation of the aromatic carbon network, where a low ratio would indicate a relatively high degree of condensation of aromatic humic components. On the other hand, Chen *et al.* (1977) concluded that, although aromaticity may influence the ratios as a secondary factor, the values are primarily governed by the molecular sizes of the HS, with small ratio values indicative of large molecluar sizes, and vice versa.

The E_4/E_6 ratio values for the HS isolated from the podzol soils were lower, in general, for the HAs than for the FAs, and that might be interpreted in terms of larger molecular sizes (or greater extents of molecular associations) for the HAs. However, the highest ratio values were obtained for the HAs and the FAs isolated at pH 7. That could be interpreted in terms of relatively lower molecular sizes for these fractions, but alternatively it might be interpreted in terms of lesser associations of molecules (because of the greater repulsions between the charged functionalities), or greater aromaticity or unsaturation. As stated already, the humic fractions isolated at pH 7 are generally more aromatic than those isolated at the higher pH values (Hayes, 1996).

The lowest ratio values were obtained for the XAD-4 acids. These fractions are the least coloured. Thus, there is considerable room for error when dealing with the low absorbance values. Invariably, the XAD-4 acids have low aromaticities, and hence the low ratio values cannot be interpreted in terms of aromaticity. The XAD-4 acids contain significant amounts of carbohydrates. Such hydrophilic moieties have high hydrostatic volumes which would influence the ratios. It will be possible to provide better interpretations of E_4/E_6 ratio data when comprehensive NMR and other compositional data are available.

Neutral Sugars (NS) Analyses

With one exception, glucose (Glu) was the most abundant neutral sugar (NS) in the samples studied (Table 4), and in the majority of samples mannose (Man) and galactose (Gal) were second or third in the order of abundance. These are the three NS components

Table 3 *E_4/E_6 ratios for the humic acids (HA), fulvic acids (FA), and XAD-4 acids isolated in water (W), in 0.1M sodium pyrophosphate, pH 7 (Pyro, 7), in 0.1M sodium pyrophosphate, pH 10.6 (Pyro, 10.6), and in 0.1M Pyro + 0.1M sodium hydroxide (P + Sh, pH 12.6) from the paired podzol soils*

Sample	Absorbance at 465nm	Absorbance at 665nm	E_4/E_6 Ratio
Kerry forested A_h FA *(W)*	0.11	0.02	7.07
Kerry forested A_h FA *(Pyro 7)*	0.34	0.03	10.69
Kerry forested A_h FA *(Pyro 10.6)*	0.06	0.02	3.63
Kerry forested A_h FA *(P + Sh 12.6)*	0.28	0.04	7.54
Kerry forested A_h HA *(W)*	0.20	0.04	4.88
Kerry forested A_h HA *(Pyro 7)*	0.70	0.09	7.87
Kerry forested A_h HA *(Pyro 10.6)*	0.45	0.07	6.6
Kerry forested A_h HA *(P + Sh 12.6)*	0.17	0.03	5.31
Kerry forested A_h XAD-4 *(W)*	0.07	0.02	3.3
Kerry forested A_h XAD-4 *(P + Sh 12.6)*	0.10	0.02	4.29
Kerry forested B_h FA *(Pyro 7)*	0.24	0.02	14.18
Kerry forested B_h FA *(Pyro 10.6)*	0.14	0.02	6.81
Kerry forested B_h HA *(Pyro 7)*	0.79	0.13	6.25
Kerry forested B_h HA *(Pyro 10.6)*	0.65	0.17	3.78
Kerry forested B_h FA/HA *(P + Sh 12.6)*	0.30	0.06	4.85
Kerry forested B_h XAD-4 *(Pyro, 7)*	0.04	0.01	3.08
Kerry forested B_h XAD-4 *(Pyro 10.6)*	0.08	0.01	6.25
Kerry unforested B_h FA *(Pyro, 7)*	0.18	0.02	9.15
Kerry unforested B_h FA *(Pyro 10.6)*	0.36	0.04	9
Kerry unforested B_h FA *(P + Sh 12.6)*	0.12	0.02	5.45
Kerry unforested B_h HA *(Pyro, 7)*	1.01	0.14	7.47
Kerry unforested B_h HA *(Pyro 10.6)*	0.33	0.07	4.55
Kerry unforested B_h HA *(P + Sh 12.6)*	0.34	0.07	5.18
Kerry unforested B_h XAD-4 *(Pyro, 7)*	0.05	0.01	5.3
Kerry unforested B_h XAD-4 *(Pyro 10.6)*	0.03	0.01	2.55

most abundantly found in polysaccharides. In general, Glu is the most abundant NS in plants (because of their cellulose and starch components), and it is often the most abundant component of microbial polysaccharides. Man and Gal are found in plant (Stephen, 1983), in bacterial (Keene and Lindberg, 1983), in algal (Painter, 1983), and in fungal (Gorin and Barreto-Bergter, 1983) sources. When present in high abundance in soils, Man and Gal are often considered to be indicative of microbial sources, and though common in bacterial polysaccharides several mannans and galactomannans are associated with yeasts and fungi (Gorin and Barreto-Bergter, 1983). Rhamnose (Rha, or 6-deoxymannose, found in D- and L- configurations), which is common in the L- form in bacterial and fungal polysaccharides, composes the sidechains in rhamnomannans, and there are also several yeast/fungal galactomannans, and some xylomannans, with xylose (Xyl) as the side chain, and an arabinoxylomannan, composed of D-Man, D-Xyl, and L-arabinose (or L-Ara) has been reported as a cellular polysaccharide of yeast (Gorin and Barrreto-Berger, 1983). However, Xyl and Ara are considered to be derived predominantly from plant sources. Rha can occur as a phenolic glycoside in tannin and flavonoid structures, and is contained in

Table 4 *Sugar contents and ratios [(Gal + Man)/(Ara+Xyl) and (Rha+Fuc)/(Ara+Xyl); Man/Rha and Gal/Fuc] in humic acids (HAs), fulvic acids (FAs), and XAD-4 acids isolated in water (W), at pH 7 in pyrophosphate, at pH 10.6 in pyrophosphate, and pH 12.6 in pyrophosphate/hydroxide from the A_h and B_h horizons of a forested (F) podzol, and from the B_h horizon of a neighbouring cleared soil site (UnF)*

		Relative		Molar		Percentage			nmol/mg	µg/mol	Weight	Gal+Man	Rha+Fuc	Man	Gal
Sample	Rha	Fuc	Rib	Ara	Xyl	Man	Gal	Glu	total	total	%	/Ara+Xyl	/Ara+Xyl	/Rha	/Fuc
F A*h* FA (W)	10.7	3.0	1.3	15.5	9.9	11.1	15.1	33.3	122.2	18.6	1.9	1.3	0.5	1.0	5.0
F A*h* FA (pH 7)	8.3	3.8	0.5	13.6	9.7	19.0	14.7	30.5	289.4	44.3	4.4	1.8	0.5	2.3	3.9
F Ah FA (pH 10.6)	7.2	4.6	0.5	10.0	13.2	20.4	12.9	31.3	661.0	101.3	10.1	1.8	0.5	2.8	2.8
F A*h* FA (pH 12.6)	6.3	4.8	0.6	9.5	15.6	20.1	10.9	32.2	733.6	112.0	11.2	1.5	0.4	3.2	2.3
F A*h* HA (W)	13.7	2.0	1.8	5.8	7.6	14.2	13.1	41.8	169.6	26.3	2.6	2.5	1.2	1.0	6.5
F A*h* HA (pH 7)	7.2	2.1	2.0	24.9	6.6	16.0	17.1	24.1	227.8	34.3	3.4	1.3	0.3	2.2	8.1
F A*h* HA (pH 10.6)	8.0	3.2	1.7	20.2	8.7	15.4	17.2	25.8	409.4	61.9	6.2	1.4	0.4	1.9	5.4
F A*h* HA (pH 12.6)	6.5	3.4	1.4	20.0	10.7	11.9	16.0	30.0	366.5	55.3	5.5	1.1	0.3	1.8	4.8
F A*h* XAD-4 (pH 7)	9.7	3.2	1.1	7.9	10.7	17.8	14.8	34.7	1135.2	175.0	17.5	2.1	0.7	1.8	4.6
F A*h* XAD-4 (pH 12.6)	5.7	2.3	1.6	7.9	28.0	13.4	9.2	31.8	2317.2	346.7	34.7	0.8	0.2	2.4	4.0
F B*h* FA (pH 7)	10.6	2.7	1.5	13.2	8.8	17.9	16.7	28.7	51.6	7.9	0.8	1.9	0.6	1.7	6.3
F B*h* FA (pH 10.6)	7.1	3.4	1.1	8.6	12.0	17.4	11.6	38.8	202.0	31.1	3.1	1.7	0.5	2.4	3.4
F B*h* HA (pH 7)	9.3	3.2	0.9	9.4	9.1	18.1	13.6	36.5	258.5	39.9	4.0	2.1	0.7	2.0	4.3
F B*h* HA (pH 10.6)	11.8	3.8	0.8	9.7	8.8	16.8	13.0	35.3	180.0	27.7	2.8	2.0	0.8	1.4	3.4
F B*h* FA/HA (pH 12.6)	8.6	2.6	4.8	10.6	15.0	15.5	11.9	31.0	444.3	67.2	6.7	1.3	0.4	1.8	4.6
F B*h* XAD-4 (pH 7)	8.8	6.3	0.6	5.3	9.5	19.6	13.9	36.0	822.6	127.6	12.8	2.8	1.0	2.2	2.2
F B*h* XAD-4 (pH 10.6)	7.8	4.7	0.9	5.4	8.0	19.0	14.0	40.1	724.5	112.9	11.3	3.0	0.9	2.4	3.0
UnF B*h* FA (pH 7)	9.8	2.8	1.0	9.3	10.9	18.1	15.6	32.5	146.3	22.5	2.3	2.1	0.6	1.8	5.6
UnF B*h*FA (pH 10.6)	9.3	2.6	1.3	9.3	12.7	16.7	13.6	34.5	340.0	52.1	5.2	1.7	0.5	1.8	5.2
UnF B*h* FA (pH 12.6)	6.5	2.5	0.9	11.4	18.3	12.6	12.6	35.1	430.3	65.2	6.5	1.0	0.3	1.9	5.0
UnF B*h* HA (pH 7)	7.9	2.4	0.7	10.0	14.1	18.6	20.2	26.2	113.0	17.3	1.7	2.0	0.4	2.4	8.3
UnF B*h* HA (pH 10.6)	9.8	2.3	2.0	11.1	9.7	16.0	13.5	35.7	266.0	40.8	4.1	1.7	0.6	1.6	5.8
UnF B*h* HA (pH 12.6)	8.2	3.7	0.3	12.7	9.8	18.2	15.6	31.5	292.0	44.8	4.5	1.8	0.5	2.2	4.2
UnF B*h* XAD-4 (pH 7)	8.4	3.3	1.5	5.5	8.8	16.8	16.4	39.1	1728.3	268.8	26.9	2.9	0.8	2.0	5.0
UnF B*h* XAD-4 (pH 10.6)	7.8	2.4	1.6	8.1	9.2	13.6	15.8	41.4	140.1	21.7	2.2	2.1	0.6	1.7	6.5

Rha = rhamnose; Fuc = fucose; Rib = ribose; Ara = arabinose; Xyl = xylose; Man = mannose; Gal = galactose; and Glu = glucose.

quercetin in oak bark. Fucose (Fuc, or 6-deoxygalactose), which can occur in the D- and L-configurations, is relatively rare in plants, but common (in the L-form and rare in the D-form) in bacteria. There is a link with D-galactose in the biosynthesis of L-Fuc, but the link with Man is less clear in the synthesis of Rha. Ribose (Rib) is common in bacterial polysaccharides, and will of course be a component of the ribonucleic acids of all cells.

Fungi will proliferate under the acidic conditions that prevail in the podzolic profile, and so it is reasonable to assume that Gal, Man, and Rha would primarily be derived from fungal sources. Fuc would also be likely to be derived from microbial sources. Although some D- and L-Ara are present in bacterial and in fungal polysaccharides, it is more likely that plants would provide the major source of the Ara detected in the present study. The same would apply also for Xyl.

Oades (1984) has used the ratio (Man + Gal)/(Xyl + Ara) to suggest origins (plant or microbial) for sugars in soils. He considered that a ratio value < 0.5 would suggest origins in plants, and a value > 2 would indicate microbial sources. The same reasoning could apply for the ratios of (Rha + Fuc)/(Ara + Xyl). However, it could also be meaningful to compare the ratio of Man/Rha in each sample, and also that of Gal/Fuc.

The total NS content of the FAs isolated in water from the A_h horizon was low, and the relative abundances of the sugars (Glu >> Ara = Gal > Man ≥ Rha > Xyl > Rib) were different from those for the other fractions studied. The relative contents of Ara and of Xyl might indicate inputs from plant sources, although the (Man + Gal)/(Xyl + Ara) ratio, but not necessarily that of (Rha + Fuc)/(Ara + Xyl), suggests inputs from microbial sources (Table 4). The order of the relative abundances was different (Glu >>> Man = Rha ≥ Gal > Xyl > Fuc > Rib) for the HAs isolated in water, as was the (Man + Gal)/(Xyl + Ara) ratio (2.5), and that of (Rha + Fuc)/(Ara + Xyl) (1.2). The data might be interpreted in terms of the involvement of microbes in the transformations to the HAs.

Sufficient materials were not recovered from the water soluble extracts from the B_h horizons to allow meaningful analyses to be carried out (Table 1).

There were exceptions, of course, but in general the order of the abundances of the NS in the fractions in the HS isolated in the same solvent systems from the A_h horizon and from the two B_h horizons was along the lines Glu > Man ≥ Gal > Rha ≥ Xyl ≥ Ara > Fuc > Rib. The relative abundances of the NS in the FAs isolated at pH 7 were similar.

The NS contents of the FAs and HAs isolated at pH 10 and at pH 12.6 from the A_h horizon were of the order of two or more times greater than those isolated at pH 7. The most highly (biologically) oxidized (transformed) fractions were isolated at pH 7, and those isolated at higher pH values are considered to be less humified substances. The abundances of the NS in the A_h FA fraction at pH 10.6 and at pH 12.6 were significantly greater than those in the corresponding HA fractions, and that is contrary to expectations. The relatively high contents of Ara in these HAs suggest inputs from plant sources. However, that is not substantiated by the abundances of Xyl, and so the (Man + Gal)/(Xyl + Ara) and (Rha + Fuc)/(Ara + Xyl) ratios, though lower than those for the corresponding FAs, would suggest that microbial as well as plant derived sugars contributed to the genesis of the carbohydrate contents. The contribution of Xyl to the total NS contents of the A_h FAs increased at pH 7, and that would suggest higher imputs from plant sources.

The total NS contents of the FAs in the isolates from the B_h horizon of the forested soil were significantly less than those in the A_h horizon, and though the amounts in the HAs isolated at pH 10.6 were less than in those isolated at the same pH from the A_h horizon, the quantities isolated at pH 7 and at pH 12.6 were similar. The Xyl contents also increased in the extracts at pH values > 7. In general, the NS contents of HS in drainage waters are up to 10 times less than those for soil HS. The lesser amounts of NS in the FAs of the B_h (compared with the A_h) horizon might be regarded as an indication that these were transported in solution in water. However, the data for the sugars of the HAs in the forested B_h (when compared with those in the A_h) might indicate that HAs were transported as dispersed colloidal systems.

The order of abundance of sugars in the FAs from the B_h horizon was similar for the forested and unforested soils, and there was an increase in the abundance of sugars of plant origins in the extracts in solvents at the higher pH values. Both extracts at pH 12.6 were enriched in Xyl. However, the total sugar contents in the unforested FA samples were higher than in the forested. With the exception of the isolates at pH 10.6, these contents were reversed in the cases of HAs, and the trends with regard to the possible origins of the sugars that applied for the FAs applied also for the HAs.

The XAD-4 acids are invariably most enriched in sugars, and this can be expected because it is the more hydrophobic components of HS that are retained by the XAD-8 resin. It is probable that the materials retained by XAD-4 are a mixture of saccharides (including polysaccharides), saccharide related substances, and other very polar FA-type substances. The order of abundances of the sugars in these acids in the extracts at pH 7 are the same, and their total contents are similar in the cases of the A_h and B_h horizons of the forested soil. The order is the same but the abundances are greater in the case of the same extract from the unforested soil. The order of abundance of sugars in the XAD-4 extract from the A_h horizon at pH 12.6 was similar to that for other extracts (FA or FA/HA) at the same pH, but the total sugar content (34.7%) was significantly greater than that of any other humic fraction isolated. The enrichment in Xyl points to origins in plant sources.

Although the order of the abundances of the sugars in the XAD-4 isolate of the unforested B_h horizon extracted at pH 10.6 was similar to the same isolate and fraction from the forested soil, the sugar content in the unforested sample was low. Most of the sugars from the XAD-4 isolate from the unforested soil were contained in the pH 7 extract.

4 Conclusions

The HAs, FAs, and XAD-4 acids isolated from the Uragh podzol profiles demonstrate different characteristics. Elemental analyses show a decrease in carbon contents from the HAs to the FAs, to the XAD-4 acids. NS analyses show that the XAD-4 acids contain up to 35% NS, and those in the B_h horizons are likely to have origins in microbial processes. The hydrophilic XAD-4 acids are likely to consist of products of microbial synthesis formed during the humification process.

A decrease in the NS contents is observed in the B_h (compared with the A_h) horizon within the forest soil profile. This is paralleled by an increase in contributions to synthesis from microbial sources. The NS contents of the less polar HS extracted at the higher pH values are indicative of plant origins. These results suggest that, as humification proceeds, microbes utilise plant NS, and microbially synthesized NS become incorporated in the HS.

Comparisons of the data obtained from the HS isolated from the forested and cleared B_h horizons show fewer differences. The sugar and E_4/E_6 ratio data do not suggest any apparent variations between the cleared and forested sites, although the elemental analyses data show a slight increase in carbon content in the cleared site.

Acknowledgements

This work is part of a project supported by the Natural Environment Research Council for studies of humic substances in paired (forested and cleared) sites.

References

Blakeny, B, P.J. Harris, R.J. Henry, and B.A. Stone. 1983. A simple and rapid preparation of alditol acetates for monosaccharide analysis. *Carbohydrate Research* **113**:219-299.

Chen, Y., N. Senesi, and M. Schnitzer. 1977. Information provided on humic substances by E_4/E_6 ratios. *Soil Sci. Soc. Am. J.* **41**:352-358.

Gorin, P.A.J. and E. Bareto-Bergter. 1983. The chemistry of polysaccharides of fungi and lichens. p. 365-409. *In* G.O. Aspinall (ed.), *The Polysaccharides* Vol. 2. Academic Press, Orlando.

Hayes, T.M. 1996. Study of the humic substances from soils and waters and their interactions with anthropogenic organic chemicals. PhD thesis, The University of Birmingham.

Keene, L. and B. Lindberg. 1983. Bacterial polysaccharides. p. 287-363. *In* G.O. Aspinall (ed), *The Polysaccharides* Vol. 2. Academic Press, Orlando.

Kononova, M.M. 1966. *Soil Organic Matter*. p. 400-404. Pergamon Press, Oxford.

Little, D.J. 1994. Occurrence and characteristics of podzols under oak woodland in Ireland. PhD thesis, The National University of Ireland.

Malcolm, R.L. and P. MacCarthy. 1992. Quantitative evaluation of XAD-8 and XAD-4 resins used in tandem for removing organic solutes from water. *Environment Internat.* **18**:597-607.

Oades, J.M. 1984. Soil organic matter and structural stability: mechanisms and implications for management. *Plant and Soil* **76**:319-337.

Painter, T.J. 1983. Algal polysaccharides. p. 195-285. *In* G.O. Aspinall (ed), *The Polysaccharides* Vol. 2. Academic Press, Orlando.

Stephen, A.M. 1983. Other plant polysaccharides. p. 97-193. *In* G.O. Aspinall (ed), *The Polysaccharides* Vol. 2. Academic Press, Orlando.

Humic Substances from Podzols Under Oak Forest and a Cleared Forest Site II. Spectroscopic Studies

A.J. Simpson, J. Burdon, C.L. Graham, and M.H.B. Hayes

THE UNIVERSITY OF BIRMINGHAM, SCHOOL OF CHEMISTRY, EDGBASTON, BIRMINGHAM B15 2TT, ENGLAND

Abstract

Fluorescence spectra demonstrate at least two fluorophores in the XAD-4 acids that were not apparent in the humic (HAs) or fulvic acids (FAs), and also the presence of fluorescent units in the FAs that were not apparent in the HAs. The FT-IR spectra were similar, as were the proton (^1H) NMR spectra of the humic substances (HS) isolated from the B_h horizons of a forested site, and of a site where the forest was cleared 400 years ago. The ^1H NMR data, however, showed a large increase in aliphatic functionalities in the HS from the B_h horizon compared with those from the A_h horizon, and there was considerable evidence for exchangeable protons in the HS fractions from the A_h horizon. The XAD-4 acids, and to a lesser extent the FAs, contained a high diversity of exchangable protons. The HAs isolated at the higher pH values exhibited resonances in the exchangeable proton region that were relatively well resolved, and that might be interpreted as evidence for protons associated with ether-, ester-, and lactone-type functionalities in the less well humified fractions that are extractable only when the aqueous media are alkaline. The data would suggest greater homogeneity in the fractions within the A_h horizon, and could indicate that considerable changes take place in the compositions of the HS before they are transported in the podzolization process, and/or that significant changes take place in the compositions of the HS during their residence in the B_h horizon.

1 Introduction

In Part I of this series, humic acids (HAs), fulvic acids (FAs) and XAD-4 acids were isolated from the A_h and B_h horizons of a podzol under oak forest and from the B_h horizon of a neighbouring site where the forest was cleared 400 years ago. Data were given for the yields obtained from extracts in water, sodium pyrophosphate (Pyro, 0.1M) at pH 7, Pyro (0.1M) at pH 10.6, and Pyro (0.1M) plus sodium hydroxide (0.1M) at pH 12.6, for the E_4/E_6 ratio values, and for the sugar contents and various sugar ratios of the different fractions. The data indicated compositional differences between the humic fractions from the A_h and B_h horizons, but not between those from the B_h horizons of the forested and cleared sites. This paper provides data for the FT-IR, fluorescence, and ^{1}H NMR spectroscopy of the samples.

2 Materials and Methods

Sources of Samples

Soil samples were collected from the forested and cleared sites at Uragh Wood, Uragh townland, Lough Inchiquin, Kenmare, Co. Kerry, Ireland (Grid reference: V.83.62). A description of the sites is given in Part I of this series.

Isolation and Fractionation of Humic Substances

The sequential extraction and the fractionation procedures used are described in Part I of this series.

Infrared Spectroscopy

KBr discs were prepared using 100 mg of dry KBr (spectroscopic grade) and 1 mg of sample. Spectra were recorded on a Perkin Elmer Paragon 1000 FT-IR spectrometer.

Fluorescence Spectroscopy

Samples (2 mg) were dissolved in 0.05M $NaHCO_3$ (10 mL) and the solutions (pH 8.2) were synchronously scanned on a Fluoromax 9000 series spectrometer using, a slit width of 1 nm, an integration time of 0.05 seconds, and a Stokes shift of 20 nm.

^{1}H NMR Spectroscopy

Samples (75 mg) were dissolved in DMSO d_6, and the spectra were recorded on a Bruker AC 300 NMR Spectrometer. A 30° flip angle, 2.6 ms pulse width, 4 s pulse delay, a 2.72 s aquisition time, and a 6024 Hz sprectral width were used as parameters.

3 Results and Discussion

FT-IR Spectroscopy

The FT-IR spectra of all samples were broadly similar, and displayed all of the major absorption bands that are characteristic of humic substances (MacCarthy and Rice, 1985).

There was a greater aliphatic character (2920 - 2860 cm^{-1}) in the humic fractions isolated from the B_h than from the A_h horizon. The spectra did not reveal any major differences between the fractions from the two B_h horizons. However, the C-O stretch at 1050 cm^{-1} was more intense for the sample isolated from the unforested (compared with the forested) site.

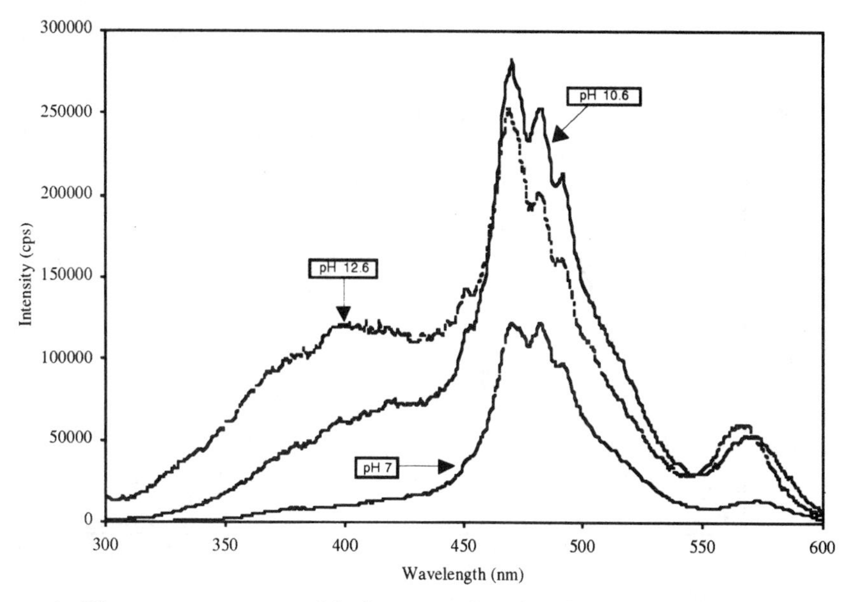

Figure 1 *Fluorescence spectra of the humic acids isolated at pH 7, pH 10.6, and pH 12.6 from the A_h horizon of the forested soil*

Fluorescence Spectroscopy

Fluorescing units are considered to be minor components of HS. Miano and Senesi (1992) have suggested that the most efficient fluorophores in HS include variously substituted, condensed aromatic rings and/or highly unsaturated aliphatic chains. There is not, however, convincing evidence from chemical degradation studies for condensed aromatic components in humic structures. Bloom and Leenheer (1989) have suggested the presence of two main types of fluorophores, one at an excitation between 315 and 390 nm, which might be attributable to carboxyphenol, and a second, whose source was not assigned, at an excitation between 415 and 470 nm. Improvements in synchronous scanning techniques allow fluorescence excitation to be measured in increments as low as 1 nm, and the more detailed spectra obtained indicate the presence of many fluorophores (with up to seven detectable bands in some spectra).

In general, the HA spectra provided three main excitation bands, at 465, 480, and at 490 nm (Figure 1), and these were least intense for the fractions isolated at the lower pH values (water and 0.1M Pyro, pH 7). The FAs show excitations in the same regions, and with intensities that are generally even greater than those of their HA analogues extracted at the same pH (as an example see Figure 2). The FAs, however, show additional excitation bands at 400, 420, and 450 nm, and these increase in intensity as the pH of the extraction solvent is raised. The excitation shifts to lower wavelengths in the cases of the XAD-4 acids, with the maximum absorbance shifting from 480 nm to 400 nm, and with new bands appearing at 330 and 370 nm (Figure 3). It is not possible at this stage to assign specific excitation bands to particular functionalities or building units.

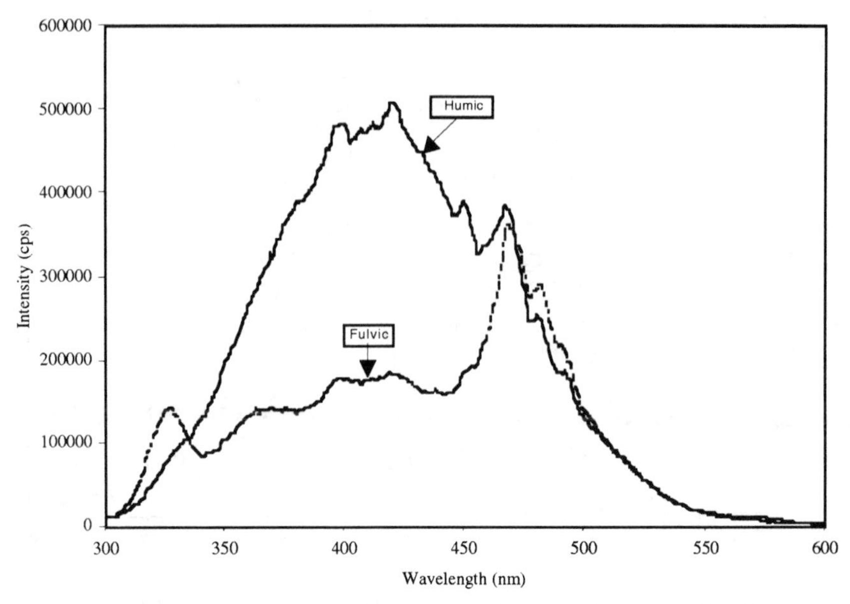

Figure 2 *Fluorescence spectra of humic acids (HAs) and fulvic acids (FAs) from the water extract of the A_h horizon of the forested soil*

Fluorescence spectra of HAs isolated at different pH values from the A_h horizon of the forested soil are shown in Figure 1. The intensities of all bands were greater for the HAs extracts at pH 10.6 and 12.6 than for those at pH 7. As the pH of the extractant increased, the intensity of the band at 360-440 nm increased, and the same trend applied for the band at 560-580 nm (which was specific for the HS of the A_h horizon, and was not apparent in spectra for the other samples studied). There were significant differences between the fluorescence spectra for the humic fractions isolated from the A_h and B_h horizons. The spectra for the HAs of the B_h horizon showed that the three excitation bands at 465, 480, and 490 nm had intensities up to 10 times greater than those for the HAs in the A_h horizon.

The spectrum for the HAs isolated in water from the A_h horizon is complex, and shows bands at 330, 370, 400, 420, 450, 480, and 490 nm. Such complexity would suggest the presence of a number of different fluorophores.

Fluorescence intensities in the spectra for the FAs from the B_h horizon decreased as the pH of extractant increased. The profiles were similar, however, indicating that the fractions contained similar fluorophores. The FAs from the A_h horizon showed greater intensities at 400 nm and at 420 nm, and these tended to increase as the pH of the extractant increased. The FAs from the water extract of the A_h horizon, like the HAs in the same extract, gave complex spectra, with at least six bands. (Figure 2).

The spectra for the XAD-4 acids (see Figure 3) were also complex, and the fluorescence intensity increased as the pH of the extractant was raised. The XAD-4 acids extracted in water from the A_h horizon had bands at 330, 360, and 420 nm that were better defined than those for the same fraction from the B_h horizon (Figure 3). Bands were evident at 330 nm and at 420 nm for the XAD-4 acids isolated from the A_h horizon at the higher pH values. These were not prominent for the corresponding isolates from the B_h

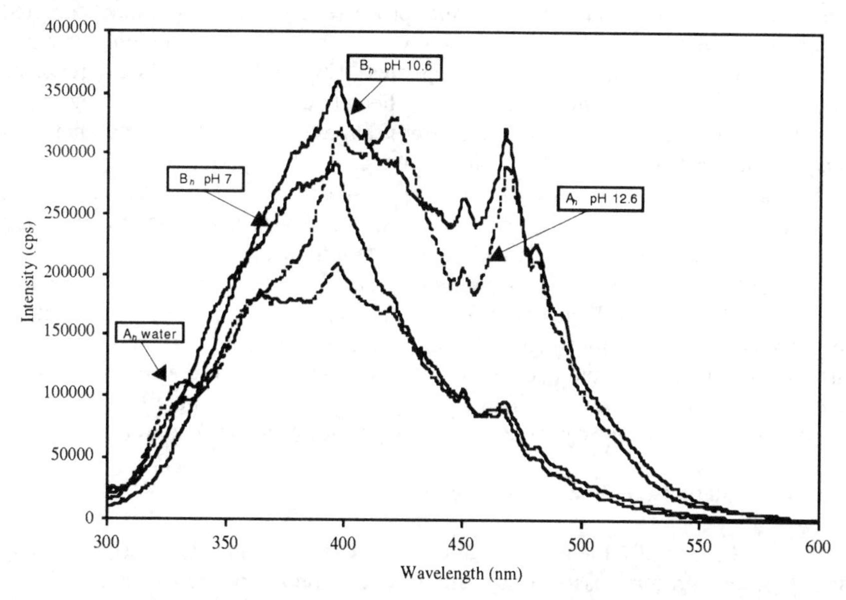

Figure 3 *Fluorescence spectra for the XAD-4 acids isolated at pH 7, 10.6, and 12.6 from the Ah and Bh horizons of the forested soil. The pH refers to that of the extractant used.*

horizon. The surface (A_h) horizon contains a diversity of plant components at different stages of humification, and a diversity of microorganisms and microbial processes. Thus, it is reasonable to assume that the humic fractions will be more diverse and more complex than those that were transported to the B_h horizon. The HS in the A_h horizon will contain a mixture of 'young' (with compositional properties related to the materials from which they were derived) and of older HS. The 'young' HS are especially prominent in the fractions extracted by the more alkaline solvents. The older HS will be more highly (biologically) transformed (oxidized), and these will be the more mobile, and are likely to contribute most to the materials transported to give rise to the HS in the B_h horizon.

There were strong similarities in both the intensities and profiles of the fluorescence spectra for the HS isolated at the different pH values from the B_h horizon inside the forest and for those for the B_h horizon of the cleared site. The spectra for the HAs had the three bands at 465, 480, and 490 nm, and the band at 465 nm became more predominant as the pH of the extractant increased. Fluorescence was most intense for the isolates at pH 7 (0.1M Pyro), and it decreased for the HAs extracted at pH 10.6, and then rose to an intermediate level for the HS extracted at pH 12.6. A similar trend was seen for the FAs, although the band in the 380-420 nm range was more significant for the FAs.

The spectra for the XAD-4 acids from the B_h horizons of the forested and cleared sites were also similar, and these had a large excitation in the 340-450 nm range. A more defined absorbance at 465 nm was seen in the case of the XAD-4 acids isolated at pH 10.6 from the B_h horizon of the forested site.

The similarities in the spectra of the HS from the two sites indicate that the same fluorophores may be present in the HS of both B_h horizons, and at roughly the same concentrations. That might indicate that similar processes are occurring or have occurred in the formation of the HS, and that these have involved the production of highly complex conjugated systems which have fluorescing properties. The similarities in the fluorescence properties of the HS in the B_h horizons that have been under different climax vegetation for the past 400 years (since the forest was cleared) suggest that the HS in the B_h horizons have retained the fluorophores which they had when formed/transported.

It will be appropriate to compare the fluorescence spectra of the HS in the B_h horizons of soils with different climax vegetation systems, and where that vegetation has not changed since podzolisation began. It will be important to be able to assign fluorescence bands to particular fluorophores, and to compare spectra for lignins, tannins, suberins, and other plant residues, as well as the skeletal residues and humic-type products of microorganisms, and especially fungi. The fluorescence tool could thus lead to a better understanding of the origins, and even to an awareness of aspects of the compositions of structural units of HS, and of the changes that take place during humification. It is possible also that the technique can be developed to provide a 'humic fingerprint' system.

^{1}H Nuclear Magnetic Resonance Spectroscopy

The proton NMR spectra for the HS isolated from the Uragh site indicated the presence of aliphatic protons, protons associated with oxygen containing functionalities, and aromatic/amide protons. Chemical shifts below 2.6 ppm can be assigned to the aliphatic protons, and the band between 6.5 and 8.5 ppm to aromatic/amide protons (Oka *et al.* 1969). There is debate about the resonances between 3 and 4.5, and these have variously been assigned to protons of methoxyl, and of lactones (Oka *et al.* 1969), and to exchangeable protons (Wershaw 1985).

Typical spectra obtained for HAs, FAs, and XAD-4 acids isolated from the B_h horizons are shown in Figure 4. The spectra for the samples from the forested site were similar to those from the cleared ground. Two main bands, assignable to CH_3-R (where R represents an alkyl group), and to CH_3-C-O protons, respectively (Williams and Fleming, 1989), were evident in the aliphatic region at 0.9 ppm and 1.3 ppm in the case of HAs isolated at pH 7. The lack of a large peak at 1.4 ppm in this spectrum, and in the spectra of all the samples isolated from the Uragh site, indicates a lack of R-CH_2-R protons. That would suggest that long chain aliphatic structures were not major components of the samples. The broad band at 3 - 4.5 ppm for the HAs isolated at pH 7 may be indicative of a dominance of exchangeable protons, and the finer resolutions, characteristic of protons associated with lactone, ether, ester, or carbonyl functionalities, were missing. The aromatic resonance covers a wide chemical shift range, indicating the presence of variously substituted aromatic functionalities.

There was somewhat better resolution for the spectrum of the FAs isolated at pH 7 from the B_h horizon. Bands were evident at 0.9 and 1.3 ppm, and additional bands were seen at 2.2 ppm and at 2.3 ppm which are likely to be attributable to CH_3 protons bonded to aromatic rings, and to CH_3-CO-R protons (Figure 4). The resonance in the 3 - 4.5 ppm region was still broad and characteristic of a high exchangeable proton content, although three bands were evident at 3.7, 4.3, and 4.5 ppm. Three distinct peaks, at 6.9, 7.1 and 7.3 ppm, were evident in the aromatic resonances. These can be assigned to amide protons or to protons on substituted aromatic rings, but assignations are not possible because of the number of combinations of substituents and substitution patterns that could apply.

The spectra for the XAD-4 acids from the B_h horizon of the podzol show lesser total

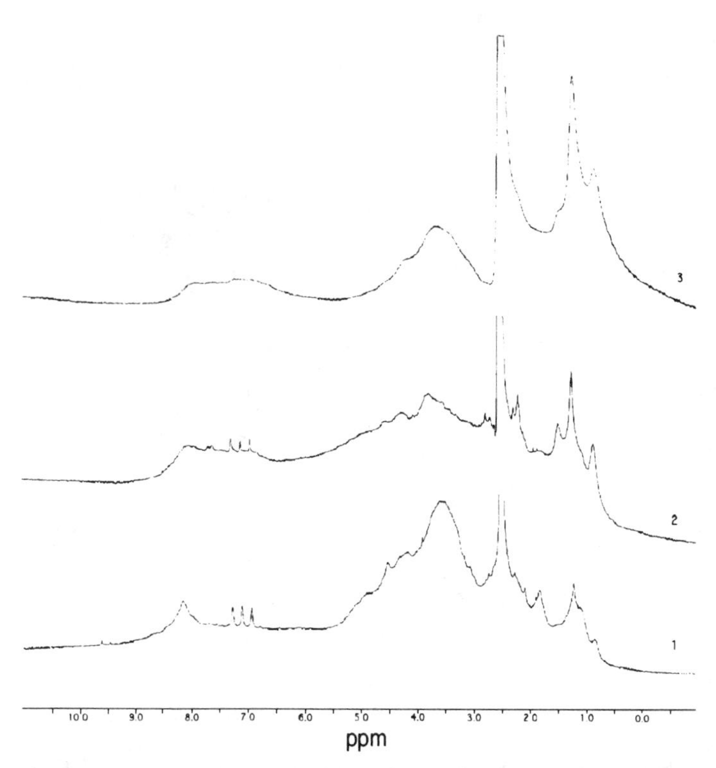

Figure 4 ^{1}H NMR spectra of the XAD-4 acids (1), the fulvic acids (2), and the humic acids (3) isolated at pH 7 from the B_h horizon of the forested soil

aliphatic functionality than is seen for the HAs and FAs, but with an additional band at 1.9 ppm that is consistent with R-CH_2-C-O protons. The broad and intense resonance from 3 - 4.5 ppm suggests a high exchangeable proton content (Figure 4). Characteristic peaks at 6.9, 7.1, and 7.3 ppm were evident in the aromatic/amide resonance, and there was an additional resonance at 8.1 ppm. As the pH of the extractant was increased, the proton composition of the B_h horizon HS became more aliphatic in character, and the resonance assigned to exchangeable protons in the HS isolated at pH 7 became less prominent.

Although the spectra were quantitative, integration was difficult because of the undulating base line and the overlapping bands. For most of the HS isolated from the B_h horizon at pH 7, rough estimates were obtained for aliphatic, aromatic, and exchangeable protons. Integration indicated that the HAs from the B_h horizons at pH 7 were ~60% aliphatic, ~20% aromatic, and had ~20% exchangeable protons. For the FAs, the data indicated ~40% aliphatic, ~10% aromatic, and ~50% exchangeable protons, and for the XAD-4 acids the values were ~10% aliphatic, ~10% aromatic, and ~70% exchangeable protons. Large errors can arise in the phase correction (used to obtain a straight baseline) and from peak overlap, and the results must be interpreted with caution.

The spectra for the HS from the A_h horizon were better resolved (cf Figure 5) than

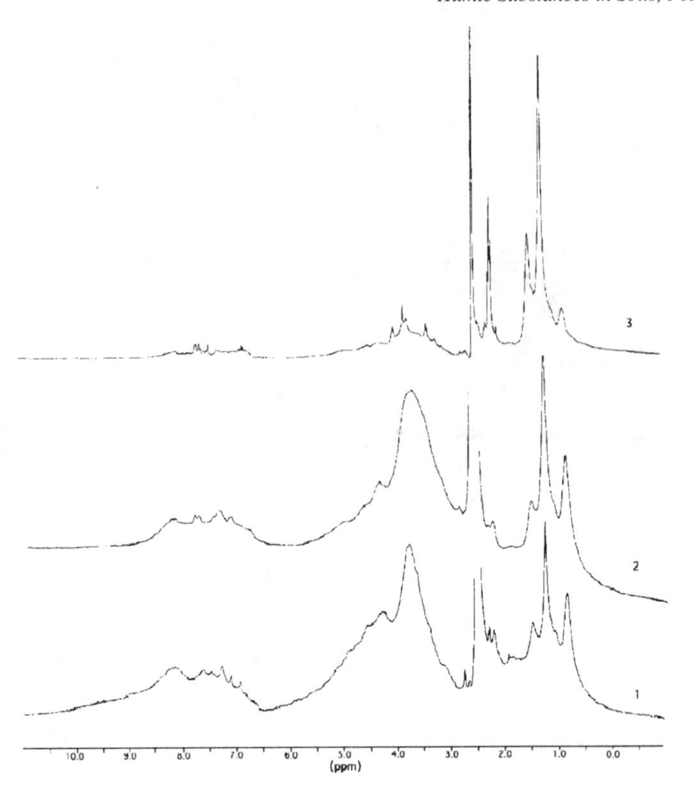

Figure 5 *Humic acids isolated from the A_h horizon of the forested soil at pH 7 (1), pH 10.6 (2) and pH 12.6 (3)*

those from the B_h. Resonances were evident at 0.9, 1.3, 1.5, 2.2 and 2.3 ppm in the spectra of the FAs isolated from the A_h horizon at pH 7, and the band at 2.2 ppm (which might be attributable to CH_3-CO-R protons) was more pronounced for the FAs than for the HAs. The band at 3 - 4.5 ppm was broad and characteristic of a high exchangeable proton content. The aromatic region was better resolved and had the characteristic peaks at 6.9, 7.1, and 7.3 ppm, and with additional broader resonances at 7.4, 7.6, and 8.0 ppm.

Spectral noise was less, and a higher level of resolution, along with a decrease in the broadness and intensity of the 3 - 4.5 ppm resonance (characteristic of exchangeable protons), was observed for HS isolated using the higher pH extractants. Proton NMR spectra are shown in Figure 5 for HAs extracted from the A_h horizon at pH values 7, 10.6, and 12.6. The spectra for HAs and FAs (not shown) isolated at pH 12.6 were very similar, except that the detail seen in the 3 - 4.5 ppm region for the FAs was superimposed on the broad shoulder indicative of exchangeable protons. (That shoulder was largely absent for the HAs from the pH 12.6 extract; Figure 5). The 3 - 4.5 ppm region shows fine detail not observed in many of the other samples. The peaks at 3.4, 3.7, 3.8 and 4.0 ppm are not inconsistent with protons of R-CH_2-OR, CH_3-O-CO-R, CH_3-OAr and R-CH_2-O-CO-R functionalities, and the region can no longer be said to be dominated by exchangeable protons. Fine resolution is evident for the 6.3 - 8.0 ppm resonances. However, the characteristic peaks at 6.9, 7.1, and 7.3, seen in the spectra for many of the HS fractions, were not prominent. On expansion, especially in the case of the 7.1 - 7.3 ppm resonances, some of the peaks were split into doublets. That would indicate that the protons in question were neighbouring one other aromatic proton. Such will only occur for certain aromatic substitutions, such as 1,4 disubstituted, 1,2,4 trisubstituted, or 1,2,3,4 tetrasubstituted six membered rings.

Intergrations of the spectra suggested that in the cases of the fractions isolated from the A_h horizon at pH 7, the HAs and FAs both contained 25% aliphatic protons, 50% predominantly exchangeable protons, and 25% aromatic protons. The HS fractions isolated from the A_h horizon at pH 12.6 were, however, more aliphatic, and the HAs contained ca 70% aliphatic protons, ca 25% protons associated with oxygen containing functionalities, and ca 5% aliphatic protons. (The detail evident in the 3 - 4.5 ppm resonances of the fractions isolated at pH 12.6 would suggest that this resonance was not dominated by exchangeable protons, and it is likely that protons associated with ether, ester, carbonyl and lactone functionalities were major contributors to the resonances.)

A comparison of the spectra for HS isolated at pH 7 from the A_h and B_h horizons would suggest that the HS from the B_h horizons were more aliphatic in nature and contained less exchangeable and aromatic protons than did the HS from the A_h horizon. However, the proton NMR spectra did not reveal any apparent differences between the HS isolated from the B_h horizons of the forested and the cleared sites.

4 Conclusions

Although FT-IR spectroscopy failed to show clear differences between the humic fractions isolated from the podzol horizons, differences were evident from the fluorescence and the ^1H NMR spectra. The fluoresence spectra demonstrated the presence of at least two fluorophores in the XAD-4 acids that were not evident in the spectra for the HAs or the FAs, and there were also fluorescent units in the FAs that were not apparent in the HAs. ^1H NMR spectroscopy data showed the presence of aliphatic protons, protons associated with oxygen containing functionalities, and aromatic protons in all the samples. The data suggest that the HAs tend to contain more aliphatic and aromatic protons, and the fulvic and XAD-4 acids (in line with their known higher contents of acidic functionalities) tend to show the presence of more exchangeable protons. The HS from the B_h horizon had more aliphatic protons than those from the A_h horizon, and these in turn (in the A_h) tended to be richer in aromatic and in exchangeable protons. Spectra for the A_h horizon HS showed better resolution, and the bands were less broad than those of the spectra for the HS of the B_h horizon. That might suggest greater homogeneity in the fractions within the A_h horizon, and could indicate that considerable changes take place in the compositions of the HS before they are transported in the podzolization process. Alternatively, it might indicate that significant changes take place in the compositions of the HS during their residence in the B_h horizon.

Acknowledgements

This work is part of a project supported by the Natural Environment Research Council for studies of humic substances in paired (forested and cleared) sites.

References

Bloom, P.R. and J.A. Leenheer. 1989. Vibrational, electronic, and high-energy spectroscopic methods for characterizing humic substances. p. 409 - 446. *In* M.H.B. Hayes, P. MacCarthy, R.L. Malcolm, and R.S. Swift (eds), *Humic Substances II. In Search of Structure.* Wiley, Chichester.

MacCarthy P. and J.A. Rice. 1985. Spectroscopic methods (other than NMR) for determining functionality in humic substances. p 527-559. *In* G.R. Aitken, D.M. McKnight, R.L. Wershaw, and P. MacCarthy (eds), *Humic Substances in Soil Sediment and Water.* Wiley New York.

Miano T.M, and N. Sensi. 1992. Synchronous excitation fluorescence spectroscopy applied to soil humic substances chemistry. *The Science of the Total Environment* **117/118**:41-51.

Oka, H., M.. Sasaki, M. Itho, and A. Suzuku. 1969. Study on the chemical structure of peat humic acids by high resolution nuclear magnetic resonance spectroscopy. *Nenryo Kyokai Shi (J. Japanese Fuel Soc.)* **48**:295-302.

Wershaw R.L. 1985. The application of nuclear magnetic resonance spectroscopy for determining functionality in humic substances. p 561-582.. In G.R. Aitken, D.M. McKnight, R.L. Wershaw, and P. MacCarthy. *Humic Substances in Soil Sediment and Water*, Wiley New York.

Williams D.H, and I. Fleming. 1989. *Spectroscopic Methods in Organic Chemistry.* McGraw-Hill, Maidenhead, Berkshire, England.

Studies of Humic Substances at Different pH Values using Scanning Electron Microscopy, Scanning Tunnelling Electron Microscopy, and Electron Probe X-Ray Micro Analysis

A.J. Simpson, C.L. Graham, M.H.B. Hayes, K.A. Stagg[1], and P. Stanley[2]

THE UNIVERSITY OF BIRMINGHAM, SCHOOL OF CHEMISTRY, EDGBASTON, BIRMINGHAM, B15 2TT, ENGLAND
[1]THE UNIVERSITY OF BIRMINGHAM, SCHOOL OF EARTH SCIENCES
[2]THE UNIVERSITY OF BIRMINGHAM, SCHOOL OF MEDICINE, DEPARTMENT OF PHYSIOLOGY

Abstract

A hydrophilic resin, which is considered to enable bulk water to escape from macromolecular samples and to allow the samples to retain their solution conformations, was used to prepare humic acids (HAs) isolated from the B_h horizon of a podzol for studies of conformations using scanning tunnelling electron microscopy (STEM), scanning electron microscopy (SEM), and electron probe X-ray micro analysis (EPXMA) techniques. At low pH values the humic sample had linear/'spike-like' conformations, indicative of associations of molecules by hydrogen bonding and by van der Waals forces. At neutral pH, the conformations were spherical/globular, which could indicate random coil conformations (should the molecules be large enough), and suggests associations between the macromolecular units. Small spheroidal particles were in evidence in which the HAs were associated with metal ions and/or finely divided clay/(hydr)oxide species.

1 Introduction

Electron microscopy, coupled with techniques such as energy dispersive spectroscopy (EDS), provides a powerful tool for observations of size, morphology, and elemental compositions of samples. However, samples must be stabilized for examination of specific features under the high vacuum conditions required when techniques such as scanning tunnelling electron microscopy (STEM) and scanning electron microscopy (SEM) are used. The classical techniques of sample preparation have involved the encasing of samples in hard hydrophobic resins, the drying of dissolved samples in a desiccator over silica gel [as used by Senesi *et al.* (1996) for studies of the Summit Hill IHSS Reference humic acid (HA)], and the rapid immersion of droplets of humic solutions (on a glass slide) in molten Freon-12 (at -155 °C) followed by submersion under liquid nitrogen, and then freeze drying, as used by Chen and Schnitzer (1976). The possibility exists that these techniques can produce conformational artefacts (caused by the redistribution of some mobile structures as the result of shrinkage) giving rise to the rupture and distortion of some delicate structures, and the misleading aggregation of sub-units (Causton, 1985; Leppard *et al.*, 1990).

Nanoplast FB101 is a melamine resin (Bachhuber and Frosh, 1983; Frosh and Westphal, 1989) which has hydrophilic properties. During the polymerization/desiccation step water is produced which evaporates together with the bulk water from the sample. Thus, the polymer slowly exchanges the water of hydration, and the sample structure is considered to be unaltered when compared to the shrinkage that results from dehydration. It is considered that the solution conformations are retained in the dried samples. A procedure described by Perret *et al.* (1991), which involves the use of the Nanoplast gel, was employed for the studies outlined in this paper in which observations were made of the conformations of HA fractions isolated from the B_h horizon of a podzol.

2 Experimental

Isolation of the Humic Substances

Humic substances (HS) were sequentially extracted from the B_h horizon of a podzol in Attiapleton townland, Pontoon, Co. Mayo, Ireland (Grid reference G.20.05) that was cleared of oak forest 250 years ago (Little, 1994). The solvent sequence used was distilled water, sodium pyrophosphate (0.1M, pH 7), sodium pyrophosphate (0.1M, pH 10.6), and a sodium hydroxide (0.1M)/sodium pyrophosphate (0.1M) mixture (pH 12.6). Extraction with each successive solvent was exhaustive. Each extract was processed by the XAD-8 and XAD-4 resins in tandem procedure described by Malcolm and MacCarthy (1992) and by Hayes *et al.* (1996).

Preparation of the Aqueous Samples

Four aqueous solutions were made up. Humic acids (HAs, 10 mg) extracted at pH 7 and at pH 10.6 were dissolved in bi-distilled water to form two of the samples. HAs (10 mg) extracted at pH 7 were dissolved in 0.1M NaOH, and one part was adjusted to pH 1-2 with HCl. (The solution was very dilute.) All solvents were made up using bi-distilled water.

Sample Preparation for Electron Microscopy

Copper SEM grids were cleaned in chloroform using an ultrasonic bath and left to dry under vacuum. Filter paper was placed under the surface of bi-distilled water and the grids were placed upon it at 1 cm intervals. A clean glass rod was dipped into a solution of 0.5% collodion (cellulose nitrate) in amylacetate (isopentylacetate) and then touched onto the surface of the bi-distilled water to form a thin film. The bi-distilled water was then drained and the film coated the grids at a thickness of ca 10-50 nm. The grids were then coated with carbon under at least 5 x 10^{-5} Torr vacuum to give a film thickness of 5 to 10 nm. Melamine resin was freshly prepared by mixing 0.025 g of *p*-toluoylsulphonic acid (catalyst B25) with 1.0 g of hexamethylol-melamine-methyl ether (monomer MME7002). Sample mixtures were prepared by mixing 1 part resin with 10 parts of aqueous solution containing the sample particles, and 3 mL sample mixtures were then pipetted onto the specimen grid held on a horizontal centrifuge with double-sided tape. After a waiting time of 30 s, the grids were spun at 7000 rpm for 15 s, then placed in a desiccator at 40 °C for 12 h, followed by 60 °C for 12 h, and finally 80 °C for 12 h. All grids were made in triplicate. Samples were studied using a Joel 100 CX-II Electron Microscope with an ASID scanner attachment, and were examined in the SEM and STEM modes, at 100 kV acceleration voltage. Electron probe X-ray micro analysis (EPXMA) used a link ISIS system.

3 Results and Discussion

The HAs in bi-distilled water extracted from the soil in pyrophosphate at pH 7 and at pH 10.6 showed no apparent differences under the microscope. Both samples showed the presence of two different types of colloidal structures. The vast majority of the colloids were irregular shaped masses or spheres showing no fine structure (Figure 3, Photographs I and II). [The scanning electron micrographs produced by Chen and Schnitzer (1976) of HAs at pH values of 6 and 8 from the A_1 horizon of a Haploboroll soil consisted of fibres and bundles of fibres]. Lesser amounts of the materials in our micrographs consisted of more dense units containing what would appear to be granular components (Figure 3, Photographs III and IV). Analysis by EPXMA (Figures 1 and 2) indicated the presence of nitrogen, oxygen, fluorine, silicon, and sulphur in both types of colloid (elements less than 14 atomic mass are too light to be detected by EPXMA). Additionally, the denser granular colloidal materials showed the presence of aluminium, potassium, and iron, elements that were absent from the spherical conformational types which did not contain granular components (the peaks due to copper at 4.5, 8, and 8.9 KeV are artefacts of the copper grids used in the study). It would seem likely that the granular conformations were caused by aggregation of organic components around metal and/or fine clay/(hydr)oxide nuclei. [The silicon content could be indicative of aluminosilicates and the iron might suggest (hydr)oxides in association with the clays]. The electron micrographs by Chen and Schnitzer (1976) of B_h horizon FAs complexed with Cu^{2+}, Al^{3+}, Fe^{3+}, and Fe^{2+} at pH 5 exhibited thin, long fibres which had a good degree of orientation. The morphologies, however, were very different from the dense granular type structures associated with the metals in our micrographs (Figure 3, Photographs III and IV).

The conformations of the HA colloidal components at pH 7 were different from those at the same pH in the case of the FAs from the B_h horizon of the Armadale profile, as presented by Chen and Schnitzer (1976). Their micrographs showed reticular-type structures at pH 7, but the structures were amorphous at the higher pH value. The morphological features evident for the Summit Hill HA studied by Senesi *et al.* (1996) at pH values of 3, 4, 5, 6, and 7 were broadly similar, and comparable to those obtained for FAs from the Armadale B_h horizon at pH values 8, 9, and 10 (Chen and Schnitzer, 1976). In our study, the morphologies of the HS that dissolved in 0.1M NaOH were not different from those that dissolved in distilled water. There were, however, significant differences in the morphologies of the HAs which had been subjected to the acid treatment (Figure 3, Photographs V and VI). These had 'spike-like' conformations distributed throughout the micrograph. These colloids would appear to be composed of aggregated fine 'hair-like' structures, and this type of conformation is also reflected in the electron micrographs of the FAs from the B_h horizon of the sample of Chen and Schnitzer at pH 2.

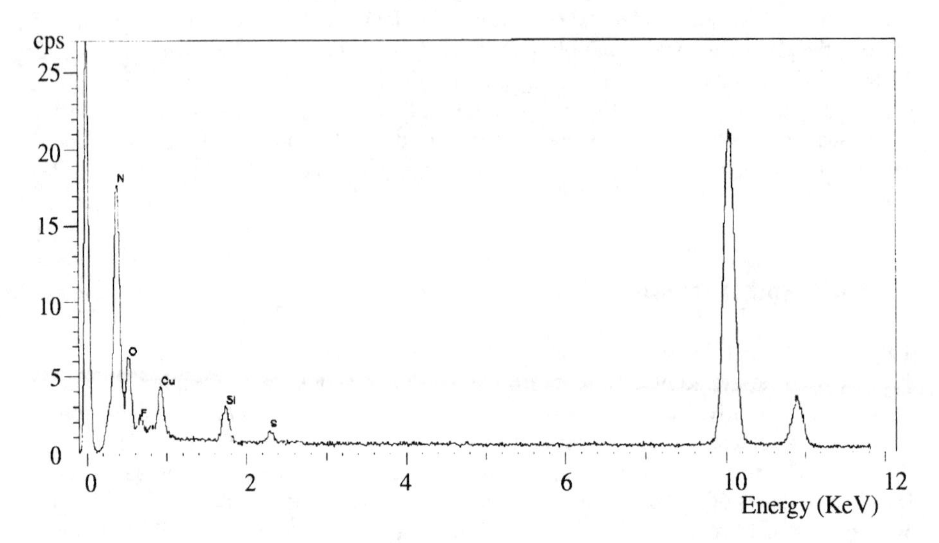

Figure 1 *The EPXMA of the colloid types (shown in Figure 3, Photographs I and II) which did not contain granular components*

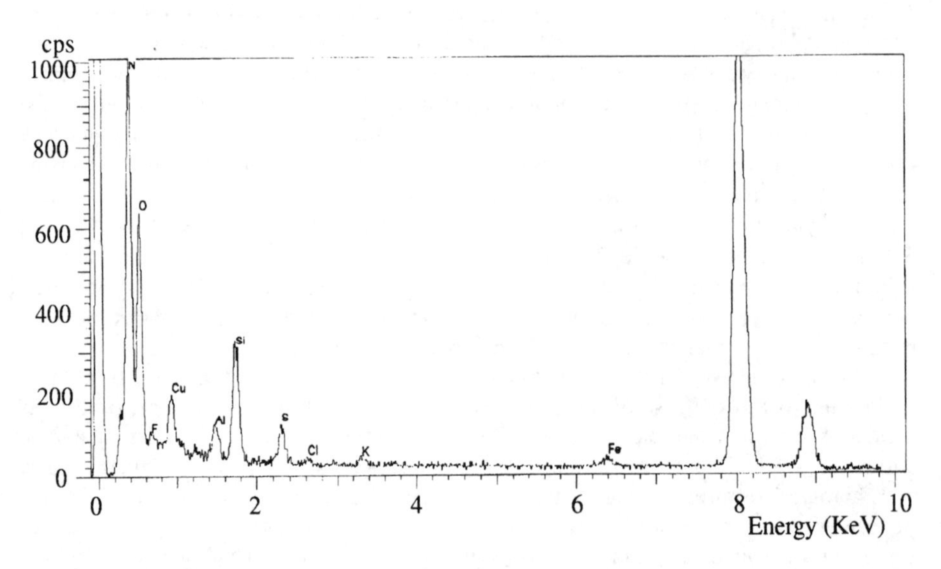

Figure 2 *The EPXMA of the colloid types (shown in Figure 3, Photographs III and IV) which contained granular components. Note the peaks for Fe, K, and Al*

We consider the fine 'hair-like' structures seen at the low end of the nanometer scale to be associations (through hydrogen bonding and van der Waals forces) of humic macromolecules.

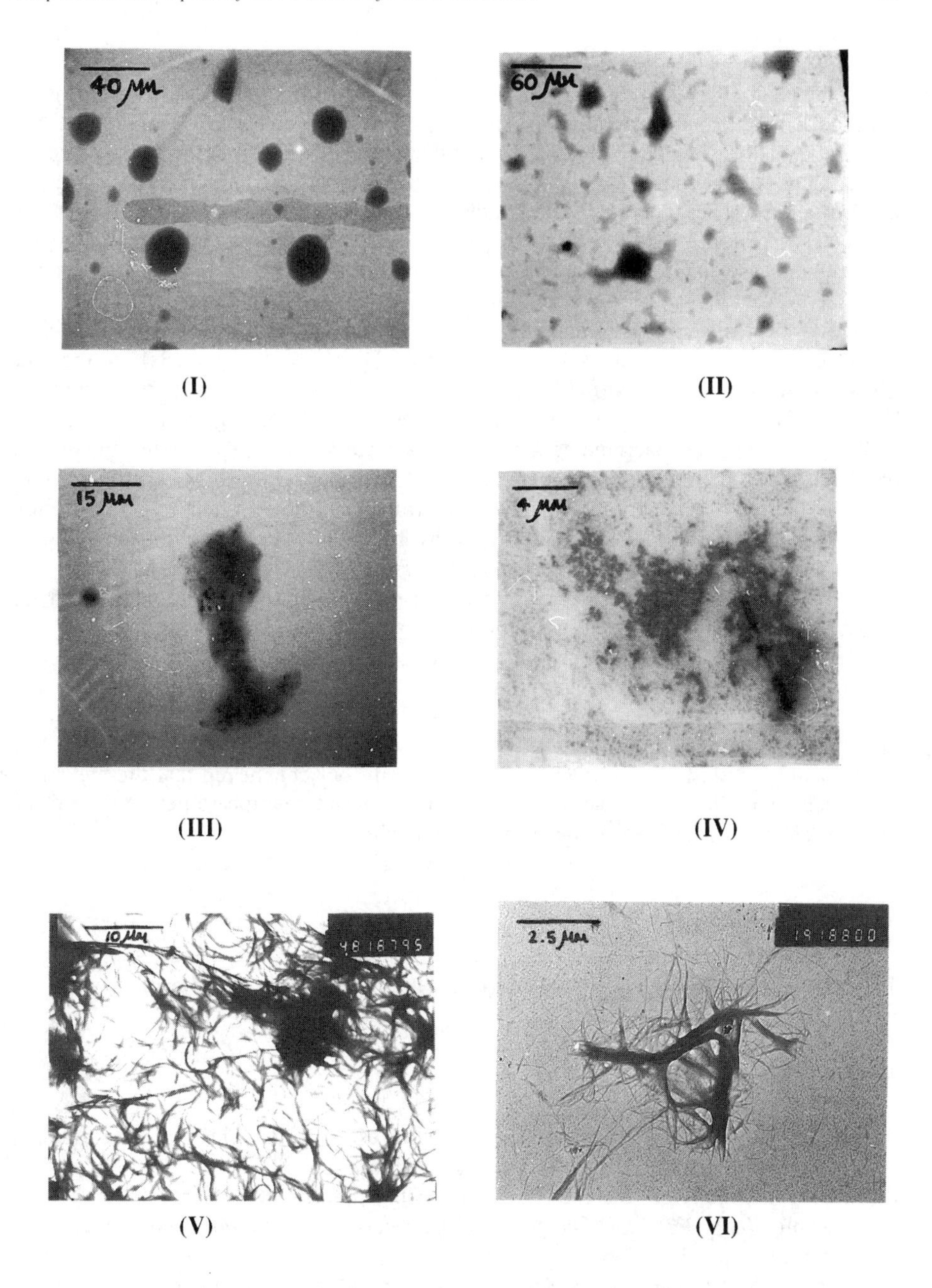

Figure 3 *Scanning electron micrographs and scanning tunnelling electron micrographs of humic acids (HAs) from the* B_h *horizon of a podzol. Photographs I and II, scanning electron micrographs of a solvated HA; III and IV, scanning electron micrographs of HAs showing granular-type components; V and VI, scanning tunnelling electron micrographs of HAs*

4 Conclusions

When fully solvated, as in the cases of the samples in distilled water and in 0.1M NaOH, the HAs have expanded, globular- or spherical-type conformations. We agree with the interpretations of Chen and Schnitzer (1976) that repulsions between the charged groups at pK_a values above those of the carboxyl functionalities of the HS would give expanded conformations. These could reflect random coil-type structures, assuming that the molecules were sufficiently large to give such conformations. However, the morphologies of the humic associations in our micrographs were different from those seen by Chen and Schnitzer (1976) for their HAs or FAs in the neutral and alkaline pH ranges. There were similarities between our electron micrographs of the acid soluble fractions (which could be expected for FA-type molecules with origins in the B_h horizons of podzols) and those of Chen and Schnitzer (1976) at the lower pH values, and aspects of our interpretations are broadly similar to theirs. At the low pH values even the strong organic acids would be undissociated, and the macromolecules would associate by hydrogen bonding and by van der Waals forces to give the linear/'spike-like' conformations observed. We consider that in solutions of high concentrations the 'spike-like' structures further associate, and eventually the associations of molecules have sufficient hydrophobicity to be precipitated and give the classical HA fraction. When metals and inorganic colloids are co-extracted, some will survive the 'purification' processes and remain associated with the HS. The HS which are associated with the metals/inorganic colloids will have contracted conformations.

The hydrophilic properties of the melamine resin, Neoplast FB101, would appear to have desirable properties with regard to the preservation of the solution conformations of humic fractions for electron microscopy studies. There is a need, however, to compare on the same samples the procedure we have used with that used by Chen and Schnitzer (1976) and by Senesi *et al.* (1996), in order to verify our conclusion that the melamine resin technique allows the conformations which apply in solution to persist during the preparation of samples for electron microscopy studies.

Acknowledgements

This work is part of a project supported by the Natural Environment Research Council.

References

Bachhuber, K., and D. Frosch. 1983. Melamine resins, a new class of water soluble embedding media for electron microscopy. *J. Microscopy* **130**: 1-9.

Causton, B.E. 1985. Does the embedding chemistry interact with tissues? p. 209-214. *In* M. Muller (ed), *Sci. Biol. Specimen. Prep. Microsc. Microanal.* SEM Inc./AMF O'Hare, Chicago, Ill.

Chen, Y. and M. Schnitzer. 1976. Scanning electron microscopy of a humic acid and of a fulvic acid and its metal and clay complexes. *Soil Sci. Soc. Am. J.* **40**:682-686.

Frosh, D. and C. Westphal. 1989. Melamine resins and the application in electron microscopy. *Electron. Microsc. Rev.* **2**:231-255.

Hayes, T.M., M.H.B. Hayes, J.O. Skjemstad, R.S. Swift, and R.L. Malcolm. 1996. Isolation of humic substances from soil using aqueous extractants of different pH and

XAD resins, and their characterization by ^{13}C-NMR. p.13-24. *In* C.E. Clapp, M.H.B. Hayes, N. Senesi, and S.M. Griffith (eds), *Humic Substances and Organic Matter in Soil and Water Environments.* Proc. 7th Intern. Meeting, IHSS (St. Augustine, Trinidad and Tobago, 1994). IHSS, St. Paul, Minn.

Leppard, G.G., J. Buffle, and B.K. Burnison. 1990. Transmission electron microscopy of the natural organic matter of surface waters. *Anal. Chim. Acta* **232**:107-121.

Little, D.J. 1994. Occurrence and characteristics of podzols under oak woodland in Ireland. Ph.D. Thesis, The National University of Ireland.

Malcolm, R.L. and P. MacCarthy. 1992. Quantitative evaluation of XAD-8 and XAD-4 resins in tandem for removing organic solutes from water. *Environ. Intern.* **18**:597-607.

Perret, D., G.G. Leppard, M. Muller, N. Belzile, R.D. Vitre, and J. Buffle. 1991. Electron microscopy of aquatic colloids: Non-perturbing preparation of specimens in the field. *Water Research* **25**:1331-1343.

Senesi, N., F.R. Rizzi, P. Dellino, and P. Acquafredda. 1996. Fractal dimension of humic in aqueous suspension as a function of pH and time. *Soil Sci. Soc. Am. J.* **60**:1773-1780.

Investigations Into the Nature of Phosphorus in Soil Humic Acids Using ^{31}P NMR Spectroscopy

E.C. Norman, C.L. Graham, and M.H.B. Hayes

THE UNIVERSITY OF BIRMINGHAM, SCHOOL OF CHEMISTRY, EDGBASTON, BIRMINGHAM, B15 2TT, ENGLAND

Abstract

Phosphorus-31 NMR was used to analyse humic acids (HAs) extracted using different conditions. Two soil samples from Western Australia were studied (RO1 and RO3) and both had significant iron and aluminium contents. The HAs isolated from these soils in 0.1M neutral sodium pyrophosphate solution had high ash contents. A further sample of HAs isolated from RO3, using a sodium hydroxide solution (0.1M, under N_2), had a much lower ash content. A HA sample isolated from the B_h horizon of a podzol from Co. Mayo, Ireland, using the neutral sodium pyrophosphate solution (following an acid wash of the soil) had lower ash and phosphorus contents than the samples from the RO. The nature of the phosphorus in these HA samples, and in a special NaOH/HCl extract of RO3, was studied by ^{31}P NMR. A link was observed between the ash contents and the residual pyrophosphate contents. It is concluded that when pyrophosphate is used as an extractant the HA fractions will be contaminated with the pyrophosphate unless the metal complexes and the (hydr)oxides associated with the HAs are removed.

1 Introduction

Phosphorus-31 NMR has been used in the studies of soils and of soil organic matter (SOM). Newman and Tate (1982) introduced this technique to study the phosphorus in the alkaline extracts of soils. Ogner (1983) expanded the technique to study 'purified' humic acids (HAs). Others who have worked in this area include Bedrock *et al.* (1994), and Condron *et al.* (1985), and they used sequential extraction of a soil to obtain a higher yield of extracted soil phosphorus. These experiments were all carried out in solution. There have been some NMR studies in the solid state by Lookman *et al.* (1994, 1996), and they also used ^{27}Al solid state NMR to study the sorption of phosphate onto the surface of the aluminium hydroxide. More recent work by Makarov *et al.* (1996) has used ^{31}P NMR to study the interactions of metal ions, phosphate, and HAs. These studies have shown that phosphate-metal-humic complexes are easily formed.

The main classes of phosphorus in extracts from soils are inorganic orthophosphate and orthophosphate monoesters. Most soils also contain orthophosphate diesters, and some soils have been found to contain phosphonates, which are thought to be of microbial origin (Newman and Tate, 1980). Previous studies have found that from 26 - 59% of the

Table 1 *Chemical shift values for various phosphorus species, based on values published by Newman and Tate (1980)*

P Species	Structure	Chemical shift (δ) / ppm
Inorganic orthophosphate	PO_4^{3-}	5.3
Orthophosphate monoesters†	$ROPO_3^{2-}$	3.5-5.3
Choline phosphate*	$(CH_3)N^+(CH_2)_2OPO_3^{2-}$	3.5
Orthophosphate diesters†	$(RO)(R'O)PO_2^-$	0- -1.5
Pyrophosphate	$P_2O_7^{4-}$	-5.5

† The exact values of δ depend on the structure of the substituents R and R'.

* Newman and Tate (1980) found choline phosphate present in nearly all of their alkaline soil extracts.

phosphorus in an extract of the total soil is inorganic. Most of the remainder is in organic phosphorus monoesters. The chemical shifts frequently seen for soil extracts are shown in Table 1.

Spectra for HAs by Ogner (1983) contain the same peaks as the alkaline soil extracts examined by Newman and Tate (1980). From the integration of these peaks, however, it was seen that the relative amounts of each phosphorous species in the samples were different. There was less inorganic phosphorus in some of the samples, and significant amounts of alkyl phosphonates (δ = 17 - 19 ppm) were evident in others.

The HAs used in the present study were isolated mainly from latosols, which are high in aluminium and iron. The latosols were taken from above a bauxite deposit in the Darling Range in Western Australia. The bauxite is mined and processed by Alcoa of Australia. A large proportion of the overburden latosol was removed prior to open cut mining, and this left the so-called 'residual overburden' (RO). RO1 is associated with granitic bauxite, and RO3 is a blend of ROs from granitic and doleritic bauxites.

In order to compare the phosphorus found in the HAs with that of the whole RO, an adaptation of the method of Condron *et al.* (1984) was used to extract the RO exhaustively. Newman and Tate (1982) used pyrophosphate as an internal standard. That enabled them to establish that the NMR was 'seeing' all the phosphorus nuclei, and demonstrated that the method was quantitative. However, because pyrophosphate is implicated as a possible contaminant of the materials under investigation in the present study, it was not possible to use pyrophosphate as an internal standard. Thus no attempt was made to quantify the phosphorus species present in each sample. Orthophosphoric acid, which provides a signal at δ = 0 ppm, was used as an external reference.

2 Experimental

Isolation of HS

The RO3 and RO1 samples were exhaustively extracted with solutions of 0.1M sodium pyrophosphate (Pyro) adjusted to pH 7 with orthophosphoric acid, and then with a 0.1M

Pyro/0.1M sodium hydroxide solution. The isolation procedure is described by Hayes *et al.* (1996). A sample of HA was also isolated from RO3 using sodium hydroxide (under N_2).

In order to obtain a spectrum of the phosphorus in the extract of the residual overburden, the RO3 was extracted (under N_2 gas) with 0.1M NaOH (10 mL per 10 g soil), and the supernatant (after centrifugation, 4000 rpm, 30 min) was neutralized with HCl (6M), then filtered (Sartorius, cellulose acetate membranes, 0.45 μm). The residue on the filter pad was returned to the bulk RO3 residue; the combination was extracted with HCl (0.1M, 24 h), centrifuged, and the supernatant was neutralized with NaOH (4M), and filtered as above. Again, the residue on the filter pad was returned to the bulk RO3 residue, and the alternate alkali and acid extractions were repeated until no further colour was isolated. The combined filtrates were freeze dried. The isolate was labelled as the 'Unfractionated Organic Fraction' (UOF).

HA was isolated also from RO1 using neutral Pyro (RO1 pH 7 HA). For comparison, a HA sample was isolated from the B_h horizon of the podzol soil. That soil was shaken with 0.1M HCl for 4 h prior to extraction with neutral Pyro.

Acquisition of ^{31}P NMR Spectra

A saturated solution of HS in NaOH (1M, 1 mL, Convol, BDH Laboratory Supplies), was prepared in a 2 mL glass vial. The solution was filtered (glass fibre, 0.4 μm) into an NMR tube and D_2O (0.5 mL, 99.9 atom % D, Fluorochem. UK) was added to provide a lock for the spectrometer. The spectra were acquired on a Bruker AMX 400, with a 30° pulse angle, and with an external standard of phosphoric acid ($\delta = 0$ ppm).

3 Results and Discussion

There are several difficulties associated with acquiring ^{31}P NMR spectra of HS. Although ^{31}P is the only natural isomer of phosphorus, it is present in HS in very low concentrations (generally < 2%). Therefore, if solution NMR is desired, the concentration of HS in solution must be maximized. The resolution of the spectra will be poor if the concentration is low, and assignations become difficult, especially in the $\delta = 3.5 - 5.3$ ppm region. There are few restrictions on the solvent that can be used for observations of the phosphorus nucleus. Consequently, 0.1M NaOH was the solvent used because it allowed saturated solutions of the solute to be prepared. To provide a lock for the NMR spectrometer, D_2O was added to the NaOH prior to dissolving the humic sample. Ogner (1983) reported that alkaline hydrolysis of humic acid phosphorus severely affected the spectra obtained. To prevent oxidation in the alkaline solution, the dissolution was carried out under N_2 and all the solutions were stored under N_2 while awaiting analyses (which were always carried out within 24 h of the sample preparation).

Phosphorus-31 NMR was used initially to study the phosphorus in some of the HA isolates from the residual overburdens; many of these samples had high ash contents. The analyses of the samples for phosphorus (Table 2) indicated that the levels of phosphorus present in the samples were higher than is normally seen for soil HAs. These samples were isolated using a neutral sodium pyrophosphate solution. Thus, using ^{31}P NMR enables identification of the different chemical species of phosphorus present in the high ash samples, and allows these to be compared with samples of lower ash contents. The results for the HAs isolated in pyrophosphate from the ROs were compared with a HA from the podzol (with a lower ash content), and these (from the ROs) were compared also with the 'full soil extract' (the UOF) isolated from RO3 in a manner similar to that used by Newman and Tate (1980), who did not find pyrophosphate in such extracts.

The ^{31}P NMR spectra obtained are shown in Figures 1 and 2, and the chemical shifts of the phosphorus functionalities in the samples are given in Table 2. This Table also provides the ash and phosphorus contents of the samples, where these were recorded.

The RO samples from which the HAs were isolated were low in organic content and high in iron and aluminium, and the ^{31}P NMR spectra indicate the presence of pyrophosphate in these. That pyrophosphate was an artefact from the extracting solution because there was no evidence for pyrophosphate in samples that were isolated using the sodium hydroxide solution. The results suggest an association between the pyrophosphate and the iron/aluminium (hydr)oxides co-extracted with the HS, and present in the ash. The HS could be sorbed to the (hydr)oxides.

The 'full soil extract' (the 'Unfractionated Organic Fraction', UOF), Sample 2, Figure 1, and the RO3 HA that was isolated using sodium hydroxide (Sample 3, Figure 1) did not contain pyrophosphate residues. The RO3 HA sample (Table 2) had less inorganic phosphate than the other RO HAs, and it contained traces of organic monoesters. The 'full soil extract' (UOF) had a low phosphorus content, though there was some evidence for inorganic phosphate and organic monoesters.

The HA sample from the podzol B_h horizon was isolated differently. Because the soil was acid washed (0.1M HCl, 4 h) prior to the extraction with the sodium pyrophosphate solution, the resultant HA had a lower ash content than those for the RO3 HAs. The NMR signal for the inorganic phosphorus was, however, relatively strong (Sample 2, Figure 2), as was that for the UOF extracted in sodium hydroxide (see spectrum for Sample 2, Figure 1). The NMR spectrum (Sample 2, Figure 2) clearly indicates a trace of residual pyrophosphate in the sample from the B_h horizon.

Table 2 *The ash and phosphorus contents, the solvent used in the extraction, and the chemical shift values (δ ppm), signal strengths [where (S) = strong, (W) = weak, and (VW) = very weak] and assignments for the humic acids (HA)*

Sample	% Ash	% P	Sample extraction solvent	δ /ppm	Assignment
RO3 pH 7 HA	62.0	1.9	sodium pyrophosphate	+5.3 (S) -5.5 (S)	Inorganic phosphate Pyrophosphate
RO1 pH 7 HA	36.9	3.3	sodium pyrophosphate	+5.3 (S) -5.5 (S)	Inorganic phosphate Pyrophosphate
RO3 HA	6.4	-	sodium hydroxide	+3.5-5.3 (W)	Inorganic phosphate and organic monoesters
RO3 'Full Soil Extract' (UOF)	-	-		+3.5-5.3 (W) +5.3 (S)	Organic monoesters Inorganic phosphate
Podzol HA†	5.2	-	sodium pyrophosphate	+3.7 (S) -5.5 (VW)	Inorganic phosphate Pyrophosphate

† This sample was provided by A. J. Simpson, from the Group's laboratory

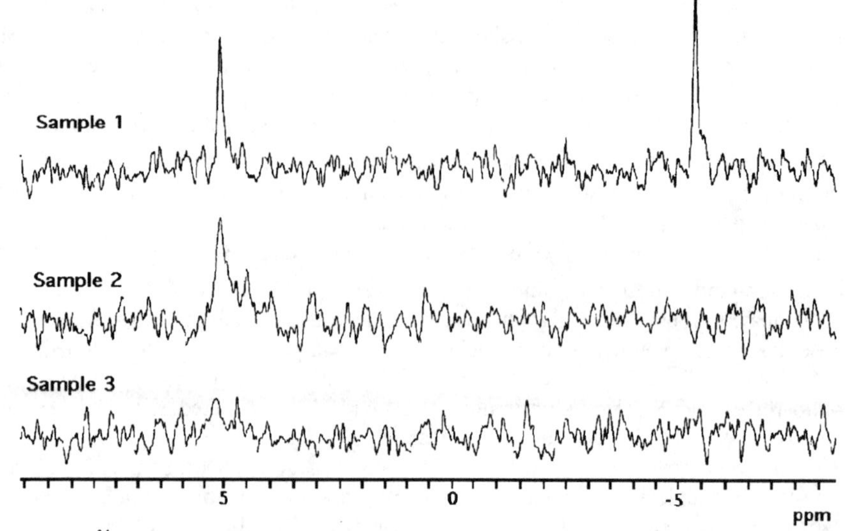

Figure 1 *^{31}P NMR spectra where Sample 1 is the RO3 pH 7 HA, Sample 2 is the 'full soil extract', or the 'Unfractionated Organic Fraction' (UOF) isolated in NaOH/HCl from RO3, and Sample 3 is the RO3 HA extracted with NaOH*

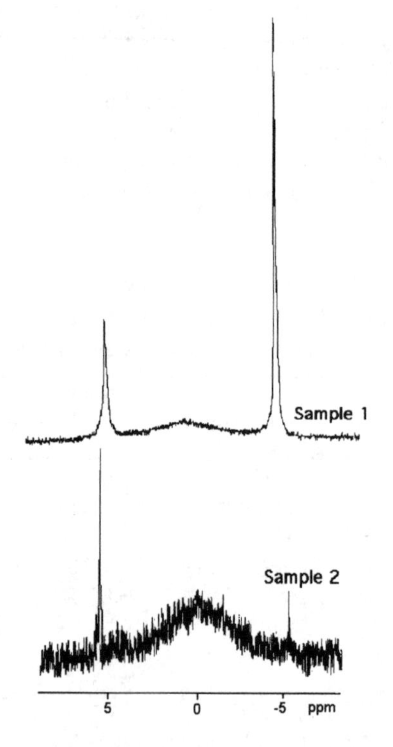

Figure 2 *^{31}P NMR spectra where Sample 1 is RO3 pH 7 HA, Sample 2 is the B_h Podzol HA, and Sample 3 is the ash extract*

The broad band centred around 0 ppm for the spectra in Figure 2 represents the orthophosphoric acid standard. That standard was run separately for the samples shown in Figure 1. The spectra in Figure 2 are made up of a large number of scans in order to provide a satisfactory signal to noise ratio.

4 Conclusions

The results show that pyrophosphate can be incorporated in the HA fraction when neutral sodium pyrophosphate is used as an extractant. The extents to which the pyrophosphate is contained in the fractions will depend on the soil type and on the methods of extraction, separation, and fractionation followed. The RO HAs were isolated from a soil low in organic content, but high in aluminium and in iron. The podzol HA was isolated from the organic rich B_h horizon which contains a high concentration of iron. However, the method of treatment of the isolate gave a HA product with a low ash content (ca 5%).

When sodium pyrophosphate was used for the isolation of the HAs it was found to be a contaminant in the fractions isolated. When sodium hydroxide was used as an extraction medium, there was no evidence for pyrophosphate in the HAs. Acid pretreatment will lower the pyrophosphate contamination, probably as the result of the removal of contaminating iron/aluminium.

Pyrophosphate solutions are used in the extraction procedure because of their abilites to complex divalent and polyvalent metals. However, it is apparent that unless the metals and the (hydr)oxide species can be removed effectively, the HA fraction will be contaminated with pyrophosphate. Makarov *et al.* (1996) have discussed the formation of phosphate-metal-humic complexes in which the metal forms a bridge between the phosphate and the humic molecule. Covalent linkages with the (hydr)oxide species, as well as ion exchange at pH values below the points of zero charge of the (hydr)oxides, are also possible. Thus, in order to decrease contamination of humic fractions with pyrophosphate, when sodium pyrophosphate is used as an extractant, it is important to remove the metal complexes and the (hydr)oxide associations from the extracts.

Acknowledgements

This study was carried out as a part of a larger project funded by Alcoa of Australia Ltd. The Podzol HA was extracted as part of a project funded by the NERC.

References

Bedrock , C.N., M.V. Cheshire, J.A. Chudek, B.A. Goodman, and C.A. Shand. 1994. Use of ^{31}P-NMR to study the forms of phosphorus in peat soils. *Sci. Total. Environ.* **152**:1-8.

Condron L. M., E. Frossard, H. Teissen, R. H. Newman and J. V. B. Stewart. 1990. Chemical nature of organic phosphorus in cultivated and uncultivated soils under different environmental conditions. *J. Soil Sci.* **41**:41-50.

Condron L. M., K. M. Goh and R. H. Newman. 1985. Nature and distribution of soil phosphorus as revealed by a sequential extraction method followed by ^{31}P NMR. *J. Soil Sci.* **36**:99-207.

Ogner G. 1983. ^{31}P NMR spectra of humic acids: A comparison of four different raw humus types in Norway. *Geoderma* **29**:215-219.

Hayes T. M., M. H. B. Hayes, J. O. Skjemstad, R. S. Swift and R. L. Malcolm. 1996. Isolation of humic substances from soil using aqueous extractants of different pH and XAD resins, and their characterisation by ^{13}C-NMR. p 13-24. *In* C. E. Clapp, M. H. B. Hayes, N. Senesi and S. M. Griffith (eds), *Humic Substances and Organic Matter in Soil and Water Environments: Characterisation, Transformations and Interactions.* IHSS, St Paul.

Lookman R., H. Geerts, P. Grobet, R. Merckx and K. Vlassak. 1996. Phosphate speciation in excessively fertilised soil: ^{27}Al and ^{31}P solid state MAS NMR spectroscopic study. *Eur. J. Soil Sci.* **47**:125-130.

Lookman R., P. Grobet, R. Merckx and K. Vlassak. 1994. Phosphate sorption by synthetic amorphous aluminium hydroxide: ^{27}Al and ^{31}P solid State MAS NMR spectroscopic study. *Eur. J. Soil Sci.* **45**:37-44.

Makarov M. I., G. Guggenberger, W. Zech and H. G. Alt. 1996. Organic phosphorus species in humic acids of mountain soils along a topsequence in the northern Caucasus. *Z. Pflanz. Bodenk.* **159**:467-470.

Newman R.H. and Tate K. R. 1980. Soil phosphorus characterisation by ^{31}P NMR. *Comm. Soil Sci. & Plant Anal.* **11**:835-842.

Tate K. R. and R.H. Newman. 1982. Phosphorus fractions of a climosequence of soils in New Zealand tussock grassland. *Soil Biol. & Biochem.* **14**:191-196.

Dissolved Humic Substances in Waters from Drained and Undrained Grazed Grassland in SW England

T.M. Hayes, B.E. Watt, M.H.B. Hayes, [1]C.E. Clapp, [2]D. Scholefield, [3]R.S. Swift, and [3]J.O. Skjemstad

THE UNIVERSITY OF BIRMINGHAM, SCHOOL OF CHEMISTRY, EDGBASTON, BIRMINGHAM B15 2TT, ENGLAND

[1]USDA-ARS & DEPARTMENT OF SOIL, WATER, AND CLIMATE, UNIVERSITY OF MINNESOTA, ST. PAUL, MINNESOTA 55108, USA

[2]INSTITUTE OF GRASSLAND AND ENVIRONMENTAL RESEARCH, NORTH WYKE, OKEHAMPTON, DEVON EX20 2SB, ENGLAND

[3]DIVISION OF SOILS, CSIRO, GLEN OSMOND, ADELAIDE, S. AUSTRALIA, 5064

Abstract

The humic acid (HA), fulvic acid (FA), and XAD-4 acid fractions of humic substances (HS) were isolated from waters flowing from a grassland soil. In one case, Devon 1 (D1), the waters were collected in March in the flow from tile drains at 0.85 m depth, and in the other, Devon 3 (D3), in November from surface run off. The humic fractions (HF) were subjected to elemental, $\delta^{13}C$, $\delta^{15}N$, neutral sugar, amino acids, NMR, and FT-IR analyses.

The analytical data showed distinct differences between the compositions of the HS fractions in waters which had passed through the soil and those in the surface run off. The differences in the sampling dates were also considered to be significant. The run-off water was sampled when the 'flush' of HS arising from the transformations of plant and microbial substrates were eluted. The drainage waters were sampled before the new growing season had contributed to the HS, and when the 'flush' of the transformation products from the previous season had ended. This showed that the samples from D1 had a longer residence time in the soil, and the analytical data indicated that these humic fractions were more extensively transformed than the fractions from D3.

1 Introduction

Soils under long-term grassland management tend to accumulate organic matter (OM), and clay-rich grassland soils in regions of high rainfall can accumulate as much as 100 and 10 t ha^{-1}, of organic C and N, respectively. This tendency is less for soils that are better aerated, such as coarser textured soils, and clay soils with field drains. There is thus a considerable potential for the leaching of soluble organics from “mature” grasslands, and especially from those that additionally receive the excretal returns from grazing livestock.

Whereas quantification of the leaching of NO_3^-, NO_2^-, NH_4^+, and molybdate-reactive P from grassland has received much attention, because of the reputed environmental impacts, little is known about the amounts and chemical compositions of soluble soil organic substances (SSOS) lost through leaching. The SSOS are of particular interest because:

1. these might subsequently release mineral nutrients to the aqueous environment, despite the fact that many of the substances are relatively intractable/recalcitrant;
2. these substances are considered to be potential "carrier" molecules for toxic elements and anthropogenic chemicals;
3. soluble humic substances (HS), major components of SSOS, can act as carbon sources and electron donors in environmentally important processes, such as denitrification and methanogenesis;
4. HS could be specific inhibitors of some soil microbial processes; and
5. HS could, as the result of the intricacies of their compositions, provide useful diagnostic tags for "fingerprinting" soil processes as influenced by different management practices and climatic factors.

This paper outlines some of the properties of HS dissolved in surface run off waters and in the waters percolating artificial field drains from intensively managed grazed grassland at an experimental site in Devon, England.

2 Details of Site, Soils, and Samplings

The samples of drainage waters were taken from hydrologically isolated plots of the Rowen Moor Drainage Experiment at the BBSRC Institute of Grassland and Environmental Research (IGER) farm at North Wyke, Okehampton, Devon. This facility was set up in 1982 on an old, unimproved, wet pasture. The average annual rainfall is 1035 mm, of which 590 mm is in excess of evapotranspiration. The existing sward was known to be at least 40 years old, and dominated by *Agrostis* spp., *Holcus lanatus*, and *Juncus* spp., and with less than 20% of *Lolium perenne*.

The soil is a Stagno-Dystric gleysol, of the Hallsworth series, which overlays the carboniferous clay/shales of the Crackington formation (culm measures). The A*p* horizon (0-27 cm) is composed of 38% clay, 50% silt, and 12% sand (mineral soil), and the clay content increases with depth to 100 cm. When the site was taken over (in 1982) the organic C and N contents in the top 30 cm were 90 t and 8 t ha^{-1}, respectively, and the pH of the A*p* horizon was 5.3.

Twelve plots, each of 1 ha, were established in two replicate blocks and six agronometric treatments were applied. These comprised two levels of fertilizer N (200 and 400 kg N ha^{-1}), with and without reseeding to *Lolium perenne* at the higher level of N input only, and with and without field drainage. On the undrained plots, > 80% of the hydrologically effective rainfall (HER) could be accounted for with "V" notch weirs sampling surface run off, whereas on the plots drained by moles (55 cm depth) and pipe drains (85 cm depth) about 85% of HER was monitored through the drainage system. Both the water content and height of the the water table were lower in the drained soil over the whole year. During the growing season (April - October), the swards were continuously grazed by beef steers (295 kg body weight at turn out) at variable stocking densities to maintain a sward height of 65 mm. Details of the sample sites are given in Table 1. The pH values of the unfiltered water samples were always in the range 7 to 8.

Table 1 *Location and description of sampling sites*

Sample	Sampling Date	Weather	Location	Plot descriptions
Devon 1 (D1)	05-03-1991	Sample taken at the end of a rainfall event, the drains were beginning to slow.	BBSERC, Institute for Grassland and Environmental Research Station, North Wyke, Devon (SX 650 995)	Permanent pipe drains at 85cm depth. Addition of 400 kg N ha^{-1} $annum^{-1}$ (Tyson *et al,* 1992)
Devon 3 (D3)	end-11-1992	Sample was taken during a downpour. The drains were flowing fast.		Surface interceptor drain at 30 cm depth. Addition of 400 kg N ha^{-1} $annum^{-1}$

Two soil drainage water samples were collected. One was from a drained (1 ha) plot, and the second was from an undrained (1 ha) plot, as indicated in Table 1.

3 Experimental

Isolation of Humic Substances from Drainage Waters

The XAD-8 [(poly)methylmethacrylate], XAD-4 (styrene divinylbenzene) and IR-120 (styrene divinylbenzene with sulphonic acid functionality) resins used were exhaustively cleaned, using a sequence of solvents (rolling in dilute NaOH, distilled water, followed by soxhlet extraction using ethanol, and acetonitrile). The resins when used were contained in glass cylindrical columns (4 L for the XAD resins and 1.5 L for the IR-120). Humic substances [humic acid (HA), fulvic acid (FA), and the XAD-4 acids] were isolated by adsorption onto XAD-8 and XAD-4 resins, using the procedure adapted by Hayes (1996) from that described by Malcolm and MacCarthy (1992).

Elemental Analyses and $\delta^{13}C$ and $\delta^{15}N$ Values

A Perkin Elmer 240 elemental analyser was used to determine C, H, and N. The precision, using standard samples, was ± 0.3% absolute of the element under consideration.

Values for $\delta^{13}C$ and $\delta^{15}N$ were determined in the USDA-ARS Laboratories, Department of Soil, Water, and Climate, University of Minnesota, using a Carlo Erba, model NA1500 elemental analyser and a stable isotope ratio mass spectrometer (Fisons, Optima model) continuous flow system. Results of the isotope analyses are expressed in terms of δ values (‰). An outline of the procedure is given by Clapp *et al.* (p. 161, this Volume).

Determinations of Neutral Sugar Contents

Monosaccharides were determined in the form of alditol acetates, using a procedure based on that of Blakeney *et al.* (1983). The humic substances were hydrolysed for 2 h in 2M trifluoroacetic acid (TFA) at 120 °C. The sugars released were reduced using sodium borohydride in dimethyl sulphoxide, and then acetylated using a mixture of acetic anhydride and 1-methylimidazole. The alditol acetate derivatives were extracted into dichloromethane and analysed by GC-FID (Pye-Unicam model 304), using a BPX70 [25 m x 0.25 mm, SGE (UK) Ltd.] column. *Myo*-inositol was used as the internal standard.

Determinations of Amino Acid Contents

Amino acids were determined using a procedure based on that of Turnell and Cooper (1982). Samples were hydrolysed for 24 h in 6M HCl at 115 °C under dinitrogen gas. Free amino acids were extracted into 0.1M perchloric acid, and separated and quantified (as *o*-phthalaldehyde derivatives) by HPLC using a 15 cm x 4.6 mm reverse phase Spherisorb ODS II 3 μm column (Phase Separations Ltd.), and fluorescence detection. Norvaline was used as the internal standard.

CPMAS ^{13}C-NMR Spectroscopy

A Varian Unity 200 spectrometer, with a 4.7T wide-bore Oxford superconducting magnet was used for CPMAS ^{13}C-NMR spectroscopy. Spectra were obtained at 50.309 MHz in a Doty probe using 7 mm diam zirconia rotors and Kel-F end caps spun at 5 kHz. The number of scans for each spectrum varied, but a typical value was 2000. A Lorentzian broadening of 50 Hz was normally used, with a 0.005 s Gaussian function.

All spectra were obtained using the methyl resonance of hexamethyl benzene as an external reference (17.36 ppm) for HAs, the acquisition time was 15 ms, with a recycle time of 0.3 s, and a contact time of 1000 ms. The recycle time was 0.5 s for XAD-4 acids; the other values were the same as for the HAs. For FAs, the acquisition time was 25 ms, with a recycle time of 1.5 s and a contact time of 1500 ms. One 'typical' sample of each type was selected to optimise the above parameters.

Infrared (IR) Spectroscopy

The FT-IR spectra were recorded using a Perkin Elmer 1600 series FT-IR instrument. A ratio of 100 mg KBr to 1 mg of humic sample was used in all cases.

4 Results and Discussion

The volumes of the samples processed and the weights of the humic fractions recovered are given in Table 2, and the percentages by weight of the humic fractions are given in Table 3. The significant differences in the amounts of the humic fractions in the two water samples could reflect the differences in the OM of the two plots (Stevenson, 1994, has also pointed out that undrained soils can contain up to 10 per cent more organic matter than their drained counterparts). However, the differences might also be partly explained by the different times of samplings. Sampling of D3 took place in November during the autumnal 'flush' when the soluble products of microbial synthesis formed during the summer months (when evapotranspiration provides the major sources of water loss from the soil) are dissolved in the drainage waters. It is likely, therefore, that lesser amounts of soluble products were available for removal in the samples (D1) taken in March. The ratio

Table 2 *Volumes of samples processed (L) and weights of humic substances recovered (g), on a dry, ash-free basis*

Sample*	Vol. Water (L)	FA (g)	HA (g)	XAD-4 acids (g)	Total (g)
Devon1 (D1)	1500	2.125	0.359	0.339	2.822
Devon3 (D3)	2000	15	7.24	1.0	23.24

* See Table 1

Table 3 *Percentage by weight of the humic (HA), fulvic (FA), and XAD-4 (XAD4) acids in the humic fractions, and concentrations (concn) of the humic substances (HS) in the water samples*

Sample	Concn of HS (mg L^{-1})	FA (%)	HA (%)	XAD-4 acids (%)
Devon1 (D1)	1.9	75.3	12.7	12
Devon3 (D3)	11.6	64.5	31.2	4.3

Table 4 *Elemental analyses data and $\delta^{13}C$ and $\delta^{15}N$ values for humic acids (HA), fulvic acids (FA) and XAD-4 acids (XAD4) in waters from the Devon drained (D1) and undrained (D3) plots. Values for C, H, N, and O are expressed as percentages of the total weight, and were calculated on a dry, ash-free basis*

Sample	Moisture	Ash	C%	N%	H%	O%	$\delta^{13}C$	$\delta^{15}N$
D1 HA	8.5	43.7	55.0	3.1	3.1	38.9	-30.8	2.8
FA	8.5	1.9	54.8	2.7	1.3	41.5	-29.8	1.2
XAD4	8.1	6.4	48.2	4.2	2.9	44.8	-28.6	2.8
D3 HA	10.4	1.1	56.0	3.4	3.0	37.6	-31.8	1.6
FA	10.5	0.2	55.8	3.2	2.9	38.1	-30.6	0.8
XAD4	10.6	4.2	55.7	4.3	3.7	36.4	-29.0	3.6

(6:1:1) of the amounts of FA:HA:XAD-4 acids in D1 was in line with that quoted by Malcolm (1991) for surface waters, and was significantly different from that (15:7:1) for the D3 sample.

The results of the moisture, ash, elemental, $\delta^{13}C$, and $\delta^{15}N$ analyses are given in Table 4. The C, H and N values for all the samples isolated are in agreement with those found by other researchers (Stevenson, 1994). All the samples show the following characteristics. In general, the carbon contents follow the order: HA > FA > XAD-4 acids, the oxygen contents follow the order XAD-4 acids > FA > HA, and generally the nitrogen contents follow the order XAD-4 > HA > FA. The hydrogen contents do not follow any discernible trend, but all of the values lie in the range 1.3 to 3.7%.

The carbon and nitrogen contents of the HAs and FAs from the two waters are similar, but the oxygen contents suggest that the FAs and XAD-4 acids were more highly oxidized in the D1 samples.

$\delta^{13}C$ Data

The $\delta^{13}C$ data indicate that all the humic materials have inputs predominantly from C3 plants which tend to have a $\delta^{13}C$ value of around -27‰ (Clapp *et al.*, p. 158 - 175 this Volume). In both cases the HAs are the most negative and the XAD-4 acids the least negative (or most ^{13}C enriched). Clapp *et al.* (p. 158 - 175, this Volume) have shown that the $\delta^{13}C$ values for HAs and FAs are similar for samples from the same sources. They also noted that XAD-4 acids from the same sources have values which are less negative than the HAs and FAs, and this might indicate that the differences arise from the microbial involvements in the processes which transform the organic materials into HS.

The ^{13}C enrichments increased (i.e. the $\delta^{13}C$ values became less negative) in the order HA < FA < XAD-4 acids for the D1 and D3 samples. That might be interpreted in terms of the microbial inputs to the genesis of the fractions. The neutral sugar data (Table 5) corroborate this hypothesis (discussed under the 'Sugar Analyses' heading below). It is worth noting that the D3 samples are less enriched in ^{13}C than the corresponding fractions from D1. That could reflect the increased biological inputs into the genesis of the humic fractions in D1. The latter fractions had been in the soil for a longer period of time and can be expected to have undergone microbial transformations during their residence in the soil.

$\delta^{15}N$ Data

A discrimination between the lighter ^{14}N and heavier ^{15}N isotopes occurs during biological and chemical processes (Delwiche and Steyn, 1970), and leads to an increase in $\delta^{15}N$ values of the unreacted fraction of the substrate. Transformations of plant and soil N are accompanied by an isotope effect in which ^{14}N is mineralized preferentially. The mineralized N is susceptible to losses through leaching and plant uptake, and thus the N in the unreacted portion of the substrate becomes enriched in ^{15}N. As a result, crops have a $\delta^{15}N$ value that is lower than that of the total nitrogen pool (Sutherland *et al.*, 1991). A number of studies have shown that nitrogen isotopes are useful for discriminating between nitrate sources in ground water (Komor and Anderson, 1993).

$\delta^{15}N$ ratio values for animal wastes range from 10 - 22‰, those for organic material in soil range from 4 - 9‰, and those for commercial fertilizers have values of -4 to +4‰. There are limitations, however, in the interpretation of the data. $\delta^{15}N$ ratios from nitrogen sources cover a range of values which do not have distinct boundaries. It is possible, for example, to have a sample with a $\delta^{15}N$ value of +6‰ which may contain 100% of its N with a $\delta^{15}N$ value of +6‰, or the N may consist of a mixture in which 50% of the N has a $\delta^{15}N$ value of 0‰ and 50% with a value of +12‰. This is an important consideration when N inputs are mixed (e.g., the mixing of commercial fertilizers with animal wastes).

The $\delta^{15}N$ data do not follow the trends noted for the $\delta^{13}C$ values for the various humic fractions, and no clear-cut trend can be described. All of the values lie in the range +0.8 to +3.6‰, and that would suggest that the nitrogenous fertilizers applied strongly influenced the ^{15}N contents of the humic fractions. The fractions isolated from the D1 sample followed the ^{15}N enrichment trend: XAD-4 = HA > FA, whereas that from the D3 fractions was XAD-4 > HA > FA. However, the differences were small. The fact that the values for the XAD-4 and HA fractions had slightly greater enrichments in ^{15}N (than the FAs) might suggest that the commercial fertilizers applied to the land made higher inputs to the water soluble FAs (than to the HAs and XAD-4 acids). The higher values for the HAs and XAD-4 acids could suggest greater inputs from animal wastes.

Results of the Neutral Sugar Analyses

The neutral sugar (NS) contents for the HS are given in Table 5. The contents of the D1 sample are similar to those quoted for HS from streams, rivers and groundwaters (Watt *et al.*, 1996a and 1996b; Thurman, 1985a and 1985b), but the D3 HS exhibit differences.

The total NS contents for the HS from the drainage waters and run off waters were significantly less than those in the corresponding soil humic fractions (Hayes, 1996). In both cases the NS contents of the FAs were least.

There were significant differences between the NS contents and in the distributions of NS in the fractions of the samples from the D1 (drained) and D3 (undrained) plots. The abundances of the NS in the fractions decreased in the order HA > XAD-4 > FA, and XAD-4 > HA > FA in the cases of samples from D1 and D3, respectively. However, the total NS contents in the humic fractions of D1 were significantly less than those in the similar fractions of D3.

In general glucose (Glu), xylose (Xyl), and arabinose (Ara) were the predominant sugars in the D1 isolates, and rhamnose (Rha) and fucose (Fuc) were least abundant. The concentrations of the other NS varied. In the cases of the D3 HA and FA samples, the order of abundances were: Xyl > Ara > Glu > galactose (Gal) > Rha > mannose (Man) > Fuc, whereas the distribution order in the XAD-4 acids was Glu > Rha > Gal > Man > Xyl > Ara > Fuc.

The variations in the contents and abundances of the NS in the humic fractions of D1 and D3 reflect the differences in the times of year when the samples were taken, and the depths in the soil to which the waters had penetrated. When adequate moisture is available high levels of microbial activity would be expected during the growing season. The products of that microbial activity will be most evident during the autumnal flush when the soluble HS are washed from the soil in run off and in drainage waters. That concept is reflected in the abundance and in the composition of the NS in the humic fractions of D3. The samples were taken when the drains had begun to flow in late November.

Oades (1984) and Murayama (1984) have suggested that the extents to which soil saccharides have their origins in plants and in microbial organisms may be deduced from the mass ratios (Gal+Man/Ara+Xyl) and (Rha+Fuc/Ara+Xyl). In general, these authors have considered that the lower ratios indicate lesser contributions from microbial sources because Ara and Xyl are important constituents of plant materials and of minor importance only in microbial sources. Thus ratios of < 0.5 and > 2 are considered to be typical of plant derived and of microbially synthesized carbohydrates, respectively. The ratios for the D1 and D3 isolates indicate that plant sources were the major contributors to the NS components of the HA and FA fractions of D1 and D3, although the evidence would suggest that contributions from microbial sources were slightly greater in the case of these fractions from D1. However, the significantly higher ratios for the XAD-4 acids reflect the importance of microbial syntheses. Also, microbial contributions to the synthesis were

Table 5 *Neutral sugar contents (as relative molar percentages of the total sugars) of the humic (HA), fulvic (FA), and XAD-4 acids (XAD) from waters from the Devon drained plot (D1) and the Devon undrained plot (D3). (Rha = Rhamnose; Fuc = Fucose; Ara = Arabinose; Xyl = Xylose; Man = Mannose; Gal = Galactose; Glu = Glucose) (values are corrected for moisture and ash)*

Sample	Rha	Fuc	Ara	Xyl	Man	Gal	Glu	Man + Gal / Ara + Xyl	Fuc + Rha / Ara + Xyl	nmol mg^{-1}	µg mg^{-1}
D1HA	5.1	2.4	18.0	27.6	10.1	9.1	27.7	0.5	0.2	114.3	**16.8**
D1 XAD	7.6	11.4	12.0	17.7	12.9	8.5	30.0	0.9	0.7	51.5	**7.7**
D1 FA	8.3	1.8	20.5	24.6	8.5	9.2	27.1	0.5	0.3	30.4	**4.5**
D3 HA	5.3	1.2	24.8	34.1	4.3	8.3	22.1	0.3	0.1	192.0	**27.5**
D3 XAD	16.0	9.9	12.4	12.7	13.5	14.7	20.8	1.4	1.1	353.5	**53.2**
D3 FA	5.9	1.6	27.7	27.7	4.9	8.7	23.5	0.3	0.2	86.2	**12.4**

*Based on the µg mg^{-1} values

Table 6 *Amino acid contents (as relative molar % of the total amino acid contents) of humic (HA), fulvic (FA), and XAD-4 acids (XAD) from water from the drained plot (D1), and from the undrained plot (D3). (Values based on a dry, ash-free basis)*

	Acidic		*Total*	Basic			*Total*	Neutral Hydrophobic (NH_O)					*Total*	Neutral Hydrophilic (NH_i)				*Total*	Other	**Total AA**	%	%
Sample	Asp	Glu	*acidic*	Arg	His	Lys	*Basic*	Val	Ile	Leu	Tyr	Phe	*NHo*	Thr	Ser	Gly	Ala	*NHi*	Met	**nmol mg^{-1}**	AA[a]	N[b]
D1 HA	12.9	14.6	27.5	1.8	1.0	5.1	7.9	8.8	5.2	6.9	2.1	3.5	26.5	9.5	5.2	16.6	11.3	42.7	0.3	**229.3**	2.3	6.7
D1 XAD	17.6	13.6	31.2	1.2	1.0	4.1	6.2	5.3	2.3	3.9	1.5	1.5	14.4	6.4	3.7	24.3	13.6	48.0	0.3	**132.5**	1.3	8.7
D1 FA	14.6	9.9	24.5	1.0	0.9	4.4	6.3	8.3	4.3	6.4	1.5	2.6	23.0	8.0	4.1	22.4	11.5	45.9	0.3	**85.6**	0.8	4.9
D3 HA	14.0	10.4	24.4	1.8	1.9	5.1	8.8	7.4	3.9	6.4	2.1	2.7	22.4	10.5	6.9	15.6	11.2	44.2	0.3	**296.6**	3.0	11.4
D3 XAD	18.1	13.5	31.6	0.7	0.8	2.7	4.2	3.9	1.4	2.3	0.6	0.8	8.8	9.0	4.8	22.7	18.6	55.2	0.3	**258.1**	2.4	11.2
D3 FA	13.2	14.6	27.8	0.9	0.9	2.0	3.9	8.4	4.3	5.8	1.5	2.5	22.5	11.1	6.7	15.6	12.0	45.4	0.4	**99.8**	1.0	4.7

[a] %AA = the percentage of the humic sample present as the amino acid [b] %N refers to the percentage of N in the sample which can be accounted for as amino acid nitrogen

greater for the D3 than for the D1 XAD-4 acids. The significantly lower NS contents in the XAD-4 acids from D1, and the lower ratio values are reflections of the transformations of the microbially derived XAD-4 acids during their residence time in the soil. That suggestion is reinforced by the significantly lower contents of NS in the hydrolysates of the samples which had 'overwintered' in the soil in the case of the D1 sample.

Results of the Amino Acid Analyses

Amino acid (AA) analyses data are given in Table 6. As observed for the NS, the contents of AAs in the fractions of D1 were less than those in D3. Generally, the contents of total acidic (TA), total basic (TB), total neutral hydrophobic (TH_O), and total neutral hydrophilic (TH_i) AAs in the different humic fractions followed the order: $TH_i >> TA > TH_O >> TB$. Some distinct trends can be observed in relation to the overall composition of the AA sub groups. The compositions (%) of the TH_i AAs followed the order, XAD-4 acids > FA > HA. In the cases of the TB% and TH_O%, the trends generally followed the order HA > FA > XAD-4 acids. Invariably the XAD-4 acids had the highest TA AA contents, but the relative abundances of these AAs in the HAs and FAs interchanged.

Total AAs in D1 and D3, in nmol mg^{-1}, followed the order HA > XAD-4 > FA. These findings are in agreement with the order found for HAs and FAs from waters (Thurman, 1985a; Watt *et al.*, 1996b). The AA contents accounted for approximately 2.3 to 3% of the HAs isolated from the Devon sample sites, and are roughly similar to the contents of NS. The AA contents of the FAs are similar for both samples, but the total AA content of the XAD-4 fraction of D3 is about twice that of the same fraction of D1. Whereas the AAs could account for nearly 12% of the N in the HA and XAD-4 acids of D3, these made a lesser contribution to the compositions of the same fractions in the D1 samples.

Ishiwatari (1985), when dealing with AAs in soil HS, considered that, in general, the relative abundances of the basic AAs and of the neutral H_i (NH_i) AAs increase in the order FA < HA < humin. The same trends are found for these AAs in the HAs and FAs from the drainage waters in the present study. Ishiwatari also observed that the acidic and the NH_i AAs decreased in the order FAs > HAs > humin. These trends did not hold in all cases for the AAs in the FAs and HAs from the drainage water samples (Table 6).

In general the abundances of the AAs decreased in the order glycine (Gly) > aspartic acid (Asp) > alanine (Ala) > glutamic acid (Glu) > threonine (Thr) > valine (Val) > leucine (Leu) > serine (Ser) > isoleucine (Ile) > lysine (Lys) > phenylalanine (Phe) > tyrosine (Tyr) > arginine (Arg) > histidine (His) > methionine (Met). This order shows two of the NH_i acids (Gly, Ala) in high abundance, and with lesser but more or less equal amounts of the (acidic) polar dicarboxylic AAs (Asp and Glu). There was an intermediate abundance of the three H_O AAs (Leu, Ile, Val), and of the (NH_i) hydroxyamino acids (Thr and Ser). The basic AAs (Lys, His, and Arg) were in relatively low abundance, as were the aromatic AAs, Phe and Tyr, and the sulphur containing compound Met. In general, the four AAs, Gly, Asp, Ala, and Glu were always present in the greatest concentrations, and generally decreased in the order Gly > Asp > Ala > Glu for all of the humic fractions studied. Phe, Tyr, Arg, His, and Met, were usually found in lowest abundances, and generally decreased in the order, Phe > Tyr > Arg > His > Met. The abundances of the other AAs in the humic fractions did not increase or decrease in any discernible order.

The NS and AA data give some interesting information about the effects of different management practices on these components of HS. The data would suggest that tile draining (for D1 samples) compared with surface run off (for D3 samples) had a significant effect on the compositions of NS and AAs. AAs and NS account for up to 5.7% of the compositions of the HAs in D3 (non-drained) and ca 4% of those in D1 (drained). The combined difference is significantly greater in the cases of the XAD-4 acids. However, emphasis has been placed on the different times of sampling, and that difference does not allow the full significance of the influences of the drains to be established.

NMR Data

The integrated areas for seven resonance bands of the CPMAS ^{13}C NMR spectra of the HA, FA, and XAD-4 acids fractions are given in Table 7.

The band at 10-45 ppm is broadly classified as aliphatic carbon, that at 45-65 ppm will contain methoxyl and also peptide-type functionalities, that at 60-110 ppm will include carbohydrate-related functionalities and the anomeric carbon resonance (95-105 ppm), that at 110-140 ppm will include aromatic functionalities, that at 140-160 ppm will include O-aromatic substituents, that at 160-190 ppm will include the carbonyl functionalities of carboxylic acids, esters, and amides, and the the resonance at 190-220 ppm will include the carbonyl functionality of aldehydes and ketones.

The data for the integrated areas in Table 7 show distinct differences between the fractions (HAs, FAs, XAD-4 acids), and between the same fractions of the D1 and D3 samples. The HAs were the most, and the XAD-4 acids the least aromatic of the three humic fractions. The integrated areas for the O-aromatic functionalities decreased in the same order. Also, the integrated areas for the carbonyl resonance at 160-190 ppm, considered most likely to be of carboxyl groups, decreased in order XAD-4 acids > FAs > HAs. The areas for the 110-65 ppm resonance increased in the order HA < FA < XAD-4 acids, and that trend does not bear a direct relationship to the total NS contents of the fractions. These decreased in the order HAs > XAD-4 acids > FAs in the case of the D1 fractions, and in the order XAD-4 acids >> HAs > FAs in the case of the D3 fractions. It is relevant to note that the D3 XAD-4 acids contained about 5.3% NS (Table 5) compared with 2.7% for the D3 HA fraction, yet the integrated area for the 110-65 ppm resonance was almost three times larger for the XAD-4 acids than for the HAs. These values can be interpreted in terms of responses to functionalities other than the NS components.

The most striking differences between the NMR spectra for the humic isolates from D1 and D3 are contained in the 10-45 ppm resonances. The data suggest that aliphatic hydrocarbon functionalities were more significant contributors to the fractions from D1 than to those from D3. That was the reverse of the trends for aromatic functionalities.

The evidence that has been presented would suggest that the humic samples from D1 had undergone more extensive transformations than those from D3. Should that be so the NMR data would indicate that in the transformation processes the aliphatic residues increase and the aromatic components decrease, or are metabolized during the transformations of water soluble humic substances.

FT-IR Data

The interpretations of the FT-IR spectra for the humic fractions were based on those of MacCarthy and Rice (1985) and Stevenson (1994). Although the FT-IR spectra for the fractions from the D1 and D3 sites were qualitatively similar, some differences in detail were evident. A comparison of the spectra for the HAs, FAs and XAD-4 acids showed that the same fractions exhibited a similar series of bands.

Table 7 *Integrated areas and the aromatic fraction (f_a)* of seven regions in the CPMAS ^{13}C NMR spectra of the humic acids (HA), fulvic acids (FA), and XAD-4 acids (XAD4) isolated from the Devon drainage water (D1) and surface water (D3)*

Sample	¶220-190	190-160	160-140	140-110	110-65	65-45	45-10	*fa
HA								
D1	4	14	5	23	10	7	37	28
D3	3	15	9	27	13	13	21	36
FA								
D1	2	16	4	11	15	-	51	15
D3	4	20	5	21	13	7	30	26
XAD4								
D1	1	24	2	6	29	-	39	8
D3	3	19	2	9	36	11	21	11

*fa = fraction aromatic (the percentage of the fraction that is aromatic). ¶ = Chemical shift (ppm)

The band at 3400-3300 cm^{-1}, due to O-H stretching, was difficult to interpret for the HS because some moisture was present even though the samples were carefully dried.

Carboxylic acid functionality dominated the spectra in the cases of all of the humic fractions from the D1 and D3 samples. That was indicated by absorption due to C=O stretch in undissociated COOH and COOR groups at 1720-1730 cm^{-1}. The absence of strong bands between 1785 cm^{-1} and 1735 cm^{-1} would suggest that esters were not major contributors to the carbonyl functionality. The bands at 1620-1630 cm^{-1} arose from COO^- groups, and/or aromatic functionalities with decreased benzene symmetry, and/or olefins. The H- bonding of OH to the C=O of quinones (quinones normally absorb between 1690 cm^{-1} and 1635 cm^{-1}) causes the C=O bonds to shift to lower frequency. The band at around 1510 cm^{-1} is also attributed to the stretching vibration of aromatic C=C bonds (MacCarthy and Rice, 1985), and it was noticeable only in the spectra of the HAs.

A broad absorption band was centred at 1212 and 1217 cm^{-1} originating from C-O stretch and/or O-H deformation vibrations of carboxyl groups. The signals in the region 1380 to 1440 cm^{-1} may be due to COO^- groups, O-H groups of alcohols, and/or C-O of phenolics. (The amide II band of proteins can also contribute to the absorption in this region.) Absorption bands and shoulders due to aliphatic compounds (C-H stretch at 2920 and 2850 cm^{-1} in methyl and methylene groups; C-H deformations of $C-CH_3$ and $C-CH_2$ at 1450 cm^{-1}), were present in all of the samples. The C-O stretch at 1000-1150 cm^{-1}, which can in some instances be attributable to polysaccharides, was evident in all spectra, though less pronounced in the FAs, and in the D1 XAD-4 acids. However, based on the relatively small NS contents (Table 5), the magnitude of the 'polysaccharides stretch' was greater than the analytical data would indicate. We do not have data for uronic acids which could make significant contributions to the polysaccharides.

The FT-IR spectrum for the D1 HAs, which deviated from the trends shown by the other samples, suggested that the acid groups were predominately in the COO^- form.

The FA spectra for the samples were very similar, especially in relation to the OH bands at 3400 cm^{-1}, the aliphatic bands at 2900 cm^{-1}, and the carboxyl bands at 1720 cm^{-1}.

There was a strong and well-defined shoulder at 1380 to 1470 cm^{-1}. This band was more intense at around 1400 cm^{-1}, indicating OH deformation and C-O stretching due to phenolic OH and carboxylic acids. The band at around 1220 cm^{-1} is typical of the others in this series. There was a noticeable absence of a shoulder between 1000 and 1100 cm^{-1}, but a slight one at around 1000 cm^{-1}, which is typical of the C-O stretching of polysaccharides.

The spectra for the XAD-4 acids were similar to those for the FAs. The main difference was in the region of 1000 to 1100 cm^{-1}. It is unlikely that these differences are related to the NS contents, or to the contents of modified carbohydrate-type materials. The integrated areas in the range 60-110 ppm (see Table 7) of the NMR spectra did not show variations on the scale shown by the FT-IR spectra. There was a distinct peak at 1400 cm^{-1} indicative a high carboxylic acid and alcohol functionalities. Similar indications were provided by a strong band at 1220 cm^{-1} .

5 Conclusions

The data have shown that there are some similarities and some distinct differences between the humic fractions from waters that had drained through soil and those from waters collected as surface run off.

There are two possible explanations for the differences observed. Firstly these may be related to the soil management practices. One of the soils is tile drained, and the other is allowed to drain naturally. The undrained soil would have a higher carbon content than the drained soil. That would result in a higher concentration of HS in the run off water, and could explain the larger amounts obtained from the D3 sample. The site of the D3 sample was less well aerated than the D1 sample, and the consequent lesser aerobic microbial activity could account for some of the structural and compositional differences observed. The most significant differences were seen in the NS and AA contents, with the greatest abundances in the D3 (sampled in November) fractions. However, the $\delta^{13}C$ data and NS ratios indicate that the D1 (sampled in March) HAs and FAs had higher inputs from microbial origins than those from D3. There were differences in the cases of the XAD-4 acids, and the microbial inputs to that fraction were greater for the D3 than for the D1 sample. That might indicate a build up of untransformed or of transforming organic material in the parent soil of the D3 sample.

Secondly, and perhaps more significantly, the differences in the results may be related to the differences in the times of sampling. D3 samples were taken at the autumnal 'flush' when the products formed during the summer were flowing as run off from the surface soil. The 'flush' was finishing, or had ended when the D1 samples were taken (March). The higher yields of the D3 fractions can be attributed to the build up of water soluble HS during the growing season in the soil, and these are then washed out in the 'flush' period. The D1 sample was collected in March, and it is likely that most of the soluble HS would have been 'flushed' out by then. That would explain the lower yields obtained for the D1 sample. The AA and NS data lend strong support for that hypothesis. The higher AA and sugar contents would be representative of a build up of (microbially) transforming organic material that was in the process of being flushed (D3 sample).

The $\delta^{13}C$ data also indicate contributions from microbial sources to the compositions of the D3 humic tractions. The implications are that the HS in the D1 sample had undergone more extensive transformations than the those in D3, and that these transformations were accompanied by an increase in the aliphatic content coupled with a decrease in the aromatic functionalities (from the NMR data).

In summary the available data indicate that the time of sampling and the extents to which the waters had penetrated into the soil have significant influences on the compositions of the humic materials isolated from soil drainage waters.

Acknowledgements

The work described in this study was part of a project supported by BBSRC

REFERENCES

Armstrong, A.C. and E.A Garwood. 1991. Hydrological consequences of artificial drainage of grassland. *Hydrol. Process* **5**:157-174.

Balsedent, J., A. Mariotti, and B. Guillet. 1987. Natural ^{13}C abundance as a tracer for studies of soil organic matter dynamics. *Soil Biol. Biochem.* **19**:25-30.

Blakeney, A.B., P.J. Harris, R.J. Henry, and B.A. Stone. 1983. A simple and rapid preparation of alditol acetates for monosaccharide analysis. *Carbohyd. Res.* **113**:291-299.

Delwiche, C.C. and P.L. Steyn. 1970. Nitrogen fractionation in soils and microbial reactions. *Environ. Sci. Technol.* **4**:929-935.

Hayes T.M. 1996. Study of the humic substances from soils and waters and their interactions with anthropogenic organic chemicals. PhD thesis, Univ. of Birmingham.

Ishiwatari, R. 1985. Geochemistry of humic substances in lake sediments. p. 147-180. *In* G.R. Aiken, D.M. McKnight, R.L. Wershaw and P. MacCarthy (eds). *Humic Substances in Soil, Sediment and Water, Geochemistry, Isolation and Characterisation.* Wiley, New York.

Komor, S.C. and H.W. Anderson Jr. 1993. Nitrogen isotopes as indicators of of nitrate sources in Minnesota sand-plain aquifers. *Ground Water* **31**:260-270.

MacCarthy, P. and J.A. Rice. 1985. Spectroscopic methods (other than NMR) for determining functionality in humic substances. p. 527-560. *In* G.R. Aiken, D.M. McKnight, R.L. Wershaw and P. MacCarthy (eds). *Humic Substances in Soil, Sediment and Water, Geochemistry, Isolation and Characterisation.* Wiley, New York.

Malcolm, R.L. 1991. Factors to be considered in the isolation and characterisation of aquatic humic substances. p. 369-391. *In* H. Boren and B. Allard (eds). *Humic Substances in the Aquatic and Terrestrial Environment.* Wiley, Chichester.

Malcolm, R.L. and P. MacCarthy. 1992. Quantitative evaluation of XAD-8 and XAD-4 resins used in tandem for removing organic solutes from water. *Environ. Int.* **18**:597-607.

Murayama, S. 1984. Changes in the monosaccharide composition during the decomposition of straws under field conditions. *Soil Sci. Plant Nutr.* **30**:367-381.

Oades, J.M. 1984. Soil organic matter and structural stability: Mechanisms and implications for management. *Plant and Soil* **76**:319-337.

Stevenson, F.J. 1994. *Humus Chemistry, Genesis, Composition, Reactions.* Wiley, NY..

Sutherland, R.A., C. van Kessel, and D. J. Pennock. 1991. Spatial variability of nitrogen - 15 natural abundance. *Soil Sci. Soc. Am. J.* **55**:1339-1347.

Thurman, E.M. 1985a. Humic substances in groundwater. p. 87-104. In: G.R. Aiken, D.M. McKnight, R.L. Wershaw and P. MacCarthy (eds), *Humic Substances in Soil, Sediment and Water, Geochemistry, Isolation and Characterisation.* Wiley, NY.

Thurman, E.M. 1985b. *Organic Geochemistry of Natural Waters.* Martinus Nijhoff/Dr W. Junk Publishers, Dordrecht, The Netherlands.

Turnell, D.C. and J.D.H. Cooper. 1982. Rapid assay for amino-acids in serum or urine by precolumn derivatization and reversed-phase liquid chromatography. *Clinical Chem.* **28**:527-531.

Tyson, K.C., E.A. Garwood, A.C. Armstrong, and D. Schofield. 1992. Effects of field drainage on the growth of herbage and the liveweight gain of grazing beef cattle. *Grass and Forage Sci.* **47**:290-301.

Watt, B.E., R.L. Malcolm, M.H.B. Hayes, N.W.E. Clark, and J.K. Chipman. 1996a. The chemistry and potential mutagenicity of humic substances in waters from different watersheds in Britain and Ireland. *Water Res.* **6**:1502-1516.

Watt, B.E., T.M. Hayes, M.H.B. Hayes, R.T. Price, R.L. Malcolm, and P. Jakeman, 1996b. Sugars and amino acids in humic substances isolated from British and Irish Waters. p. 389-398. *In* C.E. Clapp, M.H.B. Hayes, N. Senesi, and S.M. Griffith (eds), Humic Substances and Organic matter in Soil and Water Environments. *Proceedings of the Seventh International Conference, IHSS, Trinidad and Tobago.* IHSS, University of Minnesota, St. Paul.

Soil Organic Matter: Does Physical or Chemical Stabilization Predominate?

J.A. Meredith

DEPARTMENT OF SOIL SCIENCE, THE UNIVERSITY OF READING, WHITEKNIGHTS, PO BOX 233, READING, RG6 6DW, ENGLAND

Abstract

The stability of the organic matter was studied in a Vertisol from Kenya and in an Oxisol from Zambia, two tropical soils with different mineralogies and climates, and in three British soils, a stagnohumic gley, a gleyic brown earth, and a stagnopodzol with a stagnogley intergrade. The British soils had different mineralogy and organic matter contents, but with a similar climate. The results suggest that neither the tropical nor the British soils have large amounts of chemically stable soil organic matter. The soil mineralogy was the main determinant of the rate of carbon turnover, and chemical composition was a secondary consideration. Organo-mineral interactions are determined by the nature of the mineral and organic components that enable the preservation of particular organic species in certain soils and conditions.

1 Introduction

Soil organic matter (SOM) contains all of the carbon in the soil other than that in inorganic forms (e.g. calcium carbonate), and it includes the organic carbon in living soil flora and fauna. Agriculturalists, chemists, and soil scientists have endeavoured to understand the structure and the roles of SOM in the soil environment since the pioneering work of Lawes and Gilbert focused on the importance of SOM in promoting soil fertility (Schulten *et al.*, 1991). SOM has its origin in the living structures of animals and plants. These are ultimately the products of the photochemical fixing of atmospheric carbon dioxide by plants to form simple sugars and polysaccharides. The death of the animal and plant cells exposes their contents to breakdown by the microbial soil population. Animal excretion products, leaf fall, and exudates from plant roots are other sources which give rise to SOM. The mixture of chemical substances is rich and complex, and are products of genetically-controlled reactions. The range of substances, and their structures are defined by the metabolic processes of the organisms, and these in turn are supervised by the genetic codes of the organisms. After entry into the soil, most of the organic matter (OM) is rapidly utilized for energy by the soil biomass, and the carbon dioxide produced is returned to the atmosphere. (In an aerobic topsoil, 85-95% of the added SOM is thus lost within 6-12 months (Jenkinson, 1991; Jenkinson and Rayner, 1977). Some of the added OM acts as a carbon source to create more biomass, and this in turn grows, reproduces, and dies. A small proportion (5-10%) of the SOM remains in the soil for periods ranging from a few decades to thousands of years. This recalcitrant SOM may be very different from the SOM entering the soil. It is the product of the chemical and biochemical transformations of the original organic matter. The transformations involve random degradation reactions, modifications by enzymes, and the accumulation of secondary products from the biomass. These degradative and synthetic processes are not genetically controlled and their random nature produces humic substances (HS) of infinite variety in terms of molecular structure. It is thus unrealistic to consider discrete chemical structures for these HS.

Any uncultivated soil type that experiences more or less the same annual weather conditions will tend to reach a state of dynamic equilibrium where SOM inputs into the soil (plant matter, root exudates) will equal SOM losses from the soil (carbon dioxide to atmosphere, soluble organic materials leached out). When a virgin soil is brought into agricultural production the equilibrium is disturbed and the SOM levels fall rapidly (Beare *et al.*, 1994; Cambandella and Elliot, 1992). Conversely, a water-logged anaerobic soil in which SOM breakdown by the biomass is inhibited, will accumulate SOM as OM inputs exceed OM losses. Should these soils be drained, aerobic transformations of the OM will take place, and a new equilibrium will eventually be attained that is characteristic of the soil, its aeration, and its flora and fauna. Several hundred years may elapse for a new SOM equilibrium to be established when, for example, grassland soil is brought into agricultural production (Jenkinson *et al.*, 1992).

SOM is important for maintaining soil structure and productivity. In addition to providing carbon and energy sources for soil organisms, it acts as a reservoir of nutrients which may be essential or rate limiting for plant growth. The SOM contains organic bound nitrogen, sulphur, and phosphorus which are released and made available to plants by the process of mineralization. Its carboxylic acid groups and phenolic hydroxyl groups can hold cations (by ion exchange and by complexation processes) that are essential for plant nutrition. Polysaccharides, and gums produced by bacteria that utilise SOM contribute much to the stabilities of soil aggregates, and to the pore structures. Stable soil aggregates and good pore structures are important for drainage, soil aeration, and the containment of erosion resulting from the impacts of rain, and of water run-off. Larger soil aggregates are stabilized by fungal hyphae whose growth is promoted by SOM. SOM plays an important role in determining the bioactivities of many anthropogenic and xenobiotic organic compounds, and the heavy metals which enter the soil environment.

The quantity of organic carbon (OC) stored in the world's soils is estimated to be of the order of 1500 x 10^{15} g (Schlesinger, 1995). This is more than twice that in the earth's atmosphere (720 x 10^{15} g), and nearly three times that in land plants (560 x 10^{15} g). Should global temperatures rise, and as a result the rate of degradation of SOM increase, and should this not be matched by an increasing rate of carbon dioxide removal by various sinks (eg. plant photosynthesis, dissolution in the oceans as carbonate), then the carbon dioxide emissions from the SOM can be expected to contribute further to global warming. It is thus important to understand the factors that determine the stability of SOM in order to estimate the losses of SOM under changing environmental conditions, and thereby to determine whether it is significant either as a source of increasing atmospheric carbon dioxide or as an ecological and agricultural problem in its own right.

Clays are among the most important components of soil that affect the rate of decomposition of SOM. Readily decomposable materials, such as amino acids, can persist for over a year in soils with high clay contents (Sorensen, 1975). In such cases the stabilization arises from the physical protection provided by the sorption of the OM on the mineral surfaces, or its containment in a site that is inaccessible to the microbial biomass (Buyanovsky *et al.*, 1994). Particle size fractionation of soil has shown that much of the OM is bound to the clay- and silt-size fractions (Skjemstad *et al.*, 1994).

The chemical nature of the OM entering the soil is important for considerations of its transformations. For example, lignin, the highly aromatic cross-linked structural component of woody plants, can be broken down only by enzymes from certain fungi (Ambles *et al.*, 1993; Fox *et al.*, 1994; Hempfling and Schulten, 1989). OM with a high lignin content is therefore less decomposable in soil than that which is low in lignin (e.g. cellulose from herbaceous plants). Material with a high carbon to nitrogen (C:N) ratio may also be slow to breakdown. The fact that the C:N ratio is roughly constant (ca 8-15:1) across a wide range of soil types suggests that the nitrogen supply is a limiting factor. The presence of polyphenols may inhibit the activity of the soil microorganisms (deMontigmy *et al.*, 1993). Long-chain alkyl molecules, such as those from plant and microbial waxes, have been detected in soil by ^{13}C nuclear magnetic resonance (NMR) spectroscopy (Preston *et al.*, 1994), and by pyrolysis gas-chromatography mass-spectrometry (Py-GC/MS) (Tegelaar *et al.*, 1989), and by Py field-ionization MS (Sorge *et al.*, 1994). Because of their low reactivity, these are considered to be an important class of compounds that persist in soil. These various types of organic substances, large, highly cross-linked macro-molecules, stable aromatic compounds, unreactive alkyl species, are examples of how SOM may be stabilized by its chemical nature as opposed to physical protection within the soil matrix. Physical and chemical stabilization of OM may, of course, operate concurrently. Possibly the oldest characterized SOM known (6700 y, as determined by radiocarbon dating) is a long chain alkane molecule protected by intercalation between the clay layers in a New Zealand spodasol (Theng *et al.*, 1992)

Considerable effort has focused on fitting together the major soil factors to give conceptual and computer models of carbon turnover (Hsieh, 1989, 1993; van Veen and Paul, 1981). These include SOM inputs from plant debris and root exudates, SOM losses due to mineralization and leaching, the soil clay content (or its surrogate cation exchange capacity), temperature, and rainfall. The models that are most widely used are the *Century Model* (Parton and Rasmussen, 1994), and the *Rothamsted Model* (Jenkinson and Rayner, 1977; Tate *et al.*, 1993). In these models, the OM entering the soil is partitioned between different carbon "pools". The greater part of the added SOM (decomposable plant material) is mineralized to carbon dioxide which returns to the atmosphere, but the remainder (resistant plant material) is split between the active biomass and other carbon pools with different turnover times. The biomass and carbon pools each undergo mineralization at different rates with the greater part of the carbon lost as carbon dioxide, while the rest is redistributed amongst the carbon pools. These models differ in terms of the number of carbon pools in which the SOM is allocated. The Century Model has pools that increase in turnover times by a factor of ten, i.e. one year, ten years, 100 years, 1000 years, etc. The Rothamsted Model splits the OM into physically protected OM, chemically stable OM, and "inert" OM pools. These models can give very good fits of total OM, soil biomass, and overall OM turnover times for soil types and climatic conditions to which they are properly "fine-tuned", and can be used to a degree to model changes imposed by changing climates. However, the models are philosophically unsatisfying because these carbon pools are somewhat arbitrary, and there is little matching of particular SOM types or fractions to a particular pool. Without such a correlation the models cannot be considered to be truly general; extrapolations to soils and climates for which they have not been developed (or) fitted should be done with caution.

Global carbon cycle models must take account of the carbon sources and sinks for the different soil types in the different climatic environments. Ideally these SOM "black boxes" need to have their chemical and organo-mineral properties defined. SOM and soil organo-mineral species form continua in terms of chemical and physical properties, and hence in terms of turnover times. These "black boxes" are a mathematical or operational convenience. Much effort has been focused on the characterization of SOM both in forms obtained when isolated by chemical means, and when isolated from different particle-size fractions, and from soil aggregates. Modern instrumentation techniques, such as ^{13}C NMR, mass-spectrometry (MS), electron spin resonance (ESR), accelerator mass spectrometry (AMS), electron microscopy (EM), and X-ray diffraction (XRD) makes it possible to take an integrated approach to the correlation of mineralogy, chemical characterization, and the turnover of SOM. In the cases of soils with a certain mineral and climatic status, where the SOM inputs are from characteristic sources, it is possible to "open" the "black boxes" and to "look inside" at the sets of smaller "black boxes", and then to "sketch in" some broad physico-chemical characteristics. As yet, these data sets are too few to be brought into a unified SOM turnover model, but they do provide a beginning.

This study set out to contribute towards an improved understanding of the stability of SOM. It was decided to take the broad view, and to attempt to determine the relative importance of the chemical as opposed to the physical stabilization of SOM. This is seen as a potentially critical factor in determining the impact of global warming on the world's store of SOM. The decomposition rate of SOM that is stabilized chemically would be expected to increase in line with increasing temperature (according to the Arrhenius equation). Inaccessible, or physically-protected OM could be far less sensitive to temperature change. The procedure involved chemical fractionation of SOM from selected soils. Fractions were characterized by solid-state ^{13}C NMR and then were C-14 "dated".

2 Materials and Methods

Soils

Two groups of soils were chosen:

(1), a Kenyan vertisol from the Nairobi National Park, an example of a tropical grassland soil, and
a Zambian oxisol from Misamfu, an example of a tropical forest soil;

(2), three upland soils. These were:
a cambic stagnohumic gley;
a gleyic brown earth; and
a ferric stagnopodzol with a stagnogley intergrade.

All three were upland soils from Great Dun Fell, Cumbria, U.K.

The first group of two soils are from different climates and have different mineralogies. The second group of three British upland soils represent different soil types under essentially the same climatic conditions (2.5 °C at 800 m above sea level). These soils have a number of horizons having different proportions of OM, minerals, mean OM radiocarbon ages, and ^{13}C isotopic profiles. The horizons selected for study were:

a Cambic Stagnohumic Gley O_h (4-9 cm depth) with a carbon content of 35 per cent;
a Gleic Brown earth A_h (5-10 cm) with a carbon content of 6 per cent: and
a Ferric Stagnopodzol with Stagnogley intergrade B (0-8 cm), with 2 per cent carbon.

Isolation and Fractionation of the Soil Organic Matter

Isolation of SOM is a prerequisite to fractionation. Hayes (1985) has provided an extensive review of the principles and procedures used in the extraction of HS. The principal agents used are various solution strengths of sodium hydroxide, neutral or alkaline solutions of pyrophosphates, and organic solvents. Solutions of pyrophosphate displace the polyvalent ions such as calcium and aluminium that bind HS together by bridging carboxylate and phenolate functionalities. At alkaline pH values sodium ions replace the acidic protons, and the negative charges on the conjugate bases mutually repel each other, the HS tend to disperse, the charged species solvate, and the macromolecules go into solution. Although organic solvents such as dimethyl sulphoxide (DMSO) and dimethyl formamide (DMF) can offer improved selectivity for certain HS, these were not acceptable for the present study because such solvents are difficult to remove without trace and could thus lead to false radiocarbon dating values. Sodium hydroxide (0.5M) was the reagent chosen because it offered the highest yields of HS from a given quantity of soil, and it is considered not to alter significantly the functionalities of the HS as indicated by the NMR signals of extracted HS, compared with the spectra for the whole soil (Piccolo *et al.*, 1990). The humic acid (HA) fraction, precipitated when the pH of the extract is lowered to 1.0, was collected, washed, and freeze dried. The fulvic acid (FA) fraction in solution also contains carbohydrates, proteins, and inorganic salts, as well as the FAs. The major part of the true FAs, the hydrophobic fulvic acids, are isolated by selective sorption onto purified XAD-8 [(poly)methylmethacrylate] resin. The residual (non-base labile) HS is the "humin" fraction. This can contain the major part of the SOM in mineral soils when there are strong associations between the OM and the mineral matrix.

The oldest radiocarbon dates for SOM were obtained for soils that had been treated with strong mineral acids under reflux conditions. These hydrolysis conditions destroy carbohydrate and proteinaceous components of the OM, which are generally much younger than the HS. These younger substances may be physically or chemically (or both) attached to, or entrained within the older HS, and would thus give a younger mean age to the extracted materials. For that reason HAs, hydrolysed HAs, humin, hydrolysed humin, hydrolysed soil, and the original "whole" soil samples were selected for chemical analysis and for radiocarbon dating. The hydrophobic FAs were isolated from the alkaline extracts in the cases of a couple of the selected soils (see the Experimental Section).

The fractions are, of course, operationally rather than chemically defined. There are chemical differences between the fractions (which is why they separate), but the cuts-off between fractions are not sharp, and a continuum of chemical structures and molecular size distributions are the norm. Thus FAs may be considered to be more oxygenated (more carboxyl groups), and lower molecular weight analogues of the HAs. These might not originate within the same regions of the soil matrix as the HAs, however, and may undergo different degrees of interaction within the soil environment.

Fractionation of the SOM. Soil sub-samples for extraction were air-dried on poly(ethene) sheets for several weeks at room temperature, then gently ground by hand in a pristine ceramic pestle and mortar and sieved to < 2 mm mesh size.

Extraction of the Kenyan Vertisol soil. Samples of the air-dried, sieved soil (135 g) were placed in each of six 1 L "Nalgene" polypropylene centrifuge bottles fitted with air-tight screw-cap lids. These samples were decalcified with 600 mL of 0.2M hydrochloric acid in the cold for 18 h. Roots and debris floating on the surface were discarded.

After decalcification, sodium hydroxide pellets were added to produce a 0.5M NaOH solution. The bottles were immediately shaken, and flushed out with low-oxygen (BOC "white-spot") nitrogen for 10-15 minutes. Each tube was sealed and shaken vigorously for at least five minutes. The flasks were left to stand overnight then centrifuged at 2100 rpm for 25-30 minutes. The supernatant was further centrifuged for 20 minutes at 14 000 rpm, then filtered on a Number 3 sintered-glass filter to retain any particulates disturbed on decanting. The filtrate was acidified to pH 1.0 to initiate precipitation of the HA fraction (*vide infra*). For the Kenyan vertisol, five extractions were required before the colour of the extract had faded to a very pale straw-yellow (was brown-black initially), and the quantity of HA precipitated was minimal.

Extraction of the SOM of the Zambian oxisol soil. Six 135 g samples of the Oxisol from the 15 - 30-cm layer were used. No initial decalcification was required, and only three extractions with base were required to remove the coloured substances.

Isolation of humic acids (HAs) from the extracts. Each batch of extract (in 0.5M NaOH) was slowly acidified by adding 6M hydrochloric acid to the stirred extract. The acidified solutions (pH 1.0) and the precipitating HAs were allowed to stand overnight. The bulk of the HAs were separated by low-speed centrifugation (2100 rpm for 25 minutes) and the FA fraction (FAF) was decanted. The FAF was then centrifuged at high speed (14 000 rpm for 20 min) to remove all remaining HAs, decanted, filtered (No. 3 glass sinter), and transferred to glass bottles pending further processing. The HAs obtained by low and high-speed centrifugation were combined onto a No. 3 sintered-glass funnel. The filtrate was discarded. The collected HAs were washed with 2M hydrochloric acid, then typically with 3 x 200 mL distilled, de-ionised water. The washing with distilled water was continued until the filtrate was free of chloride ion (to acidified silver nitrate). The salt-free HAs (in the H^+ form) were then transferred with distilled, de-ionised water to round-bottomed flasks and freeze dried.

Collection of the Humin residues. Hydrochloric acid (6M) was added to the residues after extraction with sodium hydroxide and the mixture was stirred. The slurry pH was lowered to 3.0 - 3.5 and the mixture was left to stand overnight. Then, after further stirring, the pH was adjusted to 6.0 in the case of the Kenyan humin, and to 5.7 for the Zambian humin. The humins were isolated by freeze-drying.

Acid hydrolysis of soils and of SOM fractions. A large excess of 6M hydrochloric acid was added to weighed samples of air-dried soil or of freeze-dried SOM fractions (humin, HAs) in round-bottomed flasks, and the mixtures were refluxed for 18 h. The hydrolysed soils and humins were recovered by centrifugation, and the supernatants were discarded. The sediment was repeatedly washed with distilled, de-ionised water. When the sediment began to disperse (at pH 5 - 6) the hydrolysed material was isolated by freeze-drying.

Isolation of the hydrophobic FAs from the Kenyan and Zambian soils. The FA fraction (FAF) from the soil extracts, after centrifugation and filtering on No. 3 sinters, contained a mixture of the hydrophobic and hydrophilic FAs, peptide substances, carbohydrates, and inorganic ions. To obtain purified, desalted, hydrophobic FAs from these solutions, a column of purified XAD-8 resin was used to selectively absorb this fraction. After preconditioning the column of 4 L of XAD-8 resin, the solution of the FAF, at pH 1.0, was pumped onto the column at a rate of 100 mL min^{-1}. The solution can be added until the coloured front reaches 1/3 of the length of the column (this condition was not reached in these experiments, and all of the solution was added by the time the front was 1/6 down the column). The column was further pumped with distilled, de-ionised water to desalt and remove the unwanted hydrophilic substances from the column. The absorbed hydrophobic HS eventually began to emerge from the column. At this point the desired fraction was recovered by rapid back-elution with fresh 0.1M sodium hydroxide. About 10 L of the basic solution was required to desorb the hydrophobic FAs (some coloured material remained strongly sorbed to the resin). The FA solution was passed through a bed of purified (H^+-form) IR-120 cation exchange resin to decrease its pH to 4 - 5. The acidified solution was passed through the resin column at a fast drop rate in order to ensure that the FAs were H^+-exchanged. The samples were then freeze dried.

No decalcification was required for the Great Dun Fell Soils. Otherwise these were extracted as for the Zambian oxisol.

The Radiocarbon Dating of the Soils and of the Soil Fractions

Gilet-Blein *et al.* (1980) and Scharpenseel *et al.* (1989) have discussed the difficulties of applying radiocarbon dating to SOM. In practice, the production of C-14 in the atmosphere is not constant, and calibration curves of radiocarbon age against calendar age, based on wood samples of known age, are employed. Further complications arise from isotopic fractionation in living tissues, and in sample preparation. Contamination of samples by foreign organic material is another practical problem. However, the after effects of hydrogen bomb tests carried out in the atmosphere in the 1950's are the greatest problem in the accurate radiocarbon dating of SOM. These tests released large quantities of C-14 into the atmosphere, and thus into the carbon cycle (Schiff *et al.*, 1990). Trumbore *et al.* (1990) stated that the only sure way to counter the problems from bomb tests is to correct using corresponding samples taken before 1950. Such samples are rarely available, and not infrequently these too are contaminated. Another approach is to model the C-14 "diffusion" down the soil profile (as done, for example by O'Brien and Stout, 1978). However, this involves simplifying assumptions. Thus, this "bomb C-14" problem has to be accepted, and the limitations it imposes duly acknowledged.

SOM C-14 analyses are reported in terms of "% modern", i.e. as a percentage of the 95.5% activity of an oxalic acid standard of 1950. Thus an "old" sample will be less than 100% modern, and a sample contaminated by enough bomb C-14 to swamp the old material can be greater than 100% modern. The age can also be reported in terms of a radiocarbon age (in years) before present (BP), with the "present" being 1950. This is calculated using the "old" C-14 half-life of 5,680 years rather the new half life cited above; it should not, however, be confused with calender age. Therefore the "true" radiocarbon age, i.e. that which would prevail had bomb C-14 not been introduced, cannot usually be known.

The costs of radiocarbon analyses by AMS are high, and because of that it was feasible only to fully fractionate and to analyse for radiocarbon a single horizon or layer from each soil.

3 Results and Discussion

The Tropical Soils

There were considerable differences (Table 1) between the soil carbon contents for the vertisol and oxisol soil layers selected (0-15 cm, 15-30 cm, and 30-50 cm depths). There is no history of cultivation of these soils, and so it is likely that the SOM contents of the soils are at a steady state. Particle size analyses (Table 1) show a far greater content of clay-sized separates in the Kenyan vertisol than in the Zambian oxisol. Radiocarbon analyses (Table 2) of the whole soils clearly indicate the effect of bomb C-14 on the top layers of both soils (the analyses are greater than 100% modern). For the Kenyan vertisol, the 15-30 cm and 30-50 cm layers show a preponderence of old SOM; for these layers, mean radiocarbon ages of ca 300 and 1000 years, respectively, can be calculated. In this case sufficient "old" SOM exists at depth to overwhelm any bomb C-14 containing material that has migrated down the soil profile. In contrast, all the layers of the Zambian oxisol have been penetrated by bomb C-14 labelled SOM to such an extent that the overall analyses are greater than modern. This appears to indicate that little old material persists in this sandy soil.

The radiocarbon data for the acid hydrolysed soils (Table 3) indicate that about half the original SOM has been removed from the sample and that this material is younger than the average for the whole soil since the residual OM has a greater mean radiocarbon age. In the case of the oxisol, the upper two layers are still more than modern indicating the influence of modern OM in these horizons. Only the lowest layer (30-50 cm) of the oxisol posseses sufficient older SOM to produce a net pre-modern analysis corresponding to a radiocarbon age of 365 years.

The 15-30 cm layers of the two soils were chosen for fractionation into the HA, FA, humin, and hydrolysed humin fractions. This was a compromise between the desirability of a high SOM content (in the upper layers) to maximise the yields of humic substances and to minimise the influence of the bomb C-14 signal (greatest in the upper layers). The results of the radiocarbon analyses are given in Table 4.

Again the main observation is that the oxisol fractions are overwhelmed by the bomb C-14 signal (Table 4). The pattern of ages across the fractions are typical of the trends found in the literature, with the residual OM in the humin being the oldest material. The hydrolyzed humin is older still for both soils, but the difference on hydrolysis is much greater for the Kenyan than for the Zambian humin. That suggests that the clay of the Kenyan vertisol had retained much more effectively than the sandy oxisol what would otherwise be labile young material.

Table 1 *Aspects of the compositions of the Kenyan vertisol and of the Zambian oxisol*

Layer	C(%)	C/N	C(%)	C/N	Particle size distributions (15-30 cm.):					
	Vertisol.		Oxisol.		Clay (%)		Silt (%)		Sand (%)	
					Vertisol	Oxisol	Vertisol	Oxisol	Vertisol	Oxisol
0-15	2.36	14.9	0.93	16.7						
15-30	1.3	10.8	0.30	13.2	67.6	11.96	12.3	13.65	20.1	74.39
30-50	1.15	10.6	0.17	9.2						

Table 2 *Radiocarbon analyses of the whole soil*

	Vertisol		Oxisol
Soil depth (cm)	%Modern	Age (Years).	%Modern
0-15cm.	103	n/a	117
15-30cm.	95.6	320	107.3
30-50cm.	87.1	1070	101.6

Table 3 *Radiocarbon analyses of hydrolyzed soil samples*

		Vertisol			Oxisol	
Soil depth (cm)	%Modern	Age (years).	%C	%Modern	Age (years).	%C
0-15	93.8	475	1.3	109.1	n/a	0.46
15-30	n/a	n/a	0.4	105.5	n/a	0.125
30-50	77.9	1965	0.48	95.6	365	0.107

Table 4 *Radiocarbon analyses of organic matter fractions from the 15 - 30 cm layers*

		Vertisol			Oxisol	
Fraction	% Modern	Age	$\delta^{13}C$	Fraction	% Modern	Age
Fulvic acid	n/a	n/a	n/a	Fulvic acid	113.2	n/a
Humic acid	97.4	170	-10.9	Humic acid	105.1	n/a
Humin	91.6	660	-10.7	Humin	102.3	n/a
*Hydr. Humin	80.6	1690	-12.1	Hydr. Humin	102.0	n/a

*Hydr = Hydrolyzed

The data suggest that the clay in the Kenyan vertisol gives a more effective means of stabilizing SOM than the sand predominating in the Zambian oxisol. The relative ages of the fractions indicate a longer-term persistence of old SOM in the case of the Kenyan soil, and the faster mean turnover in the Zambian soil allowing the bomb C-14 signal to predominate. This probably explains the overall higher SOM content of the Kenyan over the Zambian soil. It should be noted that the Zambian SOM must nevertheless include some old HS to explain the data obtained.

Cross-polarization magic angle spinning (CPMAS) ^{13}C NMR spectra were obtained for the Kenyan and Zambian HAs from the 15 - 30 cm layer. The signal strengths were not necessarily directly comparable because of differing experimental conditions, and because of inevitable differences in aspects of the compositions of the samples. In the case of the humins the carbon contents of the samples were generally too low to give good spectra. The presence of iron, and of any other paramagnetic species in the soils or humins can broaden and distort the NMR spectra.

Bar charts are shown in Figure 1 for the integrated areas of the spectra for chemical shift regions which correspond to particular functionalities of the Kenyan and Zambian HAs. Some minor differences are evident in the relative proportions of functional groups in the two samples. The Zambian samples have slight relative enrichments of alkyl, carbonyl, and carboxyl functionalities, whereas the Kenyan HAs exhibit slightly enhanced

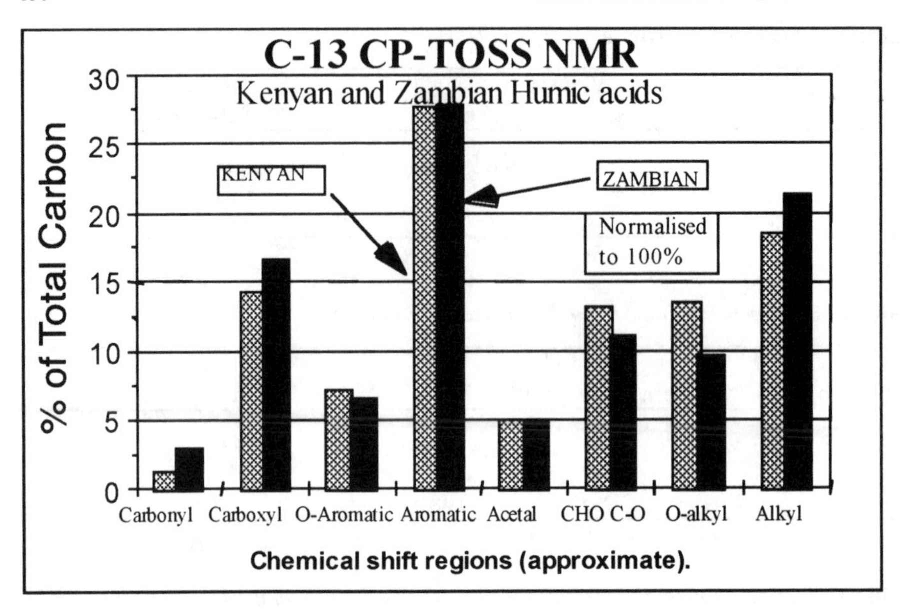

Figure 1 *Integrated areas of different resonances in the CPMAS ^{13}C NMR spectra of the humic acids from the Kenyan and Zambian soils*

signals for the carbohydrate or carbohydrate related C-O, and the O-alkyl resonances. The ^{13}C NMR data do not, of course, give the molecular connectivities or the molecular size ranges of the molecules. The enhanced carbohydrate-related functionalities in the Kenyan samples, coupled with the radiocarbon analyses data (97.4% modern for the Kenyan HA and 105% modern for the Zambian sample) would point to the physical stabilization of the OM on or within the clay mineral aggregates as the major cause of the increased stability of the Kenyan HAs.

The Great Dun Fell Soils

These soils comprise a number of horizons with different proportions of OM, mineral, and mean OM radiocarbon ages and C-13 isotopic profiles. With only a limited number of radiocarbon analyses available, it becomes imperative to select with care the horizons or layers from which to isolate the SOM. It was therefore decided to select horizons or layers with very similar gross radiocarbon ages, ca 2400-3000 years, but with very different ratios of SOM to mineral matter. The fact that these soils have very different organic carbon contents should imply different stabilities of SOM (assuming that the OM inputs and outputs are similar, and that likewise, climate and vegetation are also comparable). Thus the mean radiocarbon ages could be masking different distributions of radiocarbon across the various SOM fractions. The horizons selected were the Cambic Stagnohumic Gley O_h (4 - 9 cm depth, as measured from the upper boundary of this horizon) with a carbon content of 35%, the Gleyic Brown Earth A_h (5 - 10 cm, and with 6% carbon), and the Ferric Stagnopodzol with a Stagnogley Intergrade B (0 - 8 cm, and with 2% carbon).

Table 5 *Radiocarbon analyses of the Great Dun Fell soil fractions*

	Cambic Stagnohumic Gley (y)	Gleyic Brown Earth (y)	Ferric Stagnopodzol w. Stagnogley (y)
Whole Soil	2350	2990	2500
Humic acid	1865	2440	5050
Humic acid (hydrolyzed)	2040	2775	7380
Humin	2490	3295	4080
Hydrolyzed soil	2785		n/a
	8855		

The radiocarbon data (Table 5) show that the spread of radiocarbon dates varied considerably between the soil types. The most highly organic soil, the Cambic Stagnohumic gley, which approaches the composition of a peat, exhibited the smallest spread between the fractions. Both the HAs and hydrolyzed HAs from that soil are younger than the whole SOM, and the humin is seen to be older than the whole soil and the HA fractions from the soil. For the Gleyic Brown Earth, similar relationships exist, but the spread of ages is broader. This is best seen by calculating the fraction ages relative to the whole soil ages, as shown in Table 6.

The most mineral-based soil, the Stagnopodzol, has the greatest spread of radiocarbon ages. In this case the HA fraction is older than the whole SOM, and the hydrolyzed HAs are far older still. The hydrolyzed soil is very much older in relation to the whole soil. Account should be taken of the very low concentrations of OM, and especially in the cases of the hydrolyzed samples.

It would appear, even without considerations of the exact nature of the SOM in these soils, that it is the interaction between the OM and the mineral content that differentiates the SOM fractions by age. Whether this is by selective sorption or by selective preservation of different OM species, or indeed both, is not clear.

In the cases of the highly organic soils, it was feasible to obtain ^{13}C NMR spectra from all the fractions. However, in the case of the Ferric Stagnopodzol, sufficient carbon was present only in the whole soil sample and in the HA, and the hydrolyzed HA fractions to allow adequate NMR spectra to be obtained that would enable quantitative comparisons to be made.

Table 6 *Ratios of the ages of humic acids, hydrolyzed humic acids, humin, and hydrolyzed soil to those of the whole soil organic matter*

	Cambic Stagnohumic Gley	Gleyic Brown Earth.	Ferric Stagnopodzol w. Stagnogley.
Whole Soil	1.000	1.000	1.000
Humic acid	0.794	0.816	2.020
Humic acid (hydrolyzed)	0.868	0.928	2.952
Humin	1.060	1.102	1.632
Hydrolyzed soil	1.185	n/a	3.542

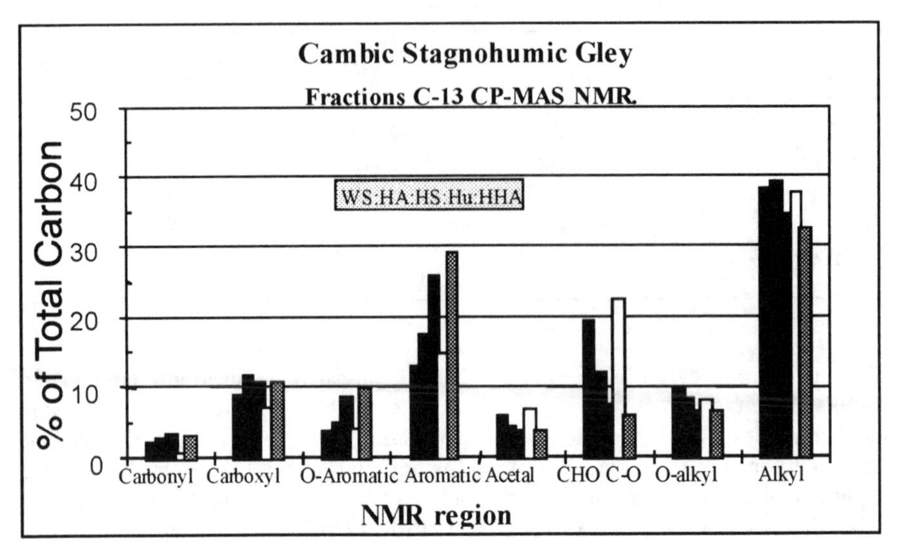

Figure 2 *Integrated areas for different resonance bands of the CPMAS ^{13}C NMR spectra of the whole soil (WS), and of the humic acid (HA), the hydrolyzed soil (HS), the humin (Hu), and the hydrolyzed HA (HHA) fractions of the Cambic Stagnohumic Gley*

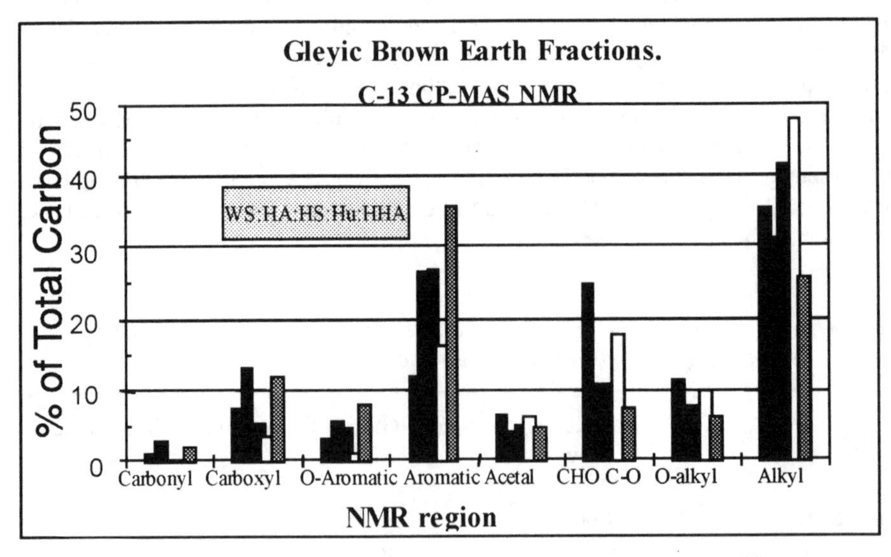

Figure 3 *Integrated areas for different resonance bands of the CPMAS ^{13}C NMR spectra of the whole soil (WS), and of the humic acid (HA), the hydrolyzed soil (HS), the humin (Hu), and the hydrolyzed HA (HHA) fractions of the Gleyic Brown Earth*

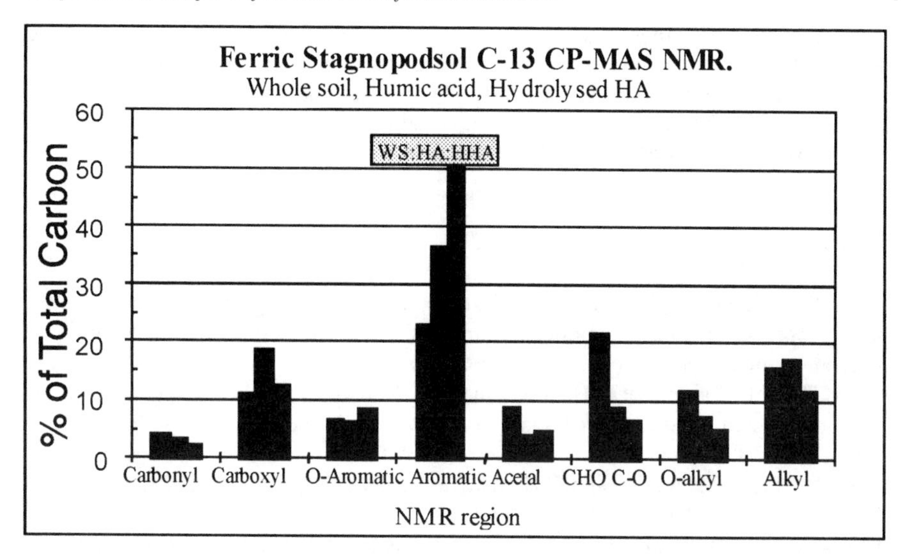

Figure 4 *Integrated areas for different resonance bands of the CPMAS ^{13}C NMR spectra of the whole soil (WS), the humic acid (HA), and the hydrolyzed HA (HHA) of the Ferric Stagnopodzol*

Data for the integration of the resonance bands corresponding to various functional groups are shown as bar charts in Figures 2, 3, and 4.

It is clear that alkali extraction of SOM tends to selectively remove aromatic and carboxyl bearing OM and to leave behind saccharide or saccharide-related materials. These saccharide related substances are largely removed on hydrolysis, indicating that this is a reasonable C-13 signal assignment, at least for these soils. Strikingly, the oldest soil, the Ferric Stagnopodzol, has a lower proportion of alkyl carbon than the more organic soils; it is highly aromatic in character. That does not support the contention that unreactive alkyl groups are selectively preserved and contribute the greatest part to the SOM age. Attempts to correlate fraction age with changes in the NMR spectroscopy of the fractions were not successful.

Overall, the data for these Great Dun Fell soils seem to indicate that SOM chemical structure is not controlling the carbon turnover. The high levels of OM at this site are due in the first instance to the cold, water-logged conditions. Differentiation of OM by age seems to be a function of the mineral content; this also is presumably selecting and controlling the preservation or formation of HS on the basis of their chemical structure. What is not seen is evidence for the existence of chemically stable OM. Some selection of SOM on the basis of innate chemical stability possibly operates in these soils, but it must be a third order effect, dominated by the climate and mineralogy.

4 Conclusions

Neither the tropical soils nor the U.K. upland soils give support to the notion that large amounts of innately chemically stable SOM are present in the soils. Both support the view that the soil mineralogy is the predominant determinant of the rate of carbon turnover. The organic chemical structure is a secondary consideration. Nevertheless, the organo-mineral interactions will be determined both by the nature of the mineral and of the organic components, and the preservation of certain organic species in certain soils and conditions will be determined by these interactions.

References.

Ambles, A., P. Jambu, J-C. Jacquesy, E. Parlanti, and B. Secouet. 1993. Changes in the ketone portion of lipidic components during the decomposition of plant debris in a hydromorphic forest-podzol. *Soil Sci.* **156**:49-56.

Beare, M.H., M.L. Cabrera, P.F. Hendrix, and D.C. Coleman. 1994. Aggregate-protected and unprotected organic matter pools in conventional and no-tillage soils. *Soil Sci. Soc. Am. J.* **58**:787-795.

Buyanovsky, G.A., M. Aslam, and G.H. Wagner. 1994. Carbon turnover in soil physical fractions. *Soil Sci. Soc. Am. J.* **58**:1167-1173.

Cambardella, C.A., and E.T. Elliot. 1992. Particulate soil organic matter changes across a grassland cultivation sequence. *Soil Sci. Soc. Am. J.* **56**:777-783.

deMontigny, L.E., C.M. Preston, P.G. Hatcher, and I. Kogel-Knabner. 1993. Comparison of humus horizons from two ecosystem phases on northern Vancouver Island using C-13 CPMAS NMR spectroscopy and CuO oxidation. *Can. J. Soil Sci.* **73:**9-25.

Fox, C.A, C.M. Preston, and C.A. Fyfe. 1994. Micromorphological and C-13 NMR characterisation of a Humic, Lignic, and Histic Folisol from British Columbia. *Can. J. Soil Sci.* **74**:1-15.

Gilet-Blein, N., G. Marien, and J. Evin. 1980. Unreliability of ^{14}C dates from organic matter of soils. *Radiocarbon* **22**:919-929.

Hayes, M.H.B. 1985. Extraction of humic substances from soil. p. 329-362. *In* G.R. Aiken, D.M. McKnight, R.L. Wershaw, and P. MacCarthy (eds), *Humic Substances in Soil, Sediment, and Water*. Wiley, New York.

Hempfling, R. and H.-R. Schulten. 1989. Selective preservation of biomolecules during humification of forest litter by pyrolysis field-ionisation mass spectrometry. *Sci. Total Environ.* **81/82**:31-40.

Hsieh, Y.P. 1989. Dynamics of soil organic matter formation in croplands - conceptual analysis. *Sci. Total Environ.* **81/82**:381-390.

Hsieh, Y.P. 1993. Radiocarbon signatures of turnover rates in active soil organic carbon pools. *Soil Sci. Soc. Am. J.* **57**:020-1022.

Jenkinson, D.S., D.E. Adams and A. Wild. 1991. Model estimates of CO_2 emissions from soil in response to global warming. Nature **351**:304-306.

Jenkinson, D.S. and J.H. Rayner. 1977. The turnover of soil organic matter in some of the Rothamsted classical experiments. *Soil Science* **123**:298-305.

Jenkinson, D.S., D.D. Harkness, E.D. Vance, D.E. Adams, and A.F. Harrison. 1992. Calculating net primary production (NPP) and annual input or organic matter to soil from the amount and radiocarbon content of soil organic matter. *Soil Biol. Biochem.* **24**:295-308.

O'Brien, B.J. and J.D. Stout. 1978. Movement and turnover of soil organic matter as indicated by carbon isotope measurements. *Soil Biol. Biochem.* **10**:309-317.

Parton, W.J., and P.E. Rasmussen. 1994. Long-term effects of crop management in wheat-fallow: II. *Soil Sci. Soc. Am. J.* **58**:530-536.

Piccolo, A., L. Campanella, and B.M. Petronio. 1990. ^{13}C NMR of soil humic substances extracted by different mechanisms. *Soil Sci. Soc. Am. J.* **54**:750-756.

Preston, C.M., R.H. Newman, and P. Rother. 1994. Using ^{13}C NMR to assess the effects of cultivation on the organic matter of particle size fractions in a grassland soil. *Soil Sci.* **157**:26-35.

Scharpenseel, H.W., P. Becker-Heidemann, H.U. Neue, and K. Tsutsuki. 1989. Bomb-carbon, ^{14}C-dating and ^{13}C measurements as tracers of organic matter dynamics as well as of morphogenetic and turbation processes. *Sci. Total Environ.* **81/82**:99-110.

Schiff, S.L., R. Aravena, S.E. Trumbore, and P.J. Dillon. 1990. Dissolved organic carbon cycling in forested watersheds: a carbon isotope approach. *Water Resources Res.* **26**: 2949-2957.

Schlessinger, W.H. 1995. An overview on the carbon cycle. p.9-25. *In* R. Lal, J. Kimble, and B.A. Stewart (eds), *Soils and Global Change*. Lewis Publishers, Boca Raton.

Schnitzer, M., and H.R. Schulten. 1992. The analysis of soil organic matter by pyrolysis field-ionisation mass spectrometry. *Soil Sci. Soc. Am. J.* **56**:1811-1817.

Schulten, H.-R., B. Plage, and M. Schnitzer. 1991. A chemical structure for humic substances. *Naturwissenschaften* **78**:311-312.

Skjemstad, J.O., P. Clarke, J.A. Taylor, J.M. Oades, and R.H. Newman. 1994. The removal of magnetic materials from surface soils . A solid state ^{13}C CP-MAS study. *Aust. J. Soil Res.* **32**:1215-29.

Sorensen, L.H. 1975. The influence of clay on the rate of decay of amino acids metabolites synthesised in soils during decomposition of cellulose. *Soil. Biol. Biochem.* **7**:171-177.

Sorge, C., M. Schnitzer, P. Leinweber, and H.-R. Schulten. 1994. Molecular-chemical characterisation of organic matter in whole soil and particle-size fractions of a spodosol by pyrolysis field-ionisation mass spectrometry. *Soil Sci.* **158**:189-203.

Tate, K.R., D.J. Ross, B.J. O'Brien and F.M. Kelliher. 1993. Carbon storage and turnover, and respiratory activity, in the litter and soil of an old-growth southern beech (Nothofagus) forest. *Soil. Biol. Biochem.* **25**:1601-1612.

Tegelaar, E.W., J.W. de Leeuw, and C. Saiz-Jimenez. 1989. Possible origin of aliphatic moieties in humic substances. *Sci. Total Environ.* **81/82**:1-17.

Theng, B.K.G., K.R. Tate, and P. Becker-Heidmann. 1992. Towards establishing the age, location and identity of the inert soil organic matter of a spodosol. *Z. Pflanzenernahr dungung Bodenk.* **155**, 181-184.

Trumbore, S.E., G. Bonani, and W. Wolfli. 1990. The rates of carbon cycling in several soils from AMS 14-C measurements of fractionated soil organic matter. *In* A.F. Bouwman (ed.), *Soils and the Greenhouse Effect*. Wiley, Chichester.

van Veen, J.H. and E.A. Paul. 1981. Organic carbon dynamics in grassland soils. I: Background information and computer simulation. *Can. J. Soil Sci.* **61**:185-201.

Humic Substances from Interment Sites I. Isolation and Characterization

M.H.B. Hayes, Lorraine J. Stewart, and P. Bethel[1]

THE UNIVERSITY OF BIRMINGHAM, SCHOOL OF CHEMISTRY, EDGBASTON, BIRMINGHAM B15 2TT, ENGLAND
[1]THE UNIVERSITY OF BIRMINGHAM, FIELD ARCHAEOLOGY UNIT

Abstract

Humic (HA) and fulvic (FA) acid fractions were isolated from grave sites in the Anglo-Saxon burial site at Sutton Hoo using neutral salt (pyrophosphate), base (NaOH), and acidified organic (dimethylsulphoxide/HCl) solvent systems. Properties, such as elemental, sugar, and amino acid compositions, cation exchange capacity, and E_4/E_6 ratio values of the fractions were compared in the cases of samples isolated from soils which clearly showed evidence for the imprints of interred bodies, and from soils in the gravefill some distance from the body 'imprints'.

Differences were observed in the compositions of the isolates from the different solvent systems, and between the humic fractions from the body 'imprint' and the gravefill sites. The data would suggest that that differences in the compositions of the HAs, arising from the protoplasm substrates of the interred bodies, were maintained throughout the centuries.

1 Introduction

This study has involved investigations of the humic substances (HS) associated with body 'imprint' and gravefill soils from the Anglo Saxon burial sites at Sutton Hoo (near Woodbridge, Suffolk, England) and at Snape (by the river Alde), Suffolk, which is about 15 km NE of Sutton Hoo. Excavation at Sutton Hoo began in 1939. Initially the remains of a ship were discovered, followed by the unearthing of grave goods, including exotic jewellery, decorated shields, bronze helmets, gold coins, and the traces of bodies.

The soils in the burial sites have sandy textures. Careful removal of the soil reveals the location of body remains as darker colorations in the light coloured sandy soil.

The objectives of the study were to compare aspects of the compositions of HS isolated from soils which had imprints of interred bodies with those of the HS from the gravefill soil well removed from the body remains.

2 Materials and Methods

Figure 1 gives a diagrammatic representation of the outline in the soil of a body as seen in

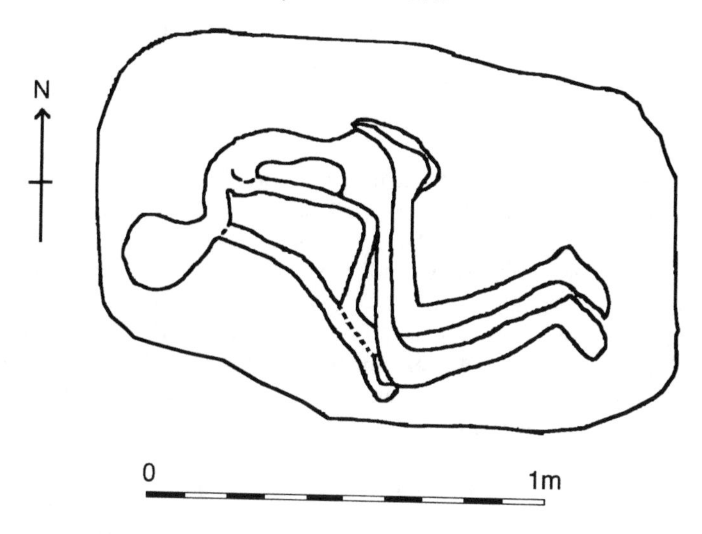

Figure 1 *Diagrammatic representation of the outline of a human body as seen in one of the internment sites at Sutton Hoo*

the graves. Sampling was achieved by dividing the outline of the body into two equal parts. The outline was then sectioned into four parts and samples were taken at 5 cm intervals along the lines of division. This pattern was repeated every 5 cm in depth. Samples were also taken from the gravefill soils removed from the locations of the body. Sampling was carried out by the Field Archaelogy Unit of the University of Birmingham.

Extractions With Sodium Pyrophosphate (pH 7)

Soil in sodium pyrophosphate (0.1M), adjusted to pH 7 with 6M HCl (1:10 ratio) was stirred for 3 h , and then centrifuged (13 000 g, 30 min). The supernatant was adjusted to pH 1 using 6M HCl. The residues were further extracted in the same way until the supernatants were virtually clear, and the combined acidified supernatants were allowed to stand overnight at 4 °C, then centrifuged (13 000 g, 30 min) to sediment the humic acids (HA). The fulvic acids (FA) fractions were contained in the supernatants.

Extractions With Sodium Hydroxide

Soils in plastic bottles were mixed with 0.1M HCl in 1:10 ratios and equilibrated (using a roller) for 12 h, then centrifuged (13 000 g, 30 min). The residue was washed with distilled water, centrifuged, and the supernatant was combined with that from the HCl extract. The soil residue, in a ratio of 1:10, was mixed with a 0.1M NaOH (carbonate-free) solution (prepared under an atmosphere of dinitrogen gas) and stirred for 3 h (under an atmosphere of N_2), then centrifuged, and the supernatant was immediately adjusted to pH 1.0 (using 6M HCl). This extraction process was repeated until the supernatants were virtually clear. The acidified supernatants were combined, let stand overnight at 4 °C (to allow flocculation to take place), then centrifuged to give the HA and the FA fractions. The FA solutions were combined with the acid washings.

Extractions With Acidified Dimethyl Sulphoxide (DMSO)

Soil was mixed (in a ratio of 1:10) with a mixture of 94 parts DMSO and 6 parts concentrated HCl (Fagbenro *et al*., 1985), rolled in plastic bottles for at least 12 h, and then centrifuged (13 000 g, 30 min). Extraction of the residue was continued in this way until the supernatant was virtually clear. All supernatant solutions were combined, the pH was adjusted to 2.0, and the mixture was passed onto a column of XAD-8 [(poly) methylmethacrylate] resin. The column was washed free of DMSO using 0.1M HCl, and the sorbed HS were recovered by back eluting with 0.1M NaOH. The pH of the back eluate was immediately adjusted to 1.0 (using HCl), and after settling overnight (at 4 ^{o}C) the HA fraction was recovered by centrifugation.

Treatment of the Fulvic Acid Fractions

All FA fractions were passed onto XAD-8 resin columns. The FAs sorbed to the resin and the polar fractions, which included sugar- and amino acid-containing residues, were washed through. The resin was washed with water till the effluents were chloride free (silver nitrate test), the FAs were desorbed from the resin columns by back eluting with 0.1M NaOH, and the eluate was immediately passed through H^+-exchanged IR-120 resin to remove the exchangeable sodium ions and to avoid oxidation in the alkaline medium. The H^+-exchanged FAs were freeze dried.

Treatment of the Humic Acid Fractions

HA fractions were dissolved in 99 parts DMSO and one part concentrated HCl (Law *et al*., 1985), then passed onto XAD-8 resin, and subsequently treated as described for the FAs.

Potentiometric Titrations

The procedure involved acid-base titrations carried out in the presence and in the absence of the HS (Borggaard, 1974). The distilled water was carbonate free (prolonged boiling), and Carbosorb was used to render the N_2 gas free from CO_2. HS samples (20 mg) were each dissolved in 10 mL of NaOH (0.1M) and 10 mL of NaCl (0.2M) under an atmosphere of dinitrogen gas, and the mixture was covered with cling film (perforated to allow the escape of gases) and stirred using a magnetic follower. The pH was recorded after equilibrium had been reached following additions of 0.1M HCl, using a PT 1-20 (Fisons) Digital Water Analyser pH meter fitted with a Russel CETL combination glass/reference electrode. Titration was continued to pH 2. Cation exchange capacity (CEC) data at different pH values were calculated from

$$\text{CEC (cmol H}^+\text{ kg}^{-1}) = \text{M} \times 100(V_b - V_s)/W \qquad \text{where}$$

M is the molarity of the sodium hydroxide solution;
V_b is the volume of HCl required to neutralise 20 mL of the blank solution;
V_s is the volume of HCl required to neutralise 20 mL of the HS solution; and
W is the weight of the sample (g).

E4/E6 Ratio Values

Humic samples (2 mg), dissolved in sodium bicarbonate solution (0.025M) adjusted to pH 7, were made up to 10 mL and the absorbances were measured at 465 nm and at 665 nm using a linear UV Spectrophotometer CE 272 (Cecil Instruments).

Sugar Analyses

HA samples were hydrolyzed in sealed tubes using 2M trifluoroacetic acid at 100 °C for 6 h. The hydrolysate was then injected onto Aminex A-28, a strong anion-exchange resin, and the solution phase was 0.13M boric acid. The sugars were eluted as their borate complexes using a borate buffer gradient of increasing pH and ionic strength, and the neutral sugars were detected colorimetrically (at 425 nm) using orcinol/concentrated sulphuric acid reagent, and heating to 90 °C for 5 min (Kennedy and Fox, 1980).

Amino Acid Analyses

Prior to amino acid analyses of the HAs, the samples were degassed by alternately freezing and thawing under vacuum, and O_2-free N_2 gas was bubbled through to remove O_2. Hydrolysis was carried out in evacuated sealed tubes using boiling HCl at 110 °C for 24 h. Solutions of the amino acids at pH 2.0 were injected onto a strong cation exchange column, and the amino acids were eluted stepwise by increasing the pH and ionic strength of the eluent. The amino acids were detected colorimetrically using ninhydrin.

Humic Substances Associated With the Soil Mineral Components

The soil components were separated on the basis of particle size differences by mixing soil with water in a ratio of 1:2.5, agitating for 30 min using a magnetic follower, ultrasonicating for 30 min, and then filtering through a 50 micron sieve. The sand material was retained on the sieve, and the silt and clay fractions were washed through and separated by sedimentation. Sedimentation was allowed to take place for 8.5 h at 20 °C, and the top 10 cm (containing clay) was siphoned off. This process was repeated until the top 10 cm was clear after 8.5 h of sedimentation time. Clay samples were subjected to X-ray diffraction for qualitative identifications of the clays present. Samples were treated with 30 per cent hydrogen peroxide to remove the organic matter, then washed with a 1M magnesium chloride solution, and centrifuged. For exchange with potassium, 1M potassium chloride was used, followed by centrifugation and washing with 50 per cent methanol, 95 per cent ethanol, and finally with 95 per cent acetone until the samples were chloride free. Then samples, suspended in acetone, were transferred to glass slides, allowed to dry, and subjected to X-ray diffraction analysis using a PW 1720 X-ray generator attached to a PW 1390 Channel Control and a PW 1373 Goniometer Supply. This instrument uses a fully collimated X-ray beam from a Cu K_α source. Sample treatments included air drying, equilibration of the Mg^{2+}-exchanged clays with glycerol in the vapour phase (McEwan, 1961), and heating the K^+-exchanged clays.

The silt (residue after removing the clay)- and clay-size fractions were freeze dried. The ash contents of the sand, silt, and clay-size fractions were determined from the loss in weight from heating the samples to constant weight in a furnace at 660 °C. C, H, and N analyses were carried out on the clay-size fractions.

Aqueous suspensions (3% w/v, in distilled water) of the silt- and clay- size fractions were subjected to particle size analyses using a Micrometric Sedigraph 5000ET Particle Size Analyser.

3 Results and Discussion

Three samples from the body 'imprint' and two from the gravefill of Sutton Hoo burial sites were combined for analyses of the sand, silt, and clay contents of the soils. The data in Table 1 show that, although the total contents were small, there were enrichments of the fine fractions (silt- and clay- sized) in the body 'imprint' soils compared with those from the gravefill. Also, measurements made using the Sedigraph indicated that the proportions of clays in the fine clay range were significantly greater in the cases of the body 'imprint' soils (73% and 53%, respectively, and 53% and 35%, respectively, were less than 2 μm and 1 μm in the cases of the 'imprint' and gravefill clays). All of the clays should fall in the < 2 μm range. However, some associations of clay size particles take place in the drying process for the recovery of the clays following isolation by the sedimentation process, and it is probable that all of these associations were not dispersed by the stirring process used in preparing the samples for the Sedigraph measurements. Because the same procedures were used for the 'imprint' and gravefill samples, it can be considered that the differences observed were meaningful. X-ray diffraction analyses of the clays gave similar but complex diffraction patterns, indicating mixtures of minerals in the clay size fractions. From Table 1 it is seen that the organic matter (OM) was primarily associated with the clay sized materials, and lesser amounts were associated with the silt sized components. It is likely that most of the OM in the silt sized materials was associated with clay aggregates or domains that had not been fully dispersed. The elemental analyses of the clays from the 'imprint' and gravefill samples show that a significant enrichment of N was contained in the 'imprint' clays.

Yields, and the E_4/E_6 ratio values are given in Table 2 for the HAs and FAs isolated from soils with body 'imprints' from the Sutton Hoo (SHB) and the Snape (SN) sites, and from gravefill (SHG) soils from Sutton Hoo. Extractions were with NaOH, DMSO, and neutral pyrophosphate solvent systems. Yields were significantly greater for the Snape than for the Sutton Hoo samples. Where comparisons can be made, yields of HAs decreased in the solvent order NaOH > DMSO > neutral pyrophosphate, and the abundances of the FAs from the different solvents were less diverse (than those for the HAs). A complete set of data are available for the E_4/E_6 ratio values only in the case of the humic isolates from the Snape soil, and the values in that case are surprisingly close for all of the fractions from all of the solvents.

Valid comparisons can be made between the data for the SHB and the SHG fractions. With the exception of the FA content of the NaOH isolate of SHB1, it is obvious that the body 'imprint' samples are significantly enriched in HAs and FAs compared with the SHG samples. Although there are some exceptions, the E_4/E_6 ratio values are less for the samples from the gravefill than for those from the body 'imprint' soils.

There is not a consensus with regard to the interpretations of the compositional or macromolecular conformational significances of E_4/E_6 ratio values, and we are avoiding assigning chemical meaning to the values obtained. We present the data only to show similarities and differences (in the data) for the different samples.

The carbon contents (Table 3) of the HAs isolated in NaOH were lower than might be

Table 1 *Sand, silt, clay, and organic matter (OM) contents (%) in the soil components isolated by sieving and sedimentation processes from body 'imprint' (BI) and gravefill (GF) soils from Sutton Hoo, and the C, H and N contents of the BI and GF clay fractions*

Sample	Sand (%)	OM (%)	Silt (%)	OM (%)	Clay (%)	OM (%)	C (%)	H (%)	N (%)
							Clay Samples		
BI	93.6	1.2	5.5	16.9	0.9	37.9	36.9	5.3	6.0
GF	94.9	0.7	4.4	7.8	0.7	30.6	32.1	6.5	3.9

expected, when compared with the data for the samples isolated in DMSO and neutral pyrophosphate. With two exceptions, the N contents are in the range of 3% to 4% for the 'imprint' and gravefill HAs. Based on the data in Table 1, it might be expected that the N contents of the body 'imprint' HAs would be greater than those for the gravefill HAs.

Comparisons of the cation exchange capacity (CEC) data (Table 4) show that the HAs from SHB1 isolated in DMSO and in neutral pyrophosphate have higher total carboxyl contents than those extracted in NaOH. The carboxyl groups will have, for the most part, dissociated at pH 7, but HS contain very strong acid groups which are dissociated at pH values even below 3 (Leenheer *et al.*, 1995).

The differences in the CEC values can be attributed to higher contents of very strong acid groups in the samples isolated in DMSO and in neutral pyrophosphate. A full set of data are given for these values only in cases of the HA and FA components isolated from the Snape soil in the three solvent systems. The trends seen for the CEC values for the HAs from the Snape soil at pH 4 have some similarities with those from SHB1. However, the carboxyl contents, as indicated by the CEC values at pH 6 and 7, are relatively low for the HAs extracted with NaOH. The abundance of strong acids is greatest for the material isolated in DMSO, and it is these strong acids which are responsible for the differences at the higher pH values in the CEC values of the HAs from the DMSO and pyrophosphate extracted materials.

The data for the Snape soil show that the CEC values for the FAs isolated with the different solvents are very similar at the different pH values, and that carboxyl functionalities are the major contributors to the acidities. As expected, the total acidic functionalities are more abundant for the FAs than for the HAs, but this difference can be attributed entirely to the carboxyl functionalities. For the most part, contributions to the CEC values above pH 7 can be attributed to activated and to non-activated phenolic functionalities. Comparisons of the differences in the CEC values for the HAs and FAs at pH 7 and at pH 10 show that the differences are always greatest in the cases of the HAs. The data for the SHB2 and SHG samples are similar.

In summary, the CEC data do not give clear indications with regard to the origins (body 'imprint' or gravefill) of the HAs. Data for SHB1 and SN would suggest that the 'imprint' HAs isolated in base are low in strong acid functionalities, but this is not reflected in the data for SHB2. Also, the data for the HAs isolated in DMSO and in neutral pyrophosphate from the SHB1 and SN indicate that the HAs isolated in these solvents have contents of strong acids similar to those in the base extracts of SHB2 and of the gravefill HAs.

The data in Table 5 show that the six neutral sugars (NS) analysed account for 0.79 to 8.63 per cent of the masses of the HAs. The DMSO extract was least enriched in sugars,

Table 2 *Yields (*mg g^{-1} of soil), and some E_4/E_6 ratio values of humic (HA) and fulvic (FA) acids from Sutton Hoo body 'imprint' (SHB), gravefill (SHG), and Snape (SN) soil samples. Extractions were with 0.1M sodium hydroxide (NaOH), dimethylsulphoxide [DMSO, (94%)/HCl (6%)], and with 0.1M Na^+-pyrophosphate (Pyro)*

Sample	Solvent used for extraction											
	NaOH				DMSO				Pyro			
	*HA	E_4/E_6	*FA	E_4/E_6	*HA	E_4/E_6	*FA	E_4/E_6	*HA	E_4/E_6	*FA	E4/E6
SHB1	3.6		0.9		1.3		1.3		0.6		1.0	
SN	15.0	7.3	6.8	8.3	9.5	7.4	6.7	8.4	6.3	7.9	8.8	8.7
SHB2	5.0	6.3	2.3									
SHB3	4.9	5.7	1.7									
SHB4	4.2	5.3	6.2									
SHB5	4.3	3.0	4.9									
SHG1	2.3	3.6	0.6	4.6								
SHG2	0.7	3.0	1.4	4.0								
SHG3	0.8	----	0.2	3.8								
SHG4	0.5	----	0.3	----								
SHG5	0.7	----	0.4	----								

Table 3 *Carbon (C), hydrogen (H), and nitrogen (N) compositions (%) of the H^+-exchanged humic acid (HA) samples isolated, using 0.1M sodium hydroxide (NaOH), dimethylsulphoxide [DMSO, (94%)/HCl (6%)], and 0.1M Na^+-pyrophosphate (Pyro) from Snape (SN) and Sutton Hoo body 'imprint' (SHB) and gravefill (SHG) soil samples*

Sample	Extractant								
	NaOH			DMSO			Pyro		
	C(%)	H(%)	N(%)	C(%)	H(%)	N(%)	C(%)	H(%)	N(%)
SN	47.3	4.9	3.2	48.3	3.2	3.0	48.7	4.7	3.8
SHB1	47.9	---	---	51.9	5.0	4.8	51.2	3.6	3.6
SHB2	48.1	4.3	1.8						
SHB3	49.0	3.9	3.5						
SHB4	48.8	3.9	3.2						
SHB5	48.9	3.9	3.2						
SHG1	44.1	4.0	1.9						
SHG2	47.7	4.5	3.6						
SHG3	47.0	4.5	3.4						

Table 4 *Cation exchange capacity (CEC) data (cmol H^+ kg^{-1}) for humic (HA) and fulvic (FA) acids from the NaOH (0.1M), dimethylsulphoxide [DMSO, (94%)/HCl (6%)], and sodium pyrophosphate (Pyro, 0.1M) extracts of the body 'imprint' Sutton Hoo (SHB) and Snape (SN) soils*

Soil	pH	CEC (cmol kg^{-1})					
		NaOH		DMSO		Pyro	
		HA	FA	HA	FA	HA	FA
SHB1	4	1.9	---	3.5	----	3.6	---
	5	2.9	---	4.2	---	4.2	---
	6	4.1	---	5.1	---	4.7	---
	7	4.6	---	5.3	---	5.3	---
	8	4.8	---	5.5	---	5.7	---
SHB2	4	3.7					
	6	5.3					
	8	6.6					
SHG1	4	4.1					
	6	5.2					
	8	5.8					
	10	6.6					
SHG2	4	3.5					
	6	5.0					
	8	6.2					
SN	4	1.4	6.0	5.0	5.7	3.7	5.7
	5	2.2	7.2	5.9	7.0	4.7	6.9
	6	2.8	8.0	6.5	7.8	5.5	7.7
	7	3.1	8.5	6.7	8.3	6.0	8.4
	8	3.5	8.8	7.0	8.7	6.3	8.9
	9	4.0	9.2	7.4	8.9	6.7	9.2
	10	5.9	9.8	8.4	9.3	7.2	9.5

and surprisingly perhaps, the gravefill soil HAs were more highly enriched than the HAs from the body 'imprint' Snape soil samples. DMSO is a good solvent for saccharides. However, the polar saccharide materials would not be retained by XAD-8 resin used to recover the HS from the DMSO unless these were bound to the more hydrophobic HAs. Häusler and Hayes (1996) have shown that the sugar contents of HAs are decreased when these fractions are subjected to the DMSO-XAD-8 treatment. That would indicate that saccharide materials that are not covalently linked to the HAs are separated by the treatment. Because all of the HAs listed in Table 5 had been subjected to the DMSO-resin treatment, it is plausible to assume that the saccharides detected were covalently linked to the HAs. The fact that the lowest sugar contents were in the DMSO extract might suggest that HA-carbohydrate associations were least in the HA fractions soluble in DMSO. Alternatively, carbohydrate materials co-extracted with the HAs in the pyrophosphate and

Table 5 *Neutral sugar (NS) contents (μg mg^{-1}), and contribution (%) of each NS to the total NS contents of humic acids extracted with sodium pyrophosphate (0.1M, pH 7), dimethyl sulphoxide (DMSO), and NaOH (0.1M) from Snape soil, and with 0.1M NaOH from Sutton Hoo gravefill soil samples SHG2, SHG4, and SHG5*

Sugars	Snape Humic Acid samples						Gravefill Humic Acid samples					
	Pyrophosphate		NaOH		DMSO		SH G2		SHG4		SHG5	
	(μg mg^{-1})	(%)	(μg mg^{-1})	(%)	(μg mg^{-1})	(%)	(μg mg^{-1})	(%)	(μg mg^{-1})	(%)	(μg mg^{-1})	(%)
Rhamnose	1.8	*9.3*	1.1	*8.8*	0.6	*7.6*	8.7	*10.1*	4.9	*9.4*	3.4	*8.7*
Fucose	2.9	*15.2*	2.2	*17.6*	1.0	*12.6*	16.0	*18.5*	9.4	*17.9*	6.8	*17.5*
Xylose	0.9	*4.7*	0.8	*6.4*	0.5	*6.3*	6.6	*7.6*	3.5	*7.4*	2.6	*6.7*
Mannose	4.8	*24.9*	2.7	*21.6*	1.9	*24.0*	20.0	*23.2*	8.9	*17.0*	7.1	*18.2*
Galactose	3.1	*16.1*	1.9	*15.2*	1.1	*13.9*	11.0	*12.7*	6.7	*10.7*	6.0	*15.4*
Glucose	5.8	*30.5*	3.8	*30.4*	2.8	*35.4*	24.0	*27.8*	19.0	*36.2*	13.0	*33.4*
Total	19.3		12.5		7.9		86.3		52.4		38.9	

Table 6 *Amino acid (AA) contents (μg mg^{-1}) of humic acid (HAs) samples isolated from body 'imprint' (SHB) and grave fill (SHG) soil samples from Sutton Hoo, and of HAs isolated from the Snape soil in 0.1M sodium hydroxide (SN NaOH), in dimethyl sulphoxide (SN DMSO), and in neutral (0.1M) sodium pyrophosphate (SN Pyro)*

Sample	Acidic			Basic				Neutral Hydophilic (NH_o)								Neutral Hydrophilic (NH_i)					
	Asp	Glu	*Total Acidic*	Arg	His	Lys	*Total Basic*	Pro	Val	Ile	Leu	Tyr	Phe	*Total NH_o*	Thr	Ser	Gly	Ala	*Total NH_i*	**Total AA**
SHB3	17.4	14.8	*32.2*	5.0	4.7	4.1	*13.8*	6.5	7.2	4.6	7.4	2.4	9.0	*37.1*	9.1	7.8	9.9	9.0	*35.8*	**118.9**
SHB4	18.8	17.8	*36.6*	5.7	5.5	5.0	*16.2*	6.6	8.0	5.5	9.8	2.7	9.9	*42.5*	9.9	8.8	11.4	9.9	*40.0*	**135.3**
SHB5	12.5	11.1	*23.6*	3.7	4.7	3.4	*11.8*	3.8	5.0	3.2	5.7	1.8	7.5	*27.0*	5.6	6.0	7.3	5.9	*24.8*	**87.2**
SHG3	12.2	10.5	*22.7*	4.3	5.4	4.0	*13.7*	3.4	4.8	2.9	5.1	-	4.8	*21.0*	6.1	6.7	7.3	8.3	*28.4*	**85.8**
SHG4	12.2	12,0	*24.2*	4.6	6.0	4.6	*15.2*	3.5	5.0	2.9	5.2	-	4.5	*21.1*	6.4	6.5	8.4	6.6	*27.9*	**88.4**
SHG5	12.5	11.1	*23.6*	4.3	5.5	3.9	*13.7*	3.8	5.1	3.0	5.5	-	5.0	*22.4*	6.5	6.5	7.5	6.6	*27.1*	**86.8**
SN NaOH	6.7	5.6	*12.3*	1.2	2.0	1.6	*4.8*	1.3	2.4	1.5	2.0	-	-	*7.2*	2.4	2.7	3.2	2.6	*10.9*	**35.2**
SN DMSO	3.4	3.4	*6.8*	0.3	0.7	0.8	*1.8*	1.3	1.2	0.7	1.0	-	-	*4.2*	1.0	0.9	1.4	1.3	*4.6*	**17.4**
SN Pyro	8.8	7.4	*16.2*	2.1	2.8	3.0	*7.9*	2.0	3.0	1.7	2.4	-	-	*9.1*	3.6	3.6	4.0	3.3	*14.5*	**47.7**

sodium hydroxide extracts became associated with the HAs in the work-up process and were not separated from these by the DMSO and XAD-8 column treatment.

The order of the abundances of the NS was similar (glucose > mannose > galactose = fucose > rhamnose > xylose) for the SN HAs isolated in pyrophosphate, NaOH and DMSO/HCl solutions. The general order of abundances was slightly different for the Sutton Hoo gravefill samples (glucose > mannose ≥ fucose > galactose > rhamnose > xylose), and these orders, though similar to that for the sugars in the HAs from a Mollisol soil (Clapp and Hayes, 1996) were significantly different from the order for NS in HS from waters (Watt *et al*., 1996), and from the HAs from other soils in Britain (Häusler and Hayes, 1996; Hayes, 1996).

Xylose is primarily a sugar of plant origins, and galactose and mannose are found in animal and microbial tissues. The surprisingly low contents of xylose would suggest that the imput from plant sources to the carbohydrate contents was small. This is also reflected in the low ratios of Xyl/(Man + Gal) in a saccharide mixtures, and it can be calculated from the data in Table 5 that these ratio values are 0.114, 0.174, and 0.167 for the pyrophosphate, NaOH, and DMSO extracts of the Snape HAs, and 0.21, 0.224, and 0.198 for the gravefill samples SHG2, SHG4, and SHG5. The slightly lower values for the SN body 'imprint' samples might be indicative of the lesser contribution of plant sources to the genesis of the carbohydrates. The lowest value for the pyrophosphate isolated sample could be expected because pyrophosphate extracted the most highly (biologically) oxidized humic materials. The ratios for the NaOH and DMSO extracts of the Snape soil approach the values for the SHG samples, and these might be considered to have more inputs from plant sources than the pyrophosphate extracted Snape HAs.

Other studies within our group (Hayes, 1996) have shown that as the pH of the extracting solution is increased, the sugar contents of the HAs increase. This arises because the more highly humified materials are extracted at the lower pH values.

In general the relative order of the abundance of each amino acid in the different samples was similar, as follows (Table 6):

Aspartic acid (Asp) = Glutamic acid (Glu) > Glycine (Gly) = Alanine (Ala) = Phenylaniline (Phe) = Threonine (Thr) = Serine (Ser) = Leucine (Leu) = Valine (Val) > Proline (Pro).

The amino acids (AAs) are categorized in Table 6 into acidic, basic, neutral hydrophobic (NH_o), and neutral hydrophilic (NH_i) groups. Tyrosine (Tyr), because of the phenolic substituent on the β carbon, is sometimes placed in the acidic group, but this can be done legitimately only when the pH of the medium is such that the phenolic OH is dissociated. Only aspartic (Asp) and glutamic (Glu) acids, which bear carboxyl functionalities on the β and γ carbon atoms, are classified in the acidic group in the present study. Data are available for Tyr contents in the cases of samples SHB3, SHB4, and SHB5, and it can be seen that the abundances of Tyr were least among the AAs listed. The abundances of the S-containing AAs were very small, in the range 0.04 to 0.1 $\mu g\ mg^{-1}$.

The differences in the relative abundances of the groups of AAs are more evident than the relative abundance of each AA. In the cases of HAs from the body 'imprint' samples, the order of abundances of the AA groups was:

total neutral hyrophobic AA ≥ total neutral hydrophilic AA > total acidic AA >> total basic AA >>> S-containing AA.

Should Tyr be assigned to the acidic group, the abundances of the total AA contents in the neutral hydrophilic, neutral hydrophobic, and acidic groups would be similar.

The abundances of the AA groups in the Sutton Hoo gravefill HAs (SHG series, Table 6) followed the order:

total neutral hydrophilic AA > total acidic AA ≥ total neutral hydrophobic AA > total basic >>> S-containing AA.

Although the AA content in SHB5 was similar to the contents in SHG3, SHG4, and SHG5 (8.58% to 8.84%), the contents in these samples were significantly lower than for the HA samples from the body 'imprint' soils (SHB3 and SHB4, 11.89% and 13.53%, respectively). The sample numbers 3, 4, and 5 represent locations at increasing depths in the grave, and whereas SHB3 and SHB4 were taken from areas which were clearly 'imprinted' from the body remains, sample SHB 5 was taken immediately below the 'imprint' region; hence the similiraty in the AA compositions of this sample and SHG samples 3, 4, and 5. However, the similarity in the relative abundances of the different AA groups in SHB5 was similar to that in SHB3 and SHB4, indicating that the body remains had influenced the AA composition of the HAs in SHB5.

The AA contents of the Snape soil HAs extracted in sodium pyrophosphate (pH 7), sodium hydroxide, and DMSO have the same order of relative abundances (in each extract) as the neutral sugars (Table 6), and the contributions of the AAs to the composition of the Snape soil HAs (maximum 4.77% for the amino acids in the pyrophosphate extracts) were significantly lower than for the HAs in the NaOH extract of the Sutton Hoo body (SHB) and gravefill (SHG) samples. The abundances of the groups of AAs in the sequential extractions (with neutral pyrophosphate, NaOH, and DMSO) were the same:

total acidic > total neutral hydrophilic > total neutral hydrophobic > total basic.

That order was significantly different from that for the samples from the Sutton Hoo sites, but the differences could be attributed to the relatively high abundances of Asp and Glu. The order was, possibly fortuitously, more in line with that for the HAs from a stagno-dystric gleysol (Hayes, 1996). Data for phenylaniline (Phe) are not available for this series, but that data, if available, would be likely to alter the order in the case of the HAs isolated from DMSO only.

We are not in a position to draw conclusions about the origins of the AAs from the data presented. The AA contents of the HAs from the Sutton Hoo 'imprint' and gravefill soils were in line with the abundances expected for extracts from soils, and up to 10 times greater than the amounts in aquatic HAs (Watt *et al.*, 1996). The abundance in the 'imprint' samples was greater than that in the gravefill samples (Sutton Hoo) and that can reflect the additional contributions from the body sources. The AA contents of the Snape samples were significantly less. Unfortunately we were unable to compare the compositions of gravefill and 'imprint' HA samples in the case of the Snape site.

4 Conclusions

There are definite differences between the compositions of the humic substances (HS) fractions isolated from body 'imprint' and gravefill sources at Sutton Hoo. However, there are also significant differences between the compositions of the 'imprint' samples from

Snape and those from Sutton Hoo. The HS formed were associated with the soil mineral colloids, and that association will have been important for their preservation. The data indicate that, where extensive soil mixing does not take place, the properties of the substrates which give rise to the HS, and the soil inorganic colloidal components will influence the compositions of samples of archaeological interest.

Acknowledgements

We are grarteful for the support which the Leverhulme Trust provided for this work.

References

Borggaard, O.K. 1974. Experimental conditions concerning potentiometric titrations of humic acid. *J. Soil Sci.* **25**:185-195.

Clapp, C.E. and M.H.B. Hayes. 1996. Isolation of humic substances from an agricultural soil using a sequential and exhaustive extraction process. p. 3-11. In C.E. Clapp, M.H.B. Hayes, N. Senesi, and S.M. Griffith (eds), *Humic Substances and Organic Matter in Soil and Water Environments.* IHSS, Univ. of Minnesota, St. Paul.

Fagbenro, J., M.H.B.Hayes, I.A. Law, and A.A. Agboola. 1985. Extraction of soil organic matter and humic substances from two Nigerian soils using three solvent mixtures. p. 22-26. *In* M.H.B. Hayes and R.S. Swift (eds), *Volunteered Papers 2nd Intern Conf. International Humic Substances Society* (Birmingham, 1984). Published by IHSS.

Häusler, M.J. and M.H.B. Hayes. 1996. Uses of the XAD-8 resin and acidified dimethylsulfoxide in studies of humic acids. p. 25-32. *In* C.E. Clapp, M.H.B. Hayes, and N. Senesi (eds), *Humic Substances and Organic Matter in Soil and Water Environments.* Proc. 7th Intern. Conf. IHSS (St. Augustine, Trinidad, 1994).

Hayes, T.M. 1996. A study of the humic substances from soils and waters and their interactions with anthropogenic organic chemicals. PhD Thesis, The University of Birmingham

Kennedy, J.F. and J.E. Fox. 1980. Fully automatic ion-exchange chromatographic analysis of neutral monosaccharides and oligosaccharides. *Methods in Carbohydrate Chemistry* **8**:3-12.

Law, I.A., M.H.B. Hayes, and J.J. Tuck. 1985. Extraction of humic substances from soil using acidified dimethyl sulphoxide. p. 22-26. *In* M.H.B. Hayes and R.S. Swift (eds), *Volunteered Papers 2nd Intern Conf. International Humic Substances Society* (Birmingham, 1984). Published by IHSS.

Leenheer, J.A., R.L. Wershaw, and M.M. Reddy. 1995. Strong-acid, carboxyl-group structures in fulvic acid from the Suwannee River, Georgia. 1. Minor structures. *Environmental Sci. and Technol.* **29**:393-398.

McEwan, D.M.C. 1961. Montmorillonite minerals. p. 143-207. *In* G. Brown (ed.), *The X-ray Identification and Crystal Structures of Clay Minerals.* Mineralogical Soc. London.

Watt, B.E., R.L. Malcolm, M.H.B. Hayes, N.W.E Clark, and J.K. Chipman. 1996. Chemistry and potential mutagenicity of humic substances in waters from different watersheds in Britain and Ireland. *Water Res.* **30**:1502-1516.

Humic Substances from Interment Sites II. Digest Products of Sequential Degradation Reactions

Lorraine J. Stewart, J. Burdon, and M.H.B. Hayes

THE UNIVERSITY OF BIRMINGHAM, SCHOOL OF CHEMISTRY, EDGBASTON, BIRMINGHAM B15 2TT, ENGLAND

Abstract

Humic acids (HAs), isolated from the sodium hydroxide extracts of gravefill and body imprint soil samples from the medieval burial site at Sutton Hoo, were subjected to sequential degradations using sodium amalgam (NaA), sodium in liquid ammonia (NaLA), and alkaline permanganate (AP). Ether soluble products were isolated from the digests, methylated, and identified by GCMS. Considerations of the degradation mechanisms that would apply led to indications of the types of structures in the HAs that could give rise to the volatile methylated products identified. A comparison was made between the digest products of HAs isolated using sodium hydroxide, dimethylsulphoxide and hydrochloric acid, and neutral sodium pyrophosphate solutions from a body 'imprint' soil, and the HAs isolated in sodium hydroxide from a sapric histosol. The results indicated differences in the origins of some of the volatile compounds, and in the ratios of these compounds in the gravefill (which had more lignin derived structures) and body 'imprint' samples (with more microbially derived components). The compounds identified in the digests of the sapric histosol were more in line with the conventional degradation products of soil HAs than with the HAs from the grave sites.

1 Introduction

The linkages between the component molecules in the 'core' or 'backbone' structures of humic substances (HS) are difficult to cleave. Hydrolyzable components, such as carbohydrate and peptide moieties, with origins in genetically or biologically controlled species, can form covalent linkages with the 'core' structures of HS (Hayes and Swift, 1978), and the amino acid and sugar components in the HS hydrolyzates can account for as much as 15 to 20 per cent of the humic isolates. The use of resins, such as XAD-8, a (poly) methyl methacrylate, allows saccharide and peptide structures that are not covalently linked to the humic 'core' to be separated (Häusler and Hayes, 1996), and that can give rise to significant decreases in the carbohydrate and peptide components in HS fractions. The majority of the linkages between the structural units, or 'building blocks' of HS are composed of carbon to carbon bonds, or of ether-type linkages which are cleaved only when high inputs of energy are applied. Such inputs inevitably give rise to changes in the compositions of the molecules released, and changes will continue to take place during the residence time in the degradation digests of the components released. When highly energetic degradative procedures are applied, such as those needed to break the C-C or the C-H bonds of hydrocarbons, the products released can be meaningless in terms of structural information and, in the extreme, only CO_2 and H_2O are given off. However, the energy requirements to cleave carbon to carbon double and triple bonds are significantly less, and some substituents can lower the activation energy necessary to cleave carbon to carbon single bonds alpha to the substituents. In general it can be said that the greater the energy applied to cleave structures, the farther removed will be the compounds in the digests from the molecules which compose the macromolecular structures. However, an understanding of the mechanisms of the degradation processes allows predictions to be made of the types of molecules which could be the precursors of the compounds identified.

It is appropriate to consider the uses of degradation processes which employ sequences of reagents and conditions that provide increasing energy inputs. In the present study, a sequence of relatively mild reductive degradation processes was followed by a more vigorous oxidative procedure for studies of the compositions of humic acids (HAs) isolated from burial sites at Sutton Hoo (near Woodbridge, Suffolk, England; see the preceding paper (Part I) in this Volume). Degradation with sodium amalgam, introduced for the degradation of HS by Burges *et al.* (1964), was followed by treatments of the non-ether soluble residues with sodium in liquid ammonia (Maximov and Krasovskaya, 1977), and finally the residues from that treatment were further degraded using alkaline permanganate [this procedure is discussed in some detail by Hayes and Swift (1978)].

2 Materials and Methods

Isolation of Humic Acids

HAs were isolated from the sodium pyrophosphate (pH 7), sodium hydroxide (0.1M), and DMSO/HCl soil extracts, as described in the previous paper (Part I in this series by Hayes, Stewart, and Bethel, p. 136 - 147 in this Volume).

Degradation with Sodium Amalgam (NaA)

Clean sodium (7.5 g) in dry toluene (45 mL) was stirred (magnetically) and heated (hot plate) till melting occurred. After removing the heat, mercury (150 g) was added slowly, dropwise, and a vigorous reaction ensued. When the reaction had subsided the amalgam was poured into a mortar (sealed with a rubber cap) and pulverized using a pestle. Humic acid (HA, 250 mg) was dissolved in sodium hydroxide (0.5M, 75 mL) in a round bottom flask and sodium amalgam (ca 30 g) was added. The mixture was stirred and the initial vigour of the reaction was allowed to subside before heating to 100 - 110 ^{o}C. The remainder of the amalgam was then added in 30 g portions over 1 h, and the reaction was continued for a further 3 h. The reduced HAs were separated from the residual amalgam and acidified to pH 1.0 with 6M HCl, and the mixture was allowed to stand for 1 h with intermittent stirring. Then the mixture was extracted with at least 3 x 75 mL portions of dry diethyl ether. The ethereal solution was dried over anhydrous magnesium sulphate, concentrated (rotary vacuum evaporator), and the product was methylated (vide infra) prior to analysis using gas-liquid chromatography, mass spectrometry (GC MS)

Degradation with Sodium in Liquid Ammonia (NaLA)

The procedure used was an adaptation of that by Sartoretto and Sowa (1937). N_2 gas was passed for 15 min through a three-necked round bottom flask with an ammonia condenser attachment. The flask was cooled to -33 ^oC by immersing in a methylated spirit/dry ice bath, and liquid ammonia (50 mL) was carefully added. Clean sodium (1.44 g) was added carefully in small portions and HA (200 mg) was added to the dark solution. The reaction was allowed to proceed for 3 h at -33 ^oC. After 3 h the liquid ammonia was allowed to evaporate at room temperature. The flask was placed in an ice bath and the mixture was carefully hydrolyzed with water, then acidified with dilute hydrochloric acid, left to stand at room temperature for 1 h, extracted with ether, dried with magnesium sulphate, and methylated (vide infra) prior to analysis using GC MS.

Degradation with Alkaline Permanganate (AP)

Methylated (vide infra) HA (1 g) was refluxed with potassium permanganate (250 mL, 4% aqueous, pH 10) for 8 h. Excess $KMnO_4$ was destroyed by careful addition of small volumes of methanol (to give MnO_2). The MnO_2 (insoluble) was separated by filtration, the filtrate was acidified to pH 2 with 6M HCl and allowed to stand for 1 h, then extracted (using a liquid/liquid extractor) for 24 h with 500 mL ethyl acetate. The extract was dried (anhydrous sodium sulphate), the EtOAc was removed (rotary evaporator), and the extracted oxidation products were methylated (vide infra) prior to analysis using GC MS.

Methylation Procedure

Potassium hydride, under N_2 gas in a conical flask stoppered with a rubber septum, was washed with 5 mL portions of hexane and ca 5 mL of dimethylsulphoxide (DMSO) was slowly added over 5 min to give a yellow-green potassium dimsyl solution. This solution (2-3 mL) was slowly added over 5 min to dried HAs in DMSO, and the viscous solution was stirred intermittently for 1 h at room temperature. The solution was cooled to 0 ^oC, and methyl iodide (1 mL) was slowly added. After reaching room temperature the solution was stirred intermittently for ca 2 h. Then it was partitioned between a mixture of 5 mL chloroform and 5 mL water. The aqueous solution was removed and more water (5 mL) was added. The extraction was repeated until the water layer remained clear. The chloroform layer was dried with anhydrous $MgSO_4$, concentrated under vacuum on a rotary evaporator, and then subjected to GC MS.

3. Results and Discussion

Aspects of the method used by Piper and Posner (1972a) were modified in the present study. In order to improve the yields of ether soluble materials following the reduction process, the amounts of HA degraded were increased from 50 (as used by Piper and Posner) to 250 mg, and the amounts of amalgam used were also increased fivefold to 150 g. In the present study, the sodium amalgam was added in batches to sodium humate solutions (Piper and Posner added their HA to the sodium amalgam preparation). Because the amalgam is readily degraded in water (as well as in air) more of the reactive reagent was maintained in the digest by following our addition procedure. The acidified digest was left to stand for at least 1 h before commencement of extraction with ether. That allowed a more extensive exchange of hydrogen for sodium to take place and yields of the order of 50 per cent of the starting HA materials were obtained as ether soluble products.

It is probable that the methyl esters associated with the aliphatic and aromatic functionalities identified were present as the carboxylic acids in the H^+-exchanged HAs. Piper and Posner (1972a; 1972b) considered that cleavages of ether linkages provided a major mechanism for the sodium amalgam degradation of HAs, and they proposed, from model studies, that the cleavages involved atomic hydrogen (H·). Their degradation products were benzene and phenol when diphenylether was subjected to reduction in sodium amalgam. We obtained the same products in low yields when 1,3-diphenoxybenzene was so degraded. However, the involvement of the hydrogen radical in the major degradation mechanism would seem improbable. Should, as was suggested, atomic hydrogen arise from the supply of protons from the aqueous medium and electrons from the sodium, the hydrogen atoms would readily combine to form H_2. The modern view of sodium amalgam reactions is that electron transfer is involved.

In order to investigate further the reaction mechanism, *p*-methoxycinnamic acid was subjected to degradation by the sodium amalgam process. Should the hydrogen radical only be involved, one product (*p*-methoxyphenylpropanoic acid) would be obtained. However, the degradation products included *p*-methoxyphenylpropanoic acid, *p*-methoxybenzoic acid, and *p*-methoxybenzyl alcohol. That would suggest that the reaction mechanism involved electron transfer (from ionization of the sodium) to the olefinic carbons, and protons from water would then give the major product, *p*-methoxyphenyl-propanoic acid. The same mechanism (electron transfer) can be invoked for cleavages of the phenylether structures. The basic medium also plays a part in the degradation process. Attack by the OH^- species on the β-carbon of *p*-methoxycinnamic acid would give rise to the β-hydroxy-*p*-methoxyphenylpropanoic acid (via an α-carbanion intermediate). A reverse aldol reaction would give rise to *p*-methoxybenzaldehyde and ethanoic acid. The former compound would undergo a Cannizarro reaction to give the minor products *p*-methoxybenzoic acid, and *p*-methoxybenzyl alcohol. Because *p*-methoxybenzyl alcohol was not detected among the reaction products it is likely that the Cannizarro reaction was not a major mechanism, and in the presence of OH^- the aldehyde functionality gave rise to the *p*-methoxybenzoate, and to hydride which reacted with water to give H_2 and OH^- (Hayes and O' Callaghan, 1989).

The data in Table 1 indicate that benzoic acid (**I**) was the most abundant degradation product in the NaA digest of the gravefill HA sample, though it was only third in the order of abundance in the case of SH 253 (body 'imprint' HA sample). It is surprising that there are not additional substituents on the aromatic nucleus of the benzoate. Their absence could suggest that the parent structure was an aromatic ether containing one carboxyl group (with the OH functionality being left on the second component of the ether). Compound **II** (1,2,3-trimethoxybenzene) is the most abundant of the volatile methylation products in the NaA digest of the body imprint HA, and is second in abundance in the gravefill sample. It is likely that this compound is the residue from the decarboxylation of gallic acid (3,4,5-trihydroxybenzoic acid) which would be decarboxylated under the digest conditions. Gallic acid would suggest origins in syringyl components of lignin. Compound **III** (methyl 2,4 or 2,6-dimethoxybenzoate - these two compounds cannot be comprehensively distinguished by MS) is not a lignin residue and could have formed from ether substituents in the 2,4- or 2,6-positions. (It is assumed that the digest product, before methylation, was the sodium salt of 2,4- or 2,6-dihydroxybenzoic acid.) Compound **IV**

Table 1 *Volatile compounds identified in the methylated degradation products of humic acids extracted in NaOH (0.5M) from gravefill (SHG 213) and body 'imprint' (SHB 253) soils. Degradations were sequential, using sodium amalgam (NaA), sodium in liquid ammonia (NaLA), and alkaline permanganate (AP) in order. R_1 to R_6 represent the substituents on benzene*

Degradation Regent	Compound Number	Compositions, and percentages (%), by peak areas, of the methylated volatile substances in the digest products
		SH 213
NaA	**I**	$R_1 = COOCH_3$; $R_2 = R_3 = R_4 = R_5 = R_6 = H$ (53.2%);
	II	$R_1 = R_2 = R_3 = OCH_3$; $R_3 = R_4 = R_5 = H$ (24.4%);
	III	$R_1 = COOCH_3$; $R_2 = R_6 = OCH_3$; $R_3 = R_4 = R_5 = H$; or $R_2 = R_4 = OCH_3$; $R_3 = R_5 = R_6 = H$ (7.0%);
	IV	$R_1 = COOCH_3$; $R_3 = R_4 = OCH_3$; $R_2 = R_5 = R_6 = H$ (2.1%);
	V	$R_1 = CH_2COOCH_3$; $R_2 = R_5 = CH_3$; $R_3 = CH(CH_3)_2$; $R_4 = OCH_3$; $R_6 = H$ (3.1%);
	VI	$CH_3(CH_2)_{14}COOCH_3$ (1.4%).
NaLA	**I**	as above (1.6%); **VI** as above (6.0%);
	VII	$R_1 = COOCH_3$; $R_3 = OCH_3$; $R_2 = R_4 = R_5 = R_6 = H$ (0.9%);
	VIII	$R_1 = COOCH_3$; $R_3 = R_5 = OCH_3$; $R_2 = R_4 = R_6 = H$ (1.9%);
	IX	$R_1 = R_2 = COOCH_3$; $R_3 = R_4 = R_5 = R_6 = H$ (1.9%);
	X	$R_1 = COCH_3$; $R_2 = R_3 = R_4 = R_5 = R_6 = H$ (37.6%);
	XI	$CH_3(CH_2)_{16}COOCH_3$ (7.5%).
AP	**VI**	as above (4.5%); **IX** as above (2.2%); **X** as above (20.3%);
	XII	$R_1 = R_4 = COOCH_3$; $R_2 = R_3 = R_5 = R_6 = H$ (2.5%);
	XIII	$R_1 = OCH_3$; $R_2 = R_3 = R_4 = R_5 = R_6 = H$ (1.8%);
	XIV	$R_1 = R_3 = OC_6H_5$; $R_2 = R_4 = R_5 = R_6 = H$ (6.6%);
	XV	$CH_3(CH_2)_{18}COOCH_3$ (4.3%);
		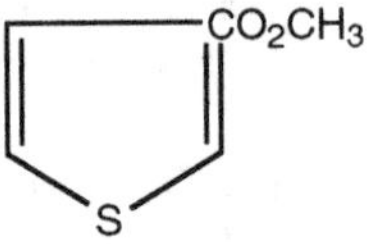 CO_2CH_3, S — **XVI** (0.8%); Unidentified **XVII** (3.9%); Unidentified **XVIII** (2.1%)
		SH 253
NaA	**I**	(8.4%); **II** (13.1%); **III** (1.2%); **IV** (1.5%);
	VI	(2.2%); **VII** (9.4%); **XI** (0.7%);
	XIX	$R_1 = COOCH_3$; $R_2 = R_4 = OCH_3$; $R_3 = R_5 = R_6 = H$ (3.7%) or $R_2 = R_6 = OCH_3$; $R_3 = R_4 = R_5 = H$
	XX	$R_1 = R_3 = OCH_3$; $R_2 = R_4 = R_5 = R_6 = H$ (3.7%)
	XXI	$R_1 = R_4 = OCH_3$; $R_2 = R_3 = R_5 = R_6 = H$ (4.9%)
NaLA	**I**	(2.3%); **X** (80.5%)
AP	**X**	(41.8%); **XVI** (24.1%); **XVII** (4.6%); **XVIII** (4.2%)

Table 2 *Volatile compounds identified in the methylated degradation products of humic acids (HA) extracted in NaOH (0.5M), DMSO (94%) plus HCl (6%), and sodium pyrophosphate (Pyro, 0.1M, pH 7) from Snape soil. Degradations were sequential, using sodium amalgam (NaA), sodium in liquid ammonia (NaLA), and alkaline permanganate (AP) in order. R_1 to R_6 represent the substituents on benzene*

Degradation Regent	Compound number, and percentages (%), by peak areas, of the methylated volatile substances in the digest products
	HA extracted in NaOH (0.5M)
NaA	**I** (7.3%); **II** (10.6%); **IV** (1.0%); **XXII** $R_1 = COOCH_3$; $R_2 = OCH_3$; $R_3 = R_4 = R_5 = R_6 = H$ (4.7%); **XXIII** $R_1 = OCH_3$; $R_4 = CH_3$; $R_2 = R_3 = R_5 = R_6 = H$ (11.9%); **XXIV** $R_1 = OH$; $R_2 = CH_3$; $R_3 = R_4 = R_5 = R_6 = H$ (20.2%).
NaLA	**I** (5.5%); **II** (0.9%); **VI** (2.5%); **VII** (0.9%); **IX** (1.3%); **X** (13.6%); **XI** (3.9%); **XIII** (14.6%); **XV** (4.7%); **XXV** $R_1 = COOCH_3$; $R_3 = R_6 = OCH_3$; $R_2 = R_4 = R_5 = H$ (0.7%).
AP	**XV** (4.8%).
	HA extracted in DMSO + HCl
NaA	**I** (15.6%); **II** (8.0%); **VII** (1.3%); **XI** (1.0%); **XXII** (4.1%); **XXIII** (30.3%) **XXVI** $R_1 = COOCH_3$; $R_4 = OCH_3$; $R_2 = R_3 = R_5 = R_6 = H$ (2.9%).
NaLA	**I** (11.4%); **X** (34.0%); **XII** (2.9%); **XXVII** $R_1 = COOCH_3$; $R_3 = R_4 = R_5 = OCH_3$; $R_2 = R_6 = H$ (1.5%).
AP	**VI** (2.1%); **IX** (2.0%); **X** (22.6%); **XI** (3.1%); **XV** (3.2%); **XXIII** (9.5%).
	HA extracted in Pyro (pH 7)
NaA	**I** (16.9%); **II** (18%); **III** (6.5%); **VI** (3.7%); **XX** (2.2%); **XXII** (6.2%); **XXVI** (5.2%).
NaLA	**I** (0.4%); **II** (1.4%); **III** (0.5%); **VI** (4.8%); **VII** (0.5%); **IX** (0.5%); **X** (19.7%); **XIII** (7.1%); **XXVII** (1.9%).
AP	**VI** (2.5%); **IX** (1.5%); **X** (9.1%); **XI** (2.9%); **XIII** (3.4%); **XV** (3.0%); **XXIII** (6.7%); **XXVIII** $R_1 = OCH_3$; $R_2 = CH_3$; $R_3 = R_4 = R_5 = R_6 = H$ (9.1%).

Table 3 *Volatile compounds identified in the methylated degradation products of humic acids extracted in NaOH (0.5M) from a sapric histosol. Degradations were sequential, using sodium amalgam (NaA), sodium in liquid ammonia (NaLA), and alkaline permanganate (AP) in order. R_1 to R_6 represent the substituents on benzene*

Degradation Regent	Compound number, and percentages (%), by peak areas, of the methylated volatile substances in the digest products
NaA	**I** (10.8%); **II** (3.9%) **IV** (0.9%); **VI** (3.1%); **VII** (3%); **XIII** (8%); **XXII** (2.5%); **XXIII** (6.4%); **XXVI** (1.8%); **XXVII** (1.1%); **XXIX** $R_1 = CH_2CH_2COOCH_3$; $R_4 = OCH_3$; $R_2 = R_3 = R_5 = R_6 = H$ (1.1%); **XXX** $CH_3(CH_2)_{12}COOCH_3$ (1.1%).
NaLA	**I** (14.6%); **IV** (2.0%); **VII** (4.8%); **IX** (0.4%); **XII** (0.5); **XXVI** (1.7%); **XXVII** (1.5%); **XXX** (0.7%); **XXXI** $R_1 = CH_2COCH_3$; $R_2 = CH_3$; $R = R_4 = R_5 = OCH_3$; $R_6 = H$ (0.4%).
AP	**I** (3.7%); **VI** (13.8%); **X** (33.1%); **XVI** (2.5%); **XVII** (1.6%); **XVIII** (3.8%); **XXXII** $R_1 = R_2 = R_4 = R_5 = COOCH_3$; $R_3 = R_6 = H$ (13.4%); **XXXIII** $(CH_2)_{10}(COOCH_3)_2$ (1.1%).

CO_2CH_3
S
XXXIV (1%)

(methyl-3,4-dimethoxybenzoate) can be considered to be derived from a lignin precursor. Compound **V** is particular to the SH 213 (gravefill) sample; the structure is highly speculative, but the compound is certainly a highly substituted aromatic, and without connections to lignin-type precursors. Compound **VI** (methyl hexadecanoate) could be contributed by microbial residues, and it could be also released in the alkaline medium from the phenolic esters of the fatty acid.

There are differences between the compositions of the NaA digest products of the HAs from the gravefill (SH 213) and body imprint (SH 253) soils. Twice as much (91%) of the volatile methylated products were identified in the gravefill HA digests compared to those in the body 'imprint' soil. Compounds **I** to **IV** were present in both HA materials, but the relative abundances of these were different in the two samples. Compound **VI** was significantly in greater abundance in the body 'imprint' digests, and additionally methyl octadecanoate (compound **XI**, which was released only in the NaLA digest in the case of the gravefill sample), presumably with origins similar to those of **VI**, was released in the NaA digest in this instance.

Compound **VII** (methyl-*m*-methoxybenzoate) was a significant componenent of the NaA digest of the body 'imprint' samples. This compound was a minor component of the

NaLA digest products, and its origins are more likely to be microbial than plant. Compound **XIX** (methyl 2,6- or 2,4-dimethoxybenzoate), like compound **III**, is unlikely to be of plant origin, and was likely to have been released through cleavages of the relevant ether linkages. Compound **XX** (1,3-dimethoxybenzene) could arise from the decarboxylation of a carboxyl group substituent, but its origins are more likely to be microbial than plant, and release was likely to have involved the cleavages of ether linkages. The same logic applies for the origins of compound **XXI** (1,4-dimethoxybenzene).

The NaLA degradation was carried out on the non-ether soluble residues after degradation with NaA. Kraus (1908) considered that a solution of sodium in ammonia contains tha Na^+ cation and electrons surrounded by an envelope of ammonia..

Diphenyl ethers react quantitatively with sodium in liquid ammonia (Sartoretta and Sowa, 1937) and two moles of sodium react with one mole of diphenyl ether to give one mole of sodium phenolate and one mole of benzene.

In the case of the degradation of the gravefill HAs with NaLA, only 56 per cent of the volatile methylated products were identified (in contrast to the results for the NaA degradations). Five of these were not released when NaA was applied to the gravefill HA. Two of the five were released by NaA from the body 'imprint' HA sample.

Compound **X** (phenyl methyl ketone) was the major product identified in the digest of the NaLA degradation of the gravefill and body 'imprint' HAs (Table 1). This compound could have formed by cleavage by NaLA of a phenolic ether structure (with the linkage likely to be in the *para* position), and the methyl ketone substituent would arise from cleavage of a side chain in the strongly basic medium. The keto group would be attached to the aromatic nucleus (a mechanism is outlined in Hayes and Swift, 1978, p. 220).

Substantial amounts of the fatty acid, compound **XI**, were released by NaLA from the gravefill HA. This acid was released by NaA in the case of the body 'imprint' HAs. Compound **VIII** from the gravefill HAs was not identified in the body 'imprint' sample. Cleavage of an ether linkage at the *para* position could give rise to this structure, or it could be indicative of ether linkages involving the 3- and 5-ring positions.

The methyl ester of benzene-1,2-dicarboxylic acid (Compound **IX**) is likely to have arisen from mechanisms similar to those outlined for compound **I**, or it could have been an artefact (phthalate esters are very widespread in the general environment).

The digest products of the alkaline permanganate (AP) degradation of the gravefill and body 'imprint' HAs are different from the range of digest products from soil humic acids (see Hayes and Swift, 1978). However, the degradation in the present study had followed prior degradation with NaA and NaLA.

Compound **X** was the major volatile degradation product in the alkaline permanganate digest of the gravefill and body 'imprint' HAs. The keto functionality would resist oxidation, and the aliphatic side chain could have given rise to the methyl ketone by mechanisms along the lines of those referred to by Hayes and Swift (1978, p. 220). Mono- and dicarboxylic acid functionalities on the aromatic nuclei would arise from oxidation of aliphatic side chains, and compounds such as **XIII** (methoxybenzene), present in relatively small amounts, is likely to be a product of decarboxylation. Compound **XV** (eicosanoic acid) is likely to have arisen from the hydrolysis of an ester functionality, or from mechanisms such as the oxidative cleavage of a double bond.

There was a significant contribution of the thiophene structure (**XVI**) in the volatile methylated derivatives of the body 'imprint' HAs. It is possible that this compound was derived from the S-containing amino acids (methionine, cystine and cysteine), or other S-containing organics. Its greater abundance in the body imprint HAs would suggest origins in protoplasmic tissues. Compounds **XVII** and **XVIII** appeared to contain iodine and were almost certainly artefacts of the degradation and work up processes.

The data in Table 2 compare the degradation products in the digests of HAs isolated in sodium hydroxide, in DMSO and HCl, and in neutral sodium pyrophosphate from the Snape (body 'imprint') HAs. Degradations were sequential, using NaA, NaLA, and AP.

Many of the digest products of the HAs isolated in the different solvent systems are the same, but there are differences as well. Significant amounts of methylbenzene structures (compounds **XXIII**, **XXIV**) were identified in the digests, and the methyl substituent was *ortho* or *para* to a methoxy or hydroxy substituent. That might suggest that the methyl groups were artifacts of the methylation process (arising from canonical forms of the phenolate anion). If the methyl substituent is discounted, compound **XXIII** would be compound **XIII**, and **XXIV** would be phenol which had resisted O-methylation.

Compound **XXVII**, methyl-3,4,5-trimethoxybenzoate, was detected in low abundance in the NaLA digest (Table 2). It is likely that the majority of this compound had decarboxylated, giving rise to compound **II** which was present in significant abundances in the NaA and NaLA digests, and as the major product in the NaA digest of the HA extracted in pyrophosphate.

Methyl benzoate (from benzoic acid) was not the major degradation product in any of the digests of the extracts from the Snape (body 'imprint') soil samples. Compound **XVI** was not detected in any of the digests, and that could result from the relatively low abundances of amino acids in the Snape HA samples (see Part I. p. 136 - 147, this Volume). Compounds **XVII** and **XVIII**, considered to be artifacts, were not detected.

The data in Table 3 give the identifications of volatile methylated derivatives isolated from the sequential degradation (in NaA, NaLA, and AP) of the HAs isolated in a sodium hydroxide solution from a sapric histosol. The list of volatile compounds identified in the digests includes six (**XXIX**, **XXX**, **XXXI**, **XXXII**, **XXXIII,** and **XXXIV**) that had not been identified in the same digests of the Sutton Hoo and Snape samples. Methyl benzoate was a prominent component in each of the digests, and it is surprising to find this compound among the products of permanganate degradation. Tetramethyl benzene-1,2,4,5-tetracarboxylate (compound **XXXII)** was a significant volatile component in the permanganate digest, and that is characteristic of the oxidation products of soil HAs (Hayes and Swift, 1978). Aliphatic dicarboxylic acids, which could have origins in the cleavages of olefinic bonds separated by -$(CH_2)_n$- groups (where n in the case of compound **XXXIII** is 10), are also characteristic of the oxidation products of soil humic substances (Hayes and Swift, 1978). Compound **XXXIV**, like **XVI**, could have its origins in S-containing amino acids, or in other organosulphur compounds, and although compounds **XVII** and **XVIII** were again detected in the permanganate digests, these were likely to have formed as artifacts. Compounds **III**, **V**, **VIII**, **XI**, **XIV**, **XIX**, **XXI**, **XXIV**, **XXV**, and **XXVIII**, which had been identified in the degradation digests of the Sutton Hoo and Snape samples, were missing from the digests of the histosol. These compounds are likely to have originated in microbial processes rather than in plant (lignin) components.

4 Conclusions

The differences between the degradation products identified in the digests of the HAs from the sapric histosol and from the gravefill and body 'imprint' soils are indicative of differences in the origins of the samples. The origins of many of the major digest products from the sapric histosol could be traced to lignin-type structures. Several of the products from the grave sites are more likely to have their origins in microbial (fungal/bacterial) processes, and the non-lignin derived products were more concentrated in the HAs from the body 'imprint' soils than from the gravefill samples.

This study shows that sequential degradation procedures, in which reagents providing increasing energy inputs to the degradation processes, can provide appropriate procedures for studies of the 'building blocks' of humic macromolecules. The sequential process gives a more complete degradation to structurally meaningful digest products than can be achieved by using any single degradation procedure.

Acknowledgements

We are grateful for the support provided by the Leverhulme Trust for this project.

References

Burges, N.A., H.M. Hurst, and S.B. Walkden. 1964. The phenolic constituents of humic acid and their relation to lignin of the plant cover. *Geochim. Cosmochim. Acta* **28**:1547-1554.

Häusler, M.J. and M.H.B. Hayes. 1996. Uses of the XAD-8 resin and acidified dimethylsulfoxide in studies of humic acids. p. 25-32. *In* C.E. Clapp, M.H.B. Hayes, and N. Senesi (eds), *Humic Substances and Organic Matter in Soil and Water Environments.* Proc. 7th Intern Conf. IHSS (St. Augustine, Trinidad, 1994).

Hayes, M.H.B. and M.R. O'Callaghan 1989. Degradations with sodium sulfide and phenol. p. 143-180. *In* M.H.B. Hayes, P. MacCarthy, R.L. Malcolm, and R.S. Swift (eds), *Humic Substances II. In Search of Structure.* Wiley, Chichester.

Hayes, M.H.B. and R.S. Swift. 1978. The chemistry of soil organic colloids. p. 179-320. *In* D.J. Greenland and M.H.B. Hayes (eds), The Chemistry of Soil Constituents. Wiley, Chichester.

Kraus, C.A. 1908. Solutions of metals in non-metallic solvents IV. Material effects accompanying the passage of an electric current through solutions of metals in liquid ammonia. Migration experiments. *J. Am. Chem. Soc.* **30**:1323-1344.

Maximov, O.B. and N.P. Krasovskaya. 1977. Action of metallic sodium on humic acids in liquid ammonia. *Geoderma* **18**:227-228.

Piper, T.J. and A.M. Posner. 1972a. Sodium amalgam reduction of humic acid-I. Evaluation of the method. *Soil Biol. Biochem.* **4**:513-543.

Piper, T.J. and A.M. Posner. 1972b. Sodium amalgam reduction of humic acid-II. Application of the method. *Soil Biol. Biochem.* **4**:525-531.

Sartoretto, P.A. and F.J. Sowa. 1937. The cleavage of diphenyl ethers by sodium in liquid ammonia I. Ortho and para substituted diphenyl ethers. *J. Am. Chem. Soc.* **59**:603-606.

Natural Abundances of ^{13}C in Soils and Waters

C.E. Clapp, M.F. Layese[1], M.H.B. Hayes[2], D.R. Huggins[1], and R.R. Alimaras

[1]USDA-ARS, & DEPARTMENT OF SOIL, WATER, AND CLIMATE, UNIVERSITY OF MINNESOTA, ST. PAUL,MINNESOTA 55108, USA
[2]THE UNIVERSITY OF BIRMINGHAM, SCHOOL OF CHEMISTRY, EDGBASTON, BIRMINGHAM, B15 2TT, ENGLAND

Abstract

The $\delta^{13}C$ values of soils and humic substances derived from soils, peats, coals, and different waters (soil waters, surface waters, streams, rivers, lakes and an ocean sample) are presented. In addition, data are given that involve soil from an 11-year experiment on corn (*Zea mays L.)*/soybean (*Glycine max* L.) cropping sequences at Lamberton, Minnesota. Data are also presented for $\delta^{13}C$ values of different particle-size fractions of the soil cropped under continuous corn and continuous soybean in the same experiment, of two soils from Israel, and a Waukegan soil from Minnesota. Data from the Waukegan soil provided an opportunity to observe the effects of residue amendments on ^{13}C contents.

1 Introduction

The isotopic composition of carbon is governed, for the most part, by the process of photosynthesis. For higher plants, those that follow the C3 pathway (Calvin cycle) discriminate against $^{13}CO_2$ during photosynthesis to a greater extent than do plants that follow the C4 pathway (Hatch-Slack cycle) (Deines, 1980). As a result, C4 plants such as corn *(Zea mays* L.), and warm season grasses have $\delta^{13}C$ values between -9 and -17‰ with most values averaging -12‰; C3 plants such as soybean [*Glycine max* L. (Merr)], wheat *(Triticum aestivum* L.), cool season grasses, and forest trees have $\delta^{13}C$ values ranging from -23 to -34‰, with most having mean values of -26‰ (Deines, 1980). The range of $\delta^{13}C$ values with different sources of C in different environments is shown in Figure 1.

Differences in the natural abundance of ^{13}C between C3 and C4 plants have been used as *in situ* labelling of organic matter for the determination of organic C turnover, or of soil organic matter (SOM) dynamics (Balesdent *et al.*, 1987, 1988; Martin *et al.*, 1989). The premise of this approach is that SOM originates from plant residues, and so changes from C3 and C4 type plants (and vice versa) will result in corresponding changes in the $\delta^{13}C$ of the SOM. The main criteria of this approach is that one type of vegetation or crop that has been growing for a long time is replaced by another having a significantly different ^{13}C isotopic composition. This could be done in a natural environmental setting, or in an

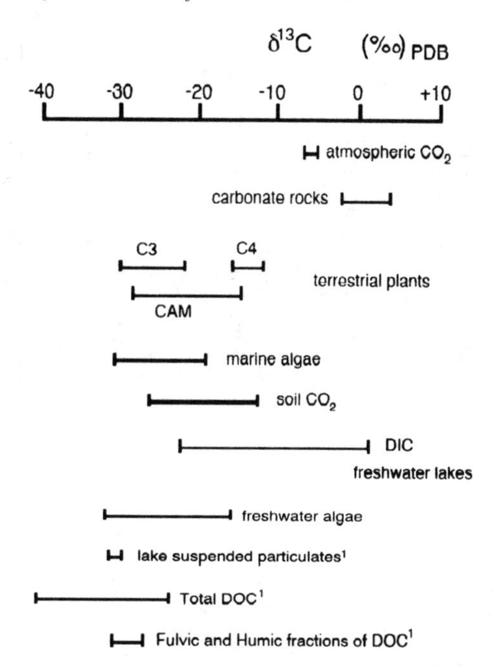

Figure 1 *Ranges of $\delta\,^{13}C$ values found in terrestrial and aquatic environments, based on the data of Deines (1980) and of Schiff et al. (1990)*

artificial setting, as may be achieved by the introduction of corn (maize), a C4 plant, into a soil that had developed under soybean, a C3 plant, or vice versa. Another example would be the cultivation of C3 crops in a soil developed under C4 prairie grasses. In the natural environment, ^{13}C data have been used as indicators of vegetation changes between two shrub species, a C3 and a C4, in a salt desert community (Djurec *et al.*, 1985); between tropical forest and savannah (Schwartz *et al.*, 1986); and between C3 woodland/C4 grassland (prairie) boundaries (Steuter *et al.*, 1990; Wang *et al.*, 1993).

When interpreting data using the ^{13}C natural abundance to estimate the rate of SOM turnover, or to detect changes in plant species composition, care should be taken to take into consideration other factors that may affect the $\delta^{13}C$ values, e.g., spatial variability (van Kessel *et al.*, 1994).

In an aquatic environment, interest is focused mainly on the humic substances (HS) in the dissolved organic matter (DOM). Dissolved organic carbon (DOC) has a wide range of $\delta^{13}C$ values, ranging from -24 to -40‰) whereas the humic acid (HA) and fulvic acid (FA) fractions of the DOC have a narrow range of -26 to -31‰ (Schiff *et al.*, 1990; see Figure 1). At issue is whether the source of HS in streams, surface waters, rivers, and ground waters is soil humic material, or whether it is synthesized *in situ*. If soil humic material is indeed a source, the question is the extent of its contribution. A critical overview of the geochemistry of stream humic substances relative to occurrence, origin, and theories of formation is provided by Malcolm (1985). A discussion is also provided of the geochemistry of HS in lakewater (Steinberg and Muenster, 1985), in estuarine environments (Mayer, 1985), and in sea water (Harvey and Boran, 1985).

Table 1 *Descriptions of the soils*

Soil series	Classification‡	Horizon	Location	pH	Vegetation§
Hallsworth	Typic loam	A	Devon, UK	6.2	permanent pasture
Denchworth	Typic Haplaquept	A	Oxfordshire, UK	6.0	arable (wheat)
Kachemak	Andic Humicryod	B_{hs}	Alaska, USA	4.7	dominated by bluejoint grass
Talkeetna	Andic Humicryod	B_{hs}	Alaska, USA	4.4	forest vegetation dominated by white spruce
Seward	Typic Cryorthod	B_h	Alaska, USA	4.8	mixed deciduous and evergreen forest (sitka spruce and mountain hemlock)
Elliot‡	Aquic Argiudoll	mollic epipedon	Illinois, USA	6.0	big bluestem grass
Harps	Typic Calciaquoll	A_p	Minnesota, USA	8.0	corn/soybean
Webster	Typic Haplaquoll	mollic epipedon	Iowa, USA	6.8	tall-grass prairie (mixed grasses)
Sanhedrin	Ultic Haploxeralf	AB	California, USA	6.2	redwood, Pacific madrone, and ponderosa pine
Armadale	Spodosol	B_h	Prince Edward Is., Can	4.4	pine forest
Summit Hill#	Typic Dystrochrept	A	New Zealand	5.4	native tussock grassland
Palmahim	Clay	A	Israel	7.7	field crops
Mevo Horon	Clay loam	A	Israel	7.7	field crops

†Descriptions of Seward, Webster, and Sanhedrin taken from Malcolm (1990). ‡Soil is classified at subgroup level; if data are not available, order or texture is given. §Bluejoint grass[*Calamagrostis canadensis* (Michx.) Beauv]; white spruce [*Picea glauca* (Moench) Voss]; sitka spruce [*Picea sitchensis* (Bong) Carr] mountain hemlock [*Tsuga mertensiana* (Bong) Carr]; big bluestem (*Andropogon gerardii Vitman var. gerardii*); corn (*Zea mays* L.), soybean [*Glycine max* (L.) merr.]; redwood [*Sequoia sempervirens* (D. Don) End I.]; Pacific madrone (*Arbutus menziesii* Pursh); sponderosa pine(*Pinus ponderosa* Laws.)
¶ IHSS standard. #IHSS reference

This contribution presents the results of C and of $\delta^{13}C$ analyses of soils and HS derived from a variety of soils, peats, coals, and different sources of water. The authors also have looked at the C and $\delta^{13}C$ values of a soil under different corn (or maize)/soybean cropping sequences, as well as at the values for different size fractions from four soils.

2 Experimental

Samples

Samples of soils, peats, coal, and waters (streams, rivers, lakes) from different locations around the world were collected and prepared for analyses. Soil materials were air-dried or freeze-dried; waters were filtered, concentrated, and stored at 4 °C prior to processing.

Isolation of Humic Substances

Soils, Peats, and Coals. Humic acids (HAs) were extracted using 0.1M NaOH or 0.1M $NaOH/Na_4P_2O_7$ under dinitrogen gas. Extracts were centrifuged and the pH values of the supernatant solutions were adjusted to 1.0 to precipitate the HAs; the fulvic acid (FA) fraction remained in solution. If not previously filtered, the HAs and supernatant FAs were treated with HCl/HF to lower the ash contents. All samples were dialyzed against distilled or deionized water and freeze-dried.

In the cases of the Kachemak, Talkeetna, Hallsworth, and Denchworth soils, the filtrates were acidified to pH 2 and then passed in tandem through XAD-8 [(poly)methylmethacrylate] and XAD-4 (styrene divinylbenzene) resins (vide infra).

Waters. The HAs from surface and stream waters, lakes, and other waters were isolated using the XAD-8 resin method, following the technique of Thurman and Malcolm (1981). Waters were filtered through 0.2 or 0.45 µm membrane filters. The filtrates were adjusted to pH 2 and then passed through XAD-8, or in tandem through XAD-8 and XAD-4 resins. The HAs and FAs were recovered by back elution with NaOH from XAD-8; the eluate was acidified to pH 1 to yield HA in the precipitate and FA in the supernatant. The material sorbed on the XAD-4 was back eluted with NaOH to yield the XAD-4 acids (XA). Samples were desalted, H^+ exchanged using Amberlite IR-120 resin, and freeze dried.

Analyses

Analyses for C and $\delta^{13}C$ were carried out, at least in triplicate, on solid samples of about 2 mg for soils and about 100 µg for humic fractions. Analyses were done on an elemental analyzer (Carlo Erba• model NA 1500) and a stable isotope ratio mass spectrometer (Fisons•, Optima model) continuous flow system. Results of the isotope analyses are expressed in terms of δ values (‰):

$$\delta^{13}C = 1000(R_{sample}/R_{standard} - 1)$$

where R = the ratio of $^{13}C/^{12}C$. The $\delta^{13}C$ values were calculated relative to the Pee Dee Beleminite (PDB) as the original standard. Urea, with a d 13C value of -18.2‰, served as a working standard.

•Mention of trade names is for reader convenience only and does not imply endorsement by the USDA-ARS or the University of Minnesota over similar products of companies not mentioned.

3 Results and Discussion

Soils

A description of the 13 soils, including classification, location, pH and vegetation is given in Table 1. The values of C and $\delta^{13}C$ of the whole soils, and of the HAs and FAs from different soils are shown in Table 2. Values ($\delta^{13}C$) are given for the whole soil and the soil humic fractions in the cases of four soils. The $\delta^{13}C$ values ranged from -19 to -29‰. Mollisols (those that have a last syllable *'ol'* in the classification column in Table 1) are less depleted (-19 to -29‰, indicating greater contribution of C4 plants) than the Spodosols (including the Cryorthod) and other soils (-25 to -29‰., indicating the contribution of C3 plants). Mollisols were developed under grassland and it is possible that the grasses were predominantly of the C4 type. Steuter *et al.* (1990) and Wang *et al.* (1993), in their studies of vegetation changes indicated that the prairies they studied were dominantly C4 types. No consistent differences in $\delta^{13}C$ were found between the HAs and the FAs for all soils. In some soils the values for HAs were lower than for FAs; in other cases, the HA values were higher; and in a few cases there were no differences. Nissenbaum and Schallinger (1974), who studied the C isotope ratios of HAs and FAs from a wide variety of soils in Israel, found that the FAs were always less depleted in ^{13}C than the HAs, on the average by 0.9960. It is probable that the FAs in these studies could be described more accurately in terms of the FA fraction, as described by Malcolm (1990). In the cases of the four soils which were completely fractionated and analyzed, the whole soil was less depleted in ^{13}C than the HA or FA fractions, and the XA fractions were always less depleted than either the HA or FA and had the lowest C content of the HS.

The values of C and of $\delta^{13}C$ in Table 3 were obtained for soil cropped to corn (maize) and soybean at Lamberton, Minnesota. There were 14 combinations of crop sequences for the period from 1981 to 1991 (Huggins *et al.*, 1995). The experiment was carried out on a Webster clay loam, classified as a Typic Haplaquoll, and treatments were replicated four times. As expected, the highest values (-17.4‰. in the 0-15 cm depth; -16.9‰ in the 15 - 30 cm depth) were found for continuous corn (maize), and the lowest (-18.3‰ in the 0-15 cm depth; -18.0‰ in the 15-30 cm depth) for continuous soybean. The difference in enrichments between the two continuous cropping systems is about 1‰. The other corn/corn-soybean combinations had values intermediate between the values for corn and soybean; however, the most recent crops in the sequence are reflected in the $\delta^{13}C$ values.

The samples from the 15- to 30-cm depths were less depleted in ^{13}C than those from the 0 - to 15- cm depth (Table 3). This was attributed to the mixing of C4 derived-C from depths below 30 cm during tillage, or from isotopic discrimination during microbial decomposition. The general tendency of $\delta^{13}C$ values to increase with depth was reported by Balesdent *et al.* (1988), Natelhoffer and Fry (1988), and Vitorello *et al.* (1989). The latter authors attributed this change to increased humification, and possibly to deposition of organic matter from a former ^{13}C-rich vegetation.

Quantifying SOM dynamics as related to agrosystem management is fundamental in identifying pathways for C sequestration in soils. Fractionating the soil according to particle size (sand, silt, and clay) yields organo-mineral fractions that could provide a more functional characterization of SOM related to biological turnover (Tiessen and Stewart, 1983). Natural ^{13}C abundance can complement physical fractionation analyses and further assess the fate of recent organic inputs into SOM.

The values for C and $\delta^{13}C$ shown in Table 4 are for different fractions of the same soil

Table 2 *Values of C and of $\delta^{13}C$ of soils and of soil-derived humic (HA), fulvic (FA), and XAD-4 acids*

Source	Whole soil		HA		FA		XAD-4 acids	
	C (%)	$\delta^{13}C$ (‰)	C (%)	$\delta^{13}C$ (‰)	C (%)	$\delta^{13}C$ (‰)	C (%)	$\delta^{13}C$ (‰)
Hallsworth†	3.6 ± 0.1	-28.2 ± 0.0	50.8 ± 0.7	-28.7 ± 0.1	51.2 ± 3.2	-29.4 ± 0.2	38.1 ± 0.4	-27.3 ± 0.1
Denchworth†	4.1 ± 0.0	-26.7 ± 0.0	38.4 ± 0.1	-29.0 ± 0.2	52.1 ± 0.6	-28.5 ± 0.1	25.5 ± 0.0	-27.3 ± 0.2
Kachemak‡	10.7 ± 0.4	-25.3 ± 0.1	50.3 ± 0.6	-26.6 ± 0.1	47.5 ± 0.6	-26.7 ± 0.1	37.2 ± 0.4	-25.3 ± 0.2
Talkeetna‡	17.1 ± 1.2	-25.6 ± 0.0	55.4 ± 0.7	-26.9 ± 0.0	51.0 ± 0.7	-26.8 ± 0.1	42.0 ± 0.3	-25.2 ± 0.1
Seward	-	-	51.9 ± 1.4	-26.4 ± 0.0	50.4 ± 1.4	-26.1 ± 0.1	-	-
Elliot§	-	-	58.4 ± 0.8	-22.6 ± 0.2	49.2 ± 0.9	-25.4 ± 0.2	-	-
Harps		-	40.7 ± 0.9	-19.0 ± 0.3	49.4 ± 1.0	-17.8 ± 0.2	-	-
Webster	-	-	55.8 ± 0.3	-18.9 ± 0.1	43.1 ± 0.2	-16.8 ± 0.2	-	-
Sanhedrin	-	-	57.2 ± 0.3	-26.5 ± 0.1	49.0 ± 0.3	-25.5 ± 0.1	-	-
Armadale	-	-	61.6 ± 2.8	-27.4 ± 0.2	48.9 ± 1.3	-26.3 ± 0.1	-	-
Summit	-	-	53.6 ± 0.8	-26.3 ± 0.2	-	-	-	-
Hill1¶								
Palmahim	-	-	57.1 ± 1.1	-24.7 ± 0.1	50.2 ± 0.3	-25.1 ± 0.1	-	-
Mevo Horon	-	-	57.9 ± 0.3	-25.2 ± 0.3	52.3 ± 2.0	-25.8 ± 0.1	-	-

† Samples provided by T.M. Hayes. (1996)
‡Samples provided by C.L. Ping *et al.* (1995)
§IHSS standard
¶IHSS reference

Table 3 *Values of C and of $\delta^{13}C$ for Webster clay loam soil at two depths under different corn (Zea mays)/soybean (Glycine max) cropping sequences at Lamberton, Minnesota*

Treatment	Crop sequence*	0-15 cm		15-30 cm	
		C(%)	δC^{13}(%)	C(%)	$\delta^{13}C$(%)
1	CCCCCSSSSSC	2.2 ± 0.1	-17.6 ± 02	1.9 ± 0.2	-17.2 ± 0.1
2	SCCCCSSSSSS	2.1 ± 0.2	-17.6 ± 0.1	2.0 ± 0.1	-17.5 ± 0.3
3	SSCCCCCSSSS	2.1 ± 0.2	-17.5 ± 0.1	2.0 ± 0.2	-17.4 ± 0.2
4	SSSCCCCCSSS	2.2 ± 0.1	-17.5 ± 0.0	2.1 ± 0.1	-17.4 ± 0.1
5	SSSSCCCCCSS	2.1 ± 0.1	-17.8 ± 0.2	1.9 ± 0.1	-17.2 ± 0.1
6	SSSSSCCCCCS	2.4 ± 0.1	-17.8 ± 0.4	2.2 ± 0.1	-17.2 ± 0.2
7	CSSSSSCCCCC	2.2 ± 0.2	-17.5 ± 0.1	2.1 ± 0.1	-17.3 ± 0.2
8	CCSSSSSCCCC	2.2 ± 0.1	-17.6 ± 0.3	2.2 ± 0.1	-17.5 ± 0.1
9	CCCSSSSSCCC	2.3 ± 0.2	-17.5 ± 0.2	2.2 ± 0.1	-17.4 ± 0.1
10	CCCCSSSSSCC	2.1 ± 0.1	-17.9 ± 0.1	2.2 ± 0.3	-17.6 ± 0.1
11	CCCCCCCCCCC	2.3 ± 0.1	-17.4 ± 0.1	2.2 ± 0.0	-16.9 ± 0.2
12	SSSSSSSSSSSS	2.1 ± 0.2	-18.4 ± 0.1	2.1 ± 0.1	-18.0 ± 0.2
13	CSCSCSCSCSC	2.3 ± 0.2	-17.6 ± 0.1	2.1 ± 0.2	-17.5 ± 0.1
14	SCSCSCSCSCS	2.3 ± 0.2	-17.9 ± 0.1	2.1 ± 0.1	-17.2 ± 0.2
15	FFFFFFFFFFF	2.2 ± 0.2	-18.0 ± 0.2	2.0 ± 0.1	-17.7 ± 0.2

C = corn (*Zea mays* L); S = soybean [*Glycine max* L. (Merr)]; F = fallow

as in Table 3, under continuous corn (maize) (treatment 11) and continuous soybean (treatment 12). The high standard deviations in some of the values are a reflection of the heterogeneity of the replicates in the field. The $\delta^{13}C$ of the corn residue averaged -12.0‰. and that of the soybean residue averaged -26.4‰, representing C4 and C3 plant material, respectively. As expected, the $\delta^{13}C$ values of the particulate organic matter (POM) closely resembled those of the residues. While the difference in $\delta^{13}C$ values in plant residue (and POM) between continuous corn and soybean is high (about -14‰ for plant residue and -8‰ for POM), the differences in the fractions and for the whole soil between the two cropping systems was much less (about 1‰). Stout *et al.* (1981) reported that the transformation of plant residue into SOM is associated with a very small ^{13}C enrichment, usually in the range of 0.5 to 1.5‰. It is evident that each particle size fraction under soybean was more depleted than the corresponding fraction under corn (maize). The fine clay fraction of either continuous corn (maize) or soybean was the most enriched.

HS derived from the whole soil and from different fractions (Table 5) from two Israeli (Tarchiyzky, 1994) soils (Palmahim and Mevo Horon) showed that the finer fractions (< 2 μm) had HAs which were more enriched in ^{13}C than the coarser fractions (> 250 μm). The FAs seemed to follow the trend of the HAs; however, the FAs were slightly more depleted than the HAs. In other studies, Vitorello *et al.* (1989) and Bonde *et al.* (1992) working independently with the same soils, found that the enrichment of SOM was associated with the clay particles in a forest soil, while the sand - sized fractions were more enriched in the case of a sugar cane soil.

The data in Table 6 indicate contributions of residues to the ^{13}C in the clay- and silt-sized fractions of a Waukegan soil (Typic Hapludoll) where corn (maize) was grown continuously for 9 years. Soils were tilled (T) and given two residue treatments. In the residue treatment (R), crop residues were left after harvest, and in the no-residue treatment

Table 5 *Values of C and of $\delta^{13}C$ of the humic acid (HA) and fulvic acid (FA) fractions from Israeli whole soils and different size fractions of these soils*

Source	Fraction	HA		FA	
		C (%)	$\delta^{13}C$ (‰)	C (%)	$\delta^{13}C$ (‰)
Palmahim†	Whole soil	57.1 ± 1.1	-24.7 ± 0.1	50.2 ± 0.3	-25.1 ± 0.0
	>250 μm	53.5 ± 1.4	-26.0 ± 0.3	49.7 ± 0.4	-26.3 ± 0.2
	50 - 250 μm	55.5 ± 0.8	-24.9 ± 0.1	51.9 ± 0.6	-27.1 ± 0.0
	20 - 50 μm	57.2 ± 1.4	-25.2 ± 0.2	51.3 ± 0.9	-25.9 ± 0.2
	5 - 20 μm	53.8 ± 4.2	-25.2 ± 0.3	53.9 ± 1.3	-25.8 ± 0.0
	2 - 5 μm	54.1 ± 0.2	-24.5 ± 0.1	50.9 ± 1.5	-25.4 ± 0.1
	<2 μm	57.2 ± 0.1	-23.9 ± 0.1	48.3 ± 1.3	-25.6 ± 0.1
Mevo Horon†	Whole soil	57.9 ± 0.3	-25.2 ± 0.3	52.3 ± 4.0	-25.8 ± 0.1
	>250 μm	60.9 ± 0.2	-27.1 ± 0.2	54.5 ± 2.5	-27.9 ± 0.2
	50 - 250 μm	59.2 ± 0.2	-26.4 ± 0.0	53.9 ± 1.4	-27.0 ± 0.1
	20 - 50 μm	58.9 ± 0.3	-25.2 ± 0.1	52.0 ± 0.3	-26.4 ± 0.2
	5 - 20 μm	57.1 ± 1.2	-24.6 ± 0.1	53.7 ± 1.9	-26.7 ± 0.4
	2 - 5 μm	54.9 ± 1.2	-24.3 ± 0.1	50.4 ± 0.6	-25.0 ± 0.0
	< 2 μm	59.6 ± 2.8	-24.6 ± 0.1	48.8 ± 1.2	-25.5 ± 0.1

†Samples provided by J. Tarchitzk (1994)

Table 4 *Values of C and of $\delta^{13}C$ of different soil fractions under continuous corn (Zea mays) and continuous soybean (Glycine max) at Lamberton, Minnesota, USA*

Crop Sequence†	Fraction	C (%)	$\delta^{13}C$ (‰)
Continuous corn	Plant residue	40.4 ± 0.6	-12.0 ± 0.2
	POM‡	21.2 ± 3.6	-14.6 ± 0.5
	Sand (>50 μm)	0.12 ± 0.03	-19.4 ± 0.6
	Co Si (20 - 50 μm)	0.43 ± 0.22	-19.0 ± 1.4
	F Si (2 - 20 μm)	2.96 ± 1.03	-17.6 ± 0.3
	Co Cl (0.2 - 2 μm)	5.92 ± 0.28	-17.6 ± 0.1
	F Cl (<0.2 μm)	4.10 ± 0.66	-16.4 ± 0.2
	Whole soil	2.27 ± 0.05	-17.3 ± 0.1
Continuous soybean	Plant residue	42.3 ± 0.5	-26.4 ± 0.1
	POM‡	24.0 ± 5.2	-22.2 ± 1.2
	Sand (>50 μm)	0.09 ± 0.02	-21.7 ± 0.1
	Co Si (20 - 50 μm)	0.60 ± 0.52	-19.9 ± 0.1
	F Si (2 - 20 μm)	3.97 ± 1.19	-18.4 ± 0.3
	Co Cl (0.2 - 2 μm)	4.59 ± 0.52	-17.8 ± 0.4
	F Cl (<0.2 μm)	4.44 ± 1.02	-17.4 ± 0.3
	Whole soil	2.18 ± 0.20	-18.4 ± 0.2

†See Table 3. ‡Particulate organic matter. Co = coarse; F = fine; Si = silt; Cl = clay

(NR) crop residues were removed after harvest. The residues added contributed to an enrichment of $\delta^{13}C$ (see data for TR vs TNR). This trend is observed on the $\delta^{13}C$ values of the whole soil and in both the clay and silt fractions, as well as in the HAs derived from the clay and silt fractions. The HA of the clay fraction was, however, more enriched than that of the silt, and that could indicate a greater contribution from the maize crop, and suggesting that the clay HA was younger. Balesdent *et al.* (1988) reported that the size fractions coarser than 50 μm and those finer than 2 μm contained the youngest OM. The

Table 6 *Values of C and $\delta^{13}C$ of Waukegan soil, of its silt and clay fractions, and of their humic acids (HA) with and without residues*

Soil fraction/residue†	Soil		HA	
	C (%)	$\delta^{13}C$ (‰)	C (%)	$\delta^{13}C$ (‰)
Whole soil TR	3.0 ± 0.0	-18.4 ± 0.1	-	-
Clay TR	6.8 ± 0.1	-17.6 ± 0.1	58.0 ± 0.8	-18.7 ± 0.1
Silt TR	1.3 ± 0.0	-19.2 ± 0.1	57.5 ± 0.5	-20.8 ± 0.1
Whole soil TNR	2.8 ± 0.0	-18.9 ± 0.1	-	-
Clay TNR	6.9 ± 0.1	-18.3 ± 0.1	59.2 ± 1.9	-18.9 ± 0.1
Silt TNR	1.1 ± 0.0	-20.3 ± 0.1	58.6 ± 1.6	-21.7 ± 0.1

†TR = tilled with residues returned; TNR = tilled with residues removed

Table 7 *Values of C and of $\delta^{13}C$ of peat-derived humic (HA) and fulvic (FA) acids*

Source	Location	HA		FA	
		C (%)	$\delta^{13}C$ (‰)	C (%)	$\delta^{13}C$ (‰)
High moor peat:	Minnesota, USA				
Fibric; Sphagnum moss		58.3 ± 0.7	-27.0 ± 0.1	52.1 ± 0.6	-27.0 ± 0.2
Hemic; reed/sedge		58.9 ± 0.9	-27.5 ± 0.4	49.2 ± 0.6	-26.7 ± 0.2
Sapric; forest/ mixed grasses		54.2 ± 1.2	-26.9 ± 0.0	49.7 ± 0.6	-26.3 ± 0.0
Everglades†; Sapric Histosol	Florida, USA	57.3 ± 1.0	-26.0 ± 0.4	47.6 ± 0.3	-25.8 ± 0.1
Everglades‡	Florida, USA	56.4 ± 2.9	-26.2 ± 0.1	45.2 ± 0.4	-26.2 ± 0.2
Woody peat; Sapric Histosol	New York, USA	49.1 ± 0.4	-24.4 ± 0.2	-	-
Fenland; Sapric Histosol	Norfolk, England	54.3 ± 1.0	-26.2 ± 0.1	-	-
Fenland Polysaccharide	Norfolk, England	13.6 ± 0.2	-28.1 ± 0.2	-	-

†IHSS standard ‡IHSS reference

OM bound to the silt-sized components had a slower turnover rate (Balesdent *et al.*, 1989; Christiansen, 1987).

Peats

The values for C and $\delta^{13}C$ of some peat-derived HAs and FAs are shown in Table 7. Samples for the Minnesota peat, a high moor peat, were taken from fibric, hemic, and sapric layers of the same peat. Fibric peat consists largely of plant remains that are little decomposed and their botanical origins can be readily determined; hemic peat contains organic matter which has been sufficiently decomposed so that the botanical origins of as much as two-thirds of the material cannot be readily determined; and sapric peat consists of products which have been sufficiently transformed that the botanical origins of the plants from which they were derived cannot be determined. Fibric, hemic, and sapric peats contain HAs which have similar $\delta^{13}C$ values, indicating that the $\delta^{13}C$ values of the HAs did not reflect the different extents of organic matter decomposition. However, the $\delta^{13}C$ values of the FAs seemed to indicate that, as the decomposition of the organic material progressed, ^{13}C enrichment increased. The $\delta^{13}C$ values for HAs and FAs were similar for samples from the same sources (the Everglades or Minnesota). The humic fractions from a low moor Fenland peat from Norfolk, England, had $\delta^{13}C$ values similar to those for the IHSS standard and reference samples from the Everglades, Florida. Fenland polysaccharide, which was isolated from the Fenland peat, had a lower C content, and a much lower $\delta^{13}C$ value than the HA from the same peat. This polysaccharide material was of microbial origin and the values indicate that the microbial processes which gave rise to these discriminated against ^{13}C inclusion. All the values suggest a dominant input from C3 plants. The composition of the woody peat from New York, however, appeared to be somewhat influenced by C4 plants.

Coals

Coals arise from the accumulation of vast quantities of plant remains and their subsequent decomposition and consolidation (Lawson and Stewart, 1989). The first phase of the decomposition involves microbial activity and the formation of peat in waterlogged environments. The peat then becomes overlain by sediments and subjected to moderate temperatures (up to 200 ^{o}C) and very high tectonic pressures. Over geological time this process results in the successive formation of lignite and brown coal, bituminous coals of increasing maturity, and ultimately anthracite. The ages of mature coals range from 2×10^6 to 250×10^6 years.

Coal HAs occur naturally in some lignites and brown coals, but little or no alkali-soluble material is contained in bituminous coals. Mild oxidation of mature coals gives gives rise to the formation of 'regenerated' HAs, or 'ulmins'. Thus, the degree of maturity of the coal has a marked effect on the quantity and on the composition of the HAs associated with it. HS in immature coals take part in the process of coal formation, while HAs derived from mature coals, by oxidation, will relate to the structure of the parent coal.

The data in Table 8 deal with three classes of coal, based on C contents, and therefore age. The low C (40%) and relatively young coals include the Thailand lignite and Canadian leonardite; the moderate C (about 55%) coals include the Dalton Ulmins and Mitchell Main Ulmin; and the high C (> 58%) include the relatively old coals, represented by Dexter coal and the British coal. The $\delta^{13}C$ values of the different coals ranged from -22.8

Table **8** *Values of C and of $\delta^{13}C$ of coals and of coal-derived humic acids (HA)*

Source	Location	Coal		HA	
		C (%)	$\delta^{13}C$ (‰)	C (%)	$\delta^{13}C$ (‰)
British coal	Central England	60.8 ± 0.1	-22.8 ± 0.1	66.8 ± 1.4	-23.3 ± 0.1
Dexter coal OX iSOC	Central England	58.0 ± 0.8	-23.0 ± 0.2	-	-
Daltons Ulmins	Central England	55.0 ± 0.4	-23.1 ± 0.0	62.3 ± 0.4	-23.0 ± 0.1
Mitchell Main Ulmin	Central England	55.9 ± 1.4	-23.4 ± 0.0	-	-
Lignite	Thailand	40.6 ± 0.7	-23.9 ± 0.1	-	-
Leonardite	Canada	40.5 ± 1.2	-23.5 ± 0.4	61.9 ± 0.5	-23.2 ± 0.1
Leonardite HA†	USA			61.2 ± 1.0	-23.8 ± 0.2

†IHSS standard

to -23.9‰. There was a very slight enrichment in ^{13}C with maturation of the coals. The $\delta^{13}C$ values of the HAs were not different from those of the coals.

HS in peatlands are considered to be precursors of coals. Comparisons of peat with some lignites in Germany showed that there were no drastic shifts in $\delta^{13}C$ in going from peats to lignites (Deines, 1980). Comparisons of $\delta^{13}C$ values of peats (-24 to -27‰, Table 7) and coals (-23 to -24‰, Table 8) indicate that the coals are more enriched in ^{13}C. Direct comparison may not be valid because the peats and coals were not restricted to a particular set of environmental conditions.

Waters

Soil waters are represented as two types, pore water and surface drainage water (Table 9). Pore water was obtained by leaching bulk soil samples with minimum volumes of water. Values of $\delta^{13}C$ were lower than -26.0‰ indicating C3 contributions. The FAs were more enriched with ^{13}C than the HAs, and the XAs (XAD-4 acids) were more enriched than the FAs giving an enrichment sequence of HA < FA < XA. The C contents of the XAs were lower than those either the FAs or the HAs, and it is considered that these fractions had undergone more extensive biological oxidation processes than the HAs and FAs. These transformation processes had given rise to enrichments of ^{13}C. The $\delta^{13}C$ values showed no difference between Hallsworth A and Hallsworth B samples, water flowing from tile drains at 50 to 60 cm, or from surface run off, and sampled at different times of the year. Data from soils, and from drainage waters from the soils (the Kachemak and Talkeetna; Hallsworth and Denchworth soils) showed that the $\delta^{13}C$ values of the water humics were much more depleted in ^{13}C than the soil humics (Tables 2 and 9).

Water HAs from peat watersheds in different locations, and in some instances formed from different plant sources, had similar organic C and $\delta^{13}C$ values (Table 10). The values for the HAs and the the FAs were similar. Again, the XAs had lower C contents, and were more enriched in ^{13}C. The $\delta^{13}C$ values for the HS from waters from watersheds of shallow silt soils were similar to those for waters from peat and Sapric Histosol watersheds. Where comparisons can be made between the data in Tables 7 and 10, it is evident that the $\delta^{13}C$ values for the HS from peat soils (Table 7) were more enriched in

Table 9 *Values of C and of $\delta^{13}C$ of the humic acids (HA), fulvic acids (FA), and XAD-4 acids of soil waters*

Soil	Type of water	Location	HA		FA		XAD-4 acids	
			C (%)	$\delta^{13}C$ (‰)	C (%)	$\delta^{13}C$ (‰)	C (%)	$\delta^{13}C$ (‰)
Kachemak†	soil pore	Alaska, USA	-	-	44.6 ± 0.7	-27.9 ± 0.3	42.8 ± 1.5	-26.4 ± 0.1
Talkeetna†	soil pore	Alaska, USA	54.7 ± 1.4	-28.0 ± 0.4	51.5 ± 0.5	-27.3 ± 0.1	48.7 ± 0.9	-27.0 ± 0.1
Hallsworth‡	surface drainage	England	55.3 ± 1.2	-31.8 ± 0.3	52.8 ± 1.0	-30.6 ± 0.2	45.0 ± 0.9	-28.4 ± 0.3
Denchworth‡ :	surface drainage	England	54.7 ± 0.6	-29.8 ± 0.0	52.4 ± 0.6	-29.4 ± 0.0	43.9 ± 0.5	-28.3 ± 0.1
Hallsworth A§	tile drain	England	34.1 ± 0.2	-30.8 ± 0.2	54.0 ± 0.5	-29.8 ± 0.0	46.0 ± 0.8	-28.6 ± 0.2
Hallsworth B§	tile drain	England	53.9 ± 1.3	-31.0 ± 0.4	57.3 ± 0.6	-30.2 ± 0.1	49.4 ± 0.7	-28.9 ± 0.2
Denchworth‡	tile drain	England	43.8 ± 1.2	-28.9 ± 0.3	53.2 ± 0.7	-28.9 ± 0.2	48.8 ± 1.1	-28.1 ± 0.0
Norfolk	drains	England	52.2 ± 0.6	-28.3 ± 0.1	53.7 ± 0.4	-28.0 ± 0.1	46.4 ± 0.9	-27.1 ± 0.1
Edinburgh	drains	Scotland	52.4 ± 1.2	-29.8 ± 0.3	56.9 ± 1.2	-29.7 ± 0.2	49.4 ± 0.7	-29.0 ± 0.4

† Samples provided by R.L. Malcolm *et al.* (1995)
‡ Samples provided by T.M. Hayes (1996)
§ Samples from Hallsworth A and B were taken at different times of the year

Table 10 *Values of C and of $\delta^{13}C$ of humic (HA), fulvic (FA), and XAD-4 acids isolated from waters from the watersheds of Upland Peats (UP) and of Sapric Histosols (SH), and from watersheds of mainly silt soils (SS)*

Source	Location	HA		FA		XAD-4 acids	
		C (%)	$\delta^{13}C$ (‰)	C (%)	$\delta^{13}C$ (‰)	C (%)	$\delta^{13}C$ (‰)
				Peats and Sapric Histosols			
Wicklow (stream from UP)	Ireland	53.0 ± 2.0	-28.4 ± 0.3	54.5 ± 0.9	-28.1 ± 0.1	49.2 ± 1.0	-27.5 ± 0.1
Liffey (stream from UP)	Ireland	54.4 ± 0.8	-27.8 ± 0.2	53.8 ± 0.8	-28.0 ± 0.2	49.2 ± 0.6	-27.1 ± 0.2
Dodder (stream from UP)	Ireland	54.4 ± 1.0	-28.0 ± 0.1	54.2 ± 2.0	-27.9 ± 0.0	49.8 ± 0.7	27.2 ± 0.1
Yorkshire Rr (from UP)	England	53.2 ± 0.7	-28.0 ± 0.0	54.5 ± 0.6	-28.0 ± 0.1	50.9 ± 1.3	-27.3 ± 0.1
Suwannee R†	Florida	53.7 ± 1.1	-27.7 ± 0.1	53.1 ± 1.3	-27.7 ± 0.1	-	-
Nordic‡	Norway	53.6 ± 0.7	-27.8 ± 0.2	53.6 ± 1.9	-27.8 ± 0.1	-	-
Norfolk II (SH)	England	52.2 ± 0.6	-28.3 ± 0.1	53.7 ± 0.4	-28.0 ± 0.1	46.4 ± 0.9	-27.1 ± 0.1
				Silt Soils (SS)			
Vyrnwy Rr* (UP and SS)	Wales	54.0 ± 0.3	-28.5 ± 0.1	39.7 ± 0.8	-28.7 ± 0.2	45.2 ± 0.6	-27.4 ± 0.1
Elan Valley (stream to Rr from shallow SS)	Wales	54.6 ± 0.5	-28.2 ± 0.3 -	54.9 ± 0.5	-28.2 ± 0.3	49.5 ± 0.5	-27.0 ± 0.2 -
Iniscarra Rr* (from silt/clay soils)	Ireland	56.3 ± 0.5	-29.2 ± 0.0	55.8 ± 2.2	-29.0 ± 0.1	48.0 ± 0.3	-27.9 ± 0.2

† IHSS standard from Suwannee River ‡ IHSS reference Rr* = Reservoir

Table 11 *Values of C and of $\delta^{13}C$ for humic (HA), fulvic (FA) and XAD-4 acids from rivers (R), lakes (L), and from the Pacific Ocean*

Source	Location	HA		FA		XAD-4 acids	
		C	$\delta^{13}C$ (‰)	C	$\delta^{13}C$ (‰)	C	$\delta^{13}C$ (‰)
Trent R.	England	55.1 ± 1.5	-28.2 ± 0.0	53.8 ± 0.2	-28.1 ± 0.1	-	-
Stratford (R. Avon)	England	49.0 ± 0.6	-28.2 ± 0.1	55.2 ± 0.9	-28.0 ± 0.1	45.3 ± 0.8	-27.5 ± 0.2
Thames R	England	48.7 ± 0.7	-28.6 ± 0.0	55.3 ± 1.3	-28.6 ± 0.1	49.2 ± 1.2	-27.4 ± 0.1
Aveiro R	Portugal	57.8 ± 2.6	-28.6 ± 0.2	57.7 ± 0.8	-28.0 ± 0.1	50.0 ± 0.7	-26.8 ± 0.0
Skjervatjern L	Norway	54.0 ± 0.5	-28.2 ± 0.0	55.4 ± 0.4	-28.0 ± 0.0	49.1 ± 1.3	-27.2 ± 0.2
Pacific Ocean	Hawaii, USA	34.2 ± 0.3	-23.8 ± 0.1	55.3 ± 0.6	-22.2 ± 0.0	58.3 ± 0.3	-23.0 ± 0.0
Suwannee R†	Florida, USA	53.7 ± 1.0	-27.7 ± 0.1	53.1 ± 1.3	-27.6 ± 0.3	-	-
Ogeechee R	Georgia, USA	55.8 ± 0.8	-27.9 ± 0.1	57.0 ± 0.9	-28.1 ± 0.1	-	-
Ohio R	Ohio, USA	57.6 ± 0.5	-27.8 ± 0.0	57.4 ± 1.4	-27.8 ± 0.2	-	-
Fremont L	Wyoming, USA	56.0 ± 1.3	-26.3 ± 0.0	54.9 ± 0.8	-25.9 ± 0.0	-	-

† IHSS standard

^{13}C than the HS from waters draining the peats.

Values of C and $\delta^{13}C$ for waters from rivers, lakes and an ocean site are shown in Table 11. Most of the $\delta^{13}C$ values of river and lake humics were close to about -28‰. The Fremont Lake, which had a $\delta^{13}C$ value of about -26‰, is an exception. There were no differences between the HAs and FAs among rivers and lakes. The Pacific Ocean sample is uniquely different from the rivers and lakes; the HA had a much lower C value and was enriched with ^{13}C by about 4‰. As previously noted, XAs were consistently more enriched than HAs or FAs; however, this trend did not hold true for the Pacific sample. Studies by Calder and Parker, as cited by Deines (1980), of waters from the Pacific Ocean and Gulf of Mexico showed that DOC had $\delta^{13}C$ values from -20 to -23‰, with most values close to -23‰.

4 Conclusions

There was a considerable spread of $\delta^{13}C$ values for mineral soils (-19 to -29‰), indicating contributions to the organic matter from C3 and C4 plants. The $\delta^{13}C$ values in mineral soils, both for the soil itself and for the humic substances from it, can be used as indicators of the plant source of the SOM. This is not the case for organic soils or peats. Although the peats studied were from different sources, the $\delta^{13}C$ values had a limited range of -24 to -28‰, with most falling between -26 to -27‰. It is possible that all the peat samples came from C3 plants. It is possible also that fractionation of the C isotope during decomposition under waterlogged conditions is a more determinant factor than plant type.

In the corn (maize)/soybean cropping systems studied, a mere 1‰ difference existed between continuous corn (maize) and continuous soybean, and the cropping combinations had intermediate values. The fine clay fraction was the most enriched of the soil fractions, and analysis of the data for ^{13}C in the different fractions of the residue-amended (R) and the no-residue (NR) treatment of a Waukegan soil cropped to corn (maize) for 8 years suggested that organic matter associated with the clay fraction was younger than that in the silts. However, it is possible that the older organic matter associated with the silt-sized fractions was associated with clay domaines/aggregates that were not dispersed in the disaggregation process. Thus, when the types of of C3 and C4 plants grown are known, and when it is known for how long these have grown, the C contents and the $\delta^{13}C$ values of the SOM of soil fractions can be used as indicators of the turnover rates of organic C.

The $\delta^{13}C$ values for soil water HAs ranged from -28 to -32‰, and the FAs were about 1‰ less depleted than the HAs; XAs were less depleted than FAs by about 1‰. HAs from waters from peat watersheds had $\delta^{13}C$ values of about -28‰; the values for HAs from shallow silt watersheds were of the order of -28 to -29‰. River and lake water HAs had $\delta^{13}C$ values ranging from -26 to -29‰, with most falling around -28‰. The FAs were either of the same order, or were slightly less depleted than the HAs; XAD-4 acids were consistently the least depleted of the water humics. The $\delta^{13}C$ values of humic samples from the Pacific Ocean were very different from those for the rivers and lakes, indicating differences in the substrates or in the biological transformation processes.

The Kachemak, Talkeetna, Hallsworth, and Denchworth soils, and their drainage waters have provided opportunities to compare $\delta^{13}C$ values of soil humics and soil water humics from the same soil sources. The results suggest that soil water humics are depleted in ^{13}C compared with their soil humic sources. The differences varied from 0.5 to 3.1‰, with most falling around 1‰. The ^{13}C depletion can also be seen in the lower $\delta^{13}C$ values for the humics from waters from watersheds containing peats and Sapric Histosols compared with those isolated from the peats. Despite the fact that only limited comparisons were made, the data suggest that the $\delta^{13}C$ values of all water humics had about the same degree of depletion, and it can be generalized that water humic components are more highly depleted in ^{13}C than are the soil humic substances. This reasoning would also suggest that the humic substances dissolved in the soil solution have origins which are not identical to those in the solid or gel states in the soil.

References

Balesdent, J., A. Mariotti, and B. Guillet. 1987. Natural ^{13}C abundance as a tracer for studies of soil organic matter dynamics. *Soil Biol. Biochem.* **19**:25-30.

Balesdent, J., G.H. Wagner, and A. Mariotti. 1988. Soil organic matter turnover in long term field experiments as revealed by carbon-13 natural abundance. *Soil Sci. Soc. Am. J.* **52:**118-124.

Bonde, T.A., B.T. Christensen, and C.C. Cerri. 1992. Dynamics of soil organic matter as reflected by natural carbon-13 abundance in particle size fractions of forested and cultivated oxisols. *Soil Biol. Biochem.* **24**:275-277.

Christensen, B.T. 1987. Decomposability of organic matter in particle size fractions from field soils with straw incorporation. *Soil Biol. Biochem.* **19**:429-435.

Deines, P. 1980. The isotopic composition of reduced organic carbon. p.329-406. *In* P. Fritz and J.Ch. Fontes (eds), *Handbook of Environmental Isotope Geochemistry* **Vol.1**. Elsevier, New York.

Djurec, R.S., T.W. Boutton, M.M. Caidwell, and B.N. Smith. 1985. Carbon isotope ratios of soil organic matterand their use in assessing community composition changes in Curley Valley, Utah. *Oecologia* **66**:17-24.

Harvey, G.R. and D.A. Boran. 1985. Geochemistry of humic substances in seawater. p. 233-247. *In* G.R. Aiken, D.M. McKnight, R.L. Wershaw and P. MacCarthy (eds), *Humic Substances in Soil, Sediment and Water.* Wiley, New York.

Hayes, T.M. 1996. Humic substances in soil and drainage waters and their interactions with anthropogenic organic chemicals. Ph.D. Thesis, The University of Birmingham.

Huggins, D.R., C.E. Clapp, R.R. Allmaras, and J.A. Lamb. 1995. Carbon sequestration in corn-soyabean agroecosystems. p.61-68. *In* R. Lal, J. Kimble, E. Levine, and B.A. Stewart (eds), *Soil Management and Greenhouse Effect.* CRC Press, Boca Raton.

Lawson, G.J. and D. Stewart. 1989. Coal humic acids. p.641- 686. *In* M.H.B. Hayes, P. MacCarthy, R.L. Malcolm and R.S. Swift (eds), *Humic Substances II. In Search of Structure.* Wiley, Chichester.

Malcolm, R.L. 1985. Geochemistry of stream fulvic and humic substances. p. 181-209. *In* G.R. Aiken, D.M. McKnight, R.L. Wershaw and P. MacCarthy (eds), *Humic Substances in Soil, Sediment and Water.* Wiley, New York.

Malcolm, R.L. 1990. Variations between humic substances isolated from soils, stream waters, and groundwaters as revealed by ^{13}C-NMR spectroscopy. p. 13-35. *In* P. MacCarthy, C.E. Clapp, R.L. Malcolm, and P.R. Bloom (eds), *Humic Substances in Soil and Crop Sciences.* Am. Soc. of Agron., Inc., Madison, WI.

Malcolm, R.L., K. Kennedy, C.L. Ping, and G.J. Michaelson. 1995. Fractionation, characterization, and comparison of bulk soil organic substances and water-soluble soil interstitial organic constituents in selected cryosols of Alaska. p. 315-327. *In* R. Lal, J.

Kimble, E. Levine, and B.A. Stewart (eds), *Soils and Global Change*. CRC Press, Boca Raton.

Martin, A., A. Mariotti, J. Balesdent, P. Lavelle, and R. Vuattoux. 1989. Estimate of organic matter turnover rate in a savanna soil by ^{13}C natural abundance measurements. *Soil Biol. Biochem.* **22**:517-523.

Mayer, M.M. 1985. Geochemistry of humic substances in estuarine environments. p. 211-232. *In* G.R. Aiken, D.M. McKnight, R.L. Wershaw, and P. MacCarthy (eds), *Humic Substances in Soil, Sediment and Water*. Wiley, New York.

Natelhoffer, K.J. and B. Fry. 1988. Controls on natural nitrogen-15 and carbon -13 abundances in forest soil organic matter. *Soil Sci. Soc. Am. J.* **52**:1633-1640.

Nissenbaum, A. and K.M. Schallinger. 1974. The distribution of the stable carbon isotope (13C/12C) in fractions of soil organic matter. *Geoderma* **11**:137-145.

Ping, C.L. , G.J. Michaelson, and R.L. Malcolm. 1995. Fractionation and carbon balance of soil organic matter in selected cryic soils in Alaska. p. 307-314. *In* R. Lal, J. Kimble, E. Levine, and B.A. Stewart (eds), *Soils and Global Change*. CRC Press, Boca Raton.

Schiff, S.L., R. Aravena, S.E. Trumbore, and P.J. Dillon. 1990. Dissolved organic carbon cycling in forested watersheds: a carbon isotope approach. *Water Resources Res.* **26**:2949-2957.

Schwartz, D., A. Mariotti, R. Landfranci, and B. Guilet. 1986. $^{13}C/^{12}C$ ratios of soil organic matter as indicators of vegetation changes in the Congo. *Geoderma* **39**:97-103.

Steinberg, C. and U. Muenster. 1985. Geochemistry and ecological role of humic substances in lakewater. p. 105-145. In G.R. Aiken, D.M. McKnight, R.L. Wershaw and P. MacCarthy (eds), *Humic Substances in Soil, Sediment and Water*. Wiley NY.

Steuter, A.A., B. Jasch, J. Ihnen, and L. L. Tieszen. 1990. Woodland grassland boundary changes in the middle Niobrara valley of Nebraska identified by delta C-13 values of soil organic matter. *Am. Midi. Nat.* **124**:301-308.

Stout, J.D., K.M. Goh, and T.A. Rafter. 1981. Chemistry and turnover of naturally occurring resistant organic compounds in soil. p.1-73. *In* E.A. Paul and J.N. Ladd (eds), *Soil Biochemistry* **Vol.5.** Marcel Dekker, New York.

Tarchitkky, J. 1994. Interaction between humic substances, polysaccharides, and clay minerals and their effect on soil structure. Ph.D. Thesis, The Hebrew University of Jerusalem.

Thurman, E.M. and R.L. Malcolm. 1981. Preparative isolation of aquatic humic substances. *Environ. Sci. Technol.* **15**:463-466.

Tiessen, H. and J.W.B. Stewart. 1983. Particle size fractions and their use in studies of soil organic matter. II. Cultivation effects on organic matter composition in size fractions. *Soil Sci. Soc. Am. J.* **47**:509-514.

van Kessel, C., R.E. Farrell, and D.J. Pennock. 1994. Carbon-13 and and nitrogen-15 natural abundance in crop residues and soil organic matter. *Soil Sci. Soc. Am. J.* **58**:382-389.

Vitorello, V.A., C.C. Cerri, F. Andreux, C. Feller, and R.L. Victoria. 1989. Organic matter and natural carbon-13 distribution in forested and cultivated Oxisols. *Soil Sci. Soc. Am. J.* **53**:773-778.

Volkoff, B. and C.C. Cerri. 1987. Carbon isotope fractionation in subtropical Brazilian grassland soils. Comparison with tropical forese soils. *Plant and Soil* **102**:27-31.

Wang, Y., T.E. Cerling, and W.R. Effland. 1993. Stable isotope ratios of soil carbonate and soil organic matter as indicators of forest invasion of prairie near Ames, Iowa. *Oecologia* **95**:365-369.

Humic Substances from a Tropical Soil

Thomas B. Rick Yormah and Michael H.B. Hayes[1]

DEPARTMENT OF CHEMISTRY, THE UNIVERSITY OF SIERRA LEONE, FOURAH BAY, SIERRA LEONE
[1]THE UNIVERSITY OF BIRMINGHAM, SCHOOL OF CHEMISTRY, EDGBASTON, BIRMINGHAM B15 2TT, ENGLAND

Abstract

Soil samples taken from the 0-20, 20-50, and 50-90 cm depths in the profile of a Tropical soil were analysed for clay and (hydr)oxide compositions, and for total organic matter (OM), for the light fraction of the OM, for OM in different soil size fractions (including the clay-size fractions), and for the humic acids (HAs) in the combined sodium pyrophosphate (0.1M, pH 7) and sodium hydroxide (0.5M) extracts. The inorganic colloids were composed mainly of 1:1 layer clays (kaolinite with some halloysite), and the (hydr)oxides gibbsite and goethite. The clay-size fractions had significant associations (11-13%) of OM.

Although the greatest amounts of OM were associated with the finer soil components in the 0-20 cm surface soil, the contents decreased only by small amounts at deeper depths in the 0-90 cm profile studied. The OM was a major contributor to the cation exchange capacity (CEC) of the soil. Although the data would suggest that humification of the OM increased with depth in the profile, the CEC values for the 100-mesh soil samples suggested that the contribution of the OM to the CEC at the higher pH values at depth was less than for the surface soil OM. The difference may be attributable to a greater ease in accessibility to the exchange sites in the case of the surface soil OM

1 Introduction

Tropical soils, especially those in the high rainfall regions are highly weathered, and the inorganic colloids are mainly kaolinitic, and with significant amounts of (hydr)oxides, especially those of iron and aluminium. Because of the high ambient temperatures the year round, and the availability of water, the turnover of organic matter (OM) is rapid. However, the OM associated with the inorganic colloids appears to have a considerable residence time in the soil environment, and this organic matter has an important role in the reactivities of the soil colloidal constituents, and in the fertility of the soils.

The soil for this study was taken from the Rice Research Station at Rokupr, in the North-Western region of Sierra Leone. The Station is located at the border between the coastal plain (with abundant mangrove swamps), and the interior plain where the vegetation is largely composed of farm bush and savannah. There are two main soil types in this area. The first, located in the coastal plain, is used almost exclusively for the cultivation of swamp rice, and has been studied in detail (Hart *et al.*, 1963; Hesse and Jeffery, 1963; Odell *et al.*, 1974). It is classified under Province C of the soil map of Sierra Leone (Dijkerman and Odell, 1976). The second soil type, the soil source for the present study, has not been studied extensively, and belongs to Soil Province F, which is characterized by soils from acid igneous and from metamorphic rocks, and shifting cultivation is the norm for the area. The samples studied are representative of an area of 6 km^2 in the region of Marcosa village in the Rokupr area. The region has a mean annual temperature of ca 27 ^{o}C and an annual rainfall of ca 300 cm. The rainy season is May to October, followed by a relatively dry period of about six months.

The purpose of this study was to investigate the properties of the OM at different depths in the surface metre of the soil profile, and to focus especially on the properties of the humic substances (HS) associated with the soil mineral colloids.

2 Materials and Methods

The Soil Samples

Soil samples were taken from the 0 - 20 cm (MC_1), 20 - 50 cm (MC_2), and 50 - 90 cm (MC_3) depths in the profile of the soils of the Rice Research Station at Rokupr, Sierra Leone. The pH values of the samples [in 0.01M $CaCl_2$, at a soil to solution (w/w) ratio of 1:2] were 4.40, 4.60, and 5.90 for MC_1, MC_2, and MC_3, respectively. The exchangeable aluminium increased with depth from 0.97 (MC_1) to 1.39 (MC_2) to 3.25 (MC_3) cmol kg^{-1}.

Fractionation of the Soils

Soils were fractionated into materials passing through 40-mesh and 100-mesh sieves, and by sedimentation (Day, 1965) into the clay size fraction (< 2 μm).

Removal of the Light Fraction of Soil Organic Matter

Soil samples (40-mesh) were suspended in carbon tetrachloride (1:8) and ultrasonicated for 10 min using an MSE 150 Watt Ultrasonic Disintegrator pre-tuned at 20 kHz. The suspension was then centrifuged at 18 000 g for 30 min and the floating materials were removed. The light fractions were recovered from the CCl_4 by distillation under reduced pressure, and the denser materials were washed with an acetone water mixture (Turchenek and Oades, 1979), dried, and sieved into appropriate size fractions.

Determinations of Cation Exchange Capacity (CEC)

The method used (described in detail by Yormah, 1981) was based on that of Polemio and Rhodes (1977). The principle involved the saturation of the soil sample with saline solutions (a 60% ethanol solution of 0.4M sodium acetate and 0.1M NaCl), then replacing the sorbed cations from these using a 0.5M $Mg(NO_3)_2$ solution, and analysing the supernatants for sodium. The CEC is obtained from the relationship:

$$CEC = (Na_t - Na_{sol}) = Na_t - Cl_t^{-} (Na/Cl)_{saturating\ solution} \quad \text{where}$$

Na_t is the total sodium determined in the supernatant;
Na_{sol} is the amount of unadsorbed sodium trapped by the soil sediment after the final decantation;
Cl_t^- is the total chloride determined in the supernatant; and
$(Na/Cl)_{saturating\ solution}$ is the ratio of sodium to chloride in the saturating solution.

An isotope (^{22}Na) dilution technique was used to determine the CEC of the soil components. The organic components were extracted using sodium pyrophosphate and sodium hydroxide solutions (vide infra). The inorganic component was isolated after digesting the associated OM with hydrogen peroxide (Kunze, 1965).

CEC values were determined at different pH values in the range 4.0 to 10.1 using Britton Robinson buffers (compositions from Heyrovsky and Zuman, 1968). An aliquot (80 mL) of a stock solution of ^{22}Na [made by diluting 120 mL of a stock (from Amersham) solution (100 μCi) to 8 mL using deionized water] was added to 20 mL of buffer solution. To this was added 500 mg (200 mg of humic material), and after mixing and centrifuging the process was repeated. The sediment was thrice washed with spectroscopic grade ethanol. The activity remaining after the washings was determined by counting the sedimented sample after it had dried. The mean activity of 1 mL of the active buffer solution used to saturate the exchange sites was 73.041 cps. The concentration of Na in solution was 12.097 mg mL^{-1}, and 1 mg of Na could be related to 6.0411 cps. These basic data were used for the determination of the amount of Na retained by the sorbent.

Soil Inorganic Carbon

The procedure used was based on that by Allison and Moodie (1965). The apparatus used is that described by Allison (1960).

Soil Organic Carbon

The gravimetric wet combustion method (Allison, 1965) and the Walkley and Black method (Allison, 1965) were used. Soil samples (100 mesh) containing ca 15 mg of organic carbon (0.1 - 0.5 g of soil) were used. The organic carbon (OC) content was calculated from the relationship:

$$\%C_{org} = (a - b)/\text{Dry weight of soil (in g)} \times (12/4000) \times f \times 100$$

where
a = milliequivalents of $K_2Cr_2O_7$ added to the soil; and
b = milliequivalents of $(NH_4)_2\,Fe(SO_4)_2.6H_2O$ needed to neutralise the excess dichromate;
f = the correction factor; and
12/4000 is the milliequivalents of carbon

The correction factor of 1.33 used by Peech *et al.* (1947) and by Grewling and Peech (1960) was employed in this study. The OM content was estimated by multiplying the OC content by 1.724.

Isolation of Humified Soil Organic Matter

Extraction of humified organic matter. A weighed sample (ca 5 g of the 100-mesh fraction) was suspended in 100 mL of 0.1M sodium pyrophosphate (Pyro) neutralized to pH 7 with phosphoric acid, rolled for 4 h, then centrifuged at 11 000 g for 30 min at 4 °C. The extraction process was repeated three times, and the fourth extract was pale yellow in colour. The supernatants were combined and filtered through fine sintered glass crucibles. Each extract was acidified to pH 1 using 6M HCl. The combination of the fulvic acid (FA) fraction in solution and the humic acid (HA) in suspension was dialyzed against distilled water till no further chloride was released from the contents of the cellulose acetate dialysis sac (silver nitrate test), and the non-dialyzable components were freeze dried.

Freeze-dried material in neutral Pyro was used to prepare a calibration curve in which the OM concentration (in mg mL^{-1}) was plotted against optical density at 400 nm.

The residual soil (after extraction with Pyro) was exhaustively extracted (as described for the extraction with Pyro) with 0.5M NaOH. The mixture was purged with dinitrogen gas, and extraction was in an atmosphere of the gas. Filtration was carried out in a glove box in an atmosphere of N_2, and the filtrates were acidified, dialyzed, freeze dried, and calibration curves were obtained as before.

Using the calibration curves, it was possible to estimate the amounts of OM isolated in each extract.

Studies of the Soil Mineralogy

Soluble salts were removed by stirring a suspension of the soil in distilled water, allowing the suspended material to settle, and decanting the supernatant. The process was repeated until the soil dispersed. Then $MgCl_2$ (0.05M) was added to flocculate the colloids, and the supernatants were decanted. Gypsum was considered to be removed when a white floc was not obtained in a 50:50 mixture of acetone and an aliquot of the supernatant.

Carbonates were removed by the sodium acetate buffer procedure, as described by Grossman and Millet (1961).

Organic matter was removed using the sodium hypochlorite (NaOCl) solution treatment (adjusted to pH 9.5 by addition of 1M HCl) described by Anderson (1963). The process was carried out a total of four times.

Free iron oxides were removed by the 'sodium-dithionite-citrate' method of Mehra and Jackson (1960). Quantitative estimates of free iron oxides were made using the procedure described by Olson (1965).

Isolation of the clay size fraction used two procedures. The first involved the sedimentation process (Day, 1965) using the pre-treated soil samples. This (sedimentation) procedure was also used for samples which were not pre-treated (for removal of OM, carbonates, and iron oxides) but were dispersed by ultrasonication using either the tank or probe method (Edwards and Bremner, 1967). The clay fractions were flocculated by adding NaCl (1M), and the sediments were washed free of salt using distilled water. The salt-free product was resuspended in water and freeze dried.

The second method involved the continuous flow process described by Oladimeji (1976). Soil samples were suspended in an aqueous solution of sodium hexametaphosphate, and the suspension was homogenized by ultrasonication (Edwards and Bremner, 1967). The apparatus used consisted of a reservoir for distilled water connected through a non-return valve to a stirred (using a magnetic follower) cell containing the dispersed soil suspended in distilled water. Dinitrogen gas was used to push water from the reservoir into the stirred cell, and the contents of this cell passed into F1 and F2 Whatman "Gamma-12" high efficiency in-line filter units (supplied by Gallenkamp) and from the F2 filter to a collecting cylinder. The F1 and F2 filters retain particles ≥ 1.8 μm and 2 μm equivalent spherical diameter, respectively. Magnetic followers were used in F1 and in F2 to maintain the particles in suspension. Complete extraction of the clay-size particles was achieved when there was no turbidity in the effluent from F2. The clay-sized particles were isolated by centrifuging the contents of the collection cylinder, washing with distilled water, and freeze drying.

X-Ray Diffraction (using CuK_α radiation and a Picker Diffratometer), and ***Differential Thermal Analysis (DTA), and Differential Thermogravimetric Analysis (DTGA)*** techniques were used for the qualitative identification of the mineralogy of the inorganic colloids in the soil samples. Magnesium-exchanged clays were prepared by filtering magnesium acetate (0.5M) , followed by $MgCl_2$ (0.5M) and methanol and acetone through the clay-sized samples sedimented on ceramic tiles. Na^+- and K^+-exchanged clays were prepared by filtering NaCl (1M) or KCl (1M) solutions through the clays sedimented on ceramic tiles, and removing the excess salts with methanol and acetone. Intercalation was carried out using dimethyl sulphoxide (DMSO) and a modification of the technique described by Olejnik *et al.* (1968). Using a dropping pipette, a 90% solution (v/v) of DMSO was added to the clay film on a ceramic tile. The sample was initially dried under suction and then in an oven at 110 °C. Improved intercalation was obtained by soaking the clay on the ceramic tile in DMSO and heating to 65 °C for 10 min, repeating the process twice, and then adding enough of the DMSO solution to moisten the hot sample. Intercalation with hydrazine used the method of Ledoux and White (1966). Intercalation using ethylene glycol (Brunton, 1955) and glycerol (McEwan, 1946) in the vapour phase used standard techniques. For intercalation with paraquat (1,1'-dimethyl-4,4'-bipyridinium dichloride), clay-sized samples were suspended for 1 h at room temperature in 5% solutions of the compound, and the clay was then mounted as a film on a ceramic tile. Paraquat gives an (001) spacing of 12.6 Å for montmorillonite and 14.5 Å for vermiculite (Burchill *et al.*, 1981), and it does not change the layer spacings of the hydrous micas or of the 1:1- and the 2:1:1-layer clays. The intersalation procedure used followed that described by Jackson and Abdel-Kader (1978).

3 Results and Discussion

The contents of OM associated with the different size fractions studied are given in Table 1. These data show that the OM contents increased as the sizes of the particles decreased. Most of the light fraction of the OM (LFOM) was associated with the coarsest of the particle sizes (40-mesh) investigated, and the least amounts were associated with the clay-size components. After removal of the LFOM, the OM contents of the 40- and 100-mesh size particles were similar for the three depth bands sampled (Table 1). The contents of OM associated with the clay-sized materials in samples MC_2 and MC_3 were the same, though less than the amounts associated with the clays in the surface soil. These data indicate that there were significant amounts of OM in the soil to a depth of 90 cm.

Aspects of the compositions and some properties of the 100-mesh size soil samples are given in Table 2. These data show that there was a significant content of OM in the 100-mesh fractions, and though the amounts decreased with depth, the content was still relatively high at the 50 - 90 cm depth. The clay, free iron oxide, and exchangeable aluminium contents, and the pH values increased with depth, whereas the CEC values decreased. The 1:1 layer clays (kaolinite, and with some halloysite) were predominant inorganic colloids, and significant amounts of gibbsite and goethite were also present.

Table 1 *Organic matter contents of the different size fractions of the Tropical soil before and after removing the light fraction (LF), and the LF content as a % of the soil and of the soil organic matter (SOM)*

Sample Fraction	Organic matter Content (%)		Light Fraction (LF)	
	Before Removing LF	After Removing LF	As % of Soil	As % of SOM
*MC_1				
40-Mesh	4.7	3.4	1.3	28.2
100-Mesh	9.9	8.8	1.1	11.2
Clay (< 2 μm)	13.3	13.0	0.24	1.8
*MC_2				
40- Mesh	3.9	3.3	0.6	16.1
100-Mesh	8.5	8.3	0.3	3.2
Clay (< 2 μm)	10.9	10.8	0.1	0.8
*MC_3				
40- Mesh	3.4	3.1	0.3	7.7
100-Mesh	7.2	7.2	0.04	0.5
Clay (< 2 μm)	10.6	10.6	0.02	0.2

* Samples taken from the 0-20 cm (MC_1), 20-50 cm (MC_2), and 50-90 cm (MC_3) depths

Neutral pyrophosphate dissolved about 30% of the OM in the exhaustive extraction process (Table 3). Pyrophosphate complexed divalent and polyvalent cations (which immobilised the HS) allowing the more highly charged fractions (at pH 7) to be solvated (in water). These fractions would be expected to have significant carboxyl functionality. An additional 27% of the OM was isolated from each of the three samples by the subsequent application of the 0.5M NaOH solution. However, it is appropriate to note that between 41% and 44% of the OM was not extracted by the solvent sequence used. In the classical definitions, the OM which resists dissolution in aqueous solvent systems is classified as humin, and these are considered to have low acidic functionality and to be relatively inert. However, much of the humin material is extractable in a dimethylsulphoxide/HCl solvent system (Hayes, 1985; Clapp and Hayes, 1996), and the DMSO/HCl soluble extract has acidity similar to that in FA/HA systems. It would appear that the associations between the HS and the mineral colloids cause the polar groups to orientate towards the inorganic colloid surfaces [to which they are held by cation bridging, and coulombic attraction, as in the cases of (hydr)oxides below their points of zero charge, and possibly to some extent by weaker energy attractive forces]. Such associations would cause the hydrophobic faces of the macromolecules to orientate to the exterior and to provide resistance to swelling in the aqueous media. It is possible also that the HS can be held by ligand exchange [with (hydr)oxide species], and molecules that are so held would not be released in the aqueous solvents.

The medium in which the CEC was determined by the method of Polemio and Rhodes (1977) had a pH value of 8.2. That value was, of course, significantly higher than the pH of the soil, and (vide infra) gave CEC values higher than those which operate under field conditions (pH values of the soil fractions are given in Table 2). The data trends would suggest that the OM was a major contributor to the CEC, especially in the case of the

Table 2 *Compositions and some properties of the 100-mesh size Tropical soil samples*

Sample	Depth (cm)	Inorganic Carbon Content (%)*	Total Organic Matter (%)*	Cation Exchange Capacity (cmol kg^{-1})*	Clay Content (%)*	Free Iron Oxide (%)*	Exchangeable Aluminium (cmol kg^{-1})*	pH	Clay Mineralogy
MC_1	0-20	0.002	9.9	24.2	11.8	2.6	1.0	4.4	Kaolinite, Gibbsite, Goethite (Halloysites)
MC_2	20-50	0.005	8.5	18.6	15.2	2.9	1.4	4.6	Kaolinite, Gibbsite, Goethite (Halloysites)
MC_3	50-90	0.02	7.2	16.0	17.0	3.2	3.2	5.9	Kaolinite Gibbsite Goethite (Halloysites)

*Measurements were made on the 100-mesh fraction of the soil samples

Table 3 *Some properties of the humic extracts isolated in pyrophosphate (Pyro, pH 7) and in sodium hydroxide (NaOH)*

Sample	Organic matter extracted				CEC values of combined Pyro and NaOH extracts at			
	With pyrophosphate (pH 7)		*With 0.5M NaOH*					
	As % of soil	As % total SOM*	As % of soil	As % total SOM	pH 4.0	pH 5.3	pH 6.0	pH 6.9
MC_1	3.2	32.3	2.6	26.4	243	341	364	354
MC_2	2.6	30.5	2.3	27.2	209	297	360	378
MC_3	2.0	29.1	2.0	27.5	214	290	351	399

SOM = soil organic matter

Table 4 *Cation exchange capacity (CEC) data, by the ^{22}Na isotope dilution method, for a kaolinite and for the 100-mesh soil fractions taken from the Tropical soil at 0-20 cm (MC_1), 20-50 cm (MC_2), and 50-90 cm (MC_3) depths in the profile. A and B refer to samples from which the organic matter (OM) was removed and was not removed, respectively*

Sample	*CEC (cmol kg^{-1})*					
	pH 4.0	pH 5.3	pH 6.0	pH 6.9	pH 8.0	pH 9.0
Kaolinite (Georgia)	5.2	6.3	6.75	6.7	7.5	
MC_1B (100-Mesh)*	13.1	17.3	19.7	25.2	27.5	34.0
MC_1A (100-Mesh)*	11.7	14.9	16.3	18.9	21.0	22.3
MC_2B (100-Mesh)*	11.9	15.3	18.9	22.0	25.0	29.0
MC_2A (100-Mesh)*	11.5	15.0	18.8	20.0	22.3	23.4
MC_3B (100-Mesh)*	12.9	17.8	17.8	20.4	23.9	26.5
MC_3A (100-Mesh)*	11.7	16.2	17.8	19.3	22.2	24.1

* Refers to the sieve mesh size used in isolating the sample

surface soil sample. At the lower pH values (pH 4 - 6), the CEC values were low and the differences in the values for the 100-mesh fractions from which the OM had been removed and that which contained the OM were small (Table 4). However, at pH 6.9 and above, the contribution of the pH dependent charges (which includes contributions from OM) to the total CEC increased. The point of zero charge (PZC) values of the (hydr)oxide components are not known, but the data indicate that these contributed to the negative charge as the pH was increased. In the pH range 7 to 9 the CEC was a linear function of OM content of the untreated samples (the linear correlation coefficient being 0.99 at pH 6.9, 0.98 at pH 8.0, and 0.99 at pH 9.0), and that would suggest that the pH dependent charges on the OM were major contributors to the CEC at the higher pH values. This was especially so in the case of the 100-mesh size surface soil sample, where the contribution of the OM to the CEC (i.e. the difference in CEC before and after the removal of the OM) was greatest. This contribution was especially large above pH 6.0.

The contribution of the OM to the CEC measured for the subsoil was significantly less than that of the surface soil, and the values were minimal at pH 6 for the MC_2 and MC_3 samples. It would seem that at pH 6, under the experimental conditions used, the contributions to the CEC of the organic and inorganic colloids balance.

The data for the CEC determinations would suggest that there were major differences in the charge densities of the organic matter in the surface and subsurface soils at the different pH values. That was not, however, substantiated by the CEC values for the humic substances isolated from the 100-mesh size fractions from the surface and subsurface soils (Table 3). The per cent OM extracted (i.e. the sum of the extracts in Pyro and in NaOH) at the different depths was similar for the two solvent systems. Also as the pH was raised, the CEC values of the HAs (combined Pyro and NaOH extracts) increased, and the charge characteristics (as deduced from the CEC values) of the HAs from the 100-mesh size samples in the pH range 4.0 to 6.9 (Table 3) were relatively similar.

The CEC data for the HAs at the different pH values indicate the extents to which the acidic functionalities were dissociated at the different pH values. The strong acids, such as carboxyl groups having substituents alpha to the carboxyl, are ionized at pH 4 (Leenheer *et al.*, 1995) and most other carboxyl functionalities are ionized at pH 6-7. CEC values were not measured above pH 6.9 for the HAs because the HAs began to dissolve at the higher pH values. Activated phenols can begin to make contributions to the CEC at pH values > 7, and so the CEC data for the HAs in Table 3 do not take account of contributions from phenolic and weaker acidic functionalities. The data would suggest that the subsoil samples were more highly humified, and the trend indicates that these would make higher contributions to the CEC at the higher pH values.

The differences in the CEC values at the higher pH values (and especially at pH 9) cannot therefore be attributed to differences in the compositions of the OM associated with the inorganic components. It is likely that these difference can be attributed to the relative ease of accessibility to the exchange sites in the surface and sub-surface samples. That could be related to differences in the ways in which the organic macromolecules were associated with each other in the different soil layers.

4 Conclusions

The OM in the 0-90 cm depth of the Tropical soil studied can be expected to be a major contributor to the soil cation exchange reactivity that is essential for soil fertility. This OM composed from 7% to 10% of the 100-mesh size fraction, more than 3% of the 40-mesh samples, and 10-13% of the mass of the clay-sized fraction. Because the clays were kaolinitic, the organic matter would be the major contributor to the negatively charged colloids. The (hydr)oxides of iron and aluminium, positively charged at the ambient soil pH, can be expected to hold the organic polyelectrolytes by coulombic attraction and thereby limit the availabilities of exchange sites for exchangeable cations. The strong associations between the OM and the inorganic colloids are likely to provide protection against microbial breakdown. It would appear also that these associations limit the ease of penetration of aqueous solvents into the macromolecular matrix.

Acknowledgements

We are grateful for the studentship provided by the British Council that enabled the senior author to research in Birmingham

References

Allison , L.E. 1960. Wet combustion apparatus and procedure for organic and inorganic carbon in soils. *Soil Sci. Soc. Amer. Proc.* **24**:36-40.

Allison, L.E. 1965. Organic carbon. p. 1379-1396. *In* C.A. Black (ed.), *Methods of Soil Analysis, Part 2.* American Society of Agronomy, Madison, Wis., USA.

Allison, L.E. and C.D. Moodie. 1965. Carbonates. p. 1367-1378. *In* C.A. Black (ed.), *Methods of Soil Analysis, Part 2.* American Soc. Agronomy, Madison, Wis., USA.

Anderson, J.U. 1963. An improved pretreatment for mineralogical analysis of samples containing organic matter. p. 380-388. *In Proc. 10 th nat. Conf., Clays and Clay Min.*

Brunton, G. 1955. Vapour phase glycolation of oriented clay minerals. *American Mineralogist* **40**:124-126

Burchill, S., M.H.B. Hayes, and D.J. Greenland. 1981. Adsorption. p. 221-400. *In* D.J. Greenland and M.H.B. Hayes (eds), *The Chemistry of Soil Processes.* Wiley, Chichester.

Clapp, C.E. and M.H.B. Hayes. 1996. Isolation of humic substances from an agricultural soil using a sequential and exhaustive extraction process. p. 3-11. *In* C.E. Clapp, M.H.B. Hayes, N. Senesi, and S.M. Griffith (eds), *Humic Substances and Organic Matter in Soil and Water Environments.* Proc. 7th. Intern Conf. IHSS (St. Augustine, Trinidad and Tobago, 1994). IHSS, St. Paul, Minnesota.

Day, P.R. 1965. Particle fractionation and particle-size analysis. p. 545-566. *In* C.A. Black (ed.), *Methods of Soil Analysis.* Part 1. American Soc. Agronomy, Madison, Wis., USA.

Dijkerman, J.C. and R.T. Odell. 1976. Properties, classification and use of tropical soils with special reference to those in Sierra Leone. *Njala University College Bull.*, University of Sierra Leone.

Edwards, A.P. and J.M. Bremner. 1967. Dispersion of soil particles by sonic vibration. *J. Soil Sci.* **18**: 47-63.

Greweling, T. and M. Peech. 1960. Chemical Soil Tests. *Cornell Univ. Agric Exp. Sta.*, Bull **960**.

Grossman, R.B. and J.C. Millet. 1961. Carbonate removal from soils by a modification of the "Acetate Buffer method". *Soil Sci. Soc. Amer. Proc.* **25**:325-326.

Hart, M.G.R., A.J. Carpenter, and J.W.O. Jeffery. 1963. Problems in reclaiming saline mangrove soils in Sierra Leone. *L'Agronomie Tropicale* **18**:800-802.

Hayes, M.H.B. 1985. Extraction of humic substances from soil. p. 329-362. In G.R. Aiken, D.M. McKnight, R.L. Wershaw, and P. MacCarthy (eds), *Humic Substances in Soil, Sediment, and Water.* Wiley, NY.

Hesse, P.R. and J.W.O. Jeffery. 1963. Some properties of Sierra Leone mangrove soils. *L'Agronomie Tropicale* **18**:803-805.

Heyrovsky, J. and P. Zuman. 1968. *Practical Polarography.* Acadwemic Press, London.

Jackson, M.L. and F.H. Abdel-Kader. 1978. Kaolinite intercalation procedure for all sizes and types with X-ray diffraction spacing distinctive from other phyllosilicates. *Clays and Clay Minerals* **26**:81-87.

Kunze, G.W. 1965. Pretreatments for mineralogical analysis. p. 568-577. In C.A. Black (ed.), *Methods of Soil Analysis.* Part 1. Am. Soc. Agron., Inc., Madison, Wis., USA.

Ledoux, R.L. and J.L. White. 1966. Infrared studies of hydrogen bonding between kaolinite surfaces and intercalated potassium acetate, hydrazine, formamide, and urea. *J. Colloid and Interface Sci.* **21**:127-152.

Leenheer, J.A., R.L. Wershaw, and M.M. Reddy. 1995. Strong-acid, carboxyl group structures in fulvic acid from the Suwannee River, Georgia. I. Minor structures. *Env. Sci. and Technology* **29**:393-398.

McEwan, D.M.C. 1946. The identification and estimation of montmorillonite group of minerals with special reference to soil clays. *J. Soc. Chem. Ind.* **65**:298-305.

Mehra, O.P. and M.L. Jackson. 1960. Iron oxide removal from soils and clays by a "Dithionite-Citrate" system buffered with sodium bicarbonate. p. 317-327. Proc. 7th Nat. Conf. Clays and Clay Minerals.

Odell, R.T., Dijkerman, J.C., W. van Vuure, S.W. Melsted, A.H. Beavers, P.M. Sutton, L.T. Kurtz, and R. Miedema. 1974. Characteristics, classification and adaptation of soils in selected areas in Sierra Leone and West Africa. *Illinois Agric. Exp. Sta. Bull.* ***748.*** *Njala University College Bull. 4.*

Oladimeji, O.M. 1976. *Mineralogical and Chemical Studies of Tropical Soils.* Ph.D. Thesis, The University of Birmingham.

Olejnik, S., S. Aylemore, A.M. Posner, and J.P. Quirk. 1968. Infrared spectra of kaolin mineral-dimethylsulphoxide complexes. *J. Phys. Chem.* **72**:241-249.

Olson, R.V. 1965. Iron. p. 963-973. In C.A. Black (ed.), Methods of Chemical Analysis. Part 2. American Society of Agronomy, Inc.Madison, Wis.

Peech, M. 1965. Hydrogen ion activity. p. 914-925. In C.A. Black (ed.), *Methods of Soil Analysis.* Part 2. Am. Soc. Agron., Inc., Madison, Wis., USA.

Peech, M., L.A. Dean, L.T. Alexander, and J.F. Reed (1947). Methods for soil analysis for soil fertility investigations. U.S. Department of Agriculture. Circular **757.**

Polemio, M. and J.D. Rhoades. 1977. Determining cation exchange capacity - a new procedure for calcareous and gypferous soils. *Soil Sci. Soc. Am. J.* **41**:524-528.

Turchenek, L.W. and J.M. Oades. 1979. Fractionation of organo-mineral complexes by sedimentation and density techniques. *Geoderma* **21**:331-343.

Yormah, T.B.R. 1981. *The Composition and Properties of Selected Samples from a Tropical Soil Profile.* Ph.D. Thesis, The University of Birmingham.

Section 2

Interactions of Humic Substances

Reference was made in Section 1 (p. 5) to the role of humic substances (HS) in the formation and stabilization of soil aggregates, and to their influences in the binding of anthropogenic organic chemicals (AOCs). Though the role of HS in the formation and stabilization of soil aggregates is not elaborated in this Volume, considerable emphasis is placed on interactions between humic fractions and AOCs.

Pre-emergent, or soil applied herbicides were introduced in the 1950s, and it has been known since then that the biological activities of sparingly soluble and of aromatic compounds are greatly decreased, and in some cases inactivated when these are applied to soils with high organic matter (OM) contents. It was soon established that it is the humic components of OM that cause the inactivations. Thus, interest has focused on the interactions with solid state HS because it has been difficult to work with HS in solution.

The contribution by Law *et al.* (p. 189 - 198) describes an automated procedure for studies of sorption by HS using a combination of continuous-flow (CF) and flow-injection analysis (FIA) processes. The procedure describes the sorption of copper by humic acids (HAs) dispersed (but not dissolved) in water in a stirred reaction vessel. The outflow from the vessel was through a membrane permeable to water and the sorptive [Cu(II)], but not to the sorbent (the HA). To be effective, the sorptive should not interact with the membrane or with the fabric of the cell, and should not bind to the cell walls or sorb to, or be rejected by the cell membrane. Thus, using this type of apparatus it is possible to study interactions with dissolved sorbents (having molecular sizes larger than the pore sizes of the retaining membranes) of metals and of organic organic sorptive species. The article by Law *et al.* shows how data can be obtained which allow the construction of sorption and of desorption isotherms. It is possible to completely automate the system.

The aromatic functionalities of the imidazolinone herbicides suggest that these would bind to HAs in ways similar to those which bind the *s*-triazines. It is now known that the imidazolinone compounds retain considerable phytotoxicity even in soils with relatively high OM contents. This feature can be predicted from the contribution by Häusler and Hayes (p. 199 - 206) for the sorption of four imidazolinone compounds by H^+- and by Ca^{2+}-exchanged HA preparations. The data show that the shapes, pK_a values, the extents of ionization of the sorptives at different pH values, and the properties of the inorganic cations neutralizing the charges on the sorptives have roles to play in determining the binding of the imidazolinone compounds by HAs. The data indicate that there would be little binding of the herbicides in agricultural soils even when the OM content is high because the inorganic cations in soils are predominantly divalent and trivalent, and the ambient pH would be significantly above the pK_a of the sorptive species.

There is environmental interest in the movement of organic chemicals and metals from soils and from landfill sites to surface and to ground waters. HS can, as stressed above, bind both metals and certain types of organic chemicals. The view is also held that HS in solution can enhance the water solubilities of sparingly soluble AOCs. Should that be so,

HS dissolved in the soil solution, in drainage waters from soils, and in leachates from landfill sites would enhancethe solubilities of various AOCs contacted.

The contribution by Hayes *et al.* (p. 207 - 217) has examined the influences of HAs, fulvic (FAs) and XAD-4 acids on the solubilities of five biocides with low solubilities in water. Their data show that the solubilities in water at pH 4 or at pH 6 of the chemicals studied were not enhanced by the presence of the humic fractions, and in some instances the HS gave rise to some suppression of the solubilities of the AOCs in water. However, the soil solution is not distilled water, and is equivalent to ca 0.01M with respect to salts. Additions of salts (calcium or sodium) to give solutions 0.01M (with respect to these) caused, in some instances, a salting out of the AOCs in the absence of humic fractions. It was evident that in some instances the presence of HS acted as protective colloids and prevented the salting out of the AOCs. However, the salts were added after the AOCs had interacted with the humic fractions. There is a need to carry out the experiments described in situations in which the humic fractions are dissolved in water that is 0.01M with respect to the salts, and to add this mixture to the sparingly soluble AOCs.

HS can be free in soils, that is be associated only with humic molecules, or they can be associated with other organic molecules, and with the inorganic soil colloids. There is not complete agreement about the ways in which HS interact with the inorganic colloids. The usual concepts presuppose that sorption takes place from solution. Because the ambient pH in the soil environment is above the pK_a values of the strong acid functionalities in HS, it can be assumed that the HS in solution will be negatively charged. That suggests, of course, that there should be repulsion between the HS and the negatively charged clays. Sorption by clays can be rationalized by considerations of cation-bridging mechanisms involving divalent and polyvalent cations. Below their points of zero charge (PZC), the (hydr)oxides of the soil environment will have positive charges. Thus the HS will be held to these by coulombic attraction. As yet there is not a satisfactory explanation for the nature of humic-inorganic colloid complexes which give rise to the so-called humin fraction of HS. (Humins, when detached from the inorganic colloids, e.g. by dissolution in dimethylsulphoxide, are often seen to have properties similar to those of HAs and FAs.)

The contribution by Lacey *et al.* (p. 218 - 224) is based on the principle that clays pillared with iron and with aluminium will present surfaces that are comparable to those provided by (hydr)oxides in the soil environment. That by Farnworth *et al.* (p. 225 - 235) takes a similar approach for aluminium (hydr)oxide species sorbed to Beringite, a calcined aluminosilicate formed from palaeozoic schists. The greater surface area provided by the Al (hydr)oxide pillared clays led to more significant sorption of the HS than was observed for the Fe (hydr)oxide species. HS in these studies were sorbed from aqueous solution, and the results provide indications of the suitabilities of the sorbents to remove colour from water. Pillaring by iron species has proved to be difficult, but the study by Lacey *et al.* produced an appropriate product using microwave energy.

The contribution by Norman *et al.* (p. 236 -244) highlights some of the problems which HS can pose in industrial processes. The bauxite recovered from open cast mining in the Darling Range of Western Australia includes 'contaminating' soil with its indigenous organic matter, as well as dead vegetation. The alkaline conditions and the elevated temperatures of the Bayer process transform the indigenous HS and the dead organic materials to give rise to new humic-type substances (H-TS). Because the process is cyclic, considerable concentrations of the H-TS can build up in the reaction digests, and these can present problems in the winning of high quality alumina from the bauxite ore. Some of the problems that are encountered have been studied and the results are discussed.

An Automated Procedure for Studying Sorption by Humic Substances Using a Combined Continuous-Flow and Flow-Injection Analysis Process

I.A. Law, J.J. Tuck, C.L. Graham, and M.H.B. Hayes

THE UNIVERSITY OF BIRMINGHAM, SCHOOL OF CHEMISTRY, EDGBASTON, BIRMINGHAM B15 2TT, ENGLAND

Abstract

A method is described for studying sorption from solution based on a continuous-flow system. In this system a continuous-flow cell is interfaced to a flow injection analysis (FIA) step using a computer-controlled auto-injection valve. The applicability of the procedure is demonstrated using a humic acid (HA) suspension under dynamic conditions and Cu(II) as the sorptive. From the known flow rate and the initial concentration of Cu(II) in the reservoir, measurement of the concentration of Cu(II) in the eluate allowed the amount of Cu(II) sorbed by the humic suspension to be calculated. Signals from the FIA detector were monitored by computer, and this subtracted digitally the background from each 'peak', measured the maximum deflection, and stored the corrected profile on disc. A complete 'wash-in' curve was built up in this way, and the data were then available for further manipulation to construct sorption isotherms.

1 Introduction

The terms sorption, adsorption, absorption, and desorption have been defined by IUPAC (1972). Briefly the term sorption is used when it is not possible to distinguish between surface-limited adsorption and the related process of absorption. Similarly the terms ad-or absorptive (used to describe the species being ad- or absorbed), and ad- or absorbent (used to describe the species doing the ad- or absorbing), can be replaced by the terms sorptive and sorbent. The process by which a macromolecular suspension (e.g. of humic acid) accumulates smaller molecules is best described using the terms sorption, sorbent, and sorptive.

Continuous-flow techniques for studying adsorption mechanisms have been known for some time (Blatt *et al.*, 1968; Ryan and Hanna, 1971; Crawford *et al.*, 1972). Generally the method is based on the ultrafiltration technique in which the sorptive solution of a known, fixed concentration is introduced into a continuously-stirred ultrafiltration cell by means of a peristaltic pump, or by gas pressure. A suspension or solution of the sorbent is held in the cell by an ultrafiltration membrane permeable to the sorptive but not to the sorbent. Traditionally, the eluate was sampled after the passage of a known volume of sorptive, collected by use of a fraction collector, and then analysed for the sorptive by an appropriate technique. In the continuous flow method the eluate concentration can be predicted as a function of cumulative filtration volume - or of time, if the flow rate is known and constant, - by evaluating a differential mass balance around the cell.

This method has been used to study the sorption behaviour of a range of materials, such as proteins and soil colloid systems (Ryan and Hanna, 1971; Grice and Hayes, 1972; Grice *et al.*, 1973; Burchil and Hayes, 1980; Smedley, 1978). Attempts have been made to automate the system and to dispense with the fraction collection step (e.g. Hartmann and Randle, 1980; Isaacson and Hayes, 1984). This work focuses on the development of automated procedures for measurements of sorption of metal cations by a humic acid (HA) preparation using the combination of flow injection analysis (FIA) with the continuous-flow stirred-cell method (CFSC), and making use of a microprocessor controlled auto-injection valve. By adopting this approach, the sorption of Cu(II) ions by HAs extracted from freshwater sediments was studied.

2 Theory

Mathematical representations of the CFC method have been presented in earlier work on the subject (Blatt *et al.*, 1968; Ryan and Hanna, 1971). Burchill and Hayes (1980) derived equations from a consideration of mass balance around the cell. The fundamental equation for such a system is:

$$C_{max}\,dv - C_i\,dv = dn_i \tag{1}$$

where C_{max} is the concentration of sorptive in the reservoir, dv is the eluate volume increment, C_i is the concentration of sorptive in the eluate (regarded as being equal to its concentration within the cell, and n_i is the total amount of solute i within the cell, both bound to the sorbent (n_i^a) and as free sorptive in solution (VC_i).

From equation (1), the following two equations are derived:

$$C_i = C_{max}\,[1 - \exp(-v_i/V)] \tag{2}$$

$$C_i = C_{max}.\exp(-v_i/V) \tag{3}$$

where V is the cell volume, and v_i is the (cumulative) volume eluted. Equation (2) represents the theoretical 'wash-in' or dilution curve. This is obtained when a solution of sorptive, i, at concentration C_{max} is pumped into a stirred cell where the initial concentration of component i (C_i) is zero, and where no sorbent is present. Equation (3) represents the equivalent "wash out" curve, obtained when a solution free of component i is used to replace the sorptive solution after C_i has reached C_{max}. These two hypothetical curves, together with the corresponding curves obtained when a sorbent material is present in the cell, are shown in Figure 1.

When sorbent is present in the cell, the sorption equation derived from equations (1) and (2) is:

$$n_i^a = C_{max}V_i - C_iV\int_0^{v_i}(\text{sign})_i\,dv \tag{4}$$

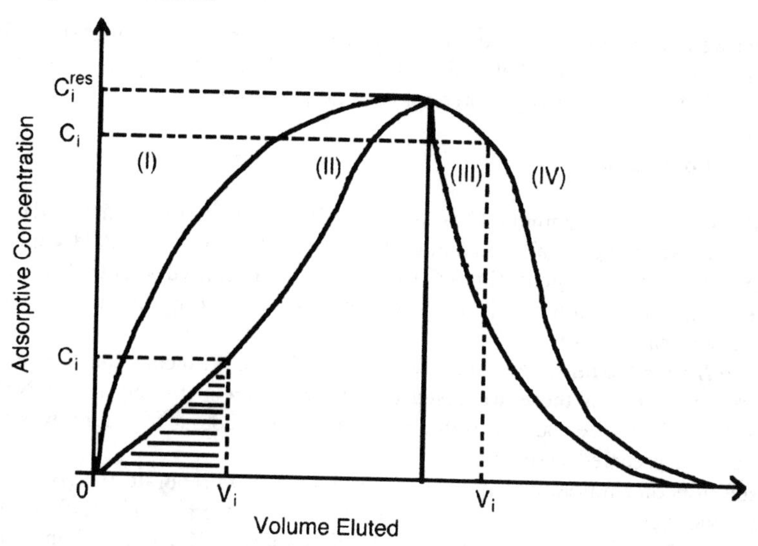

Figure 1 *Profiles obtained in practice from the CFSC technique (after Smedley, 1978)*
Curve I 'wash-in' of sorptive in the absence of sorbent
Curve II 'wash-in' of sorptive in the presence of sorbent
Curve III 'wash-out' of sorptive in the absence of sorbent
Curve IV 'wash-out' of sorptive in the presence of sorbent
The shaded area on the left hand side (lhs) represents the amount of solute that has left the cell. The 'box' on the right hs deals with the desorption mode which is not discussed

The value for the integral is obtained from the area under the wash curves for the appropriate values of C_i and v_i. The plot of n_i^a against C_i represents the sorption isotherm for the particular system under study. This form of equation (4) implies that it will be strictly applicable only to small volume increments entering the cell. The use of FIA and the sampling of the eluate by auto-injection makes the sampling of small eluate volumes possible, and allows the restrictions applied by equation (4) to be met.

3 Experimental

Extraction of Humic Acids

A freshwater sediment was extracted under dinitrogen gas using sodium pyrophosphate solution (0.1M) neutralized with phosphoric acid. Humic acids (HAs) were precipitated from the extract by acidification to pH 1.0, then dialyzed against distilled water, and freeze dried.

Fractionation of Humic Acids

The HAs were dissolved in sodium tetraborate (0.025M) and fractionated on the basis of molecular size by ultrafiltration, using a Sartorius SM11730 membrane (nominal molecular weight cut-off = 160 000 D). The fraction of nominal molecular size > 160 000 D was reprecipitated in acid, dialyzed, and freeze dried. From this a stock solution of 1 mg mL^{-1} of HA at pH 6.0 was prepared for use in the sorption studies.

Preparation of Solutions

Copper. A solution containing 1 mg mL^{-1} Cu(II) was prepared from Cu(II) sulphate pentahydrate (AR grade). This was diluted as necessary to provide standard solutions.

Cuprizone. The reagent [bis(cyclohexinone)oxalyldihydrazone; 0.15 g)] was dissolved in a hot solution of ethanol (40% v/v), cooled, and diluted to 100 mL. This reagent was prepared daily.

Citrate Buffer Mixture. Several buffer mixtures for use with cuprizone are reported in the literature. Most of these are based on citrate, either in the NH_4^+- or Na^+-forms (Marczenko, 1976). We found that the maximum rate of colour formation was obtained using a mixture prepared as follows:

A solution of ammonium citrate (0.5M) in sodium tetraborate ($Na_2B_4O_7.10H_2O$, 0.015M), was obtained by neutralizing 500 mL of 1.0M citric acid (AR grade) with concentrated ("0.88") ammonia liquor. To this was added 5.72 g of disodium tetraborate and distilled water, to a total volume of 900 mL. When the borate had dissolved, the solution was transferred to a 100 mL volumetric flask, and diluted to the mark with distilled water. The pH of the solution was adjusted to 8.8 - 9.0 by dropwise addition of

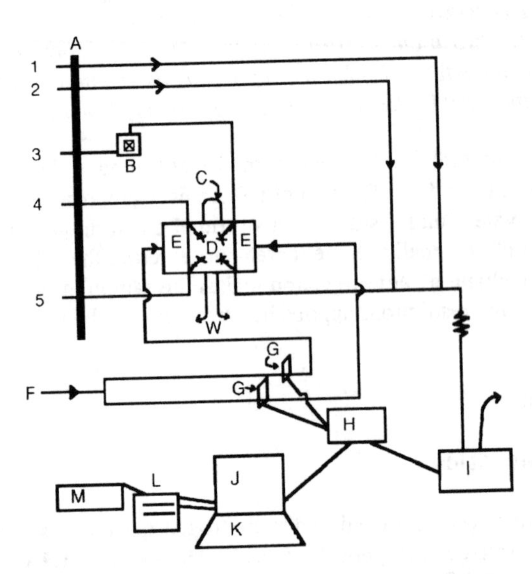

Figure 2 *A schematic representation of the CFSC-FIA equipment*
A: Peristaltic pump;
B: Continuous-flowcell;
C: 50 μL sample loop;
D: Two Altex, 4-way valves;
E: Pneumatic actuators;
F: N_2 cylinder (75 p.s.i.);
G: Solenoid valves;
H: Electronic interface and controller;
I: Spectrophotometer;
J, K: VDU and BBC microcomputer;
L: Double disk drive;
M: Printer.

10M NaOH immediately before use.

Equipment. The complete arrangement of the cell, valve, and associated hardware is shown in Figure 2. A detailed diagram of the cell assembly is shown in Figure 3. All connections to the cell were made using PTFE tubing, 0.5 mm i.d., via Omnifit gripper fittings and end pieces (Omnifit Ltd., Cambridge).

Peristaltic pump tubing of nominal flow rate 0.24 mL min^{-1}, was used for the carrier streams; the Cu(II) stream, cuprizone, and buffer were all carried in tubing of nominal flow rate 0.1 mL min^{-1}.

One Ismatec 820 peristaltic pump was used for both the CFSC and the FIA systems.

A cellulose acetate ultrafiltration membrane (Sartorius), with a nominal molecular weight cut-off of 20 000 D, was used to retain the sorbent within the the CFSC.

The auto-injection valve was constructed from two 4-way Altex valves connected as shown in Figure 4, and this was activated pneumatically when a signal was received (via a home-made interface) from the microcomputer.

The solenoids used to control the pressurized gas were purchased from RS Components, Birmingham.

A Cecil Series 2, Model 272 UV-visible spectrophotometer, fitted with a Hellma flow cell of 18 mL internal volume, was used to measure the absorbance of the Cu-cuprizone complex at 600 nm.

Adaptation of Cuprizone to FIA

Procedure. It was found from a series of batch and FIA experiments that the development of the Cu-cuprizone complex was achieved most rapidly using reagent and buffer solutions of the compositions noted above. However, the plot of colour development [concentration of Cu(II) ions] versus time was not linear at concentrations of 20.75 μg mL^{-1} Cu(II) (as shown in Figure 5) for the reaction times necessary to maintain a relatively high rate of sampling of cell eluate. This effect was overcome in the course of

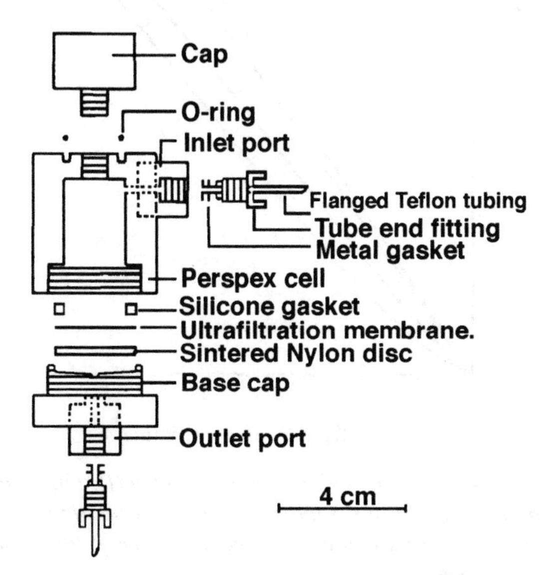

Figure 3 *Detailed diagram of the ultrafiltration cell*

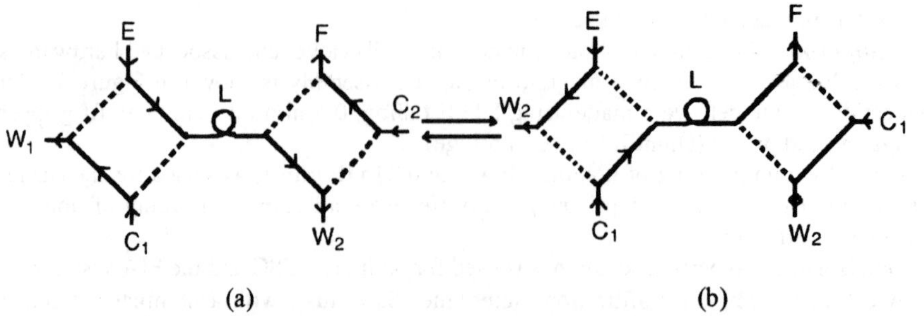

Figure 4 *A schematic representation of two "Altex" 4-way valves connected in tandem, and used to interface the CFSC system to the FIA manifold. Arrows denote directions of flow, and solutions entering or exiting from valve ports are: E = eluate from the flow cell; C1 and C2 = carrier streams; W1 and W2 = waste outlets; F = carrier or sample going to the FIA manifold. The letter L denotes the 50 μL sample loop. Position (a) represents eluate collection, while position (b) shows the 50 μL sample of eluate being introduced into a FIA manifold*

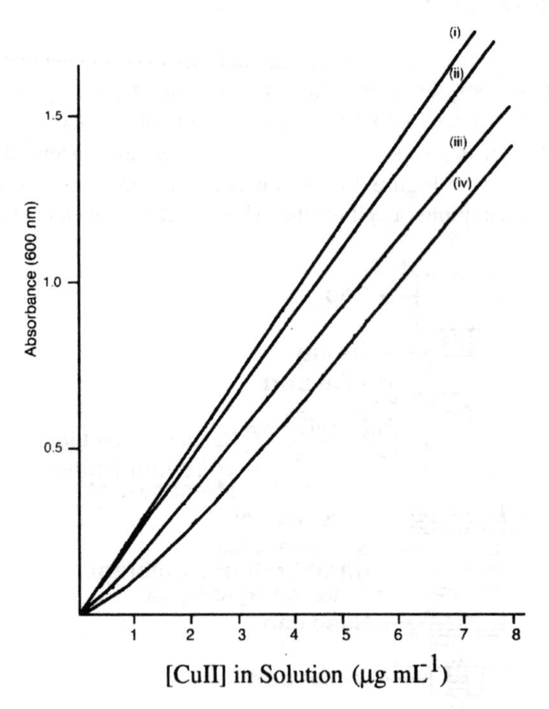

Figure 5 *The time dependence of colour formation of Cu (II) ions with cuprizone under the reaction conditions reported. (i) reaction time = 15 minutes; (ii) reactiontime = 180 seconds; (iii) reaction time = 90 seconds; (iv) reaction time = 30 seconds*

calculations of Cu(II) concentrations by using the microcomputer to fit a curve of the form $y = ax^b$ to concentration values < 0.75 μg mL^{-1}, and a straight line, $y = mx + c$, to values between 0.75 and 10 μg mL^{-1}. Values for the coefficients a, b, m, and c were calculated from the raw data obtained from calibration plots.

Investigation of the Continuous Flow Cell and Membrane

The cell volume, V, was measured by the method used by Hartmann and Randle (1980), and was found to be 8.0 mL. The total void volume of the system was calculated from a blank wash-in run, using the equation:

$$\ln\left(\frac{C_i^{res}}{C_i^{res} - C_i^{e}}\right) = \frac{v - v_o}{V} \qquad (5)$$

(where v_o is the voids volume; C_i^e is the eluate concentration [*e* denotes eluate]; v is the volume that has flowed through the cell; C_i^{res} is the reservoir concentration. A plot of v versus ln [C_i^{res}/(C_i^{res} - C_i^e)] will provide estimates of the void and cell volumes..

The behaviour of the ultrafiltration membrane towards Cu(II) ions was studied by following a blank 'wash-in' curve and comparing the result with the theoretical curve generated from equation (2). The two curves were seen to be identical (Figure 6), indicating that no sorption or rejection of Cu(II) ions occurred at the membrane surface.

The Experiment

The system as described requires that there should not be interactions between the cell (and parts leading to and from it) and the sorbent and sorptive species, and that there should not be interactions (sorption or rejection) between these species and the cell membrane.

The operational procedure used wass as follows. The cell was assembled, filled with distilled water, and connected to the peristaltic pump. An aliquot containing 1 mg HA, freshly converted to the H^+-exchanged form by the addition of H_2SO_4 (2M) to pH 1.0, was pumped into the cell, followed by distilled water until the pH of the eluate had risen to the required value. The primed cell was fitted into the apparatus and Cu(II) solution (8 μg mL^{-1}) flowed through at a rate of 0.193 ± 0.008 mL min^{-1} for 4 h (equivalent to the passage of ca six cell volumes of sorptive solution). The run was monitored in 180 sec cycles; 90 sec filling time for the sampling loop, followed by 90 sec wash period. The 50 μL sample collected and measured in that way represented the eluate from the last 16 sec of the 90 sec fill period, and can in effect be considered as a 'point value' on the 'wash-in' curve. When this is compared to a typical minimum fraction volume of 1.5 mL needed for analysis by, e.g. AAS, it can be seen that there is a significant gain in accuracy in the calculated isotherm, since 1.5 mL of eluate represents the change of concentration of sorptive within the cell over a period of 474 sec and is not a 'point value' on the 'wash-in' curve, but the average value over the period of collection

Data collected during the run were monitored by the microcomputer. This displayed the formation of each FIA signal on the VDU, converted peak heights to concentration

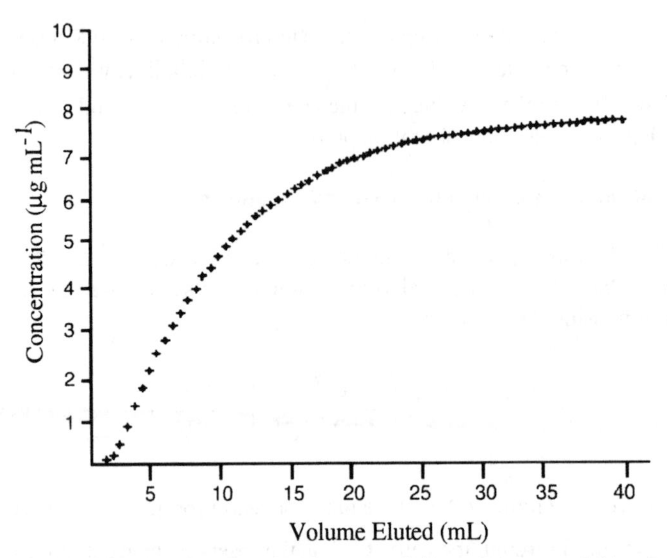

Figure 6 *'Wash in' curves showing the theoretical (-) and measured () values for Cu (II) ions in the absence of humic sorbent*

values, and calculated the amount of Cu(II) sorbed using equation 4. A plot of n_i^a versus C_i generated by the computer is shown in Figure 7.

4 Results and Discussion

'Wash-in' curves for Cu(II) at pH values of 4.5, 5.0, and 6.0 were measured for H^+-ion

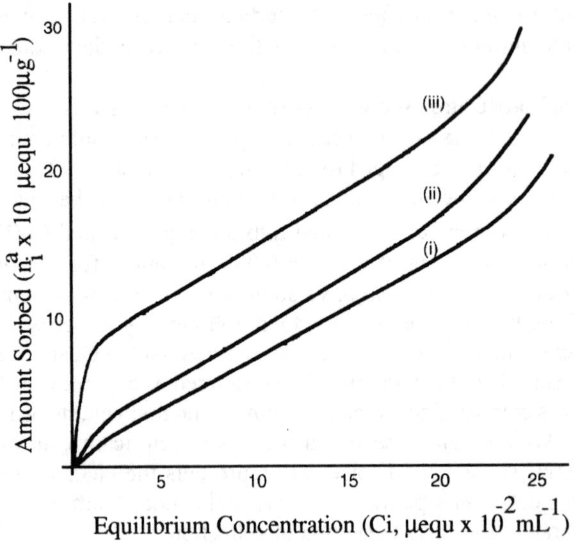

Figure 7 *Sorption isotherms for Cu (II) ions by H^+ humic acid at (i) pH 4.5, (ii) pH 5.0 and (iii) pH 6.0*

exchanged HAs and compared to theoretical "wash-curves" for the humic-free case. Isotherms for the 'wash-in' data are shown in Figure 7. Each isotherm is the average of three runs, and the variance between runs is within ± 3% over 80% of the isotherm.

The isotherms obtained by the CFSC-FIA technique are similar in shape and in the magnitude of sorption to those obtained by Hartmann (1981) for the sorption of Cu(II) by AAS. It should be noted, however, that the source materials for the HA samples were from different environments, and that the samples used by Hartmann were not fractionated. The increase in sorption with pH is consistent with the belief that copper complexes with carboxylic acid groups located throughout the macromolecules. It was noted that the pH of the cell contents at the end of each run was lowered by between 0.5 pH units (for sorption at pH 4.5) and 0.9 units (for sorption at pH 6.0). That suggests the involvement of an ion-exchange mechanism, with H^+ being released as Cu^{2+} was sorbed.

The shapes of the isotherms also support this hypothesis. The initial high affinity behaviour seen for the sample run at pH 6.0 is due to the immediate complexation of the Cu(II) ions with the ionized (COO^-) surface groups, followed by a slower sorption step where Cu(II) diffuses into the humic macromolecule and reacts by ion exchange, and by forming complexes. The absence of a rapid initial sorption of Cu at pH 4.5 is attributable to the low concentration of dissociated groups at the exposed surface of the humic colloid. The copper reacted by ion exchange, and as the sorbent structure was 'opened', the sorptive diffused to active sites within the macromolecule.

The changes in the slopes at the higher equilibrium concentrations represent errors inherent in processing data when the 'wash-in' curve obtained in the presence of the HA sorbent approaches that of the wash-in curve in the absence of the sorbent (Figure 1).

5 Conclusions

The aim of this work was to make the study of sorption and of desorption (though desorption data are not included) by continuous flow methods more rapid and more accurate by replacing the fraction collection step, and by increasing the sampling rate. These objectives were achieved by using a computer-controlled automatic-injection FIA step. The introduction of the microcomputer allowed the automation of the system, so that valve switching, calibration, background subtraction, data collection, storage, and processing were all done on line, and with a hard copy of the sorption isotherm made available at the end of the run. The use of cuprizone as a colorimetric reagent has allowed sensitive measurements of Cu(II) over the concentration range 0 to 8 $\mu g\ mL^{-1}$ to be made in aqueous media, and thereby avoiding complicated procedures such as in-line preconcentration and solvent extraction. Fitting the concentration data by microcomputer has allowed the FIA principle of measuring a reaction before the attainment of equilibrium to be applied, even though this leads to a non-linear calibration plot in the case of cuprizone. The use of stop-flow pumping or holding coils has also been avoided by this approach. The CFSC-FIA apparatus can be adapted easily to the study of sorption and desorption under dynamic conditions of a variety of sorbent-sorptive systems.

References

Blatt, W.F., S.M. Robinson, and M.J. Bixler. 1968. Membrane ultrafiltration: The diafiltration technique and its application to microsolute exchange and binding phenomena. *Anal. Biochem.* **26**:151-173.

Burchill, S. and M.H.B. Hayes. 1980. Adsorption of poly(vinyl alcohol) by clay minerals. p. 109-121. *In* A. Banin and U. Kafkafi (eds), *Agrochemicals in Soils.* Pergamon Press.

Crawford, J.S., R.L. Jones, J.M. Thompson, and W.D.E. Wells. 1972. Binding of bromosulphthalein sodium by human serum albumin using a continuous diafiltration technique. *Br. J. Pharmacology* **44**:80-88.

Grice, R.E. and M.H.B. Hayes. 1972. A continuous flow method for studying adsorption and desorption of pesticides in soil colloid systems. Proc. XI th Br. Weed Control Conf. (Brighton) **2**:784-791.

Grice, R.E., M.H.B. Hayes, P.R. Lundie, and M.H. Cardew. 1973. A continuous flow method for studying adsorption of organic chemicals by a humic acid preparation. *Chem. and Ind.* (London). p. 233-234.

Hartmann, E.H. 1981. Application of the continuous flow stirred cell technique to the study of the sorption of inorganic cations by hydrogen-ion exchanged humic preparations. Ph.D. Thesis, The University of Birmingham.

Hartmann, E.H. and K. Randle. 1980. A continuous-flow isotope dilution method for studies of adsorption behaviour of metal ions. *Anal Chem. Acta* **116**:275-287.

Isaacson, P.J. and M.H.B. Hayes. 1984. The interaction of hydrazine hydrate with humic acid preparations at pH 4. *J. Soil Sci.* **35**:79-92.

IUPAC. 1972. Division of Physical Chemistry. Manual of symbols and terminology for physicochemical quantities and units. Appendix II. Definitions, terminology and symbols in colloid and surface chemisrtry. Prepared by D.H. Everett. *Pure and Applied Chemistry* **31**:577-627.

Marczenko, Z. 1976. *Spectrophotometric Determination of Elements.* Ellis Horwood, Chichester.

Ryan, H.T. and N.S. Hanna. 1971. Investigation of equilibrium ultrafiltration as a means of measuring steroid-protein binding parameters. *Anal. Biochem.* **40**:364-379.

Smedley, R.J. 1978. Interactions of organic chemicals with clays and synthetic resins. Ph.D. Thesis, The University of Birmingham.

Sorption of Imidazolinone Herbicides by Humic Acid Preparations

Michael J. Häusler and Michael H.B. Hayes

THE UNIVERSITY OF BIRMINGHAM, SCHOOL OF CHEMISTRY, EDGBASTON, BIRMINGHAM B15 2TT, ENGLAND

Abstract

Isotherms were obtained and analysed for the sorption and desorption of a selection of imidazolinone xenobiotic organic chemicals by H^+- and Ca^{2+}-exchanged humic acid (HA) preparations isolated in a sodium hydroxide solution from a sapric histosol. The data show that the extents and the reversibilities of the sorption processes are governed by the compositions and the structures of the sorptive (sorbing) species, and by the exchangeable cation (H^+ or Ca^{2+}) satisfying the charges on the HA sorbent. To a considerable extent the sorption processes are influenced by the pK_a values of the sorptives, by their extents of ionization at the pH of the medium, and also by the different influences the cations (H^+ and Ca^{2+}) have on aspects of the structures (conformations) of the HAs. Sorption of the Ca^{2+}- exchanged HAs was small, and that would suggest that the compounds could have considerable biological activities in soils with high organic matter contents when the exchangeable inorganic cations are divalent or polyvalent.

1 Introduction

Los (1984) has described details of the synthesis of herbicidal imidazolinone compounds, discovered at American Cyanamid in 1983. The imidazolinone compounds used in this study were Compound **I**, methyl-6(4-isopropyl-4-methyl-5-oxo-2-imidazolin-2-yl)-*m*-toluate, and the *para* isomer methyl-6-(4-isopropyl-4-methyl-5-oxo-2-imidazolin-2-yl)-*p*-toluate (compound **II**). The *para* and the *meta* isomers both have a chiral carbon bearing a methyl and an isopropyl group. Each isomer has two enantiomers, and the R-enantiomer is two times more active herbicidally than the mixture. These compounds are marketed under the trade name Assert. Compound **III**, 2-(4-isopropyl-4-methyl-5-oxo-2-imidazolin-2-yl) nicotinic acid is marketed as Arsenal, and again the R-enantiomer has the highest activity. Compound **IV**, 2-(4-isopropyl-4-methyl-5-oxo-2-imidazolin-2-yl) quinoline-3-carboxylic acid has the trade name Sceptre. These compounds have solubilities in water at room temperature of the order of 50 mg L^{-1} for **I** and **II**, 250 mg L^{-1} for **III**, and 30 mg L^{-1} for **IV**, and the *n*-octanol/water partition coefficients are 66, 35, 1.3 (for the isopropylamine salt), and 2.5, respectively, for **I**, **II**, **III**, and **IV**.

Shaner *et al.* (1984) have shown that the free acids of **III** and of **IV** interfere with the

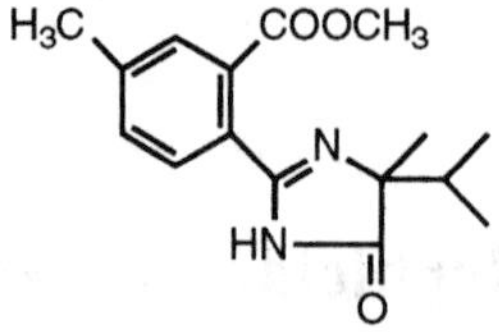

H$_3$C COOCH$_3$ N HN O

(I)

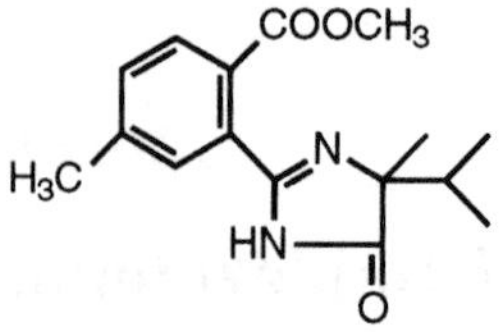

COOCH$_3$ H$_3$C N HN O

(II)

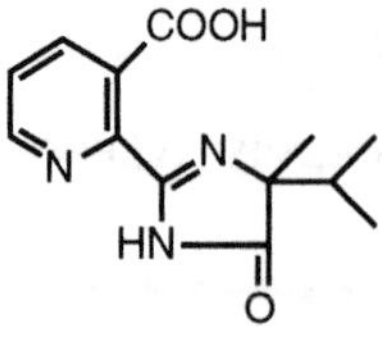

COOH N N HN O

(III)

COOH N N HN O

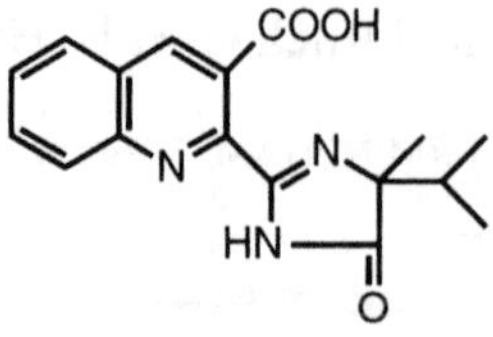

(IV)

biosynthesis of the branched chain amino acids valine, leucine, and isoleucine by inhibiting the vital enzyme acetohydroxyacid synthase. This gives rapid decreases in the pool sizes of these amino acids, followed by a decrease in protein synthesis, and that eventually leads to the death of cells. Assert (Compounds **I** and **II**) is a post emergence herbicide, and Sceptre (**IV**) can be applied to the soil as a pre-emergence herbicide, or as an early post-emergence herbicide. Arsenal (**III**), which is highly bioactive, but non-selective, can be used pre- or post-emergence, although the post-emergence application is generally considered to be most effective, and especially for the control of perennial weed species. All the herbicides are applied in an aqueous formulation containing a non-ionic surfactant.

The biological activities of aromatic anthropogenic organic chemicals are generally lost when the compounds come into contact with soil humic substances (HS), and especially with humic acids (HAs). The nature of the aromatic components of the compounds listed would suggest that these also would strongly sorb to HS, and be less effective when applied to soils rich in organic matter. The study described here set out to determine the extent and the reversibility of binding to soil HAs of the imidazolinone herbicides listed. Compounds **I** and **II** (Assert) were studied as a mixture because the compounds are marketed as such.

2 Materials and Methods

Isolation of Humic Acids

A sapric histosol was H^+-exchanged by washing with 1M HCl, and filtered. The residue was suspended in distilled water and centrifuged (20 000 g for 25 min). The washing process was repeated and the residue was extracted with 0.5M NaOH under dinitrogen gas. After centrifugation (20 000 g, 25 min) the supernatant was treated as follows:

1, Ca^{2+}-exchanged, by adjusting the pH of the supernatant to 6.5 (with HCl) and adding excess calcium acetate (1M). Excess acetate was considered to be added when the pH reached 6.8, and the colour of the suspension had changed from black to brown. After centrifugation (20 000 g, 25 min) the sediment was washed with distilled water, and centrifuged. The supernatant was Ca^{2+} free after two washings (sulphate test), and the humate was dialysed (Visking tubing, 12 000 - 14 000 MW cut off), then freeze dried; and

2, H^+-exchanged, by adjusting the alkaline extract to pH 1.0 (with 2M HCl), centrifuging (20 000 g, 25 min), washing with water, dialysing againsty distilled water till chloride free, and the H^+- HA was freeze dried.

Measurements of Sorption

The continuous flow stirred cell (CFSC) technique. The technique is described in this volume by Law *et al.* (p. 189 - 198, this Volume), and SARTORIUS Type SM (cellulose acetate) membranes were used which retained molecules of molecular weight values of 20 000 and of 5000. The concentrations of sorptive in the effluents from the stirred cell were monitored continuously using a differential refractometer, or by collecting fractions and measuring the concentrations of herbicide by UV at the isobestic point of the herbicide.

The batch-slurry method. A known amount of sorbent (10 mg) was equilibrated, with intermittent shaking for 24 h with a known volume (5 mL) of sorptive solution, and the amount sorbed was calculated from the concentration of sorptive in the supernatant after centrifugation (40 000 g for 20 min). Supernatant (1 mL) was added to scintillator (13 mL) in a counting vial, and solution concentrations were measured by liquid scintillation counting.

Sorption by non-freeze dried H^+-HA. To aliquots (1 mL) of non-freeze dried H^+-HA suspensions (containing 11.7 ± 0.2 mg HA), ^{14}C-labelled solutions (4 mL) of Assert (compound **III**) were added, and the mixtures were equilibrated for 24 h. Then the mixture was centrifuged (40 000 g, 1 h), and the concentration of sorptive in the supernatant was measured.

Analytical Procedures

Liquid scintillation counting. Stock solutions of herbicide were prepared using unlabelled compound to which small, but known amounts of the labelled materials were added. Some labelling details of the stock solutions are given in Table 1. These solutions were diluted to give a range of concentrations between ca 0.1 and 15 ppm. An Amersham PCS-II cocktail, a xylene-surfactant based liquid scintillation counting cocktail, was used for the phase combining system for the counting of the aqueous herbicide solutions. The efficiency of the counter was determined using an internal standard method, and the efficiency was 75% for the unquenched samples. The degree of quenching was monitored for every sample by the channels ratio method, and when appropriate a quench correction curve was prepared. Quench correction was always necessary for studies with HA preparations. The total counts recorded for a single sample were never less than 5000, and that value was calculated to give an uncertainty of 1.41%. All the results were reproducible.

Measurements of Desorption

This study was carried out on the materials used for the sorption studies using the batch

Table 1 *Data for the labelled herbicide preparations*

Herbicide	Compound No.	^{14}C label	Stock solution (ppm)	Specific activity (dpm)
Assert	**I, II**	Imidazolone-5	31.568	145435
Arsenal	**III**	COOH	61.890	221289
Scepter	**IV**	Pyridine ring	15.591	193628

slurry technique. For measurements of adsorption 1 mL aliquots were taken. Then additional aliquots were withdrawn, and these were replaced by one half the same volume of distilled water. The mixture was further equilibrated for 24 h, and the process was repeated several times to obtain the data for the desorption isotherm.

3 Results

Sorption by H^+- Humic Acid

The cellulose acetate membrane used for the CFSC technique gave rise to rejection of sorptives and to concentrations within the cell in excess of the concentrations in the reservoir. Thus the concentrations in the effluents could not be considered to represent the equilibrium solution concentration within the cell. Rejection by the membrane was overcome in the presence of 0.05M NaCl, but the presence of the salt made detection by refractometry unreliable. The CFSC technique has many advantages over batch-slurry procedures for studies of sorption and desorption, provided that there are no interactions between the interacting species and the cell membrane (Hayes, 1980). It would be important to use a membrane which will not carry a charge under the experimental conditions, and will not interact with the contents of the cell.

The extent of the membrane interactions was such that, although a linear isotherm was obtained for the sorption of Arsenal (Compound **III**) by H^+-HA, the amount sorbed for any 'equilibrium' concentration was ten times lower than for the batch results.

An equilibration time study indicated that deviations of 2 to 3 h at either side of a 24 h equilibration period did not affect the results significantly. Isotherms for the sorption and desorption of Assert (mixture of **I** and **II**), Sceptre (Compound **IV**), and Arsenal (Compound **III**) by H^+-HAs are given in Figure 1. The isotherms are linear, and the extents of sorption decrease in the order Assert > Scepter > Arsenal. The distribution between the HA and water (K_d value) folow the order Assert (**I** and **II**) > Scepter (**IV**) >> Arsenal (**III**). The value found for Assert, ca 2500, was relatively constant over a wide concentration range, and that for Scepter (1250) decreased steadily with increasing equilibrium concentration values. The behaviour was different for Arsenal. At low equilibrium solution concentrations (0.2 to 0.4 ppm) a K_d value of the order of 60 was obtained, and this increased to a constant value of ca 250 in the 0.6 to 10 ppm range.

Figure 2 provides linear plots of the Freundlich equation, expressed as:

$$\log (x/m) = \log K + (1/n) \log C$$

where (x/m) is the amount of sorptive sorbed per unit mass of sorbent, C is the equilibrium solution concentration.

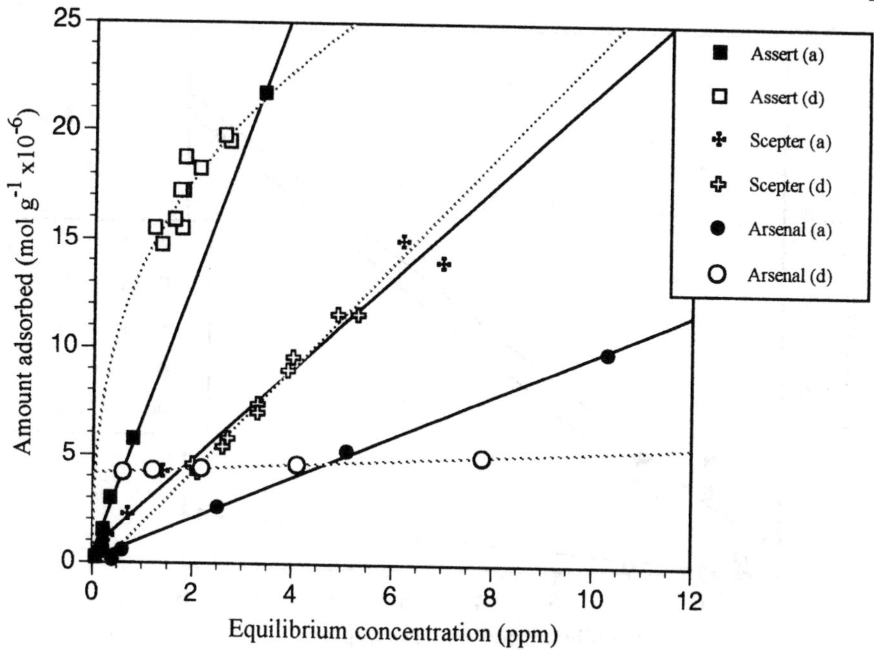

Figure 1 *Isotherms for the sorption (a, solid lines) and desorption (d, broken lines) of imidazolinone herbicides by H^+-humic acid preparations*

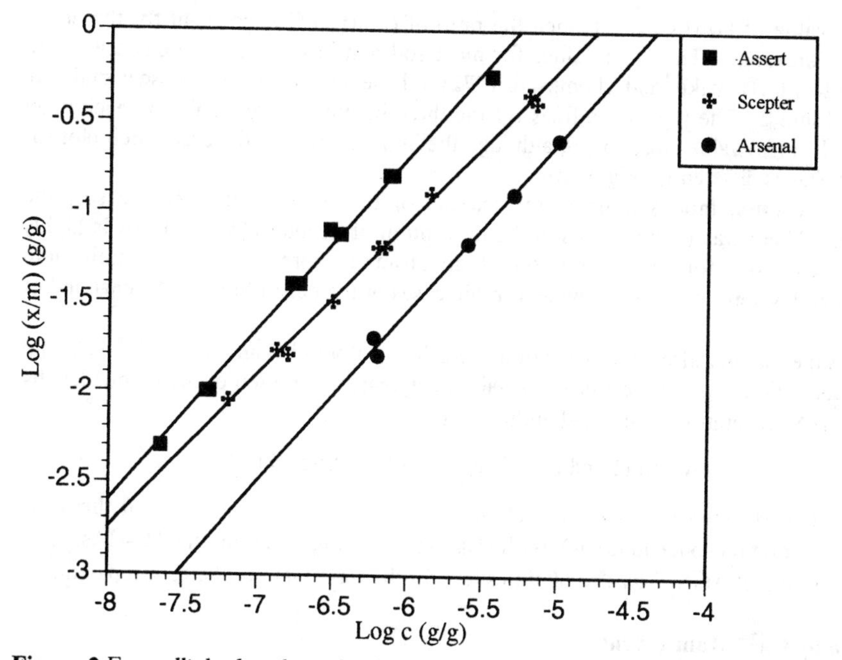

Figure 2 *Freundlich plots from the data for the sorption of imidazolinone herbicides by H^+-humic acids*

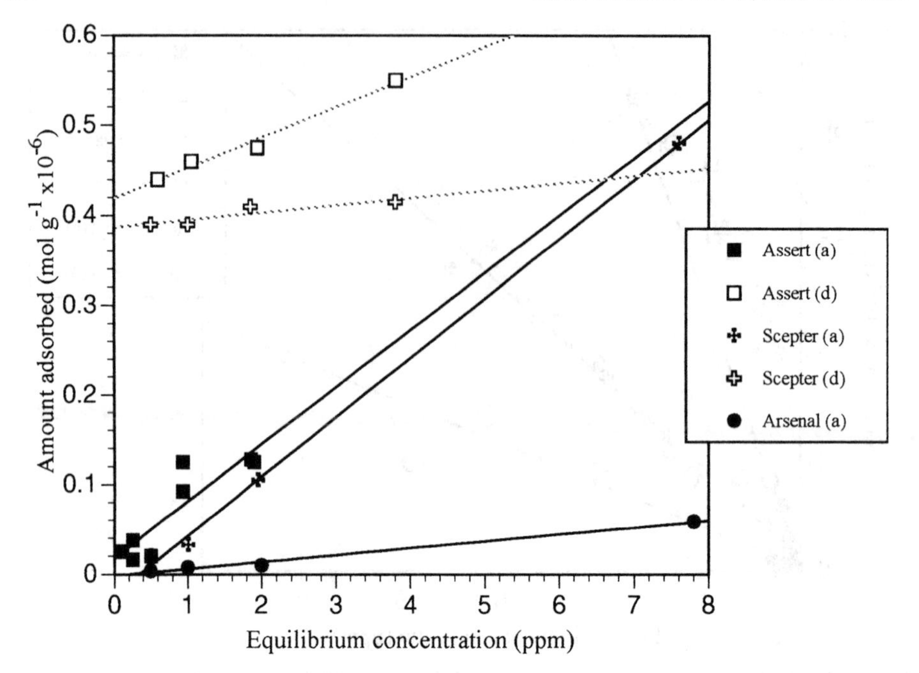

Figure 3 *Isotherms for the sorption (a, solid lines) and desorption (d, broken lines) of imidazolinone herbicides by Ca^{2+}-humic acids*

If the value of (1/n) equals 1, then the ratio of (x/m) to C is constant for the entire concentration range. The (1/n) values for most soil and biocide interactions are in the range 0.7 to 1.0 (Hamaker and Thompson, 1972). Those obtained for the present study are given in Table 2. The regression lines for the three herbicides are parallel to each other and with the (1/n) value close to 0.9, although the initial section of the Freundlich plot for Arsenal behaves differently (Figure 2).

The desorption modes indicate the extents of the reversibilities of the sorption processes. There was no hysteresis in the desorption of Scepter (**IV**), and that is highly unusual behaviour for the interaction of an aromatic sorptive with a humic acid preparation. Hysteresis was very evident in the cases of the desorptions of Arsenal and of Assert.

The supernatants after centrifugation were less coloured than for the H^+-HA, but quench correction was still necessary. Again the extents of sorption of the herbicides by the Ca^{2+}-HA preparations decreased in the order

Assert (**I** and **II**) > Scepter (IV) >> Arsenal (**III**).

The extent of sorption of Assert and of Scepter was about 20 times less (at the same equilibrium solution concentrations) for the Ca^{2+}-preparations than for the H^+-HAs, and that for sorption of Arsenal was *ca* 60 times less for the Ca^{2+} than for the H^+-preparation.

Sorption by Ca^{2+}-Humic Acid

Isotherms for the sorption/desorption of Assert and of Scepter by the Ca^{2+}- HAs are

Table 2 *Correlation coefficients and Freundlich (1/n) values for the sorption of imidazolinone herbicides by H^+-humic acids*

Herbicide	*Compound No.*	*Correlation coefficient*	*(1/n) value*
Assert	Mixture of **I** and **II**	0.998	0.93
Scepter	**IV**	0.999	0.86
Arsenal	**III**	0.966	0.92

given in Figure 3. The extents of sorption of Arsenal were too small to obtain reliable desorption data. The desorption data (Figure 3) clearly show hysteresis.

4 Discussion

It is appropriate to take account of the acid-base properties of the imidazolinone compounds when considering differences in the extents and the reversibilities of sorption by the humic preparations. The pK_a values, as determined by Dolling (1985), are listed in Table 3. These values were determined by ultraviolet spectroscopy, and in the case of Assert it was not possible to make a definite assignment of pK_a values because the compound is a mixture of *m* and of *p* isomers.

The pK_a values in the acidic range are attributed to the dissociation of the carboxyl group, or of the protonated nitrogen of the zwitterion form. These assignments are supported by the findings of Green and Tong (1956) who showed that both quinolinic and nicotinic acids are both almost entirely zwitterionic. Protonation of the imidazolone ring is likely to take place in the same pH range. This pK_a value is likely to overlap with the dissociation of the carboxyl group of Arsenal (**III**) and of Scepter (**IV**). For Assert (**I, II**), the estimated pK_a value of 3.3 is entirely attributable to the protonation of the imidazolone structure, and possible protonations of the imidazolone structure are shown in Figure 4.

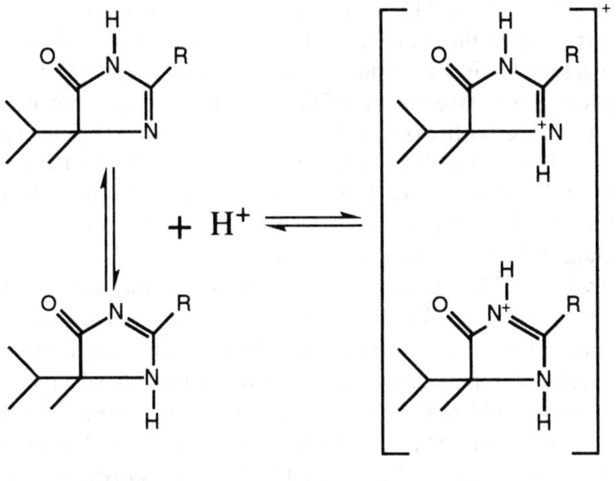

Figure 4. *Suggested protonations of the imidazolone structure*

Table 3 *pK_a values for the imidazolinone herbicides*

Herbicide	*Compound No.*	*pK_a^1*	*pK_a^2*
Arsenal	**III**	3.1	11.2
Scepter	**IV**	3.3	11.1
Assert	**I** and **II**	3.3*	11.1*

* = estimated value

The pK_a values in the alkaline region would appear to represent the dissociation of a proton from the imidazolone ring. Based on the equation

$$pK_a = pH + \log (a_A/a_B)$$

where a_A and a_B are the activities of the acid and base species, respectively, the pK_a values in Table 3, and on the fact that the pH of the aqueous H^+-HA suspension was 3.7, it is calculated that (at pH 3.7) 80% of the Arsenal molecules have shed a proton and will therefore be negatively charged. At the same pH, 71% of the Scepter molecules will be anionic, and protonation of the imidazolone ring will take place. For a pK_a value of 3.3, 24% of the imidazolinone herbicides would be protonated at the imidazolone ring at the ambient pH (3.7). The net charge would still be negative (but decreased by 24%) for Arsenal and for Scepter, but Assert would be positively charged. Arsenal would therefore carry a net positive charge at pH values below 3.2 and a net negative charge above pH 3.2. At pH 3.7, 24% of Assert molecules would be positively charged, and on raising the pH the extents of protonation of the imidazolone ring is significantly decreased. At pH 4, protonation will be 17%, at 4.5 it will be 6%, and only 2% at pH 5.

The cation-exchange capacity (CEC) for H^+-HAs can be mainly attributed to carboxylic functionalities, and these include activated carboxyls that are dissociated at pH values as low as 3.5 (Leenheer *et al.*, 1995). Adsorption of an anionic species at pH 3.7 will not therefore be favoured because of repulsion between the negative charges. Bonding may take place by ion exchange between the protonated imidazolone ring and ionized carboxyl groups on the HA. This mechanism is favourable in the pH range 3-4 where the concentration of the protonated imidazolone species is significant. There are two possible processes for ion exchange of Arsenal and Scepter in the pH 3 - 4 range. Firstly, bonding could take place via one of the protonated ring nitrogens, and secondly the protonated nitrogen of the zwitterion could take part in the ion exchange process.

The Freundlich plots for the three herbicides are linear with (1/n) values of the order of 0.9. The Freundlich K values, which may be considered to be a hypothetical index of the amount of herbicide sorbed from solutions having a unity equilibrium concentration (Biggar and Cheung, 1973), are given in Table 4.

Adsorption by Ca^{2+}-HA was only a small fraction of that for the H^+-HA (compare Freundlich K values, Table 4). The pH values of the Ca^{2+}-HA suspensions were *ca* 7.3, and at that pH value 99.99% of Scepter and > 99.99% of Arsenal molecules would be anionic, and the carboxyl functionalities of the sorbent are ionized. Inter- and intrastrand bridging by Ca^{2+} ions would provide a pseudo cross linking effect for the humic species, inhibiting diffusion of sorptive species to binding sites in the macromolecular matrix. Thus, appropriate mechanisms for the sorption of Scepter and Arsenal would involve salt formation between the ionized herbicide and hydrated Ca^{2+} ions neutralizing the charges

Table 4 *Freundlich K values for the sorption of the imidazolinone herbicides by H^+- and Ca^{2+}- humic acids*

Herbicide	Compound No.	Freundlich K (at C = 1 ppm)	
		H^+- Humic Acid	Ca^{2+}-Humic Acid
Assert	**I** and **II**	2147	19
Scepter	**IV**	954	15
Arsenal	**III**	290	10

on the humic substances.

Assert is uncharged at the ambient pH of the Ca^{2+}-humate medium, and the adsorption mechanisms are likely to involve hydrophobic bonding, charge transfer between the aromatic nucleus and aromatic moieties in the sorbent, and hydrogen bonding processes.

The mechanisms of sorption are compatible with the concept of HS as self assemblies of species, and it strengthens the view that bridging through divalent and polyvalent cations can provide one mechanism by which the humic molecules are associated.

Acknowledgement

Work in this area was supported by a studentship (to M.J. Häusler) provided by American Cyanamid.

References

Biggar, J.W. and M.W. Cheung. 1973. Adsorption of picloram (4-amino-3,5,6-trichloropicolinic acid) on panoche, Ephrata, and Palouse soils: A thermodynamic approach to the adsorption mechanism. *Soil Sci. Soc. Amer. Proc.* **37**:863-868.

Dolling, A.M. 1985. Studies of interactions of some imidazolinone herbicides with clays. M.Sc Thesis, The University of Birmingham.

Green, R.W. and H.K. Tong. 1956. The constitution of the pyridine mono-carboxylic acids in their isoelectric forms. *J. Am. Chem. Soc.* 78:4896-4900.

Hamaker, J.W. and J.M. Thompson. 1972. Adsorption. *In* C.A.I. Goring and J.M. Hamaker (eds), *Organic Chemicals in the Soil Environment.* Dekker, New York.

Hayes, M.H.B. 1980. The role of natural and synthetic polymers in stabilising soil aggregates. p. 263-296. *In* R.C.W. Berkeley, J.M. Lynch, J. Melling, P.R. Rutter, and B. Vincent (eds), *Microbial Adhesion to Surfaces.* Ellis Horwood, Chichester.

Leenheer, J.A., R.L. Wershaw, and M.M. Reddy. 1995. Strong-acid, carboxyl-group structures in fulvic acid from the Suwannee River, Georgia. 2. Major structures. *Environ. Sci. & Technology* 29:399-405.

Los, M. 1984. *o*-(5-oxo-2-imidazolin-2-yl)arylcarboxylates: A new class of herbicides. p. 30-44. *In* P.S. Magee, G.K. Kohn, and J.J. Menn (eds), *Pesticide Synthesis through Rational Approaches.* ACS Symposium Series No. 255.

Shaner, D.L., P.C. Anderson, and M.A. Stidham. 1984. Imidazolinones - Potent inhibitors of acetohydroxyacid synthase. *Plant Physiol.* **76**:545-546.

The Influence of Humic Substances from Drainage Waters on the Transportation of Anthropogenic Organic Chemicals

T.M. Hayes, M.H.B. Hayes, and L.V. Vaidyanathan

THE UNIVERSITY OF BIRMINGHAM, SCHOOL OF CHEMISTRY, EDGBASTON, BIRMINGHAM B15 2TT

Abstract

The maximum concentrations in aqueous solutions of the xenobiotic substances, alachlor [2-chloro-2',6'-diethyl-*N*-(methoxy-methyl)acetanilide], DDT [1,1,1-trichloro-2,2-bis(4-chloro-phenyl)ethane], isoproturon [3-(4-isopropylphenyl)-1,1-dimethylurea], trifluralin [2,6-dinitro-N,N-dipropyl-4-(trifluoromethyl)benzenamine], and of simazine [2-chloro-4,6-bis(ethylamino)-1,3,5-triazine] in the presence of dissolved humic acids (HAs), fulvic acids (FAs), and XAD-4 acids, isolated from soil drainage waters, were studied. Account was taken of the effects of pH (4 and 6) and of background electrolyte. Concentrations of simazine in solution were not increased or decreased by the presence (in solution) of HA, FA, or XAD-4 acids in concentrations ranging from 20 to 100 mg L^{-1}, or by the pH of the environment (4 or 6), or by the presence or absence of salt. The solubilities of alachlor, DDT, isoproturon, simazine, or of trifluralin were not enhanced by the presence of the humic fractions. The evidence indicates that in some instances, the concentrations in solution of these xenobiotics could be lowered by the presence of humic fractions. Background electrolyte, at concentrations characteristic of the soil solution (0.01M), and added to the aqueous media after equilibrium had been established between the chemical in the solid state and in solution (in water and in the HS solutions), lowered the concentrations in solution of alachlor, of DDT, of isoproturon, of simazine, and of trifluralin. When DDT or trifluralin were bound to the sorbents, protection was provided against such '*salting out*'. There was no convincing evidence to indicate that the water soluble HS enhanced the solubilization of the xenobiotic compounds.

This study highlights the need to investigate the ways in which salts in the soil solution will influence the concentrations in solution of anthropogenic chemicals with solubilities in water less than 1 mg L^{-1}.

1 Introduction

Complexation with water soluble humic substances (HS) could have a significant influence on the apparent mobility of sparingly soluble biocides. Traces of some persistent chemicals have been found in areas well removed from the treated site (Stevenson, 1994). Thus, the properties of the chemicals and of the soil colloids are important in the transportation of chemicals from the area of treatment. More comprehensive information is needed about the mechanisms involved in the transport processes.

Xenobiotic, sparingly soluble compounds used in studies of solubility enhancements by dissolved organic matter (DOM) and by HS have included DDT, polycyclic aromatic hydrocarbons, and chlorinated aromatic hydrocarbons (Calvet, 1980; Carter and Suffet, 1982; Chiou *et al.*, 1986, 1987; Kile and Chiou, 1989; Chiou, 1989; Hayes and Mingelgrin 1991; Stevenson, 1994). Stevenson (1994) has referred to the special role that fulvic acids (FAs) can play. Their relative abundance, high polarities and acidities, and relatively low molecular weight (MW) values cause these to be readily soluble in water. FAs are ca 10 times more abundant than humic acids (HAs) in aquatic environments (Malcolm, 1991).

Wershaw *et al.* (1969) were first to address the question of solubility enhancement of sparingly soluble anthropogenic chemicals by HS. Their data suggested that the solubility of DDT was enhanced by a factor of at least 20 (compared with its solubility in distilled water) when the medium used was 0.5% Na^+- humate. The 'solubilization' (or binding) of DDT in that medium was not changed significantly by the presence of 0.1M NaCl. However, the Na^+- humate used was not isolated from water, and its source was soil, lignite, or leonardite. The physicochemical properties of HAs isolated from water are significantly different from those isolated from soil, or from coal related sources.

The data in the literature would suggest that solubility enhancement by dissolved organic carbon (DOC) is greatest for sparingly soluble chemicals (Kile and Chiou, 1989). There is not, however, universal acceptance of the thesis that DOC enhances the solubility of sparingly soluble chemicals. Recently, Zsolnay (1994) showed that DOM had little effect on the solubility of the herbicide terbuthylazine (2-*tert*-butylamino-4-chloro-6-ethylamino-1,3,5-triazine; solubility in water 5 mg L^{-1}). Maaret *et al.* (1992) found no solubility increase for DDT by natural aquatic HS (DOC of 14.5 mg L^{-1}) at pH 6.5. Kulovaara (1993) found only a 4-6% increase in the solubility of DDT in the presence of DOM (DOC was 14.6 mg L^{-1} at pH 6.3), but indicated that this enhancement may be within the experimental error.

This study assessed the influence of the DOM from the drainage water of a grassland soil on the solubility of five anthropogenic organic chemicals (AOCs) with a range of functionalities and water solubilities. Salts, representative of the total concentrations, but not of the diversity of the salts in the soil solution, were included in the study.

2 Materials and Methods

The classical fractions of HS, the HAs and the FAs, as well as the more recently described XAD-4 acids (Malcolm and MacCarthy, 1992) were isolated from the drainage waters of a clay loam soil of the Hallsworth series (Stagno-dystric gleysol); (Armstrong and Garwood, 1991), using XAD-8 [poly (methylmethacrylate)] and XAD-4 (styrene divinylbenzene) resins in tandem (Malcolm and MacCarthy, 1992; Hayes, 1996). The land use is long term grassland, and the site is located at the Institute of Grassland and Environmental Research (IGER) farm at North Wyke, Okehampton, Devon, England.

The pH values of aqueous humic solutions (some at 0, 20, 40, 60, 80, and 100 mg L^{-1}, and others at 0, 40, and 100 mg L^{-1} concentrations) were adjusted to 4 or 6 using either 0.1M HCl or NaOH (Convol concentrate, from BDH), but never both (to minimise the salt

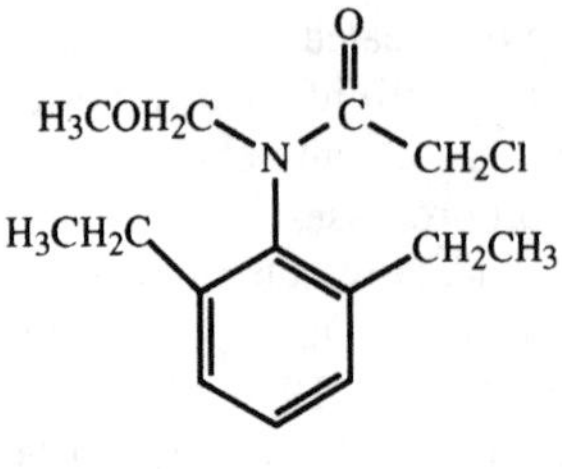

(I) 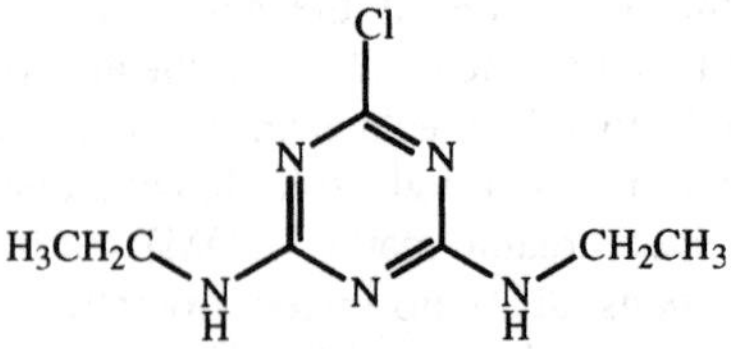 **(II)**

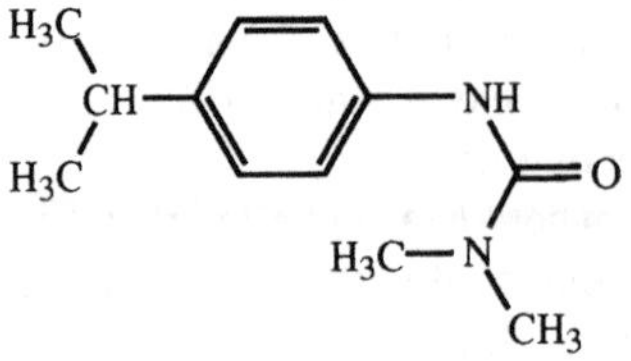

(III) 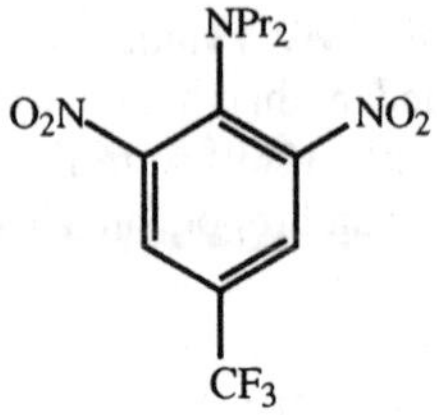 **(IV)**

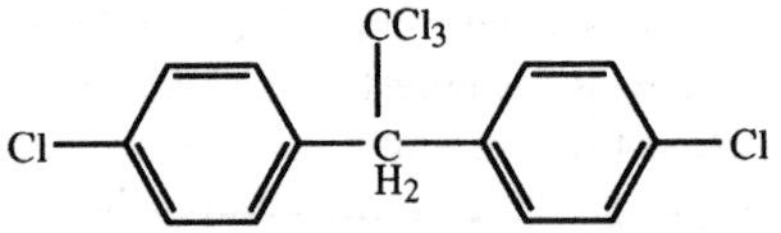

(V)

background). The blanks used were distilled water, with the pH adjusted to 4 and to 6.

Alachlor [2-chloro-2',6'-diethyl-*N*-(methoxy-methyl)acetanilide], structure (**I**), with a purity of > 99% was purchased from Greyhound UK Ltd., and the ^{14}C-labelled alachlor [specific activity (s.a.) of 2.3 mCi $mmol^{-1}$) was purchased from Sigma, USA. Simazine [2-chloro-4,6-bis(ethylamino)-1,3,5-triazine, master standard grade], structure (**II**), with a purity of 99.2%, and labelled (^{14}C) simazine, with a s.a. of 50 μCi mg^{-1} was donated by Ciba-Geigy, Switzerland. Isoproturon [3-(4-isopropylphenyl)-1,1-dimethylurea, 99% purity], structure (**III**), was purchased from Greyhound UK Ltd., and labelled isoproturon with a s.a. of 78 μCi mg^{-1} was donated by Zeneca, UK Ltd. Trifluralin [2,6-dinitro-N,N-dipropyl-4-(trifluoromethyl)benzeneamine], structure (**IV**), and *pp'*-DDT [1,1,1-trichloro-2,2-bis(4-chlorophenyl)ethane], structure (**V**), were purchased (> 99.0% pure) from Greyhound UK Ltd. ^{14}C-labelled DDT with a s.a. of 11.8 mCi $mmol^{-1}$ was purchased from Sigma UK. Dow Elanco, Indianapolis, USA donated the ^{14}C labelled trifluralin (s.a. = 77.4 μCi mg^{-1}).

A stock solution of each AOC, made up in acetone (SLR grade), was spiked with the ^{14}C-labelled compound to give a measured dpm mL^{-1}. Each stock solution (1 mL) was pipetted into a 15 mL Corning culture tube. As the acetone evaporated, the compound was deposited on, and 'lined' the walls of the tubes. The amount of AOC deposited was far in excess of that soluble in 5 mL of the 100 mg L^{-1} humic solution.

Solutions (5 mL) of each HS preparation were pipetted into the 'lined' culture tubes. These were capped with Teflon-lined screw caps, and shaken at room temperature (18 ± 3 °C) for a minimum of 24 h (the concentration of dissolved chemical was not changed by more prolonged shaking). In general, triplicate samples were prepared for each concentration of the HS. Subsequently, samples were centrifuged at 4500 rpm for 45 min. A 1 mL aliquot of the supernatant was carefully withdrawn and added to 10 mL of Ultima Gold LSC cocktail (Canberra Packard, UK). ^{14}C activities were counted on a Beckman LS 1800 series liquid scintillation system, with automatic quench correction.

Subsequently, 25 μL each of solutions of NaCl and of $CaCl_2$ were added to the culture tubes such that the solutions in the tubes were 0.01M with respect to salt (the ratio of Na^+ to Ca^{2+} was 2:1). That would simulate the concentration of salt in the drainage waters. The culture tubes were shaken for a further 24 h, and then resampled and ^{14}C activities were measured using the procedure described above.

3 Results

Table 1 gives data for the solubilities of the xenobiotic compounds in the 40 and 100 ppm humic preparations at pH 4 and at pH 6, and in the presence and in the absence of salt. The data indicate that the solubility of simazine is not affected by the pH, the humic fraction, or by the presence of salt. There are, however, significant differences in the solubilities of alachlor, DDT, isoproturon and of trifluralin that can be attributed to the humic fraction, to the salt, and to the pH of the media.

Solubilities in Water

The solubility of alachlor (**I**, ca 340 ppm) was similar at pH 4 and at pH 6, when a salt background was absent. However, the salt background had a significant effect on the solubility at pH 4 (ca 137 ppm) due to a *'salting out'* effect. The solubility at pH 6 increased slightly (ca 3%) due to a *'salting in'* effect.

In the absence of salt, the solubility of DDT (**V**) was the same at the two pH values, but 'salting out' occurred, and this effect was greater at pH 6 than at pH 4. Solubilities of isoproturon (**III**) at pH 4 and at pH 6 were similar (ca 63 ppm). 'Salting out' occured at both pH values, though the effect was much greater at pH 4 (ca 55 ppm). There was little difference in the solubility of simazine (**II**) at pH 6 (ca 4.1 ppm) and at 4 (ca 3.2 ppm). Addition of salt did not change the solubilities significantly. The solubility of trifluralin (**IV**) in water was greater at pH 6 (ca 830 ppb) than at pH 4 (ca 620 ppb). It 'salted out' at pH 6 and 4 (solubility from ca 667 to ca 421 ppb).

Influences of Humic Acids on Solubilities

The solubility of alachlor (**I**) at pH 6, with and without a salt background, was not affected significantly by the presence of dissolved HA. The HA solution did not have an effect at pH 4, but alachlor was salted out in the presence of a salt background.

Although in the absence of salt, dissolved HA slightly suppressed the solubility of DDT (**V**) at pH 4, its influence on solubility was very small. Dissolved HA did, however, diminish the 'salting out' of DDT, and this effect was greater at the 100 mg than at the 40 mg L^{-1} concentrations, and it was greater at pH 6 than at pH 4 (Table 1).

HAs did not affect the solubility of isoproturon (**III**) at pH 6. However, at pH 4, in the presence of the HA solutions, some inhibition of solubility was observed, and additions of salt did not affect the inhibition of solubility.

Table 1 *Maximum solubilities of alachlor (**I**), DDT (**V**), isoproturon (**III**), trifluralin (**IV**), and of simazine (**II**) in water, and in 40 and 100 mg L^{-1} aqueous solutions of humic acids (HA), fulvic acids (FA), and XAD-4 acids in the presence and in the absence of salt (made from additions of equal volumes of 0.01M NaCl and 0.005M $CaCl_2$)*

Sorbent		Maximum Solubility									
Concentration		Alachlor (**I**) (mg L^{-1})		DDT (**V**) (µg L^{-1})		Isoproturon (**III**) (mg L^{-1})		Simazine (**II**) (mg L^{-1})		Trifluralin (**IV**) (µg L^{-1})	
(mg L^{-1})	pH	No Salt	Salt	No Salt	Salt	No Salt	Salt	No Salt	Salt	No Salt	Salt
H_2O only	[4]	337± 2.4	136.9± 0.3	19±0.6	07±1.0	63.5± 1.1	54.5± 2.3	3.21± 0.0	3.14± 0.01	620±30	421±66
H_2O only	[6]	340± 5.3	352.4± 6.7	19±0.8	04±0.2	62.8± 1.4	60.2± 0.2	4.13± 0.1	4.04± 0.03	829±12	667±40
HA (40)	[4]	336.3± 0.7	139.8± 3.4	17±0.6	12±1.2	59.8± 0.1	57.6± 0.7	3.03± 0.06	3.21± 0.13	655±100	791±01
HA (40)	[6]	356.3± 1.9	350.6± 3.1	20±2.6	16±0.1	64.4± 1.0	62.9± 0.7	4.03± 0.0	4.0± 0.07	820±45	981±59
HA (100)	[4]	344.5± 3.6	139.0± 0.1	14±0.4	16±0.6	59.9± 0.0	57.7± 0.0	3.43± 0.04	3.46± 0.04	640±30	742±28
HA (100)	[6]	350.4± 5.4	350.9± 0	19±1.1	18±1.0	64.3± 1.1	64.4± 2.7	3.88± 0.04	4.0± 0.05	895±57	942±48
FA (40)	[4]	342.2± 2.5	140.0± 2.6	11±0.3	08±0.4	60.4± 1.0	59.1± 0.5	3.0± 0.02	2.96± 0.01	565±35	531±14
FA (40)	[6]	323.7± 0.4	317.1± 2.6	10±2.1	05±0.6	55.5± 3.9	55.4± 2.8	3.04± 0.04	3.1± 0.06	271±34	244±33
FA (100)	[4]	345.7± 0.0	140.0± 0.0	09±0.6	07±0.3	56.4± 2.5	54.9± 1.8	2.93± 0.13	2.67± 0.08	545±35	575±47
FA (100)	[6]	319.5± 3.6	321.9± 1.3	13±2.1	05±0.5	54.6± 4.5	53.1± 4.4	2.99± 0.16	2.89± 0.06	235±10	212±05
XAD-4 (40)	[4]	339.9± 2.0	139.7± 1.3	09±0.3	05±0.1	64.0± 0.0	63.1± 0..0	3.19± 0.03	3.22± 0.01	560±20	468±10
XAD-4 (40)	[6]	329.8± 9.1	325.5± 5.5	21±1.9	07±0.8	57.8± 3.2	54.5± 1.7	3.38± 0.09	3.32± 0.06	465±28	442±66
XAD-4 (100)	[4]	338.4± 4.5	139.0± 0.3	06±0.2	04±0.1	63.2± 0.0	51.4± 0.0	3.18± 0.03	3.11± 0.11	568±20	485±10
XAD-4 (100)	[6]	314.9± 6.1	329.2± 2.4	21±1.7	08±0.6	54.6± 0.3	54.2± 0.3	3.12± 0.13	3.06± 0.12	558±38	486±12

The solubility of simazine (**II**) was not affected by the dissolved HAs at pH 6 or at pH 4, in the presence or in the absence of a salt background.

HAs in solution at two different concentrations, and at the two different pH values, did not influence the concentrations of trifluralin (**IV**) in solution in the absence of salt. HAs, however, provided protection from 'salting out', and more trifluralin was in solution (and associated with the HA) in the presence than in the absence of salt. This enhanced solution concentration was greater at pH 6 than at pH 4.

Influences of Fulvic Acids on Solubilities

The solubility of alachlor (**I**) at pH 6 was inhibited slightly by FAs in solution in the absence and in the presence of a salt background. FAs did not affect the solubility at pH 4, but salting out occurred in the presence of the salt.

FAs, in solution at 40 and at 100 mg L^{-1}, and at pH 4 and pH 6, suppressed the concentration of DDT (**V**) in the solution. Suppression was less at 100 mg L^{-1} and at pH 6. The FAs did not affect the suppression of the 'salting out' at either pH.

FAs suppressed the solubility of isoproturon (**III**) at pH 4 and at pH 6, in the presence and in the absence of salt. Suppression was slightly greater in the presence of salt, and appeared to increase as the concentration of FAs in solution increased.

FAs solutions at pH 4 slightly inhibited the solubility of simazine (**II**), and addition of salt enhanced the salting out at the higher FA concentrations. Solubility was also inhibited at pH 6, with and without a salt background.

FAs, in concentrations of 40 and 100 mg L^{-1}, also suppressed the concentrations of trifluralin (**IV**) in solution, and the suppression was much greater at pH 6. At pH 4, the FAs provided protection against the 'salting out', but salt did not influence the suppression of solubilization by the FAs at pH 6.

Influences of XAD-4 Acids on Solubilities

The XAD-4 acids did not influence the solubility of alachlor (**I**) at pH 4, or provide protection from salting out. At pH 6, with and without salt, these acids inhibited the solubility of alachlor, although the effect was less in the presence of the salt background.

The influences of the XAD-4 acids on the concentrations of DDT (**V**) in solution at pH 4, in the presence and in the absence of salt, were similar to those of the FAs. However, in the cases of the 40 and 100 mg L^{-1} concentrations, a slight enhancement of the concentration of DDT in solution at pH 6 took place in the absence of salt. 'Salting out' of the solute was similar at the two different concentrations of the XAD-4 acids at pH 4, and although the extents of 'salting out' were the same for both concentrations at pH 6, the effects were slightly less than at pH 4.

The solubility of isoproturon (**III**) was slightly inhibited by the XAD-4 acids in solution at pH 6. In the absence of salt, these acids did not affect the solubility at pH 4.

The XAD-4 acids did not affect the solubility of simazine (**II**), in the presence or absence of a salt background at pH 4, but did inhibit the solubility at pH 6. Addition of a salt background did not affect the inhibition.

Suppression of the concentration of trifluralin (**IV**) in solution resulted also from the presence in solution of the XAD-4 acids at concentrations of 40 and 100 mg L^{-1}. This suppression was the same for the two concentrations at pH 4, but it was greater at pH 6 (more so for 40 mg L^{-1}). The XAD-4 acids lessened the salting out effect observed for the

Table 2 *Cation exchange capacity (CEC) data (cmol kg^{-1}) of the humic substances fractions at pH 4 and at pH 6. (Hayes, 1996)*

	CEC (cmol kg^{-1})	
	pH 4	pH 6
Humic acid (HA)	120	240
Fulvic acid (FA)	320	510
XAD-4 acids (XAD4)	420	630

trifluralin in water at pH 4, but the amounts of the compound in solution at pH 6, in the presence and in the absence of salt, were similar.

4 Discussion

The results for simazine (**II**) are straightforward. Our data indicate that the solubility is not influenced significantly by the presence of HAs, FAs, and XAD-4 acids at pH 4 or at pH 6, or by the presence or absence of the NaCl/$CaCl_2$ salt system. Solution concentrations were influenced, however, by the HS fractions, by salt, and by pH in the cases of the alachlor (**I**), DDT (**V**), isoproturon (**III**), and trifluralin (**IV**) compounds.

In order to take account of the ways that HS fractions might influence the solubilities of organic chemicals, it is appropriate to consider aspects of the compositions and structures of the fractions.

The CPMAS ^{13}C NMR data which we have for the HS isolates used indicate that aromaticity decreased in the order HA (36%) > FA (26%) >> XAD-4 acids (11%), and acidity (for carboxyl functionalities, from titration data) increased in the order HAs < FAs < XAD-4 acids (Hayes, 1996). The CEC values for the HS fractions are given in Table 2.

The HAs were the most aromatic of the fractions, had the most hydrophobic functionalities, and it is assumed that their molecular weight values were greater than for the FAs and the XAD-4 acids. The solution conformations of the HS could be expected to be more expanded (or less aggregated) at pH 6 than at pH 4; the lesser extents of dissociations of the carboxyl groups at pH 4 would allow some shrinking of the macromolecules caused by inter- and intra-strand hydrogen bonding. That would be especially true in the cases of the HAs, as judged by the difference in the CEC values at pH 4 and at 6. It can be assumed that the humic strands would be more flexible at pH 6 than at pH 4, and thus more free to rotate and to orientate randomly with respect to time and space. At pH 6, the negatively charged groups can assume conformations in which repulsive interactions are least, and hydration of the charged species would be complete. At pH 4, the macromolecules would be less flexible, less charged, and less amenable to solvation.

Some of the foregoing considerations have been used by Burns *et al.* (1973) to explain the enhanced binding of paraquat (1,1'- dimethyl-4,4'-bipyridinium dichloride) by the sodium salt of HA from a sapric histosol compared with the binding by the H^+- exchanged HA from the same source. The Na^+- humate dissolved in water, and the sorptive readily entered the sorbent matrix (in the expanded conformation) and bound to the negatively charged functionalities in the matrix. Penetration into the matrix was inhibited in the case of the H^+-HA because inter- and intra- strand hydrogen bonding restricted the entrance to the exchange (binding) sites within the matrix.

Solution concentrations were influenced, however, by the HS fractions, and/or the salt, and/or by pH in the cases of the sparingly soluble DDT (**V**) and trifluralin (**IV**). Studies of solubilities of DDT by Wershaw *et al.* (1969) used Na^+-humate and, though the source was not specified, it is presumed to be from soil or from lignite. The HA and FAs used by Hayes *et al.* (1968) and by Burns *et al.* (1973) were isolated from the NaOH extract of a H^+-exchanged sapric histosol. Chiou *et al.* (1986, 1987) variously used commercial HA preparations, a soil HA, HAs from the Suwannee and the Alcasieu rivers, FA from the Suwannee, and river waters from the Suwannee and the Sopchoppy rivers.

The Na^+-humate used by Wershaw *et al.* (1969) had a high capacity to bind DDT. Binding is not solvation, and is best described as sorption. Dissolved Na^+-humate can be regarded as an expanded structure with internal and external surfaces available to bind the sorptive. Burns *et al.* (1973) used the concept of expanded macromolecular conformations to explain how paraquat had access to the negatively charged sites contributing to the CEC of the sorbent (see also Burchill *et al.*, 1981). Alternatively, it might be considered to resemble micellar associations of molecules to give pseudo high MW properties.

The foregoing considerations can be used to explain our results for the solubility effects observed for DDT at pH 6. The HA did not affect the concentration of DDT in solution. When salt was added, however, a 'salting out' of the DDT took place in the absence of the HAs. The HAs provided protection against the 'salting out', and this protective effect increased up to a concentration of HAs of ca 80 mg L^{-1}.

The additions of salt were made after each humic fraction had equilibrated with the sorptive. Thus an equilibrium was established between the DDT in association with the HAs and that free in solution. The DDT in association with the HAs was protected from the 'salting out' effect, and it would seem that very little free DDT was present when the concentration of HAs in solution was 80 mg L^{-1}.

The linear suppression by the dissolved HAs of the solubility of DDT (**V**) detected at pH 4 would suggest that the association of water with the HAs depressed the amounts that could associate with the sorptive. Again, the data suggest that the DDT that had associated with the HAs at this pH was not 'salted out' when the salt was added.

The data in Table 1 do not indicate that the dissolved HAs influenced the concentration of trifluralin (**IV**) in solution. It is clear, however, that the solute became associated with the dissolved HAs because it provided protection against salting out. The data indicate that the concentration of the trifluralin in the solution was greater at pH 6 than at pH 4. That is surprising in view of the fact that any degree of protonation of the solute would be more likely to occur at the lower pH value. It would seem that in the presence of salt, more trifluralin was associated with the HAs than in its absence.

From the data in Table 1 it is clear that the solubility of DDT in solution was suppressed by the FAs, both at pH 4 and pH 6, and that was also evident for the XAD-4 acids at pH 4. The FAs and XAD-4 samples had high carboxyl contents, and contributed 315 and 545 cmol kg^{-1}, respectively, to the CEC at pH 4. The suppression of the solubility of DDT in the presence of FAs would suggest a diminution in the water available for the solvation of the solute. It is unlikely, however, that the solute was associated with the FAs since it was salted out at pH 4 and at pH 6.

The FAs also suppressed the solubility of trifluralin at pH 6 and at pH 4, and these suppressions were significantly greater at pH 6. The data indicate that the FA and trifluralin were associated at pH 4, and that this association prevented the 'salting out' of the solute. It can be predicted that trifluralin would be associated with the FA at pH 4 should some protonation of the solute take place.

The XAD-4 acids did not have a significant effect on the concentrations of DDT in solution at pH 6. As indicated above, the CEC of these acids at pH 6 was 630 cmol kg^{-1}, and the aromaticity was low (11%). The effects of the XAD-4 acids on the solubility of DDT might be comparable to that of poly(acrylic acid) shown by Chiou *et al.* (1986) not to influence the solubility of DDT. That can be interpreted as indicating that DDT did not interact with the highly charged and polar polymer.

The presence in solution of the XAD-4 acids at pH 4 suppressed the solubility of DDT, and again that might indicate insufficient free water in the medium to solvate the DDT. It is clear that the DDT was not associated with the XAD-4 acids because the solute was 'salted out' at pH 6 and at pH 4. Again, analogies can be drawn between the physicochemical properties of the XAD-4 acids and (poly)acrylic acid in so far as interactions with the solute are concerned.

The lowering of the concentrations in solution of trifluralin in the presence of the XAD-4 acids would suggest competition between the humic substance and the solute for water. The fact that some protection against 'salting out' was provided by the XAD-4 acids at pH 4 would suggest interactions between the acids and the solute; ion exchange is the likely mechanism. There is no evidence to indicate that protection was provided at pH 6, and protonation of the solute, and ion exchange would be less at that pH.

5 Conclusions

It is not valid to extrapolate to HAs in the soil solution and in drainage waters the possible solubility enhancements of anthropogenic organic chemicals that have been observed for HS isolated from soil in the usual solvent systems used. The Na^+- humate used by Wershaw *et al.* (1969) was extracted from soil or lignite, and the influence of that material on the solubilization of DDT was significantly different from that experienced in our studies. HS from soils are larger (or the component macromolecules are more extensively associated), and less polar than those in water, and have compositions, shapes, and sizes that are significantly different.

The presence of salt, even at the relatively low concentrations present in the soil solution, can inhibit the solubilization of sparingly soluble chemicals. However, such chemicals can be protected from salting out when the compounds are sorbed to HS.

Our data do not indicate that dissolved HS enhance the solubilization of sparingly soluble anthropogenic organic chemicals (AOCs) which have solubilities in the range of 300 mg L^{-1} to 5-20 µg L^{-1}. It is clear that some fractions of dissolved HS, and especially the HAs (the most aromatic and hydrophobic), will bind hydrophobic AOCs, but the evidence would suggest that the binding will take place when the chemicals are already in aqueous solution.

The hydrophilic fractions of DOM, with low aromaticities and high acidic functionalities, which can include the FAs and will include the XAD-4 acids, have little affinity for hydrophobic xenobiotic compounds. At concentrations as low as 40 mg L^{-1} FAs and XAD-4 acids can inhibit the solubilization of some such chemicals.

Salt, even at concentrations as low as 0.01M - the concentration that can normally be expected in the soil solution, can inhibit the solubility of sparingly soluble chemicals.

However, when such chemicals are bound to HS, the 'salting out' effect is avoided, and the HS act as protective colloids.

The salt was added to the media after the humic/xenobiotic complexes had formed. Salts are present in the soil solution, and any complexes that will form between the xenobiotic compounds and the HS will do so in the presence of salt. There is a need to investigate the interactions when sparingly soluble xenobiotics are added to HS in solution in the presence of salt.

The evidence we have does not indicate that the solubilities of the xenobiotics studied have been enhanced in the presence of HS.

HS will aid the transportation of sparingly soluble xenobiotic organic chemicals only when these xenobiotics are bound to the mobile HS. Evidence from this study would suggest that in the cases of the xenobiotics and humic fractions studied, dissolved HS will not transport more xenobiotics than will be dissolved in water in the absence of salt.

The concept of self association (or self assembly) as an explanation of the ways in which humic molecules come together to give the appearance of high molecular weight structures, and the notion that biological molecules can link humic 'core' structures is not alien to the views given here. Kenworthy and Hayes (p. 39 - 45, this issue) have speculated about the ways in which pyrene is protected (sorbed) in 'cage like' hydrophobic structures in the humic molecules. That concept can also be considered as a mechanism for the binding by the humic molecules (and especially the HAs) of the relatively hydrophobic AOCs used in the present study. However, it would appear that the AOCs would need to be in solution in order to migrate to the 'cages'.

References

Armstrong, A.C., and E.A Garwood. 1991. Hydrological consequences of artificial drainage of grassland. *Hydrological Processes* **5**:157-174.

Burchill, S., M.H.B. Hayes, and D.J. Greenland. 1981. Adsorption. p. 221–400. *In* D.J. Greenland and M.H.B. Hayes (eds), *The Chemistry of Soil Processes.* Wiley, Chichester.

Burns, I.G., M.H.B. Hayes, and M. Stacey. 1973. Studies of the adsorption of paraquat on soluble humic fractions by gel filtration and ultrafiltration techniques. *Pestic. Sci.* **4**:629–641.

Calvet, R. 1980. Adsorption - desorption phenomena. p. 1–29. *In* R.J. Hance (ed.), *Interactions Between Herbicides and the Soil.* Academic Press, London.

Carter, C.W. and I.H. Suffet. 1982. Binding of DDT to dissolved humic materials. *Environ. Sci. Technol.* **16**:735–740.

Chiou, C.T. 1989. Theoretical considerations of the partition uptake of nonionic organic compounds by soil organic matter. p 1–29. *In* B.L. Sawhney and K. Brown (eds), *Reactions and Movement of Organic Chemicals in Soils.* SSSA and ASA, Madison, WI.

Chiou, C.T., R.L. Malcolm, T.I. Brinton, and D.E. Kile. 1986. Water solubility enhancement of some organic pollutants and pesticides by dissolved humic and fulvic acids. *Environ. Sci. Tech.* **20**:502–508.

Chiou, C.T., D.E. Kile, T.I. Brinton, R.L. Malcolm, and J.A. Leenheer. 1987. A comparison of water solubility enhancements of organic solutes by aquatic humic materials and commercial humic acids. *Environ. Sci. Tech.* **21**:1231–1234.

Hayes, T.M. 1996. Isolation and characterization of humic substances from soil, and the soil solution, and their interactions with anthropogenic organic chemicals. Ph.D. Thesis. The University of Birmingham.

Hayes, M.H.B. and U. Mingelgrin. 1991. Interactions between small organic chemicals and soil colloidal constituents. p. 323–407. *In* G.H. Bolt, M.F. DeBoodt, M.H.B. Hayes, and M.B. McBride (eds), *Interactions at the Soil Colloid-Soil Solution Interface.* Kluwer, Dordrecht.

Hayes, M.H.B., M. Stacey, and J.M. Thompson. 1968. Adsorption of *s*-triazine herbicides by soil organic matter preparations. p. 75–90. *Isotopes and Radiation in Soil Organic-Matter Studies.* (Vienna), IAEA, Vienna.

Kile, D.E. and C.T. Chiou. 1989. Water solubility enhancement of nonionic organic contaminants. p 131–157. *In* I.H. Suffet, and P. MacCarthy (eds), *Aquatic Humic Substances. Influence on Fate and Treatment of Pollutants.* American Chemical Society, Washington, DC.

Kulovaara, M. 1993. Distribution of DDT and benzo[a]pyrene between water and dissolved organic matter in natural humic water. *Chemosphere* **27**:2333-2340.

Maaret, K., K. Leif, and H. Bjarne. 1992. Studies on the partition behaviour of three organic hydrophobic pollutants in natural humic water. *Chemosphere* **24**:919-925.

Malcolm, R.L. 1991. Factors to be considered in the isolation and characterization of aquatic humic substances. p. 369–391. *In* H. Boren and B. Allard (eds), *Humic Substances in the Aquatic and Terrestrial Environment.* Wiley, Chichester.

Malcolm, R.L. and P. MacCarthy. 1992. Quantitative evaluation of XAD-8 and XAD-4 resins used in tandem for removing organic solutes from water. *Environ. Int.* **18**:597–607.

Stevenson, F.J. 1994. *Humus Chemistry, Genesis, Composition, Reactions.* Wiley, NY.

Wershaw, R.L., P.J. Burcar, P.J., and M.C. Goldberg. 1969. Interactions of pesticide with natural organic material. *Environ. Sci. Tech.* **3**:271–273.

Zsolnay, A. 1994. The lack of effect of the dissolved organic material in soil on the water solubility of the herbicide, terbuthylazine. *Sci. Tot. Environ.* **152**:101–104.

Preparation of Iron Pillared Clays and Their Applications for Sorption of Humic Substances

A.L. Lacey, M.H.B. Hayes, and L.V. Vaidyanathan

THE UNIVERSITY OF BIRMINGHAM, SCHOOL OF CHEMISTRY, EDGBASTON, BIRMINGHAM B15 2TT, ENGLAND

Abstract

Pillared clays are smectites intercalated with large cations that prop clay layers apart. Such clays have uses as catalysts and as molecular sieves. A reproducible method is reported, using microwave radiation, for the preparation of Fe pillared clays with interlayer spacings of 25 Å. Iron pillared clays were ineffective for binding water soluble humic substances (HS), though Fe/Al and Fe acetate pillared products did bind HS, but to different extents.

1 Introduction

Clays are naturally occurring sheet silicates (Diddams, 1992). Smectite clays undergo swelling, and have cation exchange and intercalation properties. Montmorillonite, the smectite clay used in this study, consists of continuous two-dimensional silicon tetrahedral sheets sandwiching an aluminium octahedral sheet. Cations in the clay interlayer region will undergo exchange with cations in the bulk solution. When mono- or divalent cations in the interlayers of smectite clays are exchanged for large, thermally stable and robust cations, the layers are propped apart and the interlayer spacing is increased (Pinnavaia, 1983). This class of compound is a pillared clay structure, with the large cations acting as the pillars. A schematic diagram of a pillared clay is shown in Figure 1 (Diddams, 1992).

Compared with unpillared clays, pillared clays have an increased porosity, increased acidity, and a greater total surface area (as measured by sorption of an inert gas in the dry state). Such preparations have a two-dimensional micropore system analogous to zeolites. By regulating the synthesis processes, pore spaces can be obtained that are larger than those for zeolites.

Some of the chlorinated organic compounds which arise from the chlorination of humic substances in potable waters are mutagenic, and some of these are carcinogenic (Långvik and Holmbom, 1994). It is appropriate therefore to remove humic substances prior to the chlorination of water. Because the clays and hydr(oxides) of iron and aluminium in soils bind humic substances, it is reasonable to assume that combinations of clays and iron and aluminium (hydr)oxy species, i.e. pillared clays, could have uses for the removal of the humic substances from waters.

Alkylammonium ions, bicyclic amine cations, metal chelate complexes, and polynuclear

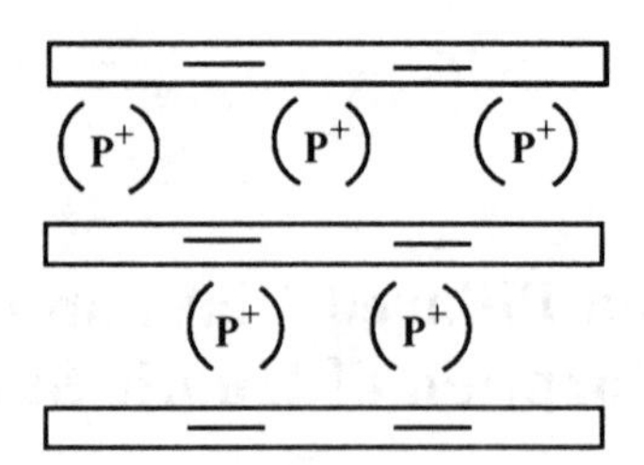

Figure 1 *Schematic representation of a pillared clay showing the negative charge on the clay balanced by the pillaring cations (P^+)*

humic substances from waters.

Alkylammonium ions, bicyclic amine cations, metal chelate complexes, and polynuclear hydroxy metal cations, as shown in Figure 2, are the cation types most commonly used in pillaring (Pinnavaia, 1983).

2 Preparation of Iron Pillared Clays

The general method for the preparation of pillared clays uses a solution of sodium hydroxide (or another base) added to an aqueous solution of the metal cation in a known base to metal ratio. The mixture is left to age, or is heated to allow the polycationic (hydr)oxy species to form. The polycation solution is added to an aqueous suspension of sodium-exchanged montmorillonite clay, and the mixture is then left to stand to allow the pillaring species to cation-exchange with the hydrated interlayer cations. Excess salts are removed from the mixture, either by dialysis or by filtration and by repeated washings of the clay. The clay is then freeze dried. Freeze drying has been shown to lead to a wider distribution of pores within the clay than any other form of drying (Pinnavaia *et al.*, 1984).

The pillared clays prepared in this study, and the interlayer spacings obtained are given in Table 1. Data for non-pillared sodium montmorillonite have been included for comparison.

The preparation of iron pillared clay is complex because of the large number of diverse species present in the hydrolyzed iron solution (the pillaring system). Procedures reported by Doff *et al.* (1988), Zhao *et al.* (1993), and Bergaya *et al.* (1993) to provide iron pillaring were followed, but these gave rise to amorphous products. The successful method used by us was based on that of Lee *et al.* (1989). Iron (III) chloride solution was mixed with sodium carbonate in an OH/Fe ratio of 2.0. The solution was aged at 95 °C for 36 h and a brick-red suspension was formed prior to addition to the clay. Sufficient pillaring solution was added to satisfy four times the cation exchange capacity of the clay. It was assumed that the iron pillaring solution contained only $[Fe_3(OH)_4]^{5+}$. The calculations used in this study have meant that sufficient pillaring solution was added to the clay to produce an interlayer spacing of 25 Å. It was necessary to use elevated temperatures because the ionic product of water is such that at 95 °C the proportion of H^+ and OH^- is two orders of magnitude higher than its value at room temperature (Tödheide, 1973). Iron has a strong tendency to bond with oxygen, in preference to hydroxide ions, to form oxides. Only at 95 °C is the concentration of OH^- sufficient to prevent this. The

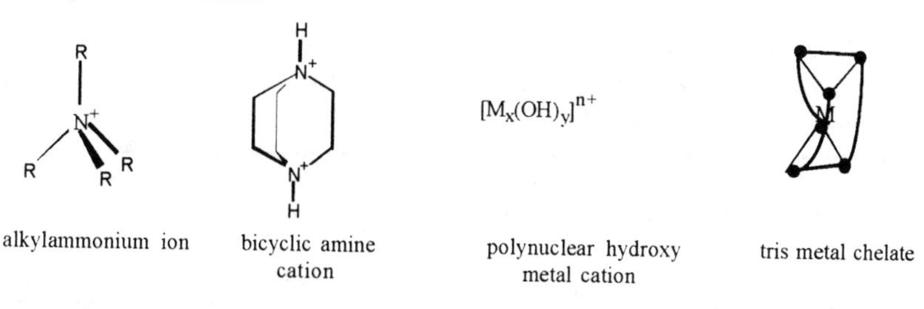

Figure 2 *Various cations used to make pillared clays*

strength of base used must be regulated because the addition of strong base to the iron (III) solution gives rise to the formation of precipitation products.

A novel preparation of iron pillared clay was achieved employing microwave energy. Using the proportions of base to iron and iron to clay outlined above, iron-pillared clay with an interlayer spacing of 25 Å was achieved in a fraction of the time required when conventional heating was used. The pillaring solution (prepared as described above) was placed in a sealed acid digestion bomb and heated for several one minute periods (cooling for 30 minutes between each minute of irradiation). Clay was then added to the solution and heating was continued for several minutes. When microwave energy was used iron pillared clay with a basal spacing of 25 Å (comparable to that for the preparation when conventional heating was used) was obtained in minutes rather than hours. The basal spacing of the clay produced by Lee *et al.* (1989) was 13.8 Å.

Initially, the pillars are held electrostatically within the clay layers. At the more elevated temperatures, however, dehydroxylation takes place and metal oxide pillars covalently bonded to the silicate layers are obtained. At a critical temperature (e.g. > 700 °C for aluminium pillared clays) the pillars decrease in height and eventually collapsed (Figueras, 1988).

The surface areas (N_2, BET) for the iron and the iron acetate pillared clays are given in Table 2. That for the Na^+-montmorillonite used in this study was measured as 13 $m^2\,g^{-1}$. In view of the large basal spacing (25 Å) for the iron pillared clay, the surface area was low. This is explained by the precipitation of iron (hydr)oxide species between the layers, and by the presence of a film of Fe_2O_3 that covered the clay surface, as was evident from

Table 1 *Summary of the pillared clays preparations used in this study*

Type of pillared clay	Pillaring species	(001) Interlayer spacing (Å)
Iron	Polymeric Fe^{3+} (hydr)oxy species	25
Iron Acetate	$[Fe_3(OCOCH_3)_7OH]^+$	amorphous
Iron/Aluminium	A mixture of the iron and aluminium pillars	14
Unpillared Na^+ montmorillonite	-	9.6

Table 2 *The differences in surface area on heating pillared clays to 400 °C using microwave and dielectric heating. The surface area of the unheated clays is included for comparison*

Type of pillared clay	Surface area of the unheated pillared clays (m^2g^{-1})	Surface area after microwave heating (m^2g^{-1})	Surface area after conventional heating (m^2g^{-1})
Iron	36	43	40
Iron Acetate	155	142	132

Mössbauer spectroscopy data. Heating decreased the surface areas of the pillared clays, and the higher the temperature of heating the greater was the decrease. At approximately 700 °C, the pillars, and thus the interlayer spacings, collapsed. Scanning electron microscopy indicated that there were significant differences between the morphologies of the clays that were conventionally heated and those that were heated with microwave energy. The microwave-treated clay had deep, narrow pores whereas the clays that were prepared by the conventional procedures had broad, shallow pores.

3 Sorption of Fulvic (FA) and of Humic (HA) Acids by Iron Pillared Clays

Fulvic acids (FAs), isolated from the drainage waters taken in 1994 from the artificially acidified watershed of Lake Skjervatjern (near Förde) in Western Norway (Hayes *et al.*, p. 298 - 309, this Volume), were used in studies of uptake by the iron pillared clays. The results of the study are shown in Figure 3, and data for Al and Al + H_3PO_3 pillared clays are included for comparison.

The data from Figure 3 show that the Al-H_3PO_3 pillared clay has the greatest capacity to sorb FAs, and this is followed by Al-pillared clay. The same trend applied for the case of sorption of humic acids (HAs). Both clays have higher capacities than the Fe acetate-pillared clay to sorb the humic substances (HS), and all of these clays had significantly greater capacities to sorb the HS than the Fe/Al pillared clay. In general, the masses of FAs/HAs sorbed by a clay increase with increasing concentrations of the FA/HA. A number of explanations can be put forward for this. As sorption proceeds, fewer sites are available to take up the remaining FA/HA molecules in solution. However, as the concentrations of HS in the medium are increased, more humic molecules are available to compete for the unfilled sites. Larger concentrations of FA/HA molcules are needed to increase the probability of a FA/HA molecule to find a free sorption site.

Another possible explanation for the concentration effect is that the pH of the fulvic/humic acid solution decreases with increasing concentration. As the pH decreases fewer acidic groups will be dissociated, and hence the humic molecules become more closely associated (through hydrogen bonding), and repulsion from the negatively charged clay surface is lessened. Associations with the pillaring species are unlikely to be influenced substantially, but sorption to the clay surfaces will be enhanced.

In the case of the sorption of fulvic acid by the Fe acetate pillared clay, however, the

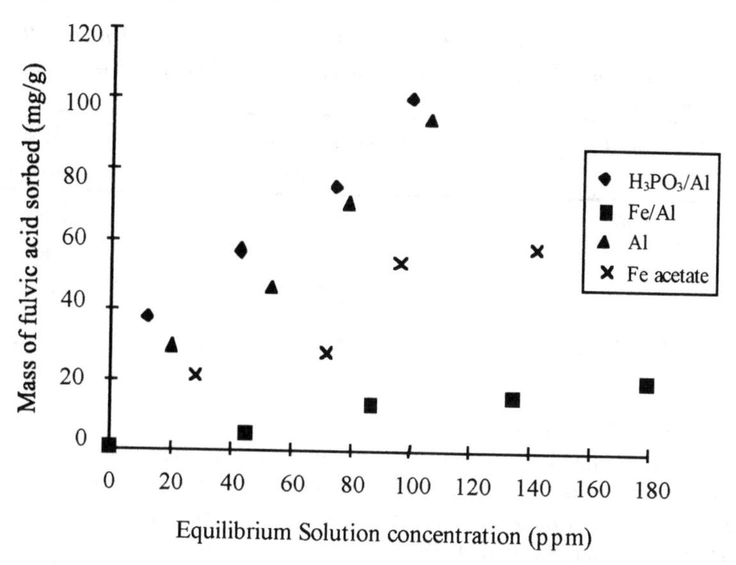

Figure 3 *The extents of sorption of fulvic acids by various pillared clay preparations after the systems had been equilibrated 24 h*

sorption isotherm tended to level off, suggesting that eventually a maximum amount of FA would be sorbed by the clay, amounting to ca 70 mg of FA g^{-1} of Fe acetate clay. A similar situation is apparent for the Fe/Al pillared clay. However, the maximum sorption for the FA will be less (ca 25 mg g^{-1}) for the Fe/Al pillared clay.

The amounts of HAs/FAs sorbed to the clays increased with time. That might be attributed to a slow diffusion of sorptive species to sorption sites.

The fact that the Fe/Al pillared clay had a lesser ability than the other clay preparations to sorb FA reveals the importance of having a large surface area (i.e. interlayer area) accessible for sorption. The extents of the sorption are measures of the availabilities of hydroxyl groups for binding to the FA/HA molecules. This is because the sorption mechanism is probably due more to FA/HA molecules sorbing onto the large number of binding sites on the pillars (and binding to water molecules surrounding the cationic pillars) than it is due to FA/HA molecules entering and filling the pore spaces. Hence, the larger the surface area, the greater the number of binding sites, and provided there is access to these the greater will be the sorption by the pillars.

The clays had larger capacities to sorb FA than HA molecules. That may reflect the generally smaller sizes of FA molecules, which would allow easier access to the sorption sites in the microporous regions of the clay. Also the pH of the fulvic acid is lower than that of the humic acid for the same concentration, and hence there are fewer negative sites on the fulvic acid molecules and, on approaching the negative clay layers, there will be less repulsion than for the humic acid molecules

It can be seen that all of the sorption isotherms are Langmuirian (Giles *et al*., 1960). This type of isotherm is typical of a surface that contains micropores and it is this sorption into micropores that limits the adsorption rate, hence the time dependent sorption. There may also be a fraction of the FA molecules that find it more difficult or

Table 3 *Sorption of humic and fulvic acids by aluminium and iron pillared clay preparations. The solution concentrations were 200 ppm and the equilibration time was 24 h*

Type of clay	% Fulvic acid sorbed by weight	% Humic acid sorbed by weight
Al-H_3PO_3	10.2	7.9
Al	9.4	4.8
Fe acetate	5.8	-
Fe/Al	2.1	-

slower to sorb onto these pillared clays.

The iron pillared clay did not, however, sorb humic substances. Instead, the HS tended to dissolve the iron pillaring species. It is known that in the presence of light and/or organic ligands, surface iron species and iron pillaring species dissolve out of the clay and into solution (Stumm and Sulzberger, 1992).

The results of this study indicate that pillared clays can be useful for the removal of humic substances from water. However, the data indicate that aluminium pillared clays are better than iron pillared structures for this purpose (Table 3).

4 Conclusions

Dielectric heating has provided a novel and exciting method for the preparation of iron pillared clay. This study has shown that such pillared clays can be synthesized also by more conventional methods. The reproducible procedure described has led to the production of iron pillared clay with a large interlayer spacing of 25 Å. Several other pillared clays were successfully prepared and characterized during the course of this study.

The effect of heat on pillared clays was studied by heating the clays conventionally (using a furnace) and by a novel method using dielectric heating (in a microwave oven). Generally, for a given temperature, the clays heated by these two different methods had similar characteristics, and the use of microwave radiation did not confer any unusual properties on the clays.

Some of the pillared clays retained HS sorbed from solution. The amounts sorbed by the clays varied with the type of species used in the pillaring process, with the type of HS used, with the equilibration time, with the temperature to which the clay had been heated and, to a small extent, with the type of heating used.

References

Bergaya, F., N. Hassoun, J. Barrault, L. Gatineau. 1993. Pillaring of synthetic hectorite by mixed $[Al_{13-x}Fe_x]$ pillars. *Clay Miner.* **28**:109-122.

Diddams, P.Chapter 1, *Solid Supports and Catalysts in Organic Synthesis.* K. Smith(Ed). Ellis Horwood Ltd., Chichester, 1992.

Doff, D.H., N. H. J. Gangas, J. E. M. Allan, and J. M. D. Coey. 1988. Preparation and characterization of iron oxide pillared montmorillonite. *Clay Miner.* **23**: 367-377.

Figueras, F. 1988. Pillared clays as catalysts. *Catal. Rev. Sci. Eng.* **3**: 457-499.

Giles, C.H., T. H. MacEwan, S. N. Nakhwa, D. Smith. 1960. Studies in adsorption, Part XI. Adsorption system of classification of solution adsorption isotherms and its use in diagnosis of adsorption mechanisms and in measurement of specific surface area of solids. *J. Chem. Soc.* 3973-3993.

Långvik, V.-A. and B. Holmbom. 1994. Formation of mutagenic organic byproducts and AOX by chlorination of fractions of humic water. *Water Res.* **28**:553-557.

Lee, W.Y., R.H. Raythatha, and B.J. Taturchuck. 1989. Pillared clay catalysts containing mixed metal species. *J. Catal.* **115**:159-179.

Pinnavaia, T. J. 1983. Intercalated clay catalysts. *Science* **220**:365-371.

Pinnavaia, T. J., M. S. Tzou, S. D. Landau, R. H. Raythatha. 1984. On the pillaring and delamination of smectitic clay catalysts by polyoxy cations of aluminium. *J. Mol. Catal.* **27**:195-212.

Stumm, W. and B. Sulzberger. 1992. The cycling of iron in natural environments: cosiderations based on laboratory studies of heterogenous redox processes. *Geochim. Cosmochim. Acta* **56**:3233-3257.

Tödheide, K. 1973. Chapter 13. *In* F. Franks (ed.), *Water: A Comprehensive Treatise*. Vol 1. Plenum, London. .

Zhao, D.Y., G.J. Wang, Y.S. Yang, X.X. Guo, Q.B. Wang, and J.Y. Ren. 1993. Preparation and characterization of hydroxy-pillared clays. *Clays and Clay Minerals* **41**:317-327.

Uses of Schists- and Clay-Derived Materials to Remove Humic Substances from Waters

J.J. Farnworth, L.V. Vaidyanathan, and M.H.B. Hayes

THE UNIVERSITY OF BIRMINGHAM, SCHOOL OF CHEMISTRY, EDGBASTON, BIRMINGHAM B15 2TT, ENGLAND

Abstract

Epidemiological and laboratory studies have raised concerns about the risks of human cancers from consumption of chlorinated waters. The data would suggest that the risks are greatest for chlorinated waters that are rich in humic substances. The work reported here sought materials for the removal of humic substances from water prior to chlorination and, thereby, to lower the risks of forming mutagenic substances in waters subsequently subjected to treatments with chlorine. The primary aim was to work with relatively inexpensive materials that would sorb water soluble humic substances. The solid phase sorbents used were all based on different schist and clay preparations that were modified with aluminium hydroxide species.

1 Introduction

A Sanskrit Lore, dating back to 2000 BC stated "It is good to keep water in copper vessels, to expose it to sunlight, and filter through charcoal", and "Impure water should be purified by being boiled on a fire, or by being heated in the sun, or by dipping a heated iron into it, or it may be purified by filtration through sand and coarse gravel and then allowed to cool" (Baker, 1949; Faust and Aly, 1983). The connection between the quality of drinking water and disease was well known to the Greek and Roman civilisations. Hippocrates, the 'father of medicine' (460-356 B.C.) stated that "water contributes much to health", and "water should be boiled before drinking or it creates hoarseness in the voice". The philosopher Pliny documented that the City of Rome had taken great steps to improve public health by operating a main sewer system in the fourth Century B.C., and the first of the aqueducts to supply the city with fresh water was built in 313 B.C. (Smith, 1874).

A scientific connection was made between the quality of water and the qualtity of health, when the causative agents of typhoid (*Salmonella typhimurium*) and of cholera (*Vibrio cholera*) were identified in waters in 1880 and in 1884, respectively. In an effort to provide water free from infectious organisms, the Reading Water Authority pioneered in 1910 the use of chlorine in continuous chemical treatments. This process was rapidly taken up by other water authorities, both in Britain and in the United States.

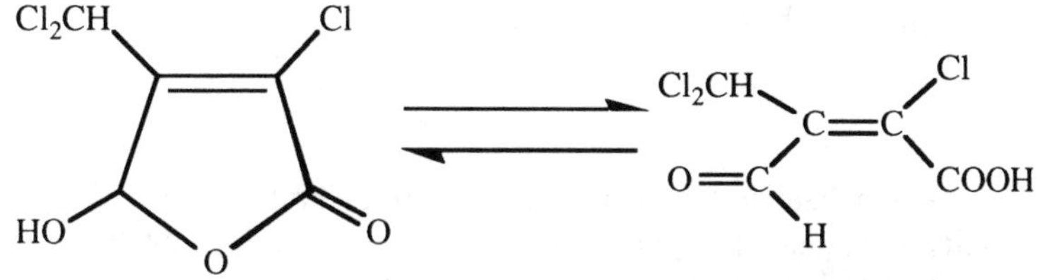

MX
3-Chloro-4-(dichloromethyl)
-5-hydroxy-2(5H) furanone

Z-MX
Z-2-Chloro-3-(dichloromethyl)
-4-oxo-butenoic acid

Figure 1 The structure of MX

Chlorination as a water treatment process was considered to be satisfactory and safe until Rook (1974), Bellar and Lichtenberg (1974), and Bellar *et al.* (1974) reported increases in the levels of trihalomethanes (THMs) when waters containing humic substances (HS) were chlorinated. Subsequent studies found that THMs were ubiquitous in chlorinated water. Hemming *et al.* (1986) found (Z)-2-chloro-3-(dichloromethyl)-4-oxobutenoic acid and its cyclic isomer 3-chloro-4- (dichloromethyl)-5- hydroxy-2(5H) furanone, known as MX (Figure 1), as chlorination products in drinking water, and MX was shown to be responsible for a significant amount of the total mutagenicity in chlorinated water samples. (See MacDonald and Chipman, p. 335 - 343, this Volume.) MX was originally identified in the chlorinated effluent from pulp mills (Rook, 1974).

The mutagenicity of chlorinated waters has led to suggestions that such waters might also be carcinogenic. A recent meta-analysis in the USA of chlorination, chlorination by-products, and cancer suggests that there is a positive association between the consumption of drinking water containing chlorinated by-products and the occurrence of bladder and rectal cancer in humans (Morris *et al.*, 1992). That supports the results of earlier studies (Meier *et al.*, 1986; Cantor *et al.*, 1987; Flaten, 1992).

Colour in water arises from the presence of naturally occurring HS. These are generally characterized as being yellow to black in colour, and are highly refractory (Aiken *et al.*, 1985). Humic acids (HAs) and fulvic acids (FAs) are the major fractions of HS found in natural waters, and in general the ratios of FAs to HAs are in large excess and of the order of 9:1 (Malcolm, 1985).

Several methods have been employed for the removal from waters of the colour attributable to aquatic HS. These include powdered activated carbon (PAC), slow sand filters, and the use of aluminium salts (Hahn and Klute, 1990; Mbwette *et al.*, 1990; Wilmanski and van Breeman, 1990; Najm *et al.*, 1991; Collins *et al.*, 1992).

Humic substances in soil systems interact with clays and with (hydr)oxides (Schulthess and Huang, 1991), especially those of the most abundant species (iron, aluminium, and manganese), and with metals (Bartoli *et al.*, 1992). It is reasonable to assume, therefore, that combinations of clays, metals, and (hydr)oxide species could have uses for the removal of HS from waters. We have been involved with studies of the effectiveness of combinations of clays and (hydr)oxides for the sorption of humic substances, metals, and anthropogenic organic chemicals. The work described in this paper investigates the uses of Beringite, a calcined aluminosilicate schist material, and of aluminium-exchanged and aluminium pillared clays for the removal of humic substances which give rise to colour in waters.

2 Materials and Methods

Beringite

Beringite (1.0-2.0 mm diameter), was provided as a gift by Professor M. F. De Boodt, The State University of Ghent, Belgium. Beringite is an aluminosilicate formed from Palaeozoic schists, which provide a porous structure on calcination, with a surface area of the order of 20 m^2 g^{-1}, and a cation exchange capacity (CEC) in the range of 0.16-0.22 meq g^{-1} (De Boodt, 1991). The schists contain hematite, magnetite, ettringite, and minerals of the formula $(Mg,Fe)Al_2O_4$. These, in the presence of water, form minerals of the pyroaurite family $[(M^{2+})_6(M^{3+})_2OH_{16}CO_3.H_2O]$ which are known to have high sorption capacities. Aluminium (hydr)oxide species are also formed by the addition, with heating, of aluminium salts and of calcium or magnesium carbonate, and these contribute to the sorption capability.

Beringite has a high buffering capacity at alkaline pH, due to the presence of magnesium and calcium oxides. The composition of Beringite is given in Table 1.

Pillared Clays

Pillared clays are formed by 'propping open' the clay lamellae using a 'pillaring agent'. In general the agent is a polymeric metal (hydr)oxide species (Lahav *et al.*, 1978), but it can be any large polymeric species (Barrer and MacLeod, 1955; Farfan-Torres and Grange, 1990). The intercalation of the polymeric species is thought to evolve from ion exchange where the pillaring species replaces the counterions found between the clay layers.

In the work reported here 'pillaring solutions' were obtained by the hydrolysis of aluminium chloride with sodium hydroxide (the *cross-linkage* method). The processes used have enabled the preparation of pillared clays with d_{001} spacings of 18-23 Å. It is widely believed that the particular polymeric aluminium species responsible for the large increase in basal spacing is an Al_{13}^{7+} species, known as the Keggin ion (Figure 4).

Two 'starter' clays were used in a study of the pillaring process. These were Fulbent 570, a sodium montmorillonite clay from bentonite rock material (obtained from Laporte Industries, UK), and hectorite (SHCa-1) from the Source Clay Mineral Repository, University of Missouri. Some aspects of the chemical compositions and properties of Fulbent 570 and of hectorite are given in Table 2.

Prior to pillaring, the clays were cleaned using H_2O_2, and size fractionated to obtain

Table 1 *Chemical composition of Beringite (deBoodt, 1991)*

Element /Oxide	Amount (%)	Element	Amount (mg kg^{-1})
SiO_2	52.00	Mn	1100.00
Al_2O_3	30.00	Cu	120.00
CaO	3.45	Zn	630.00
MgO	1.48	Cd	9.10
K_2O	2.65	Co	98.00
Na_2O	0.58	Ni	123.00
Fe_2O_3	4.72	Pb	203.00
		Cr	950.00

Table 2 *Chemical compositions and cation exchange capacity (CEC) data of clay minerals used in the preparation of sorbents ([1] Laporte, 1992; [2] Newman, 1987)*

	FULBENT 570[1]	HECTORITE[2]
Type of clay	2:1	2:1
Type of substitution	Dioctahedral	Trioctahedral
Layer causing charge deficit	octahedral	octahedral
SiO_2	56.00	55.86
Al_2O_3	15.00	0.13
Fe_2O_3	6.50	0.03
Na_2O	5.00	2.68
MgO	3.50	25.03
CaO	2.50	trace
TiO_2	0.50	---
K_2O	0.50	0.10
Li_2O	----	1.05
F	----	5.96
CEC (meq g^{-1})	0.79	0.44

particles of < 2 μm. The suspension was dispersed by mechanical stirring, and ultrasonication using an ultrasonic probe (Ultrasonics Rapid, Ultrasonics, Surrey, UK), and the final concentration of the suspension was adjusted to 5% (w/v).

The Pillaring Process

Al hydroxide (pillaring) solutions with OH/Al (B/M) molar ratios from 0.100 to 10.0 were prepared by the addition of aluminium chloride (of several different molarities) to an equal volume of sodium hydroxide (of several different molarities). Initially, a precipitate was observed in some of the the mixtures, but in most cases these cleared after aging for a minimum of 6 days at 25±1 °C.

Aliquots (10 mL) of the clay suspension (5% w/v) were transferred to glass screw top containers. The pillaring solution (10 mL) was added to each suspension, and stirred using a magnetic follower. Clays suspended in the pillaring solutions were left for 24 h, then transferred to dialysis tubing and desalted by dialysis. Dialysis was continued until the dialyzate gave a negative chloride test with silver nitrate. Some of the pillared clay suspensions were used for X-ray diffraction studies, and the remainder were freeze dried.

X-Ray Diffraction

A Picker X-ray diffractometer and generator (producing CuK_{α} radiation of 1.5406 Å), in conjunction with a Philips PW 1710 diffractometer control and Siemens Diffract-AT software, was used for X-ray diffraction (XRD) studies. The pillared clay suspensions (0.5 mL) were placed on a glass slide and allowed to dry at room temperature. The clays were then dried (100 °C) to remove intercalated water before being analysed by XRD.

Isolation of Fulvic Acids (FAs)

FAs, the most abundant of the HS in the aquatic environment, were used as the sorptive. Methods were also developed for their desorption from the Beringite sorbent.

FAs were isolated by the tandem XAD-8/XAD-4 resin procedure (Aiken *et al.*, 1992; Malcolm and MacCarthy, 1992) from water samples taken from a reservoir, near Skipton, Yorkshire. The watershed was mainly upland peat.

Measurements of Sorption by Beringite

Beringite was washed with distilled water until a clear supernatant (free from fines caused by the abrasion of the dry material) was obtained. Washed material was packed into the glass columns (Omnifit, Cambridge) by adding the granules to water in the column (to avoid inclusion of air). Washing was continued using an additional 10 column volumes of distilled water.

A solution of FAs (200 mL, 50 mg L^{-1}), adjusted to pH 7 (0.1M NaOH), was passed at a flow rate of 0.5 mL min^{-1} through Beringite (5 g; in an Omnifit 150 x 10 mm glass preparation column) until the colour of the eluate equalled that of the influent (from UV/Vis analysis). The eluate, collected in 10 mL aliquots, was acidified to pH 2 (HCl), and absorbance (400 nm) was measured using a Perkin-Elmer Lambda 2 UV/Visible spectrophotometer.

Fractionation and Reconcentration of Fulvic Acids Using a Beringite Column System

A solution of FAs (1 L, 100 mg L^{-1}, pH 7) was passed through a column (Omnifit 500 x 15 mm) of Beringite (50 g) at a flow rate of 2.5 mL min^{-1}. The eluate (pH 10), collected in 50 mL aliquots, was acidified to pH 2 (HCl). Separate columns were back eluted with sodium tetraborate (0.1M), sodium carbonate (0.1 and 1.0M), tetrasodium pyrophosphate (0.1M), and with sodium hydroxide (0.1M) solutions.

Solutions of sodium tetraborate and of sodium hydroxide failed to remove the FAs from Beringite, but some desorption was obtained using solutions of sodium pyrophosphate and of sodium carbonate. Eluates in sodium carbonate and in tetrasodium pyrophosphate were collected in 10 mL aliquots, acidified to pH 2 with HCl and with orthophosphoric acid, respectively, and the absorbance (400 nm) was measured spectrophotometrically.

Sorption of Fulvic Acids by Pillared Clays

The FAs used were isolated from drainage waters from the Norfolk Fenlands, using the Amberlite XAD-8 and XAD-4 resin in tandem method (Aiken *et al.*, 1992; Malcolm and MacCarthy, 1992).

The pillared clays used to obtain the sorption isotherms were prepared using an OH/Al ratio of 2.25. These had a basal spacing of 18 Å.

Solutions of FAs were prepared at pH 2 and at pH 3, at concentrations of 150, 300, and 450 mg L^{-1}, and 10 mL of each of these was added to the pillared clay samples (10 mg). Equilibration was allowed to take place as the samples were rotated end over end (ca 10 rpm). The absorbance (λ=400 nm) values of the supernatants were measured daily, using a Perkin-Elmer UV/Vis spectrophotometer.

3 Results and Discussion

Sorption by Beringite

Some coloured substances were contained in the eluate from Beringite even after one void volume (35 mL) of FA had passed through the column. FAs continued to be sorbed by the Beringite until the sorption capacity of the column was reached, after ca 150 mL of the solution had passed through (Figure 2).

At maximum sorption, 2.5 mg (from the total 7.5 mg applied) of FAs was retained by Beringite (5 g). The eluate was at pH 10.

Back elution with a tetrasodium pyrophosphate (0.1M) solution removed 65% of the sorbed FAs (based on the recovery of coloured material) from the Beringite. However, interactions between the pyrophosphate and the Beringite particles gave a solid plug which could be dispersed using concentrated HCl (which degraded the Beringite particles).

Sodium carbonate (0.1M and 1.0M) desorbed part of the FAs from the Beringite without interacting with the sorbent. Treatment with 0.1M sodium carbonate removed 56-60% of the coloured material, and the total removed was 72% when 1.0M sodium carbonate was used as an additional treatment.

Sorption data indicate that 1 g of Beringite sorbed 0.5 mg of FAs. However, no threshold was found for breakthrough, and some FAs were eluted even in the first void volume from the column. The data suggest that 10 g of Beringite would be required to treat 1 L of water containing 5 mg L^{-1} DOC. The process would decrease, but not totally remove the dissolved coloured substances.

The pH values of the eluates were of the order of 10 [at which the metal hydr(oxides) in the Beringite and the FAs are negatively charged] which would suggest that the sorption mechanism would involve the formation of metal-fulvate complexes. This is substantiated by the release of fulvates from Beringite when pyrophosphate was used as a desorbent.

In terms of viable, full scale methods for the removal of colour from water, Beringite has some advantages compared with existing methods because it is obtained from a naturally occurring schist. It can be regenerated by heating in a fluidized bed at 800 ^{o}C, or disposed of in landfill or in concrete. However, the high pH (10) of the eluates, and the release of metals from the sorbent are significant disadvantages for its use.

Sorption by the Pillared Fulbent

The X-ray diffraction data for the pillared Fulbent 570 samples (prepared using a OH/Al ratios of 0.1 to 10.0) alllowed determination of the optimum conditions for pillaring.

From the results (Figure 3) it can be seen that the maximum d_{001} spacing was obtained using an OH/Al ratio of 2.25 in the case of the pillared Fulbent and ca 1.4 in the case of the hectorite. These results show that a large range of basal spacings were obtained for the OH/Al ratio range studied. This is thought to result from the nature of the pillaring species. The use of Al-27 NMR has provided information on several of the pillaring species that may be in the solution (Akitt *et al.*, 1972). These include:

$Al(OH)_6^{3+}$; $Al_2(OH)_2(H_2O)_8^{4+}$; $Al_6(OH)_{12}(H_2O)_{12}^{6+}$; $Al_{13}O_4(OH)_{24}(H_2O)_{12}^{7+}$; and $Al_{28}(OH)_{70}(H_2O)_{28}^{14+}$.

$Al_6(OH)_{12}(H_2O)_{12}^{6+}$ (which has a diameter of 4.7 Å), and $Al_{13}O_4(OH)_{24}(H_2O)_{12}^{7+}$

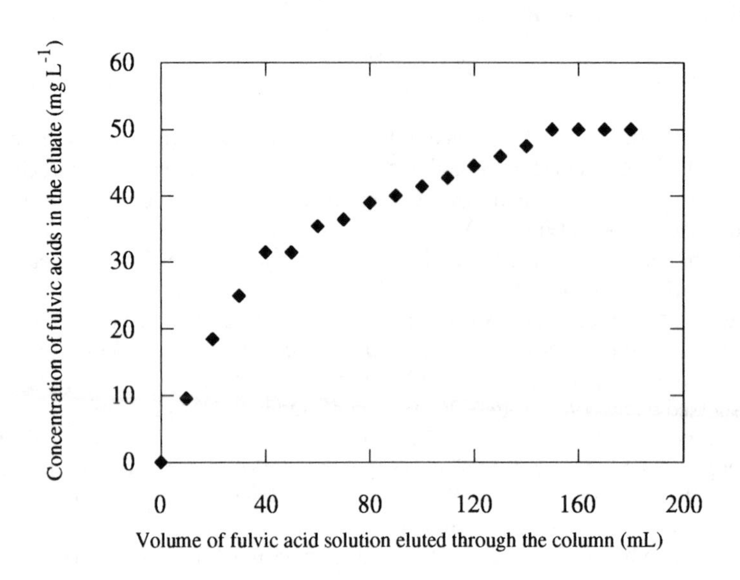

Figure 2 *Sorption by Beringite of fulvic acids (50 mg L^{-1}) from solution*

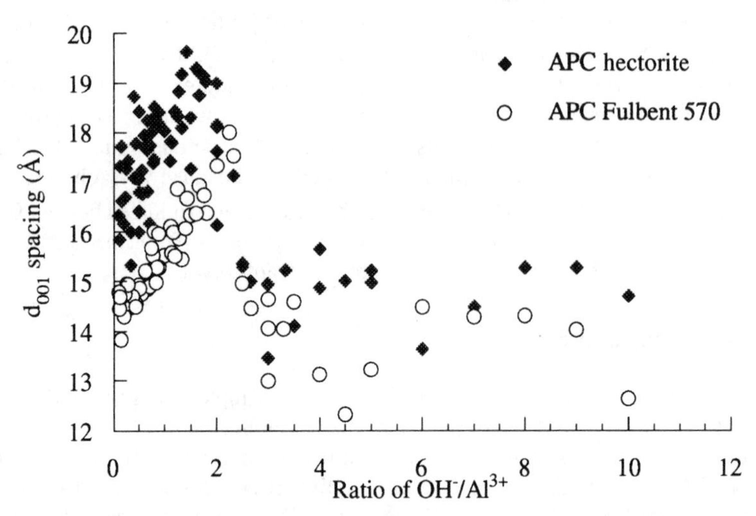

Figure 3 *d_{001} spacings of Fulbent 570 and Hectorite reacted with hydr(oxy) solutions containing a range of OH^-/Al^{3+} ratios*

with a diameter of ca 8.9-9.4 Å) are the species most likely to give rise to the larger d_{001} spacings (Lahav *et al.*, 1978).

$Al_{13}O_4(OH)_{24}(H_2O)_{12}{}^{7+}$ (Figure 4), known as $Al_{13}{}^{7+}$ or the Keggin ion, is spherical, and is thought to be largely responsible for the pillar spacings that give d_{001} spacings in the range 18-19 Å.

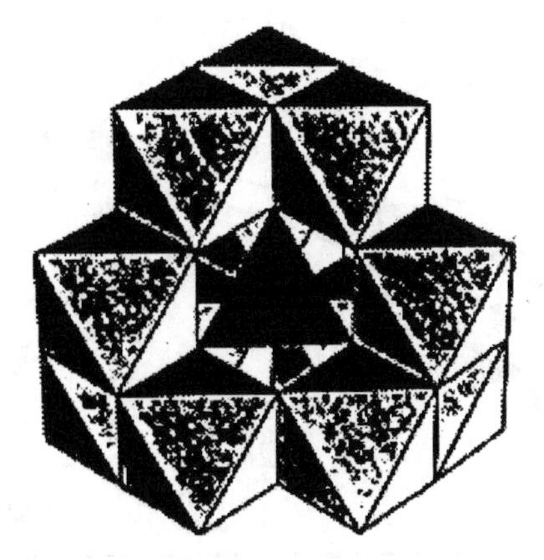

Figure 4 *The Keggin ion*

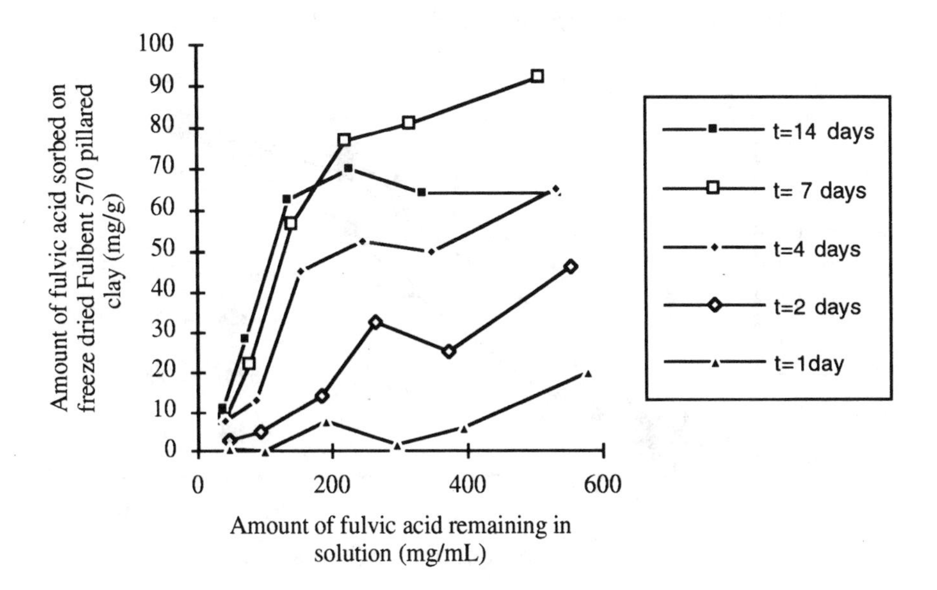

Figure 5 *Isotherm (20 ± 2 °C) for the sorption of fulvic acids by Al-pillared Fulbent 570*

Figure 5 shows that the amounts of the FAs sorbed by APC Fulbent 570, pillared at an OH/Al ratio = 2.25, increased with time, and after 48 h this pillared clay had sorbed 9% (w/w) of the FAs. The time dependence of the sorption process might be attributable to the diffusion of the sorptive solution through the pores of the pillared clay. Plots of the amounts sorbed at pH 2 and at pH 3 as a fraction of the total amount sorbed (fractional sorption) as a function of the square root of time for initial solution concentrations of 150, 300, and 450 mg L^{-1}, are given in Figures 6 and 7. The data imply that pH has a controlling influence on the rate of sorption.

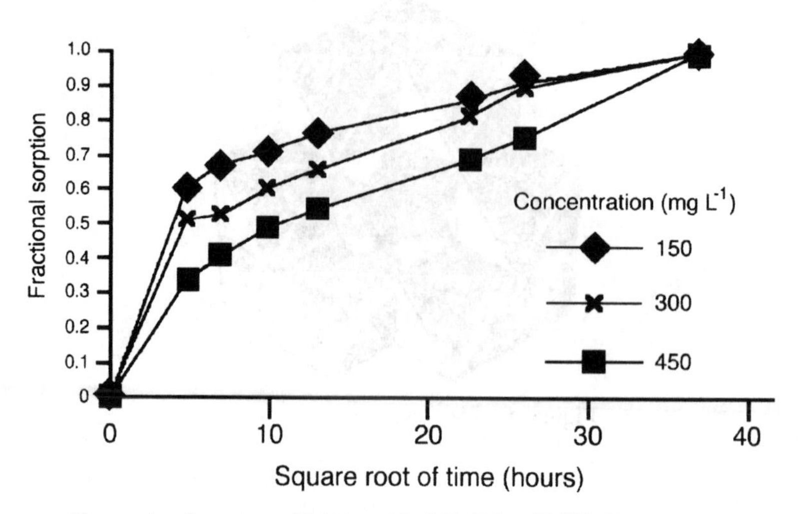

Figure 6 *Sorption of fulvic acids (pH 2) by APC Fulbent 570*

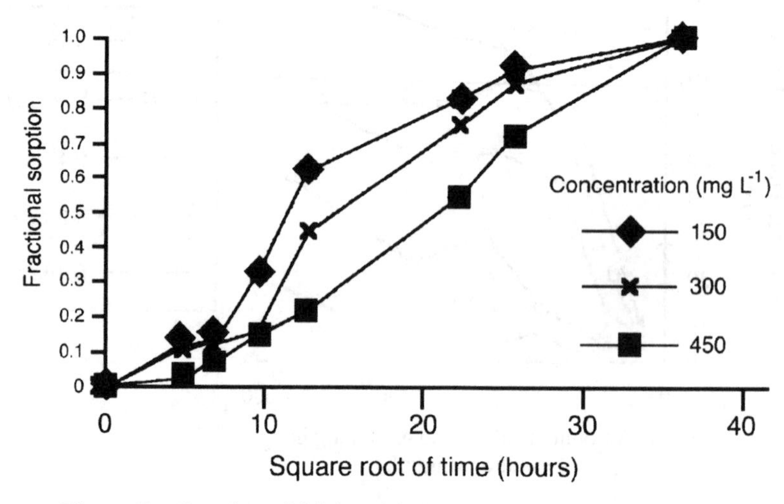

Figure 7 Sorption of fulvic acids (pH 3) by APC Fulbent 570

Investigations of the sorption of water soluble humic fractions by different pillared clays are ongoing.

4 Conclusions

1. Beringite provides a relatively inexpensive treatment for decreasing colour in waters.

2. Sodium carbonate solutions (0.1 and 1.0M) were the most effective of the reagents tested for the desorption of the FAs retained by Beringite.
3. The alkaline pH values (10) of the water after passing through the Beringite column, and the salts released to the waters are disadvantages in practical considerations for the removal of colour.
4. Initial results indicate that pillared clays can be effective sorbents for the removal of humic substances from natural waters.
5. The sorption capacity of pillared clays is greatly in excess of the sorption capacity of Beringite.

Acknowledgment

The work described here is part of that supported by a grant from the 'Safe Drinking Water Initiative' of SERC.

References

Aiken, G.R. 1985. Isolation and concentration techniques for aquatic humic substances. p. 363-386. *In*: G.R. Aiken, D.M. McKnight, R.L. Malcolm and P. MacCarthy (eds), *Humic Substances in Soil, Sediment, and Water*. Wiley New York.

Aiken, G.R., D.M. McKnight, R.L. Malcolm and P. MacCarthy. 1985. An introduction to humic substances in soil, sediment, and water. p. 1-9. In G.R. Aiken, D.M. McKnight, R.L. Malcolm and P. MacCarthy (eds), *Humic Substances in Soil, Sediment, and Water*. Wiley New York.

Aiken, G.R., D.M. McKnight, K.A Thorn,. and E.M. Thurman. 1992. Isolation of hydrophilic organic acids from water using non-ionic macroporous resins. *Org. Geochem.* **18**:567-573.

Akitt, J.W., N.M. Geenwood, B.L Khandelwal,. and G. D. Lester. 1972. ^{27}Al nuclear magnetic resonance studies of the hydrolysis and polymerisation of the hexa-aquo-aluminium(III) cation. *J.Chem. Soc.* Dalton. p. 604-610.

Baker, M.N. 1949. *The Quest for Pure Water*. The American Water Works Association Inc., New York.

Bellar, T.A. and J.J. Lichtenberg. 1974. Determining volatile organics at microgram-per-liter level by gas chromatography. *J. Am. Water Works Assoc.* **66**:739-744.

Bellar, T.A., J.J. Lichtenberg, and R.C. Kroner. 1974. The occurrence of organohalides in chlorinated drinking waters. *J. Am. Water Works Assoc.* **66:**703-706.

Barrer, R.M and D.M. MacLeod. 1955. Activation of montmorilonite by ion exchange and sorption complexes of tetra-alkyl ammonium montmorillonites. *Trans. Faraday Soc.* **51:**1290-1300.

Bartoli, F., R. Philippy, and G. Burtin. 1992. Poorly ordered hydrous Fe-oxides, colloidal dispersion and soil aggregation. II. Modification of silty soil aggregation with Fe(III) polycations and model humic materials. *J. Soil Sci.* **43**:59-75.

Cantor, K.P., R. Hoover, P. Hartge, T.J. Mason, D.T. Silverman, R. Altman, D.F.Austin, M.A. Child, C.R. Key, M.D. Marrett, M.H. Myers, A.S. Narayana, L.I. Levin, J.W. Sullivan, G.M. Swanson, D.B Thomas. and D.W. West. 1987. Bladder cancer, drinking water source and tap water consumption: A case control study. *J. Natl. Cancer Inst.* **79**:1269-1279.

Collins, M.R., T.T. Eighmy, J.M. Fenstermacher Jr. and S.K. Spanos. 1992. Removing natural organic matter by conventional slow sand filtration. *J. Am. Water Works Assoc.* **84**:80-90.

De Boodt M.F. 1991. Applications of the sorption theory to eliminate heavy metals from wastewater and contaminated soils. p. 293-329. In: G.H. Bolt, M.F. de Boodt, M.H.B. Hayes, and M.B. McBride (eds), *Interactions at the Soil Colloid-Soil Solution Interface*. NATO Series, Kluwer Academic Publishers, Dordrecht.

Farfan-Torres, E.M. and P. Grange. 1990. Pillared clays. *J. Chim Phys.* **87:**1547-1560.

Faust, S.D. and O.M. Aly. 1983. *Chemistry of Water Treatment*. Butterworths, London.

Flaten, T.P. 1992. Chlorination of drinking water and cancer incidence in Norway. *Int. J. Epidemiology* **21**: 6-15.

Hahn, H.H. and R. Klute. (eds), 1990. *Chemical Water and Wastewater Treatment* Springer Verlag, Berlin-Heidelberg.

Hemming, J., B. Holmbom, M. Reunanen and L. Kronberg. 1986. Determination of the strong mutagen 3-chloro-4-(dichloromethyl)-5-hydroxy-2(5H) furanone in chlorinated drinking waters. *Chemosphere* **15**:549-556.

Lahav, N., U. Shani, and J. Shabtai. 1978. Synthesis and properties of hydroxy-aluminium montmorillonite. *Clays and Clay Minerals* **26:**107-115.

Laporte Industries (Widnes), Technical Bulletin (1992)

Malcolm R.L. 1985. Geochemistry of stream fulvic and humic substances. p. 181-210. *In* G.R. Aiken, D.M. McKnight, R.L. Malcolm and P. MacCarthy (eds), *Humic Substances in Soil, Sediment, and Water*. Wiley, New York.

Mbwette, T.S.A., M.A.R. Stieitieh and N.J.D. Graham. 1990. Performance of fabric protected slow sand filters treating a lowland surface water. *J. Instn. Wat. & Envir. Mangt.* **4**:51-61

Meier, J.R., H.P. Ringhand, W.E. Coleman, K.M. Schenck, J.W. Munch, R.P. Streicher, W.H. Kaylor, and F.C. Kopfler. 1986. Mutagenic by-products from chlorination of humic acid. *Environ. Health. Perspect.* **69**:101-107.

Morris, R.D., A.M. Audet, I.F. Angellillo, T.C. Chalmers, and F. Mosteller. 1992. Chlorination, chlorination by-products and cancer: A meta-analysis. *Am. J. Publ. Health* **82**:955-963.

Najm, I.N., V.L. Snoeyink, B.W. Lykins Jr. and J.Q. Adams. 1991. Using powdered activated carbon: a critical review. *J. Am. Water Works Assoc.* **83**;65-76.

Newman, A.C.D. 1987. The interaction of water with clay mineral surfaces. p. 2-128. *In Chemistry of Clay and Clay Minerals*. A.C.D. Newman (ed.), Mineralogical Society, London.

Malcolm, R.L. and P. MacCarthy. 1992. Quantitative evaluation of XAD-4 and XAD-8 resins used in tandem for removing organic solutes from water. *Environment International* **18**:597-607 .

Rook, J.J. 1974. Formation of haloforms during chlorination of natural waters. *Water Treat. Exam.* **23**:234-243.

Schulthess, C.S. and C.P. Huang. 1991. Humic and fulvic acid adsorption by silicon and aluminium oxide surfaces on clay minerals. *Soil Sci. Soc. Am. J.* **55**:34-42

Smith, W. 1874. *A Smaller Classical Dictionary*. John Murray, London.

Wilmanski, K. and A.N. van Breemen. 1990. Competitive adsorption of trichloroethylene and humic substances from groundwater on activated carbon. *Water Res.* **24**:773-779.

Investigations of the Impacts of Humic-Type Substances on the Bayer Process

E.C. Norman, I.R. Dixon, C.L. Graham, M.H.B. Hayes, and S.C. Grocott[1]

THE UNIVERSITY OF BIRMINGHAM, SCHOOL OF CHEMISTRY, EDGBASTON, BIRMINGHAM B15 2TT, ENGLAND

[1]ALCOA OF AUSTRALIA LTD, KWINANA, PO BOX 161, WA 6167, AUSTRALIA

Abstract

Soil organic matter in the bauxite feedstock gives rise to humic-type substances (H-TS) in the Bayer process used to extract aluminium hydroxide from bauxite ore. Tests were carried out which studied the influences or impacts of these H-TS on:

1, the solubility of sodium oxalate (a major contaminant in Bayer liquors which arises from the oxidation of organic substances contaminating the feed stock);
2, foaming (an undesirable property) of the liquors; and
3, the agglomeration of aluminium hydroxide particles.

The H-TS taken from different stages of the Bayer cycle were fractionated into humic-type acids (H-TA) fulvic-type acids (F-TA) and XAD-4 acids. The results indicate that some of H-TS affect the stability of oxalate in the liquor. Only F-TA gave rise to stable foams, and the Bayer liquor H-TS decreased the agglomeration of aluminium hydroxide.

1 Introduction

Origins of Humic-Type Substances (H-TS) in the Bayer Process

The Bayer process (Bayer 1888) is used to isolate aluminium hydroxide from bauxite ore. It utilises the amphoteric nature of aluminium for the dissolution of aluminium hydroxide in a strong sodium hydroxide solution. The present study is part of a programme designed to take account of the unusual feed stock supplied to the Alcoa of Australia Ltd. refineries in Western Australia. There, open cut mining is used to recover bauxite from the Darling Range. In the mining process as much of the topsoil and overburden is removed as is feasible. However, the bauxite deposits are in uneven strata and it is inevitable that some residual overburden is contained in the mined ore. The alkaline conditions of the Bayer process give rise to partial oxidation of the soil organic matter, and to the formation of range of sodium carboxylates, fulvates, and humates.

To date over 1600 organic compounds have been identified in Bayer liquor. Because the Bayer process is cyclic, fresh organic matter is continually added and removed, and this leads to a steady state concentration of the order of 25 g L^{-1} organic carbon in the liquors in Alcoa's Western Australian plants. The organic species have been classified into several categories at Alcoa (Lever 1978, Guthrie *et al.* 1984), and fulvic-type acids (F-TA) and humic-type acids (H-TA) comprise 22-25% and 3-5%, respectively, of the total organic carbon. The term '-type acids' is used in order to differentiate the operationally defined humic substances (HS) found in the Bayer liquor from those isolated from soil and water samples.

The mechanisms by which the H-TS interact in the Bayer process are not well understood. Because these substances are macromolecular and anionic polyelectrolytes, the polar and non polar functionalities will have an impact on the Bayer process. Three of the features of the Bayer process that are considered to be influenced by H-TS are:

(i) The stability of sodium oxalate;
(ii) the inhibition of the aluminium hydroxide agglomeration; and
(iii) the foaming of the liquor.

Oxalate Stability

Sodium oxalate, by far the most abundant organic impurity in Bayer liquor, is formed through oxidation of organic species during the cycle (Lever 1978). The *oxalate breakpoint* is the critical oxalate concentration at which primary and/or secondary heterogeneous nucleation occurs. At that breakpoint sodium oxalate starts to precipitate from the Bayer liquor. The breakpoint is dependent on temperature, on sodium ion concentration, and on the concentration of H-TS. Sodium oxalate crystals precipitate in the shape of fine needles (acicular) and the aluminium hydroxide will then co-precipitate as a very fine product on the oxalate needles (in preference to forming pure aluminium hydroxide crystals when the aluminium hydroxide seed crystals are added). The concentration of oxalate in the liquor can reach 400% supersaturation, depending on the stabilizing properties of the H-TS and of the other organic compounds present in the Bayer liquor. Norman *et al.* (1996) have shown that some fractions of H-TS prevent to some degree crystallization of sodium oxalate. Their impact on oxalate stability was studied by measuring the changes in breakpoints for process liquors when a variety of H-TS were added. No additions were made to the Bayer liquor in the reference run.

In the present study, in order to test the impact of H-TS on the oxalate breakpoint, a series of test liquors, of increasing oxalate concentrations, were prepared from a sample of Bayer liquor, and a known concentration of H-TS was added. The volume of H-TS solution added was determined using the Bayer liquor concentration (BLC).

Bayer Liquor Foaming

It is well known that HS in aquatic environments produce foams, and foaming is a common phenomenon in Bayer liquors. A test was designed to ascertain which, if any, of the H-TS in the Bayer liquor would produce a stable foam from the liquor.

Agglomeration Tests

Agglomeration is an important part of the precipitation stage of the Bayer process.

Aluminium smelters will not operate efficiently when fine alumina is supplied. The precipitation of aluminium hydroxide (gibbsite) particles takes place over 30 h and agglomeration can be monitored in the early stages of the precipitation process. The impact of an additive on the agglomeration can be observed in 4.5 h. HS are known to interfere/interact in aluminium hydroxide precipitation in the soil solution (Mazet *et al.* 1990, Edwards and Amirtharajah 1985, and Chadik and Amy 1987). However, the interactions in the soil solution are vastly different from those in the Bayer process because the compositions of the HS and H-TS, the pH values, the ionic strengths, and the aluminium hydroxide concentrations are vastly different in the two systems.

The agglomeration process is best studied in a synthetic liquor. In this way the effects of the H-TS can be assessed independently of the other organic compounds present in the Bayer liquor. Because the agglomeration test focuses on the precipitation of aluminium hydroxide, the initial concentration of sodium aluminate must be high in order to mimic the concentrations of green liquor (the liquor from which aluminium hydroxide has not precipitated.) The liquor was dosed with the H-TS, to a concentration equivalent to the BLC. The dosed liquor was then equilibrated at 70 °C. Duplicate runs were carried out, and duplicate controls (containing no additives) were also run.

The aluminium hydroxide seed was then added to each test bottle. The seed used was 'tertiary thickener seed', or tray seed. The test bottles were rotated for 4 h. The precipitated aluminium hydroxide was separated by filtration, and washed until free from the caustic solution. The tray seed, and the aluminium hydroxide from each test were assessed for particle size density on a Malvern Mastersizer, in which a measured amount of the aluminium hydroxide sample under test was suspended in de-ionized water. The analysis was carried out using laser light diffraction measurements. The size distribution curve, and a Table for each of the particle sizes measured was obtained for each sample.

2 Materials and Methods

Isolation of H-TS from Bayer Process Samples

Isolation was from:

1 Spent Bayer Liquor, sampled after aluminium hydroxide has been precipitated;
2 Sodium Oxalate 'Cake' (40% liquor moisture), sampled directly after filtration through large fabric filters during the oxalate removal process. The sample had not been pressed, and contained a high content of Bayer liquor;
3 Sodium Oxalate 'Cake' (20% liquor moisture) taken after pressing 40% liquor moisture cake to remove Bayer liquor; and
4 Lake Water, sampled from the residue settlement 'lakes'. Because of additions of washing waters, the Lake Water was ten times more dilute than Bayer liquor.

Extraction of H-TS from Bayer Liquor

Citric acid solution (0.6M, 8 L) was added, with stirring, to Bayer liquor (8 L), and the mixture was pre-filtered through two Whatman GF/B glass microfibre filter papers. Citric acid prevents the precipitation of aluminium hydroxides on acidification. The sample was filtered under pressure through Sartorius 0.2 μm cellulose acetate membrane filters. The filtrate was diluted two hundred fold with distilled water, the pH was lowered to 1.96, and then processed using XAD-8 and XAD-4 resins in tandem as described in this Volume by Hayes *et al.* (p. 107 - 120). Humic-type acids (H-TA) and fulvic-type acids (F-TA) were isolated from the XAD-8 resin and XAD-4 acids (XAD-4-TA) were isolated from the XAD-4 resin.

Extraction of H-TS from Sodium Oxalate Cake (40% Liquor Moisture) and Sodium Oxalate Cake (20% Liquor Moisture)

Sodium oxalate cake (40% liquor moisture, 40 g) was dissolved in distilled water (1 L). The pH of the resulting solution was lowered to 6-7, then pre-filtered through a Whatman GF/B glass microfibre filter. In the pH range 6-7, amorphous aluminium hydroxide is precipitated from solution. The combined filtrate from the processing of six kg of oxalate cake was then filtered under pressure through Sartorius 0.2 μm cellulose acetate membranes, and after dilution the filtrates were subjected to treatment with XAD- resins, as was done for the Bayer liquor sample.

Sodium oxalate cake (20% liquor moisture, 180 g) was dissolved in distilled water (4.5 L), and mixed with an equal volume of 0.6M citric acid solution. The pH of the resulting solution was lowered to 6 - 7, pre-filtered through a Whatman GF/B glass microfibre filter and then through Sartorius 0.2 μm cellulose acetate membrane, and the filtrate was treated as outlined for the H-TS isolated from the Bayer liquor sample.

Extraction of H-TS from Lake Water

The pH of the Lake Water sample (50 L) was adjusted to pH 6-7, then filtered under pressure through Sartorius 0.2 μm cellulose acetate membrane filters, and treated as described for the H-TS isolated from the Bayer liquor sample.

Determination of the *Oxalate Breakpoint*

The oxalate breakpoint was determined by increasing the sodium oxalate concentration and measuring the point at which sodium oxalate crystallisation occurred. The procedure used was an Alcoa test method.

Liquor Foaming Tests

H-TS were added (at the concentrations listed in Table 2), to a filtered solution of organic and inorganic salts, mixed into a solution of NaOH and $Al(OH)_3$ (55 ^oC), and equilibration was allowed to take place for 30 min at 60 ^oC. Air, at a flow rate of 3 L min^{-1} was pumped through a glass tube with the fritted end placed in the synthetic liquor. The foam height was recorded at set time intervals. The foam was allowed to subside, and the test was repeated.

The Agglomeration Tests

Two equivalent doses of Bayer liquor H-TA and F-TA were added to a solution of $Al(OH)_3$, NaOH, and Na_2CO_3 preheated to 75 ^oC. After equilibration for an hour at 75 ^oC, the $Al(OH)_3$ start seed was added, and the assembly was allowed to equilibrate for 3.5 h. The liquor was then filtered and the solids were washed with deionized water until

Table 1 *Concentrations of the H-TS in the test liquor (300 mL)*

Sample	Abbreviation	Concn $g L^{-1}$	Bayer Liquor Equivalent Concn (BLC)
Bayer liquor H-T acid	BL H-TA	0.1095	x 0.2
Bayer liquor F-T acid	BL F-TA	1.1554	x 0.2
Bayer liquor XAD-4 acids	BL XAD-4	0.0161	x 0.2
Oxalate 20% H-T acid	Ox(20%)H-TA	0.0036	x 1
Oxalate 20% H-T acid	Ox(20%) H-TA	0.0180	x 5
Oxalate 20% F-T acid	Ox(20%) F-TA	0.0095	x 1
Oxalate 20% F-T acid	Ox(20%) F-TA	0.0499	x 5
Oxalate 20% XAD-4 acids	Ox(20%) XAD-4	0.0240	x 5
Oxalate 40% H-T acid	Ox(40%) H-TA	0.0004	x 1
Oxalate 40% F-T acid	Ox(20%) F-TA	0.0041	x 1

free of NaOH, then dried. A Malvern Mastersizer was used to measure particle size density.

3 Results

Impact of H-TS on the Oxalate Breakpoint

With the exception of the oxalate (20%) H-TA, the results from the oxalate Breakpoint Test indicate that the H-TS had, at the concentrations tested, only small influences on the stability of the sodium oxalate in the Bayer liquor. The oxalate 20% H-TA was dosed at five times its normal Bayer liquor concentration and it gave a seven per cent increase in the stability of the dissolved sodium oxalate. It produced only a small increase in the oxalate stability at its BLC. Both of the oxalate (20%) F-TA additions increased the stability of the oxalate in the Bayer liquor. However, the effect was small, and close to the error limit of this experiment. The changes in oxalate stability induced by the oxalate (40%) H-TS again were very small, and close to the error limits of the test. Both of the XAD-4 acids which were tested gave a decrease in the stability for the sodium oxalate (see Figure 1).

Impact of H-TS on Liquor Foaming

The H-TA, XAD-4 acids, and oxalate H-TS tested did not produce stable foams in the synthetic Bayer liquor. Stable foams were, however, produced by the Bayer liquor F-TA and by the Lake Water F-TA (Figures 2 and 3). These two samples were then tested at varying concentrations, and both samples showed a rapid enhancement of the amount of foam produced. The Lake Water F-TA exceeded the measuring capabilities of the equipment. After the initial rapid increase in the amount of foam produced, the Bayer liquor F-TA reached a maximum. Increasing the concentration of the F-TA increased both the rate at which the maximum was attained, and the height of that maximum. As the concentration of the Bayer liquor F-TA was increased, the maximum foam height

Table 2 *Concentration of humic-type substances in 250 mL of synthetic liquor*

Sample*	Concentration $g\ L^{-1}$	BLC*
BL H-TA	0.1168	x 0.2
BL F-TA	0.6169	x 0.1
BL F-TA	1.2328	x 0.2
BL F-TA	1.8486	x 0.3
BL F-TA	1.9654	x 0.4
BL F-TA	3.0810	x 0.5
BL XAD-4 acids	0.1390	x 0.2
Lake Water H-TA	0.0202	x 1
Lake Water F-TA	0.5289	x 1
Lake Water F-TA	1.0578	x 2
Lake Water F-TA	1.3222	x 2.5
Ox (40%) H-TA	0.0015	x 5
Ox (40%) F-TA	0.0163	x 5
Ox (20%) H-TA	0.0192	x 5
Ox (20%) F-TA	0.0713	x 5
Ox (20%) XAD-4 acids	0.0400	x 5
Refinery liquor 1:5 with synthetic liquor	-	x 0.2

* See Table 1 for abbreviations meanings

attained converged to a value of 750 - 800 mL. This was due to the limit imposed by the amount of synthetic liquor used. At higher F-TA concentrations all of the synthetic liquor was in the foam, and no liquid was available for the generation of more foam.

The characteristics of the foam generated were different in the case of the Lake Water (LW) F-TA. This F-TA produced a white foam, consisting mainly of a relatively small number of bubbles with large internal volumes. On the other hand, the Bayer liquor F-TA produced a brown coloured foam consisting of a large number of bubbles with small

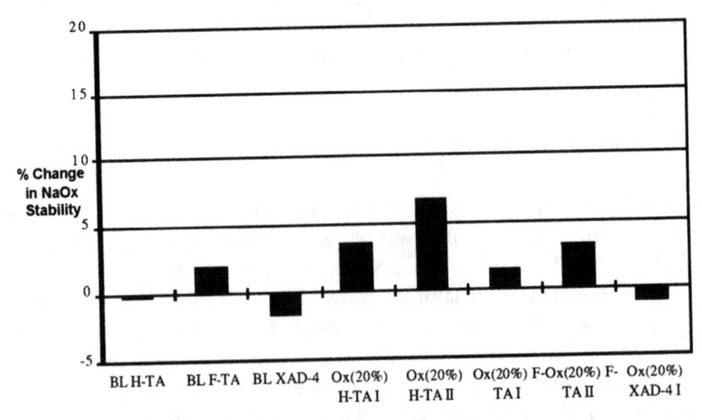

Figure 1 *Plot showing the percentage change in the stability of sodium oxalate with the addition of H-TS*

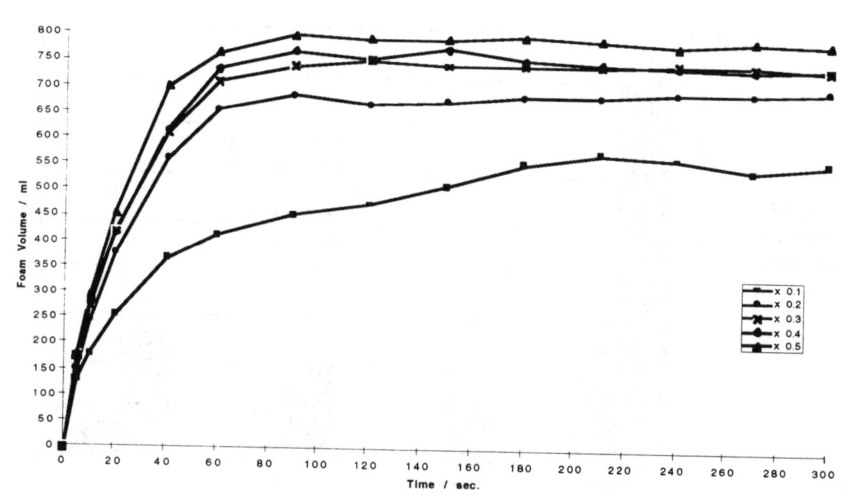

Figure 2 *Plot showing the volume of foam produced (mL) against time (sec) for Bayer liquor fulvic-type acids at x 0.1, x 0.2, x 0.3, x 0.4, and x 0.5 Bayer liquor concentration*

internal volumes. This difference in the foam characteristics enables the LW F-TA to produce enough foam to exceed the measuring capabilities of the equipment used, and still to have a significant amount of synthetic liquor available to generate additional foam.

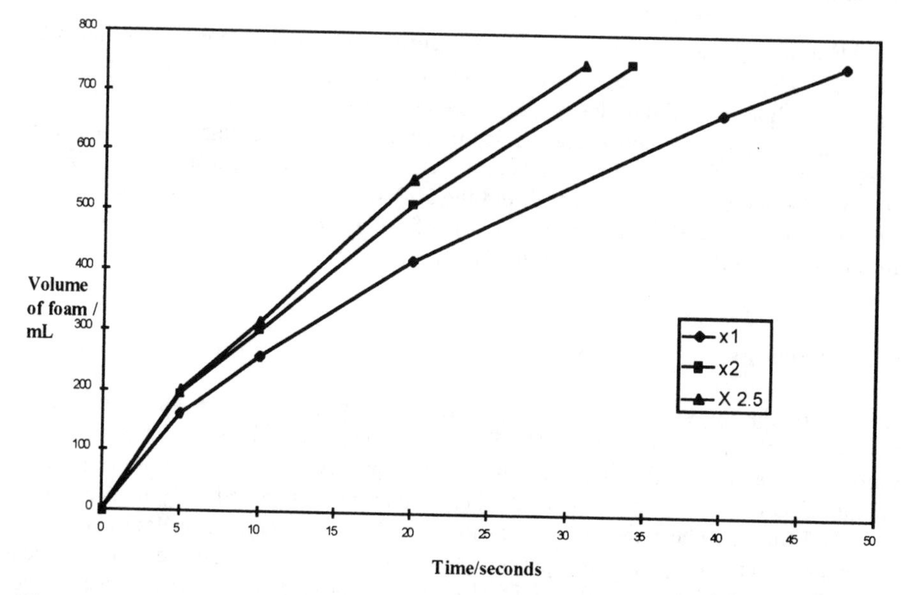

Figure 3 *Plot showing the volume of foam produced (mL) against time (sec) for lake water fulvic-type acids at x 1, x 2, and x 2.5 Bayer liquor concentration*

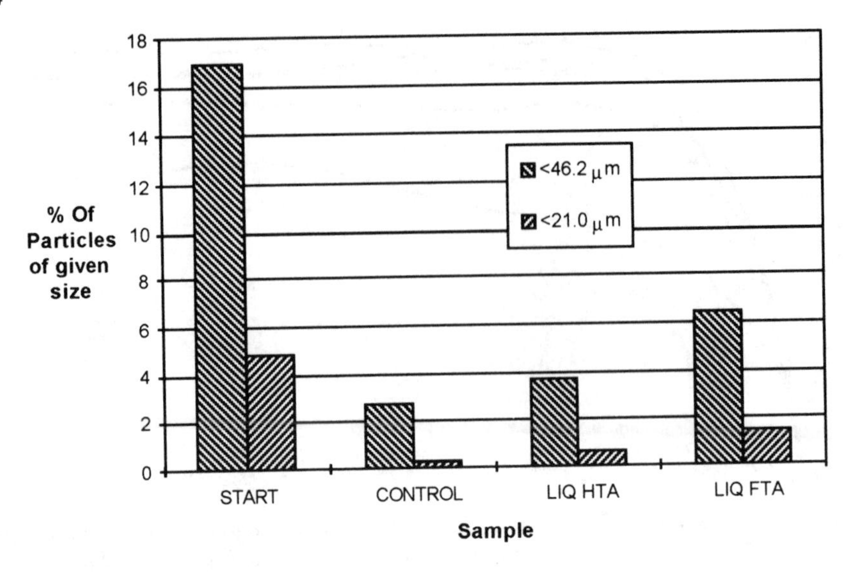

Figure 4. *The impact of Bayer liquor humic-type substances on the agglomeration of aluminium hydroxide with respect to a control containing no additive. The smaller the percentage of particles of < 46.2 μm and < 21.0 μm, the greater the agglomeration*

The Impact of Bayer Liquor H-TS on Aluminium Hydroxide Agglomeration

The extent of agglomeration is assessed by measuring the relative proportions of the < 21 μm and < 46.2 μm diameter particles present. As agglomeration takes place, the proportion of particles of this size range would be expected to decrease.

Figure 4 shows that agglomeration had taken place in all the runs. The H-TS were dosed in at their Bayer liquor concentrations. At these concentrations, the F-TA inhibited the agglomeration more than did the H-TA. The variation was greater for the duplicate runs for the H-TS compared with that for the control duplicate runs. This was attributed to the amorphous nature of the H-TS.

4 Conclusions

The aim of the investigation was to quantify the impacts of the H-TS in the Bayer process. This would imply that the H-TS should be tested at their BLC. That presented a problem when testing the impacts of the Bayer liquor H-TS on the stability of sodium oxalate. As opposed to what is generally believed in the alumina refining industry, none of the H-TS appeared to exert any statistically significant effect on the oxalate stability.

In the Bayer liquor foaming experiments the H-TA and XAD-4 acids did not produce foam. The oxalate 20% and 40% H-TS also failed to produce foam. In the case of the oxalate 20% and 40% F-TA, this failure was due to the low concentrations of material used in the test. The Bayer liquor and Lake Water F-TAs both produced foams, and in both cases production was rapid. That from the Lake Water F-TA exceeded the measuring capabilities of the equipment. The F-TAs in the Bayer process liquor are likely to be the major factor contributing to the excessive foaming of the Bayer process liquor.

Agglomeration of aluminium hydroxide is inhibited by Bayer liquor H-TS. When compared at their Bayer liquor concentrations, the F-TAs have a larger impact than the H-TAs. However, the F-TAs are present in the Bayer liquor at approximately ten times the concentration for the H-TAs.

The tests carried out show that H-TS have influences on the production of alumina by the Bayer process. It is clear that the different components of the H-TS have different influences. However, the impacts are not directly related to the concentrations of the H-TS components.

References

Bayer, K. J. 1888. German Patent. 43 977. Aug 3.

Bremner, J. M. 1949. Some observations on the oxidation of soil organic matter in the presence of alkali. *J. Soil Sci.* **1**:198-204.

Chadik, P.A. and G. L. Amy. 1987. Coagulation and adsorption of humic substances - an analysis of surrogate parameters for predicting effects on trihalomethane formation potential. *Environ. Tech. Let.* **8**(6):261-268.

Edwards, G. A. and A. Amirtharajah. 1985. Removing color caused by humic acids. *J. Am. Water Works Assoc.* **77**(3):50-57

Guthrie, J. D., P. J. The, and W. D. Imbrogno. 1984. Characterisation of organics in Bayer liquor. *Light Metals.* p.127-136.

Humski, E. 1994. Oxalate in liquors by GC. KW-LSM-74-141-LQ-009. Private Alcoa of Australia Ltd. Document.

Lever, G. 1978. Identification of organics in Bayer liquor. *Light Metals* **2**:71-83.

Malcolm, R. L., E. M. Thurman, and G. R. Aiken. 1977. The concentration and fractionation of trace organic solutes from natural and polluted waters, using XAD-8, a methylmethacrylate resin. p. 307-314. *In* D. D. Hemphill (ed.), *Trace Substances in Environmental Health-XI. A Symposium.* University of Missouri.

Mazet, M., L. Angbo, and B. Serpaud. 1990. Adsorption of humic acids onto preformed aluminium hydroxide flocs. *Water Res.* **24**(12):1509-1518.

Neyroud, J. A. and M. Schnitzer. 1975. The alkaline hydrolysis of humic substances. *Geoderma* **13**:171-188.

Tran, T., K. A. Chouzadjian, A. D. Stuart, and D. A. J. Swimbles. 1986. Oxidation of organics in simulated Bayer liquor using MnO_2 ore. *Light Metals.* p. 217-222.

Yamada, K., T. Harato, and H. Kato. 1981. Oxidation of organic substances in the Bayer process. *Light Metals.* p. 117-128.

Section 3

Environmental Impacts of Humic Substances and Organic Matter

The influences of soil organic matter (SOM) on soil structure and on soil fertility are well recognised. So too is the important role which peatlands have as a carbon sink in the higher latitudes of the Northern Hemisphere. Concerns for the environment are expressed when peats are harvested and when wetlands are drained. There is a need therefore to know more about the nature of the organic matter in peatlands and in wetlands, and about the extents to which these are transformed when management systems are altered.

The review by Little and Farrell (p. 247 - 259) has outlined the topographic features and some of the soil textural characteristics which give rise to podzols in Ireland. Podzolic soils can lead to the genesis of peats in areas of high rainfall where hardpan is formed that impedes drainage. Podzols will commonly be found in upland regions in oceanic climates where base poor parent materials and appropriate topography prevail. Podzols may also be found in lowland areas where ironpan is often a feature of their compositions.

Undisturbed podzols with a climax vegetation that has not changed for thousands of years, and which have A_h as well as B_h horizons, provide fascinating features for studies of the kinds of transformations in OM which take place with time. It is assumed that the humic substances (HS) in the B_h horizon originated in the A_o/A_h horizons These solubilized in water, or formed dispersed colloidal suspensions prior to transportation in the soil profile. The study by Simpson *et al.* (p. 73 - 82) has shown that there are significant differences in aspects of the compositions of HS in the A_h and B_h horizons where the climax vegetation was predominantly oak forest.

Blanket peats, which consist predominantly of ombrotrophic peats, and cover substantial areas of Britain and Ireland, are formed in shallow base deficient soils where the annual precipitation exceeds 1250 mm and the rain days per year exceed 225. Low moor peats can form in shallow lakes and stagnant water when the drainage waters are rich in minerals. Thus the lower horizons of such peats can be formed from wood and from vegetation requiring good supplies of cations, and a pH that is not strongly acid. As the peat level is raised above the water source, the supply of minerals is diminished, and the vegetation is changed to mosses and other species which can survive on a low mineral supply and tolerate acid conditions. These conditions give rise to raised bogs.

The substantial areas of blanket peatlands has focused attention on their economic development, and considerable investment has been made in the afforestation of such peats. The review by Byrne and Farrell (p. 260 - 270) has highlighted the ways in which afforestation affects the physical and chemical properties of peats, and the quality of the runoff waters from the forest lands. This review casts serious doubts on the environmental desirability of planting forests on blanket peats. There are abundant data to show that afforestation promotes the oxidation of the indigenous OM, and the drainage waters from afforested peats are significantly more rich in dissolved organic carbon and organic acids than those from non-forested blanket peats. Among other things, the altered water quality can have adverse effects on the biota in the water catchment, and the forest management system that must be employed can give rise to depletion of the soils by erosion following harvest, and also give rise to biological oxidation of the blanket peats as the result of drainage and the alteration of the water levels. The authors caution about the environmental impacts of afforestation of blanket peats, and consider that "rehabilitation of forested blanket peats may be the most appropriate environmental option".

Blanket and high moor peats are poor in cations, and are sensitive to inputs of acids in precipitation. White *et al.* (p. 278 - 287) have reported on extensive regional surveys carried out on Scottish moorland peats and podzols to assess the relative adverse effects of acid depositions on the soil chemistry of low base status soils. Their data show that, for 'acidification-vulnerable' soils, the relationships between soil pH and the $[H^+]:\sqrt{[Ca^{2+}]}$ for the upper highly organic surface horizons are similar to those for the peats. Thus, measurements of acid-deposition effects on peat chemistry could be used as indicators of likely acidification damage to adjacent, sensitive mineral soils. The study of Dawod and Cresser (p. 288 - 298) has compared the streamwater chemistry from three catchments of peats and peaty podzol soils with the compositions of the precipitation. The compositions of the waters from surface run off were similar to those of the precipitation, and in the case of a catchment with negligible amounts of mineral soil underlying peat, the relative cation proportions in the river water were similar to the proportions in precipitation.

An international project at Lake Skjervatjern in Norway has looked at the influence of simulated acid rain on the lake and on the watershed composed of a dystrophic peat. Hayes *et al.* (p. 299 - 310) have studied the influences of the acid rain on the neutral sugar contents in the humic fractions from the drainage waters from the acidified and control watersheds, and in the control and acidified sides of the lake. Their study has shown that, overall, the acidification process did have some effects on the totals of sugars bound to the HS fractions. The major influences, however, involved changes to the relative abundances of the sugars in the different humic fractions as the result of the acidification process,

A contribution by Ball and Bullimore (p. 311 - 318) has examined the decomposition of C3 and C4 plants grown at ambient and at elevated atmospheric CO_2 concentrations. This is relevant to the environmental influences of increasing levels of greenhouse gases from the burning of fossil fuels, and from the breakdown of SOM as the result of changes in soil management practices. In the cases of the C3 plants the decomposition rates were less in the cases of all plants grown at elevated CO_2 concentrations, presumably because under these circumstances there was a 10% decrease in plant N. The compositions and C:N ratios of C4 plants grown under ambient and elevated CO_2 did not differ significantly, and the decomposition rates were essentially the same for both CO_2 regimes.

The contribution by Lara *et al.* (p. 319 - 325) focuses on marine HS from the polar regions. Marine humus is produced by the transformation of constituents of phytoplankton which give rise to particulate (POM) and dissolved organic matter (DOM). The amount of dissolved C in the oceans is about 10 times that in particulate form; hence there is a case for detailed studies of its formation and transformations. This publication pays especial attention to the amino acid contents, and the data show that there is a selective preservation of amino acids (and especially glycine) in the hydrophobic neutral fraction of the HS.

Applications of an enhanced chemiluminescence (ECL) procedure to studies of HS are described by Watt *et al.* (p. 326 - 333). Their data indicate that the humic fractions with the greatest resemblances to functionalities in lignin- and tannin-type structures are the strongest inhibitors of ECL. Thus the amounts and origins of HS in waters should be considered when using ECL to detect and to determine pollutants in waters.

A Review of the Characteristics and Distributions of Organic Matter-Related Properties in Irish Podzols

Declan J. Little

DEPARTMENT OF ENVIRONMENTAL RESOURCE MANAGEMENT, UNIVERSITY COLLEGE DUBLIN, BELFIELD, DUBLIN 4, IRELAND

Abstract

This review of the literature on podzols in Ireland has concentrated primarily on properties associated with organic matter (OM). Most of the studies on Irish podzols have placed emphasis on morphology, and on physical and chemical properties. Others have studied their genesis, taking account of anthropogenic influences, parent materials, relief, topography and vegetation.

The distribution of OM, whereby the maximum amount occurs at the soil surface, has implications for the overall nutrient status of these soils. The soils are acidic, nutrient deficient, and have high C:N ratios. The mor humus layer contains the maximum cation exchange properties and total exchangeable bases, and is the zone where the maximum concentration of fine roots occur. The nature of the OM is transformed upon cultivation, and thereby the productivity of these soils is increased appreciably. However, such soils become base desaturated in a relatively short period after abandonment, then translocation of organo-metallic complexes is resumed, and re-podzolisation eventually takes place.

The nature of translocated, organically bound iron appears to change as podzolisation proceeds. The highest values occur at shallow depths during the earlier phases of podzolisation (humic-brown podzolic/ferric podzols), and at greater depths in the more podzolised soils (humo-ferric podzols). Well developed ironpan stagnopodzols, which are common at high altitudes, and at lower elevation in western coastal counties have lower amounts of these iron species, probably as the result of immobilization and reduction processes. Gleisation of humo-ferric podzols results in lower amounts of translocated organic and inorganic iron species.

The composition and fractionation of organic matter of Irish podzols requires further investigation, especially in relation to vegetation type and the transformation of translocated organo-metallic complexes. This would help elucidate specific podzolisation processes occurring under alternative vegetation/soil regimes.

1 Introduction

Podzols are a common occurrence in the upland regions of Ireland, especially in western coastal counties (Culleton and Gardiner, 1985). The principal factors in their formation are the oceanic nature of the climate, base-poor parent materials, topography, and the presence of vegetation communities which promote podzolisation processes. Anthropogenic impacts also have important influences in accelerating podzolisation (Cruickshank and Cruickshank, 1981; Little, 1994; Cunningham, 1996). An outline of the environmental conditions affecting pedological conditions in Ireland (geology, soil parent materials, pedogenic processes, long-term climatic data, and other contributory items are given in the Agroclimatic Atlas of Ireland (Collins and Cummins, 1996).

The General Soil Map of Ireland gives the major source of information on the distribution of soils, and on the occurrence of podzols in Ireland. Podzols are estimated to compose *ca* 11% of the Irish soil landscape (Gardiner and Radford, 1980). McAllister and McConaghy (1968) have provided general data on the soils of the northern region, and more recently specific data on podzolic soils have been provided by by Geddis (1986). Also, the completion of the mapping of soils of Northern Ireland in the past decade (scale 1:50000) has provided data on the extent of podzols in this region.

In common with the mid-latitudes in general, and as a result of Ireland's island and coastal zone position, Irish soils have been subjected to a leaching regime since their initiation following the Pleistocene glacial events. Hence leached soils, and podzols in particular, have developed not only in siliceous parent materials (granites, gneiss, shales, schists, etc.), but also on some limestone bearing materials (Gardiner and Radford, 1980).

In many areas podzolisation has probably been a dominant soil process for 5000 years or even longer (Lynch, 1981; Geddis, 1986). Anthropogenic influences in marginal upland areas, especially through clearance for agriculture, which accelerated during the Bronze Age, has accentuated podzolisation (Little *et al.*, 1996). Podzolisation increased as the result of increased soil exposure and the curtailment of nutrient recycling because of the replacement of woodland vegetation with pasture and tillage crops. As the base status was lowered, more acid-tolerant plant taxa, such as *Vaccinium* and *Erica* became dominant, especially where forest cover had been removed (Culleton and Gardiner, 1985). These changes, in turn, accentuated podzolisation, and the process probably increased westward with the increasing rainfall:evapotranspiration ratio. Thus, as in the rest of north-west Europe (Ball, 1975), the main climatically driven trend of soil development in Ireland, in freely drained soils, is for progressive leaching, with consequent acidification and, in susceptible soils, eventual podzolisation.

This paper outlines the principal features of Irish podzols based on a selection of studies, and it emphasizes features that are dominated or influenced by organic matter. Morphological and chemical characteristics of the organic matter in surface and sub-surface horizons are elucidated. The concentration of organic matter/organic carbon in horizon samples of most podzols studied in Ireland is confined to loss-on-ignition (Ball, 1964) or by wet oxidation values, based on the Walkley and Black method (Piper, 1947) and its modifications.

2 Organic Matter-Related Properties of Podzols

Review data indicate that podzols are characterised by having a raw humus surface organic layer containing up to 95% organic matter and 55% organic carbon (Little, 1994). This surface horizon may or may not possess two distinct layers, and where it does the upper layer is poorly decomposed and fibrous, in contrast with the lower layer which is humified

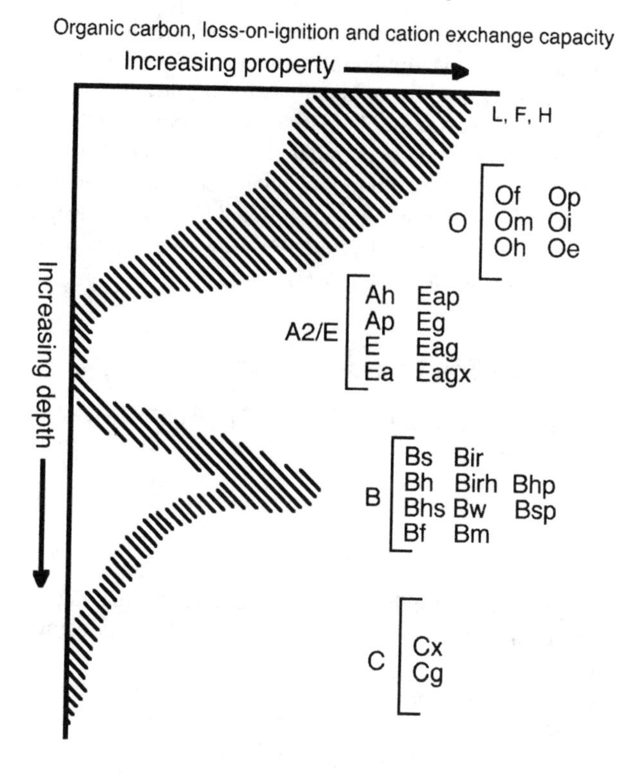

Figure 1 *Schematic representation of a typical Irish podzol with examples of commonly used master subordinate horizon designators*

amorphous, and greasy in texture. In all podzols the maximum cation exchange capacity (CEC), total exchangeable bases (TEB), and percentage base saturation values (PBS) in the soil profile occur in the surface organic layer. This layer typically overlies a leached E_a horizon which has lost its colouring iron and organic compounds, in addition to most of its clay content as a result of mobilization promoted by organic acids emanating from the surface organic horizon. The resultant eluvial horizon is typically whitish grey in colour (i.e. chromas of 2 or less on soil colour charts) with values of organic matter (and hence organic carbon) generally less than 10%. This horizon is underlain by an illuvial B horizon containing iron and aluminium sesquioxides, with or without a high humic component, and with a higher clay content and a larger CEC than the overlying E horizon. In soils, the occurrence of dark organic and iron oxide-rich subsurface layers with brownish, yellowish colours, and with dark brown to black surface organic horizons are unique to podzols. A generalized depiction of the horizon arrangements and properties with increasing depth is shown in Figure 1.

Although a range of mechanisms for podzolisation has been proposed, most authors attribute the bulk of sesquioxide mobilization and translocation to complexing organic substances. Under certain circumstances this would include translocation of other inorganic substances (McKeague *et al.*, 1983; Wang *et al.*, 1986). Where present, humic B horizons contain a large proportion of organo-metallic overlying a continuum of complexes.

These are believed to have been translocated from the horizons resulting in the deposition of a continuum of amorphous and inorganic metallic-humic compounds. The use of *'h'* as a designator in illuvial horizons was initially used in conjunction with *'ir'*, meaning iron, and subsequently with *'s'*, meaning sesquioxides to describe dark coloured orange/red humic-rich B horizons. Humic B*h* horizons may contain as much as 15% organic matter, and generally will contain the maximum amounts of amorphous, poorly crystalline and highly crystalline pedogenic iron oxides in the soil profile. Subjacent B and C horizons, deeper in the profile, have lower quantities of organo-related properties, and with the C horizon typically containing less than 5% organic matter (Little, 1994).

3 Irish Podzols - Setting and General Characteristics

Gardiner and Radford (1980) have indicated that the majority of podzols belong to three broad physiographic divisions (see Table 1):

(i) Mountain and Hill;
(ii) Hill; and
(iii) Rolling Lowland.

By referring to the profile descriptions and to published chemical and physical data, it is possible to subdivide the soils labelled 'podzol' into a number of sub types using the system outlined by Avery (1980).

Podzols developed on siliceous parent materials at altitudes greater than 500 m O.D. (Mountain and Hill), i.e. granite and sandstone, and to a lesser degree quartzite and mica schist, are typically peaty ironpan stagnopodzols, stagnogley-podzols, and humo-ferric podzols. These soils occupy approximately 6.4% of the total area of Ireland (the Republic of Ireland plus Northern Ireland), and are most often associated with lithomorphic and high level raw peat soils. The organic surface horizon is 20-40 cm thick which contains 10 to 30% organic carbon. Typically, the C:N ratio is very high and the pH is very low, i.e. 40 and 4, respectively. This horizon is described as peaty, massive, and moist friable with plentiful roots (Gardiner and Radford, 1980). The vegetation is dominated by *Calluna vulgaris, Molinia caerulea, Nardus stricta* and *Scripus caespitosus.* Organic carbon values decrease markedly beneath the surface organic horizon - values are commonly less than 10%. Beneath the eluvial horizon a cemented, indurated ironpan, which impedes further vertical root penetration, maintains very moist soil conditions in the horizons above the pan. Organic matter in the ironpan is closely associated with sesquioxides, and in particular with highly crystalline iron oxides as extracted by dithionite-citrate-bicarbonate. These soils are often the precursors to blanket peat formation (Moore, 1975; Cunningham, 1996). High level blanket peats are often underlain by ironpan stagnopodzols.

The Hill physiographic division, at altitudes between 150 and 365 m O.D., is dominated by brown podzolic soils (Table 1). These soils, formed mostly from glacial till of predominantly Devonian Sandstone composition, occupy approximately 4% of the total area of Ireland, and are characterized by a dark, organic loam surface horizon overlying a yellowish-red, iron-enriched B horizon of loam texture. This grades into a reddish-brown sandy loam parent material at about 75 cm. Occasionally a thin ironpan occurs at about 35 cm, but where the soil has been reclaimed either by deep ploughing, ripping or rotovating, only relics of the ironpan remain. Widespread reclamation of these soils has taken place in

Table 1 *Soil Associations which include podzols as their principal or associated soils (Gardiner and Radford, 1980)*

Broad Physiographic Division	Soil Association Number	Parent Material	% Podzols
Mountain and Hill	1	Mainly granite - sandstone	75
	2	Mostly mica schist, gneiss, quartzite, and sandstone	10
	4	Mostly sandstone, granite, quartzite or mica schist	5
	6	Mostly sandstone	5
	8	Mainly granite	10
Hill	9	Ordovician-Silurian, Cambrian shale, and mica schist	5
	12	Mostly granite or rhyolite glacial till	5
	16	Morainic sands and gravels, and blown sands	2
Rolling Lowland	18	Sandstone, granite, mica schist glacial till	70
	32	Mostly limestone glacial till	10

recent years (Gardiner and Radford, 1980).

Podzols on Rolling Lowland landform are derived mainly from glacial drift of acid igneous, sandstone, or of mica schist composition (Table 1). Some have a carboniferous limestone/sandstone influence in the parent material, and these podzols are found mainly at elevations below 150 m O.D. and occupy approximately 0.6% of the total land area (Gardiner and Radford, 1980). Such podzols have a surface texture which ranges from sandy loam to organic and peaty sandy loam. Organic matter contents tend to be relatively high. A leached E_a horizon overlies an iron enriched B horizon that is frequently streaked with humus, and is enriched in clay relative to the overlying E_a horizon. An ironpan may also be present. Podzol subtypes in this division include ferric, humo-ferric, iron pan stagnopodzol, ferric stagnopodzol, and gley podzols. Though not as acid, nor as poorly structured (especially in the surface layers) as podzols of Mountain and Hill, these podzols are, nevertheless, nutrient poor, and have an appreciable sesquioxide enrichment in their B horizons.

A detailed study of podzolic soils in the north of Ireland has indicated that ironpan stagnopodzols and ferric podzols are the predominant profile types (Geddis, 1986), and occur most frequently on parent materials dominated by schist and granite. Ferric podzols, which occur at low altitude, become ironpan stagnopodzols at higher altitudes. Ironpan stagnopodzols occur at altitudes of between sea level on the west coast and 533 m O.D. in the Mourne mountains. Cultivated brown podzolic soils are common - many of these soils are podzols which have been substantially modified by cultivation.

Local Studies of High Altitude Podzols

Podzolic features dominated the soils encountered in an altitudinal soil sequence between *ca* 240 and 775 m on the slopes of the Galtee mountains in south central Ireland (Lee *et al.*,1964). The study constituted part of a general investigation to examine the relationship between soils and environmental features that obtain at high altitude. Parent materials consisted mainly of Devonian sandstone conglomerates and Silurian material with strong colluvial and solifluction influences. Brown podzolic soils predominated at altitudes between 240 m and 300 m where the soils may well have been ameliorated/cultivated. These were characterized by the presence of a mull-like surface humus layer and by a strong brown B horizon containing 9.4% and 3.5% organic carbon, respectively.

Buried ironpan stagnopodzols under colluvial material predominated at altitudes between 300 m and 450 m. A mature climax soil was found below *ca* 30 cm. A surface horizon of acidic mor humus, and a distinct eluvial and well-defined ironpan were evidence for a considerable period of stability in this accretion zone before superposition of the colluvial material.

The dominant soil type between 450 m and 600 m was an ironpan stagnopodzol. A notable feature of these mature climax soils was that with increasing elevation there was also an increase in the thickness of the surface organic mor humus, from 18 cm at 450 m to 38 cm at 570 m. The depth at which ironpans occurred also increased from 28 cm to 53 cm in the same zone (Lee *et al.*, 1964). Lithomorphic soils, i.e. podzolic rankers, and climatic blanket peats dominated the landscape above 570 m.

Conry (1970) studied high altitude reclaimed podzols in southern and south-western counties. The soils were derived primarily from Devonian sandstone. The study also involved unreclaimed podzols for comparative purposes. Because of the large tracts of reclaimed stagnopodzols, it was desirable to determine the extent to which human activity has been a major factor responsible for the often advanced modified state of reclaimed ironpan stagnopodzols. The history of soil use was based mainly on literary evidence. Like similar soils described by Lee *et al.* (1964) and by Gardiner and Radford (1980), Conry found that the unreclaimed profiles studied displayed identical trends in organo-related properties, and there were substantial accumulations of humus and sesquioxides in the B_1 or/and placic (ironpan) horizons. Unreclaimed podzols at and above *ca* 200 m O.D. had ironpans, while those below this altitude did not.

Reclaimed ironpan stagnopodzols which had been cultivated for long periods had features that could allow classification as brown podzolic soils, but upon close examination traces of the diagnostic features of iron pan stagnopodzols were observed. Remnants of the ironpan were found in the A_p horizons, or/and small remnants of an E_a horizon overlay an unbroken ironpan. There was a very sharp decrease in organic matter as the result of cultivation, and this decrease was directly related to the intensity of cultivation. Thus, as cultivation proceeded organic matter content decreased despite the fact that farmyard manure and crop residues were added.

There was a distinct correlation between soil colour and the soil organic matter content. The original podzols had black peaty surface horizons. The organic matter decreased with increased cultivation, and the colour became less dark until the soil, after prolonged cultivation, had apparently reached equilibrium at 2.5% to 3% organic carbon, and had a brown to dark brown (7.5YR 4/2) colour which gave a mull-like appearance to the A_p. Though total nitrogen decreased sharply after cultivation, the mineral and readily hydrolyzable nitrogen contents increased. The level of biological activity depended largely on factors such as pH, aeration, nutrient supply, and organic matter, and it was improved by any cultural practice that tended to ameliorate these major requirements. Earthworms, which were absent in the non-reclaimed soils, were plentiful in soils which received heavy dressings of farmyard manure, and in those profiles which had been most intensively reclaimed. Conry (1970) attributed the presence of mull-like A_p horizons in a number of profiles to high earthworm populations/activity.

The genesis of a selection of closely located podzols in the same locality, and having relatively uniform morphological properties but derived from three contrasting rock types - granite, schist/shale and dolerite - at high elevation in the Dublin/Wicklow mountains was elucidated by O'Dubháin and Collins (1981). They described these podzols as having a fibrous organic surface horizon at least 15 cm thick with a placic/iron pan horizon on top of, or within, a spodic B horizon. It was found that, in virtually all pedons, organic carbon values declined rapidly with increasing depth. This reflected the sharp morphological distinction between the organic and mineral parts of the sola. The distribution patterns for both the organically-bound (pyrophosphate-soluble) and the total pedogenic iron content [dithionite-citrate-bicarbonate (DCB) soluble] exhibited extremely low values in the E_a horizons, with contrasting high values in the B_2 and B_3 horizons. Maximum DCB concentrations occurred in iron pans. The proportion of organically bound material was generally low when a ratio of pyrophosphate-soluble to DCB-soluble was calculated. Of the seven pedons examined, only two had ironpans which had a pyrophosphate/DCB ratio which exceeded 0.62. The majority of ratios were between 0.03 and 0.53. The continuity and cementation of the ironpans, as well as the aquic nature of the E_a horizons in the majority of cases suggests that these soils were at advanced stages of podzolisation. O'Dubháin and Collins (1981) concluded that the process of acidification/podzolisation had advanced to the extent that most of the exchangeable bases of all the soils studied had been replaced by H^+ and Al^{3+}. It appears that exchangeable cations were removed from the soil environment as soon as they were released by weathering.

Local Studies of Low Altitude Podzols

Studies of plaggen soils near sea-level in the south-western coastal counties indicated that these soils had very low contents of organic carbon compared to their uncultivated counterparts (Conry, 1971). The soils had been ameliorated to increase agricultural production. Included in this group of soils were a number of gley-podzols. In the plaggen layer the organic carbon content ranged from 1.2% to 2.9%, and the C:N ratios were always less than 12, irrespective of soil colour. Peaty podzolised gleys with black (5YR 3/2) A_p horizons (6.2% organic carbon) gave rise to plaggen horizons (2.2% organic carbon) with a very dark grey (10YR 3/1) matrix colour, and containing abundant grey sand grains. There was no evidence that the plaggen layer had transformed the underlying horizons of the original soil profile. However, the formation of a B_{22irhb} horizon in one of the profiles studied could be tenuously attributed to the overlying plaggen layer.

In a study of an altitudinal sequence of soils in the Sutton-Howth area of Co. Dublin, O'Flynn and Collins (1972) concentrated on the inter-relationships between soil type and the underlying geologic materials, topographic position, and vegetative cover. Annual precipitation in this area is approximately 750 mm. Of the eight profiles studied, two were podzols which had developed between *ca* 110 and 130 m O.D. on materials derived from Cambrian shales, grits, and quartzites. These occupied relatively stable topographic positions, and as a result had distinct horizonation compared to humic podzolic soils located nearby. The podzols had developed under a *Calluna/Ulex* vegetation cover, were characteristically low in pH, high in exchangeable acidity, and had increased values for total pedogenic iron, and/or organic carbon in the B horizons. One podzol profile had an incipient ironpan (ironpan stagnopodzol) which restricted root penetration, and the other (humo-ferric podzol), though not possessing an iron pan, had an indurated E_{a2} which also hindered root penetration and inflated the carbon content of the overlying $E_{a1/2}$ horizon. The E_a and $E_{a1/2}$ horizons had organic carbon values of *ca* 3% and 13%, respectively. This profile had a higher organic carbon content throughout (compared to the former), and this was reflected in higher CEC values in every horizon.

The extreme acidity (less than pH 4.2 in most cases, including the podzols), and the very strong base desaturation of all the study soils was of note. The author considers that soil pH values below 4.2 are rare, other than in acid sulphate soils. The bulk of the total exchangeable bases occurred in the surface organic and in the upper B horizons in the podzol profiles reflecting the predominance of exchange sites from an organic source.

The uses of radiocarbon dating and of pollen analyses to elucidate soil genesis and soil processes co-incident with vegetation change and/or anthropogenic influence have been a feature of podzol studies in Ireland and elsewhere. Evidence from archaeological sites such as hearths, burial cists, temporary dwellings, iron furnaces etc., is often used to augment the pedological radiocarbon data (Proudfoot, 1958; Cruickshank, 1980; Cruickshank and Cruickshank, 1981; Wilson and Bateman, 1986; Wilson, 1991; McBride and Wilson, 1991; Little *et al.*, 1996; Cunningham, 1996). Radiocarbon dating of charcoal in buried, relict podzols in sand dunes near Dundrum in Co. Down (Cruickshank, 1980), and at Goodland in north Antrim (Proudfoot, 1958) indicated that podzolisation coincided with the arrival of the first farmers *ca* 5000 BP. From radiocarbon dates obtained in the organic-rich B_1 horizons of coastal sand podzols at Portstewart (Wilson, 1991) and at Magilligan Foreland (Wilson and Bateman, 1991), both in Co. Derry, it was concluded that once podzols had formed in coastal sands in Northern Ireland, further development was restricted to profile thickening rather than to enhanced chemical differentiation. At Portstewart, the podzols were much deeper, but there were similar amounts of translocated organic matter and of iron in the B horizons at both locations. This conclusion is based on the fact that the podzol-forming interval was 1900-2900 years longer at the former than at the latter site.

Local Podzol-Ecosystems Relationships

A study on the development of humo-ferric podzols on a site which supports semi-natural oakwood at this time was undertaken near Ballycastle, County Antrim (Cruickshank and Cruickshank, 1981). The site, on glacial drift of Pre-Cambrian mica-schist, lies between 140 m and 180 m O.D., and receives approximately 1300 mm precipitation annually. Pollen, and soil profile analytical data from several sites on a centrally situated ridge in Breen Wood were examined and used to reconstruct vegetation and soil history. Radiocarbon dates provided a time scale. The humo-ferric podzols studied were mainly free draining, extremely acid (pH 3.8 in mor layers), leached profiles with low clay contents (< 10%). TEC values were very low, as were those for the PBS (always < 10%). There was a characteristic relationship between the organic carbon, loss-on-ignition, and CEC. In these very sandy (*ca* 75% sand) and stony soils, the presence of humus, particularly in the B_{hs} horizons, was the main influence on the CEC. Organic carbon in the surface mor composed *ca* 20% by weight, and it was *ca* 6% in the B_{hs} horizons. There was very little organic carbon in the E_a horizons and these were always light grey in colour.

The profile peaks of total pedogenic iron (DCB-soluble) usually coincided with the humus peak in the B_{hs} at the top of the B horizons. In some profiles an ironpan had developed, and a smaller amount of humus had accumulated subsequently on top of the

Table 2 *Organic matter-related morphological and chemical characteristics of soil profiles at wooded and non-wooded sites*

Horizon	Wooded	Non-wooded (non-cultivated)
Upper organic layer	Thinner	Deeper, more fibrous
common designator	O_f, O_m, O_h, L/F	O_f
Lower organic layer	Poorly developed, shallower	Well developed, deeper
common designator	O_m, H	O_h, O_m, H, A_h
Leached layer	Lighter colour, very low in organic carbon, often firm	Darker colour, more organic carbon, deeper often indurated
common designator	E_a, E, E_g	E_a, E_{agx}
Upper B	Drier, more friable - 1 in 14 had an ironpan less organically bound Fe	Wetter, more cemented - 2 of 6 had ironpans more organically bound Fe
common designator	B_h, B_{hs}, B_{sh}, B_s	B_h, B_f, $B_{h(f)}$, B_s

ironpan (humus-ironpan stagnopodzol), possibly representing a different sequence of pedogenesis. Sodium pyrophosphate extractions were used at Breen Wood to estimate the organically bound mobile iron fraction. This fraction accounted for between a quarter and a half of the total iron in these soils, i.e. very high pyrophosphate/DCB ratios, indicating that translocation was probably very active.

Pollen analysis at Breen Wood indicated that there were two phases of forest clearance (between *ca* 2000 and 1700 BP, and between *ca* 1400 and 600 BP) followed by a resurgence of *Calluna*, and a decline in tree pollen. Radiocarbon dating of mor layers and of B_{hs} horizons indicated when these horizons formed and, combined with pollen data, pinpointed critical events which accentuated podzolisation, and hence influenced the development of these horizons. No evidence of agricultural activity was found in this study from the pollen spectra at the base of the mor or B_{hs} horizons. It was concluded that podzols almost certainly existed in Breen Wood for the last 2000 years, possibly even from the time of the first recorded mixed oak forest. Clearance between *ca* 2000 and 1700 BP led to heath dominated vegetation which can be tenuously linked chronologically with the first phase of humus movement from the mor into the B horizon of the podzols. After the major clearance of forest, *ca* 1440 BP, there was even stronger evidence for the subsequent development of B_{hs} horizons. Hence, the development of the B_{hs} horizons at Breen Wood coincided with these phases of forest clearance. The circumstantial evidence derived from organic matter-related properties has provided valuable information on the formation of these soils.

A study of podzolised soils in semi-natural oakwood ecosystems has revealed variation in morphological and chemical soil properties, especially when oakwood podzols were compared to other podzols formed under similar environmental conditions under acidophillous vegetation located nearby (Little *et al.*, 1990; Little, 1994). Twenty three profiles were studied, and these were located at altitudes between 15 m and 250 m O.D. Generally, organic matter, organic carbon, exchangeable cations, and base saturation values displayed similar patterns to those for the soils examined by Cruickshank and Cruickshank (1980).

Surface organic horizons under oakwood were generally between 6 cm and 11 cm deep, and tended to be dark greyish brown (2.5Y 3/0) to black (10YR 2/2) in colour, and with increasing humification towards their base. At non-wooded sites, distinctive O_1 and O_2 horizons were always present, and these were usually deeper and more fibrous than their woodland counterparts (Table 2). Loss-on-ignition values were similar in paired wooded and non-wooded surface organic horizons, but organic carbon values tended to be much greater in wooded sites. This property probably reflects greater rates of decomposition due to a more palatable litter source and greater biological activity. It was notable that the colour of the eluvial horizons at the non-wooded sites tended to be darker. There appeared to be more organic matter translocated down-profile compared to the wooded sites.

Humo-ferric podzols were the podzols most frequently encountered in this study. These soils often possessed a thin black B_h horizon having between *ca* 7% and 14% organic carbon. All iron DCB-soluble, oxalate-soluble, and tetraborate-soluble iron species were generally at a maximum in the B_l horizon. With rare exceptions the iron contents decreased in the order, DCB-soluble iron > oxalate-soluble iron > tetraborate-soluble iron in all horizons of the soils analysed. Maximum values for all iron species were obtained where maximum loss on ignition and organic carbon contents occurred in mineral horizons, i.e. in B_l horizons. These horizons also contained appreciable amounts of poorly ordered oxides or "amorphous" inorganic and organic iron species (oxalate-soluble), and exclusively organically bound iron species (tetraborate-soluble) relative to DCB-soluble iron. This indicated that appreciable quantities of organically bound iron were being translocated down the profile and deposited in the B_l horizon of all these soils. It is believed that the oxalate-soluble iron species represent most of the iron involved in podzolisation (Schwertmann and Murad, 1990). Non-wooded tended to have more than wooded soils of exclusively mobile organo-metallic compounds (tetraborate-soluble) in the B_l horizons, and this reflected disturbance episodes and increased exposure of the former soils to leaching. Podzolisation was believed to have been the dominant soil processes at both sites.

With the exception of previously clearfelled areas within woodlands, which have since regenerated naturally, a consistent factor distinguishing the wooded podzols from all others was the absence of an ironpan. Pan formation may have been initiated in the period following clearfelling. It was suggested that when deciduous woodland is conserved, the humo-ferric podzols present have a friable B_h horizon as the result of higher biological activity. If, on the other hand, deciduous woodland is exploited and substituted exclusively by acidophillous species, the B_h shows progressive cementation. In addition, changes in hydrological conditions within the soil appeared to alter pedogenic processes or/and the chemistry of translocated materials in the B_l horizon. The proportion of exclusively mobile amorphous iron species (tetraborate-soluble) decreased relative to the total pedogenic iron content (DCB-soluble) as pan formation proceeded.

Pollen analysis and radiocarbon dating of mor humus (non-wooded site - Cunningham, 1996) and small hollow samples (wooded site - Little *et al.*, 1996) at Uragh, Co. Kerry, revealed that woodland on the non-wooded site was cleared *ca* 350 years ago and that the humo-ferric podzols within the wood formed prior to any major anthropogenic influence. Podzols probably formed at Uragh under forest vegetation (Little *et. al.*, 1996), particularly when *Pinus sylvestris* dominated in the region between *ca* 8500 BP and 5700 BP (Mitchell, 1990; O'Sullivan, 1990). A date of 5520 BP was obtained from fragments of pine charcoal found in the B horizons at Uragh Wood (Little *et al.*, 1996). It may be that the podzols present in Irish oakwoods developed under a pine regime and that the humus that formed at that time has changed with the replacement of pine with deciduous litter. Descriptions of typical Scottish pinewood podzols (McVean and Ratcliffe, 1962; Gauld, 1982) indicate that these tend to have a thicker, less decomposed surface organic horizon than do the Irish oakwood podzols (*ca* 30 cm versus 10 - 15 cm).

A number of podzols that were once cultivated but subsequently abandoned were compared with oakwood podzols in the Irish oakwood study (Little, 1994). Detailed land use history of one of these sites provided a timescale for pedogenetic processes which had occurred there as a result of anthropogenic influences (Little and Collins, 1995). Both sites had free draining soils with westerly aspects, at an elevation of *ca* 30 m O.D., and on slopes between 10% and 13%. The previously cultivated site was dominated by *Pteridium aquilinum*. Higher values of organic carbon were found in the B*h(p)* horizon of the previously cultivated site, and these may have resulted from the inversion of the soil at reclamation; i.e. a remnant of the old O horizon, and the subsequent translocation of organo-metallic complexes (tetraborate-soluble) after clearfelling. In addition, the inversion of charcoal, cultivation and abandonment, may also have contributed to the increase in the loss-on-ignition values. Since abandonment, *ca* fifty years ago, the soil had become base unsaturated although the correlation between CEC and organic matter was not as strong in this instance compared with that for the wooded profile. The higher values of oxalate-soluble and tetraborate-soluble iron compared to the overlying $E_{a(p)}$ horizon of the previously cultivated soil provided evidence for the mobilization and translocation since abandonment of organo-metallic complexes from the $E_{a(p)}$ to the $B_{h(p)}$ horizon. Podzolisation seemed to have been the dominant pedogenetic process in this soil.

Cunningham (1996) studied the effects of vegetation and deforestation on pedogenic processes at Uragh, Co. Kerry, which is close to the south-western seaboard. The study area ranged in altitude between 40 m and 150 m O.D., and most of the soils studied exhibited podzolic features. Soils in similar topographical positions inside and outside the wood were compared. The area outside the wood supported oakwood up to *ca* 350 BP, and hence many of the present soil features were as a consequence of woodland clearance and the colonization of the site with acidophillous vegetation communities.

Deeper, more fibrous organic layers, darker coloured eluvial horizons, gleization, and the presence of ironpans were distinguishing features of non-wooded soils. Wetter site conditions had led to a greater accumulation of organic matter. The average thickness of combined O horizons in the wooded and non-wooded site was 7 cm and 16 cm, respectively. Increased translocation of organic matter, and the mixing of surface layers due to disturbance were probably the principal factors responsible for the darker coloured E horizons. Gleization had resulted in lower amounts of oxalate-soluble extractable iron in illiuvial horizons. (These compose most of the species involved in podzolisation.) However, where ironpans had formed, these iron species were very abundant relative to the total pedogenic iron content (DCB-soluble). It was concluded that the development of surface O horizons and of ironpan layers are often two complimentary processes. The build up of a moisture retentive surface organic layer can cause reducing conditions leading to the formation of the pan. This in turn will act as a water barrier and contribute to conditions favourable for further peat development.

References

Avery, B.W. 1980. Soil classification for England and Wales [higher categories]. *Soil Survey Technical Monograph No. 14*. Bartholomew Press.

Ball, D.F. 1964. Loss-on-ignition as an estimate of organic matter and organic carbon in non-calcareous soils. *J. Soil Sci.* **15**:84-92.

Ball, D.F. 1975. Processes of soil degradation: a pedological point of view. p. 20-27. *In* J. G. Evans, S. Limbrey and H. Cleere (eds), *The Effect of Man on the Landscape: the Highland Zone.Council for British Archaeology Research Report*, No. 11.

Collins, J.F. and T. Cummins. 1996. *Agroclimatic Atlas of Ireland.* AGMET, Joint Working Group on Applied Agricultural Meteorology, Dublin.

Conry, M.J. 1970. *Mans role in soil profile modification and formation in Ireland.* Ph.D Thesis (unpublished), University College Dublin.

Conry, M.J. 1971. Irish Plaggen soils - their distribution, origin, and properties. *J. Soil Sci.* **22**:401-416.

Cruickshank, J.G. 1980. Buried, relict soils at Murlough sand dunes, Dundrum, Co. Down. *Irish Naturalists Journal* **20**:20-31.

Cruickshank, J.G. and M.M. Cruickshank. 1981. The development of humus-iron podsol profiles, linked by radiocarbon dating and pollen analysis to vegetation history. *Oikos* **36**:238-253.

Culleton, E.B. and M.J. Gardiner. 1985. *Soil Formation.* p. 133-154. *In* K.J. Edwards and W.P. Warren (eds), *The Quaternary History of Ireland.*Academic Press, London.

Cunningham, D.A. 1996. *The influence of vegetation and landuse change on soil development at Uragh.* Co. Kerry. MSc. Thesis (unpublished). University College Dublin.

Gardiner, M.J. and T. Radford. 1980. Soil associations of Ireland and their land use potential. *An Foras Talúntais Soil Survey Bulletin No. 36.* An Foras Talúntais, Dublin.

Gauld, J.H. 1982. Native pinewood soils in the northern section of Abernethy forest. Aberdeen. *Scottish Geographical Magazine* **98**:48-56.

Geddis, P.W. 1986. Some podzolic soils in the North of Ireland. *Proceedings of the Royal Irish Academy* **86B:**121-140.

Lee, J., T.F. Finch, and P. Ryan. 1964. An altitudinal sequence on the slopes of the Galtee mountains. *Irish Journal of Agricultural Research* **3**:175-187.

Little, D.J., E.P. Farrell, J.F. Collins, K. Kreutzer, and R. Schierl. 1990. Podzols and associated soils in semi-natural oak woodlands - a preliminary report. *Irish Forestry* **47**:79-89.

Little, D.J. 1994. *Occurrence and characteristics of podzols under oak woodland in Ireland.* Ph.D. Thesis (unpublished). University College Dublin.

Little, D.J. and J.F. Collins. 1995. Anthropogenic influences on soil development at a site near Pontoon, Co. Mayo. *Irish Journal of Agricultural and Food Research* **34**:151-163.

Little, D.J., F.J.G. Mitchell, S.E. von Engelbrechten, and E.P. Farrell. 1996. Assessment of the impact of past disturbance and prehistoric *Pinus sylvestris* on vegetation dynamics and soil development in Uragh Wood, SW Ireland. *The Holocene* **6:**90-99.

Lynch, A. 1981. Man and Environment in S.W. Ireland. *British Arch. Reports, British Series,* 85.

McAllister, J.S.V. and S. McConaghy. 1968. *Soils of Northern Ireland and their influence upon agriculture.* Ministry of Agriculture. p. 101-105.

McBride, N. and P. Wilson. 1991. Characteristics and development of soils at Magilligan Foreland, Northern Ireland, with emphasis on dune and beach sand soils. *Catena* **18**:367-378.

McKeague, J.A., F. DeConick, and D.P. Franzmeier. 1983. Spodosols. p. 217-252. *In* L.P. Wilding, N.E. Smeck and G.F. Hall (eds.), *Pedogenesis and Soil Taxonomy II. The Soil Orders.* Elsevier, New York.

McVean, D.N. and D. Ratcliffe. 1962. *Plant communities of the Scottish Highlands.* H.M.S.O. London.

Mitchell, F. 1990. *Reading the Irish landscape.* Criterion Press, Dublin.

Moore, P.D. 1975. Origin of blanket mires. *Nature* **256:**267-269.

O'Dubháin, T. and J.F. Collins. 1981. Morphology and genesis of podzols developed in contrasting parent materials in Ireland. *Pedologie* **31**:81-98.

O'Flynn, B. and J.F. Collins. 1972. An altitudinal sequence of soils in the Sutton-Howth area of Co. Dublin. *The Scientific Proceedings of the Royal Dublin Society* **4:**315-330.

O'Sullivan, A. 1990. *Historical and contemporary effects of fire on the native woodland vegetation of Killarney, S.W. Ireland.* Ph.D. Thesis (unpublished), University of Dublin (Trinity College).

Piper, C.S. 1947. Soil and Plant Analysis. Interscience Publishers, Inc., New York.

Proudfoot, V.B. 1958. Problems of soil history. Podzol development at Goodland and Torr townlands, Co. Antrim, Northern Ireland. *J. Soil Sci.* **9**:186-198.

Schwertmann, U. and E. Murad. 1990. Forms and translocation of Iron in podzolised soils. *In* J.M. Kimble and R.D. Yeck (eds), *Charactersiztaion, Classification, and Utilization of Spodosols.* Proceedings of the Fifth International Soil Correlation Meeting (ISCOM IV). USDA Soil Conservation Service, Lincoln, NE.

Wang, C., J.A. McKeague, and H. Kodama. 1986. Pedogenic imogolite and soil environments: case study of Spodosols in Quebec, Canada. *Soil Sci. Soc. Am. J.* **50**:711-718.

Wilson, P. 1991. Buried soils in coastal aeolian sands at Portstewart, Co. Londonderry, Northern Ireland. *Scottish Geographical Magazine* **107**:198-202.

Wilson, P. and R.M. Bateman. 1986. Nature and paleoenvironmental significance of a buried soil sequence from Magilligan Foreland, Northern Ireland. *Boreas* **15**:137-153.

The Influence of Forestry on Blanket Peatland

Kenneth A. Byrne and Edward P. Farrell

FOREST ECOSYSTEM RESEARCH GROUP, DEPARTMENT OF ENVIRONMENTAL RESOURCE MANAGEMENT, UNIVERSITY COLLEGE DUBLIN, BELFIELD, DUBLIN 4, IRELAND

Abstract

Blanket peatlands cover substantial areas of Great Britain and Ireland. The uses to which peatlands are put are governed by economic considerations without full regard for the ecological consequences.

Peat has an important role in the carbon cycle. It absorbs (fixes) and releases CO_2, and it releases methane.

This review indicates how the physical and chemical properties of peats, the ecology, carbon cycling, and the surface water quality are influenced by afforestation. Attention is focused on the ways forests influence depositions of sulphates and nitrates in blanket peats, and how these peats affect the deposition of organic acids, and the depositions of neutral salts from wind blows in the reafforested areas of the west of Ireland.

It is important to give careful consideration to the future role of blanket peats. This should involve the monitoring of various diagnostic features, including surface water quality, peat erosion, and the stability of ecosystems. Overall, the rehabilitation of forested blanket peats may be the most appropriate environmental option.

1 Introduction

Blanket peatlands, consisting predominantly of ombrotrophic peat, are a significant component of the landscape in Ireland and Great Britain. In Ireland these cover ca 10% of the land area, and are formed in areas where the annual precipitation exceeds 1250 mm and the number of rain days exceeds 225 (Hammond, 1981). The most extensive areas in Ireland are found in the western counties of Galway and Mayo and to a lesser extent in counties Cork, Donegal and Kerry. Blanket peats are also found on mountain tops throughout the rest of the country. The most significant areas in Great Britain are in the north and west of Scotland, although blanket peatland also occurs in Wales and the north of England.

Over the last 50 years extensive areas of blanket peatland have been afforested in Ireland and in Great Britain. Farrell (1990) estimated that there are some 210 000 ha of forestry on peat in Ireland, and the majority of this is on blanket peat in the west of the country. In Great Britain there are some 190 000 ha of forestry on peat that is more than 45 cm deep, and the greater part of this is on blanket peat in the uplands (Pyatt, 1993).

There has been concern in recent years about the potential impact of blanket peatland afforestation on peatland hydrology, on its physical and chemical properties, on the flora and fauna, on the ecological diversity, on the surface water quality, and on carbon cycling. The scale of the impact can vary from local, as in the case of physical and chemical properties, to regional as in the case of surface water quality where the peatland plantation forest can represent a threat to salmonid populations, and the macroinvertebrates on which they depend..

Attention has also focused on the need to conserve representative areas of blanket peatland. These conservation values have been widely discussed, e.g. by Atherden (1992), Doyle (1983, 1984, 1990a, b, c, 1991), Ryan and Cross (1984), Watts (1990), and by Foss (1991). This recognition of the value of conserving undeveloped peatlands has helped focus attention on the impact of afforestation on blanket peatlands.

This paper reviews the nature of impacts of forestry on blanket peatland in the light of Irish, British, and international experiences.

2 Forestry Impacts

Ecological Impacts

Many ecosystems demonstrate resilience in the face of land-use change. Take, for example, a mineral soil-based ecosystem such as oak woodland. Oak woodlands are characterised by a large number of plant and animal species distributed over several layers. These are remarkably resilient, and many of the ground species will persist after conversion to coniferous plantation and often reappear as the coniferous forest approaches maturity. On the other hand, ombrotrophic peatlands are simple. Only a limited number of species can survive in the harsh regime of high moisture and low nutrient supply. Peat soils are far simpler in composition and behaviour than mineral soils. Peats lack the range of soil minerals which, by their complex chemical interactions confer diversity on the processes of ionic retention and release.

When a previously undisturbed blanket peatland is converted to plantation forestry, a process of change is initiated which profoundly affects the ecosystem and impacts on neighbouring ecosystems. There is no doubt but that the impacts on the ecosystem are drastic, some are almost irreversible. Lowering of the water table, applications of fertilizers, and shading by the forest crop alter the growing conditions. As a consequence, much of the natural vegetation is eliminated for most of the rotation, and some species are eliminated permanently.

Physical and Chemical Properties of Peats

The peatland ecosystem is strongly influenced by establishment practices. Drainage and cultivation of upland areas is manifested in the increased magnitude and frequency of downstream flooding (Acerman 1985). Robinson (1980) demonstrated that ditching of a small upland peat basin prior to afforestation produced an increase in unit hydrograph peak flow of 40%, and halved the time to peak flow. Four to five years after afforestation peak flows in the basin were decreased compared with that in the years following drainage. This decrease was attributed to the establishment of the young trees and to vegetation colonisation of drains which slowed their hydraulic efficiency. Afforestation may lead to sediment losses as a result of drainage. Newson (1980) studied bed-load yields in mid-Wales and found a considerable danger of erosion in open ditches (and consequently of sedimentation) on wet upland soils. Robinson and Blyth (1982), working in a small Welsh blanket peat catchment, found that the sediment yields following drainage were equivalent to almost 50 years yield at pre-drainage rates. During the period of the study sediment yields did not return to pre-drainage levels but remained at about four times that level.

An increase in suspended solids in surface water following drainage has also been recorded. Kenttamies (1980), working in Finland, found a dramatic rise following drainage, and Burt *et al.* (1983) reported that ploughing of peaty gley soils in the Pennines produced an increase in the suspended solids in surface water. The ploughing gave rise to a significant amount of loose material which was subsequently flushed from the drains. Revegetation decreased yields significantly by protecting the soil from rainbeat, and helped to trap the sediment transported by overland flow.

Site cultivation also increases soil aeration and leads to an increase in soil oxidation and mineralization rates. The products can influence stream water chemistry through surface run-off and leaching. Hornung *et al.* (1988) found the increased oxidation and mineralisation rates of organic nitrogen and sulphur compounds in peat to produce nitrate, ammonia, and sulphate, which may in turn be leached into drainage waters. Robinson (1980) found increased concentrations of the four main basic cations, potassium, calcium, magnesiun, and sodium in surface run-off during ploughing and drainage in an upland peatland area. Significant increases in water turbidity, colour, total iron, manganese, and aluminum were reported in a similar study by Stretton (1984).

Phosphorus is the fertilizer additive most commonly applied in Irish forestry (Farrell 1990). Loss of phosphorus from soils is low except in the cases of those which contain less than 5% clay (Cooke, 1974; Kilmer *et al.*, 1974). Losses of phosphorus from drained peatlands have been reported, especially following applications of easily soluble phosphate (Henkens, 1972; Tamm *et al.*, 1974). In a study of Scottish streams, Harriman (1978) found that the phosphorus loss to streams continued three years after fertilization. Kenttamies (1980) studied losses of phosphorus following applications of fine-grained rock phosphate and muriate of potash in a drained peat basin in Finland. Concentrations of phosphorus in run-off were, at first, ten to fifteen times greater than before fertilization.

Due to its waterlogged and nutrient deficient condition peat, in its unmodified state, is unsuitable for any form of crop production. High water storage capacity and the ability to draw up water by capillary action maintains the peat at, or close to saturation. Undrained peats have low shear strength and trafficability.

Drainage leads to a loss of water and consequently to an increase in density followed by subsidence of the bog surface. Further shrinkage of the peat arises because it must bear more of its own weight. Subsidence is a permanent lowering of the surface which occurs in peat following drainage, and it involves both physical and biochemical agencies leading to a loss of mass and volume (Davis and Lucas, 1959; Terry, 1980; Tate, 1980).

Subsidence takes place rapidly in the five to ten years following drainage. The consequent increase in density, shrinkage due to drying, compression due to capillary action, and compaction due to traffic leads to consolidation of the peat mass. Farrell (1985) attributed subsidence, in part, to de-watering of the peat brought about by increased evapotranspiration by a forest canopy, and in part, to peat decomposition. The amount and rapidity of subsidence depends on drain depth and intensity as well as on peat depth and bulk density (Nesterenko, 1976; Yevdokinova *et al.*, 1976). Subsequent to this initial period, the major cause of decomposition and subsidence arises from microbial oxidation of the soil organic matter (Stephens, 1969). Farrell (1985) reported an average subsidence of 1.2 cm per year during the 15 years preceding the thinning of a forestry plantation in a blanket peat in the west of Ireland. Schothorst (1976) reported subsidence of 6-10 cm six years after drainage of an unplanted, low moor peat soil. He attributed 65% of the subsidence to shrinkage and oxidation of organic matter above the water table, and 35% to compression below the water table. Following this period, subsidence is variable and is caused directly or indirectly by biochemical processes. From here the subsidence levels off to a value related to the depth of the water table and to its influence on biochemical oxidation (Schothorst, 1979; Stephens, 1969, and Eggelsmann, 1976).

It is reasonable, therefore, to expect greater subsidence and shrinkage under afforested peat due to the additional compression caused by the effects of drainage and the weight of the the developing crop, coupled with increased oxidation rates and moisture demands.

Many studies have been carried out on the effects of cultivation, tree growth, and stand development on the physical and chemical properties of the peat (Davis and Lucas, 1959; Stephens, 1969; Nesterenko, 1976; Yevdokinova *et al.*, 1976; Eggelsman, 1976; Schothorst, 1976, 1977; Williams *et al.*, 1978; Terry, 1980; Tate, 1980, and Braekke, 1987). Binns (1968) compared unplanted peat in Scotland with peat that had been shallow drained, fertilized with rock phosphate and was under thirty two year old Lodgepole pine (*Pinus contorta* Loud.) and Scots pine (*Pinus sylvestris* L.) stands. There were significant differences in the physical and chemical properties of the planted and unplanted peat. There was a marked decrease in the moisture content of the planted peat due to evapotranspiration from the forest canopy. Subsidence was not observed due to the absorption of shrinkage by underground cracks and cavities. Potassium and phosphorus were lower under the fertilized lodgepole pine, and similar trends were found under the less vigorous Scots pine. There was a marked increase in sodium under the vigorously growing trees, and pH was decreased where the peat had dried. Tree growth had no significant effect on total nitrogen, calcium, or magnesium.

In a survey of tree crops on deep peat in Northern Scotland, Pyatt (1976) showed that drying and development of shrinkage cracks are dependent on the original depth of peat, on its degree of humification, on the tree species, and probably also on climate. Lodgepole pine had a much greater drying effect than Scots pine, Japanese larch [*Larix kaempferi* (Lamb.) Carr.] or Sitka spruce (*Picea sitchensis* (Bong.) Carr). Drying was more intense and cracking was more rapid where the peat was relatively shallow (less than 1.5 m) and well humified than it was where the peat was deep and more fibrous.

Burke (1978) studied drainage and land-use effects on blanket peat in Western Ireland. He found that a seventeen year old Sitka spruce stand in tunnel ploughed and open ditched areas had a dewatering effect on the peat mass even at depths considerably deeper than the root zone. The resulting shrinkage led to an increase in bulk density with a decrease in porosity and hydraulic conductivity.

Cracking has not been observed in afforested blanket peat in Ireland. Shrinkage and cracking occurs on bare and afforested amorphous *Phragmites* peat in the midlands of Ireland. Differences in regional climatic conditions, coupled with the lower hydraulic conductivities of blanket peats may contribute sufficiently to the peat moisture status and determine whether or not crack formation will occur.

Williams *et al.* (1978) compared acidity and exchangeable cations in the top 30 cm of peat beneath a lodgepole pine stand and unplanted peat in Scotland. Contents of nitrogen, phosphorus, and potassium were not significantly altered by the presence of the trees. Calcium and magnesium contents were lower in the peat beneath the trees.

Kelly (1993), working in the west of Ireland, studied the influence of afforestation on the physical and chemical properties of blanket peat. The drying and associated shrinkage of the forest peat resulted in increased bulk density and subsidence of the bog surface. Drainage and evapotranspiration by the forest canopy led to an accelerated rate of organic matter decomposition. This is reflected in the changed fibre content, in the pyrophosphate index, and in the volume weight of the forested peat. The pH of the forested peat was decreased significantly as the result of the combined effects of plantation establishment and tree growth. This decrease could be attributed especially to the increased oxidation of organic matter, and to the release of acidity associated with the uptake of nutrient cations. Calcium and magnesium were significantly lower in the forest floor due to the uptake of nutrients by the forest crop. Significantly, higher levels of phosphorus and iron were found in the forested peat. This may be the result of the application of rock phosphate which, in addition to phosphorus, contains a large number of trace constituents (Farrell, 1971).

Carbon Cycling

In their natural states, peatlands accumulate carbon. Peats are thus large sinks for carbon dioxide (CO_2). This carbon is usually added to the top 10 cm (Clymo 1983), and the rate of addition will depend on factors such as soil temperature, water table level, and nutrient availability. In terms of carbon cycling, peatlands are unusual ecosystems. On the one hand they sequester as peat the major "greenhouse" gas CO_2 from the atmosphere, while on the other hand they can release large amounts of both CO_2 and the second major "greenhouse" gas, methane (CH_4). Methane is produced by the decay processes in the catotelm, the anaerobic layer below the water table level. CO_2 is produced by aerobic decay in the acrotelm, the well aerated surface layer. The emission of CH_4 is controlled mainly by soil temperature and water table level (Crill *et al.*, 1992), and has been extensively studied (e.g. Svensson and Rosswall, 1984; Moore and Knowles, 1987, 1989; Martikainen *et al.*, 1992 and Bubier, 1995). CO_2 emissions have also been extensively studied (e.g. Silvola, 1986; Moore and Knowles, 1987, 1989; Nykänen *et al.*, 1996).

Peatlands form an extensive biome in the boreal and temperate zones and store an estimated 455 Pg (where 1 Pg = 10^{15} g) of carbon (Gorham, 1991). This is approximately equal to one third of the world pool of soil carbon (Post *et al.*, 1982). These peatlands play an important role in regulating the global climate and release *ca.* 24-39 Tg (1 Tg = 10^{12} g) of CH_4 annually to the atmosphere. This is 5-20% of the annual anthropogenic CH_4 emission to the atmosphere (Bartlett and Harriss, 1993; Matthews and Fung, 1987; Aselman and Crutzen, 1989).

There has developed in recent years a better awareness of the importance of peat in the global carbon cycle (e.g. Sjörs, 1981; Armentano and Menges, 1986; Gorham, 1991; Laine and Paivanen, 1992; Vasander and Starr, 1992; Bartlett and Harriss, 1993; Maltby and Immirzi, 1993; Laine *et al.*, 1996). Franzen (1994) hypothesised that peatlands may play a role in the ice-age cycle. The suggested mechanism is as follows. Peatland growth gradually decreases the atmospheric carbon supply. Consequently, the greenhouse effect is decreased and the global temperature is lowered. Temperatures decline beyond a critical point where glaciation is easily triggered (e.g., by volcanic eruptions). Glacial activity brings organic material to the ice margin where it is oxidised and brought back into circulation. The concentration of CO_2 gradually rises to give an increase in the greenhouse effect and in the global temperature. Ultimately this leads to deglaciation and the cycle is complete.

Attention has also been focused on the impact of afforestation on the carbon balance of peatlands (Vompersky and Smagina, 1984; Laine and Laiho, 1992; Sakovets and Germanova, 1992; Anderson *et al.*, 1992, 1994; Cannell *et al.*, 1993).

Some 14 400 x 10^3 ha of peatland have been drained for forestry in Europe, most of it since the Second World War (Paavalainen and Päivänen 1995). In Ireland and in the UK this has involved the establishment of forest crops on peatlands that are naturally treeless. In contrast, forestry development of peatlands throughout the rest of Europe has involved, for the most part, drainage and amelioration of peatlands with a natural tree layer.

When peatlands are drained for forestry the carbon balance is greatly changed. Drainage leads to a decrease in the water table, to increased aeration, and to greatly increased decomposition and oxidation of stored carbon. As a result, the loss of CO_2 will be greatly increased (Silvola and Alm, 1992; Moore and Knowles, 1989; Freeman *et al.*, 1993; Silvola *et al.*, 1985). Part of this loss may be caused by enhanced root respiration (Silvola *et al.*, 1992). The increased activity of methane oxidizing bacteria in the aerated surface peat will affect the amount of CH_4 being released to the atmosphere (Martikainen *et al.*, 1994; Lien *et al.*, 1992; Martikainen *et al.*, 1992; Moore and Knowles, 1989; Freeman *et al.*, 1993; Roulet *et al.*, 1992). Methane emissions have been found to decrease or to cease after drainage, and in some sites net consumption of CH_4 has been observed (Martikainen *et al.*, 1992, 1994). Emissions of N_2O from undrained peatlands are negligible (Nykänen *et al.*, 1996) but water level drawdown at nutrient-rich sites has stimulated N_2O production and fluxes (Freeman *et al.*, 1993; Martikainen *et al.*, 1993; Regina *et al.*, 1996).

These losses of carbon may be offset by increased carbon inputs from:

(i) increased growth of ground vegetation;
(ii) the development of a forest litter layer; and
(iii) the growth of the forest crop.

In studies in Finland, Laine and Laiho (1992) have calculated that the annual post-drainage above-ground tree stand biomass would compensate for about 200 years for the loss of carbon resulting from the oxidation of the peat. The calculations for the losses from peat were by Armentano and Menges (1986). Laine *et al.* (1996) use the changes caused to the carbon balance by forestry to simulate the impact of the drying caused by climate change on the global climate. This drying, they found, would decrease the impact of northern peatlands on the total radiative forcing for about 100 years by 0.1 W m^{-2}.

Vompersky and Smagina (1984) concluded that after 10 years of drainage, peats 350 km west of Moscow were still acting as a sink for carbon, although the rate of carbon accumulation was less than that for the undrained peats. They found the increased litter production from the tree layer compensated for the increased mineralization of peat, despite the fact that below-ground litter production was not included in their study.

Studies in Russian Karelia have shown inconsistent results. Sakovets and Germanova (1992) found that increased decomposition decreased the carbon store of the peat by 0.32 t ha^{-1} y^{-1}. However, if the increased phytomass production is taken into account the total carbon accumulation is 1.23 t ha^{-1} y^{-1}. In contrast to this, Makarevskii (1992) found that for tree stands on drained areas in Karelia, the average additional increment does not compensate for carbon losses from the peat.

The time scale needs to be considered, as does the life time of the forest products (how soon will the carbon in the forest products be returned to the atmosphere) when considering the changes to the carbon balance following drainage for forestry. Cannell *et al.* (1993) have carried out theoretical calculations for drained and afforested peatlands under British conditions. They assumed that a typical Sitka spruce crop growing on peat has a yield class of 12 m^3 ha^{-1} y^{-1}. Such a crop will accumulate 16.7 kg C m^{-2} in trees, wood products, litter and soil, and this is equivalent to about 35.5 cm of deep peat, or to 20.9 cm of shallow peat. Forests planted on peat deeper than this may, in the long term, be net sources of CO_2. They found that where CO_2 loss rates were 50-100 g C m^{-2} y^{-1}, an increased carbon storage within the system was likely for three rotations. Should CO_2 loss rates be 200-300 g C m^{-2} y^{-1}, this increased storage could be limited to the first rotation after which there would be a net loss of carbon.

Increased leaching of organic carbon after drainage also plays a role in the carbon balance, but this has not been considered in many studies. Laine *et al.* (1996) found leaching to be substantial when compared to carbon accumulation. Drainage increased leaching by ca 1 g C m^{-2} y^{-1}.

Surface Water Quality

The development of blanket peatland for forestry in Galway and south-west Mayo has brought forestry into contact with the upper reaches of many important sea trout and salmon fisheries. A number of these catchments are acid sensitive; i.e. the soils and the underlying rocks (e.g. granite and quarzite) are base poor and resistant to chemical weathering. Thus these soils have little buffering ability. Acidification of surface waters (i.e. an increase in the H^+ concentration or a decrease in pH), irrespective of land-use, is observed in acid sensitive areas in North-Western Europe (Battarbee and Renberg, 1990; Ormerod and Gee, 1990) and North America (Kaufmann *et al.*, 1992). This has generally been attributed to the increased consumption of fossil fuels and the associated emissions of oxides of sulphur and nitrogen leading to a consequent increase in acidic deposition.

Concern has been expressed that forestry may be causing an increase in the acidity of streamwater in upland areas. Research in Britain has shown that streams draining catchments with closed forest cover are significantly more acid than moorland streams (Harriman and Morrison, 1982; Stoner and Gee, 1985; Ormerod *et al.*, 1989).

Interception of airborne nutrients and pollutants by coniferous forest canopies can have a significant impact on forest soil and surface water chemistry. There are three types of atmospheric deposition:

(i) wet deposition (rain and snow),
(ii) dry deposition (gases, particles and aerosols); and
(iii) occult deposition (fog, cloud and mist).

Forests can 'scavenge' both occult and dry deposition very efficiently. Throughfall will contain all three plus substances leached from the forest canopy.

Nutrient concentrations in throughfall may be increased by the evaporation of water intercepted by the canopy, by the washing off of soluble compounds captured by the forest canopy as dry deposition, and by the leaching of solutes from above ground parts of plants (Velthorst and Van Breemen, 1989).

The pH of stream water is controlled by complex interactions between precipitation and the vegetation, soils, geological substrate, and the drainage characteristics of the catchment. Three main sources can be considered to be responsible for the enhancement of acidity by forests. These are:

(i) sulphate and nitrate in acidic deposition;
(ii) neutral sea salts; and
(iii) organic acids.

Sulphate and Nitrate in Deposition

The natural uptake of base cations by tree growth generates acidity. This natural process can be accelerated by increased rainfall, and by increased or concentrated acid inputs in polluted rain. Polluted rain has been responsible for accelerated soil acidification in continental Europe, and it has been implicated in a variety of growth disorders in forests (Farrell *et al.*, 1993). Polluted rain has also been responsible for the acidification of streams draining forested catchments. The throughfall is acidified in polluted environments, and this may result in an increase in acidity in the soil, and to an increase in the solubility of potentially toxic aluminum in the soil water (Ormerod *et al.*, 1989, 1991).

Neutral Sea Salts

Many authors have discussed the potential of sea salts to induce short term acidification in surface waters (Wright *et al.*, 1988; Hultberg *et al.*, 1990; Farrell *et al.*, 1993; Lydersen, 1994; Farrell, 1995a). Laboratory studies by Wiklander (1975) have demonstrated the ability of neutral salts to displace protons from the exchange complex of acid soils and thus to lead to an increase in the acidity of the soil solution or to the export of acidity from the soil. Field experiments by Wright *et al.* (1988) and by Hultberg *et al.* (1990) have confirmed this. During storm events, high salt inputs from Na^+ and possibly Mg^{2+} in the rainfall result in the displacement of hydrogen (H^+) and of aluminum (Al^{3+}) from the exchange complex of the soil. There is no aditional acidification of the soil water system. What occurs is as a transfer of acidity from the exchange complex to the soil solution, and possibly to surface waters. Such storm events often occur in winter or in early spring. This is a critical time for the spawning of salmonids, and such sea salt effects can have serious consequences for fish life (Baker and Schofield, 1982).

Farrell (1995a) has described the impact of such storm events on the chemistry of soil water in a forest plantation on deep peat in the west of Ireland. During a major storm in 1991 the deposition of marine ions was enormous compared with depositions for the rest of 1991, or for subsequent years. Farrell (1995a) demonstrated that during storms the high levels of Na^+ displace H^+ from the exchange complex. This would give rise to the generation of acidity. The slow rate of recovery of the exchange complex to its original state was further demonstrated.

Organic Acidity

Organic acids play an important role in surface water acidity. In a survey of lakes and streams of the United States (Kaufmann *et al.*, 1992) organic acids were found to be the dominant source of acidity in one third of the lakes and streams. Wigington *et al.* (1992) compared episodic acidification in lakes and streams of Canada, Europe, and the United States, and found that organic acids contribute to acidity in all streams studied.

DOC concentrations increase with the passage of water through the forest canopy and

Table 1 *Mean concentrations of DOC (mg L^{-1}) in bogwater, stemflow, humus water, and forest peat water at 25 cm, 75 cm, and 100 cm depths, and in drain water for the years 1991-1995 (Farrell et al., 1996). Estimates for precipitation and throughfall are taken from Cronan (1990).*

	Mean concentration of DOC (mg L^{-1})					
Source	1991	1992	1993	1994	1995	Mean
Precipitation (estimate)	–	–	–	–	-	1-3
Throughfall (estimate)	–	–	–	–	-	8-12
Bog water	33	32	28	31	29	31
Stemflow	–	37	38	39	-	38
Humus water	51	54	53	49	50	50
Forest peat water 25 cm	38	84	47	59	58	58
Drain water	54	52	55	61	56	56
Forest peat water 75 cm	34	40	46	54	43	43
Forest peat water 100 cm	–	32	32	–	-	32

forest floor (McDowell and Likens 1988; Cronan 1990), and is probably attributable to concentration by evapotranspiration and leaching. On mineral sites, these high DOC concentrations are decreased as the result of abiotic sorption (McDowell and Likens 1988).

Since peatlands provide a rich medium for organic matter production, their drainage waters are often high in organic carbon. In a survey of Finnish lakes, Kortelainen *et al.* (1989) associated high organic matter contents with catchments having a high proportion of peatlands and of acid organic soils under coniferous forest. Marmorek *et al.* (1989) surveyed lakes of low pH (< 6.0) in the eastern United States and found that organic acidity arising from wetlands, or from land use changes were responsible for the low pH status of one quarter of these lakes.

Working in catchments with predominantly peat soils in the south west of Scotland, Grieve (1990) found that DOC concentrations in the streams draining forested areas were approximately double those in streams draining moorland sectors.

Similar results have been reported in the west of Ireland. Allott *et al.* (1990) found that afforested blanket peatland catchments produce more organic carbon than do such unafforested catchments. In a further study Allott *et al.* (1993) found concentrations of organic acids to be high in streams draining afforested catchments. These acids contributed to the acidity in 11 out of 12 acid episodes studied. The two possible sources of these increased levels of organic acidity are decomposition of peat soils as a result of peat drainage and tree growth, or the decomposition of leaf litter. This observation is supported by the work by Farrell *et al.* (1996). They found a clear increase in the DOC content of water passing through the litter layer in a coniferous stand (Table 1). The DOC contents of the drainage waters of this stand were considerably higher than those in the surface waters of a nearby open bog.

3 Future Prospects

It is clear from what has been discussed that the development of plantation forestry on blanket peatland exerts a considerable influence on both the peatland itself and on adjacent ecosystems.

On the issue of surface water quality, continued monitoring of streams in forested areas is needed to study the processes involved. Such monitoring would also provide a baseline data set which would augment future research needs in this area.

Harvesting operations provide another aspect of blanket peatland forestry which has potential impacts on surface water quality. Many areas of forest on blanket peatland in the west of Ireland are currently being clearfelled. Given the sensitive nature of peat soils, there is considerable potential for peat erosion and for the efflux of ions with consequent negative impacts on streamwater quality and fish habitats. There is a need to quantify these impacts, should these exist, and to develop management guidelines for harvesting operations so as to minimise the impacts described.

It is possible that reforestation following clearfelling will not be carried out in certain cases, based on ecological considerations where peatland forests threaten adjacent ecosysytems, or where it is not considered to be economically viable to replant. One management option would be to allow these sites to revert naturally after removal of the first rotation trees. Many blanket peatlands contain evidence of invasion by prehistoric forests which, due to a change in climate, were eventually succeeded by a further accumulation of ombrogenous peat. However, these forests differed substantially from the forests of today, and were not sustained by artifical fertilizer or drainage.

If these blanket peatlands are to be rehabilitated it will be necessary to sustain in the future the conditions which favour their formation. That would entail blocking drains and ditches in order to achieve near constant saturation with water of low trophic status. Given the probability of climate change, it is possible that this might not happen. Should that be so, naturally regenerated offspring of the forest may well prevail. Allowing the original trees to succumb to a higher water table has been successful for the rehabilitation of raised bogs in Switzerland (Anderson *et al.*, 1995). That approach has the advantage of avoiding the potential site damage that would occur during harvesting, but it may be a realistic option only where the cost of harvesting is greater than the value of the crop. Clearfelling the trees might provide a better alternative. Increased light and exposed conditions would be restored quickly to the bog surface. The nutrients in the trees would also be removed from the site. Removal of the harvest waste would also promote site restoration. An attempt to restore a drained ombrotrophic bog in central Finland has been described by Vasander *et al.* (1992). The process was moderately successful, although restoration proved to be more difficult than was anticipated.

Damming can be very effective in raising the water table in flat areas of blanket peat. However, on sloping terrain the area affected by damming would be a small proportion of the whole.

Given that ombrotrophic peat soils are composed almost totally of organic matter, and that afforestation promotes its oxidation, it is prudent to question the long-term sustainability of forestry on blanket peatland. This question has been raised by Farrell (1995b) who points out that the development of peatlands for forestry begins a process of peat subsidence and oxidation which immediately threatens the sustainability of the peat soil.

References

Acerman, M.C. 1985. The effects of afforestation on the flood hydrology of the upper Etterick valley. *Scottish For.* **39**:89-99.

Allott, N, M. Brennan, J. Reynolds, D.Cooke, J. Gillmor, P. Mills, E.P. Farrell, Boyle, G.M., Cummins, T., M. Kelly-Quinn, J.J. Bracken, D. Tierney, and S. Coyle. 1993. Evaluation of the Effects of Forestry on Surface-Water Chemistry and Fishery Potential in Ireland. *Final Report, Volume 2, Stream Chemistry and Biota, Galway - Mayo Region.* 142pp.

Allott, N.A., W.R.P. Mills, J.R.W. Dick, A.M. Eachrett, M.T. Brennan, S. Clandillon, W.E.A. Phillips, M. Critchley, and T.E. Mullins. 1990. *Acidification of Surface Waters in Connemara and South Mayo.* DuQuesne Ltd., Dublin, 61 pp.

Anderson, A.R., H.A. Anderson, and D.G. Pyatt. 1994. Carbon loss from afforested peat. Poster paper presented to *"Greenhouse Gas Balance in Forestry"*, Forest Research Coordination Council Conference, The Royal Geographic Society, London, 9-10 November 1994.

Anderson, A.R., D.G. Pyatt, J.M. Sayers, S.R. Blackhall, H.D. Robinson. 1992. Volume and mass budgets of blanket peat in the north of Scotland. *Suo* **43**:195-198.

Anderson, A.R., D.G. Pyatt, and I.M.S. White. 1995. Impacts of conifer plantations on blanket bogs and prospects of restoration. p. 533-548. *In* B.D. Wheeler, S.C. Shaw, W.J. Fojt, and R.A. Robertson (eds), *Restoration of Temperate Wetlands.* Wiley, Chichester.

Armentano, T.V. and E.S. Menges 1986. Patterns of change in the carbon balance of organic soil-wetlands of the temperate zone. *J. Ecol.* **74**:755-774.

Aselman, I. and P.J. Crutzen. 1989. Global distribution of natural frashwater wetlands and rice paddies, their net primary productivity, seasonality, and possible methane emissions. *J. Atmos. Chem.* **8**:307-358.

Atherden, M. 1992. *Upland Britain - A Natural History.* Manchester University Press.

Baker, J.D. and C.L. Schofield. 1982. Aluminum toxicity to fish in acidic waters. *Water, Air and Soil Pollution.* **18**:289-309.

Bartlett, K.B. and R.C. Harriss. 1993. Review and assessment of methane emissions from wetlands. *Chemosphere* **26**:261-320.

Battarbee, R.W. and I. Renberg. 1990. The Surface Waters Acidification Programme. Paleolimnology Programme. *Philosophical Transactions of the Royal Society* (London) **B, 327**:227-232.

Binns, W.O. 1968. Some effects of tree growth on peat. Proc. 3rd Int. Peat Congr. (Quebec). pp 358-365.

Braekke, F.H. 1987. Nutrient relationships in forest stands: effects of drainage and fertilisation on surface peat layers. *For. Ecol. Man.* **21**:269-284.

Bubier, J.L. 1995. The relationship of vegetation to methane emission and hydrochemical gradients in northern peatlands. *J. Ecol.* **83**:403-420.

Burke, W. 1978. Long-term effects of drainage and land use on some physical properties of blanket peat. *Irish J. Agric. Res.* **17**:315-322.

Burt, T.P., M.A. Donohoe, and A.R. Vann. 1983. The effect of forestry drainage operations on upland sediment yields: the results of a storm based study. *Earth Surface Processes and Landforms* **8**:339-346.

Cannell, M.G.R., R.C. Dewar, and D.G. Pyatt. 1993. Conifer plantations on drained peatlands in Britain: a net gain or loss of carbon. *Forestry* **66**:353-369.

Clymo, R.S. 1983. Peat. p. 159-224. *In* A.J.P. Gore (ed.), *Mires: Swamp, Fen, Bog and Moor. Ecosystems of the World,* vol **4A**. Elsevier, Amsterdam.

Cooke, G.W. 1974. A review of the effects of agriculture on the chemical composition and quality of surface and underground waters. Agriculture and Water Quality. M.A.F.F. Tech. Bull.

Crill, P., K. Bartlett, and N. Roulet. 1992. Methane flux from boreal peatlands. *Suo* **43**:173-182.

Cronan, C.S. 1990. Patterns of organic acid transport from forested watersheds to aquatic ecosystems. p. 245-260. *In* E.M. Perdue and E.T. Gjessing (eds.), *Organic Acids in Aquatic Ecosystems*. Report of the Dahlem Workshop, Berlin, May 7-12, 1989. Wiley, Chichester.

Davis, J.F. and R.E. Lucas. 1959. Organic soils, their formation, distribution utilisation and management. *Mich. Agr. Expt. Sta., Special Bull. 425.*

Doyle, G.J. 1983. Conserving bogland. p. 191-202. *In* J. Blackwell and F.J. Convery (eds), *Promise and Performance - Irish Environmental Policies Analysed.* University College Dublin,

Doyle, G.J. 1984. Pollardstown fen. p. 37-48. *In* D.W. Jeffrey (ed.), *Nature Conservation in Ireland: Progress and Problems.* Royal Irish Academy, Dublin.

Doyle, G.J. 1990a. Bog conservation in Ireland. p. 45-58. *In* M.G.C. Schouten and M.J Nooren (eds), *Peatlands, Economy and Conservation*. SPB Academic Publishers, The Hague.

Doyle, G.J. 1990b. *Ecology and Conservation of Irish Peatlands* Royal Irish Academy, Dublin.

Doyle, G.J. 1990c. Future prospects for Irish bogs. p. 211-218. *In* G.J. Doyle (ed.), *Ecology and Conservation of Irish Peatlands.* Royal Irish Academy, Dublin.

Doyle, G.J. 1991. Progress and problems in the conservation of Irish peatlands, 1983 to 1991. p. 511-519. *In* J. Feehan (ed.), *Environment and Development in Ireland.* The Environmental Institute, University College Dublin.

Eggelsmann, R. 1976. Peat consumption under influence of climate, soil condition and utilisation. *Proc. 5th. Int. Peat Congr.* (Poland) **3**:233-247.

Farrell, E.P. 1971. The effects of ground North African phosphate and sulphate of ammonia on the growth and mineral composition of *Picea sitchensis* and *Pinus contorta* grown on blanket peat. Ph.D. Thesis, The National University of Ireland.

Farrell, E.P. 1985. Long term study of Sitka spruce on blanket peat. 2 Water-table depth, peat depth and nutrient mineralisation studies. *Ir. For.* **42**:92-105.

Farrell, E.P. 1990. Peatland forestry in the republic of Ireland. p. 13-18. *In* B. Hånell (ed.), Biomass production and element fluxes in forested peatland ecosystems. *Proceedings of a Seminar Held in Umeå, Sweden,* September 3-7, 1990.

Farrell, E.P. 1995a. Atmospheric deposition in maritime environments and its influence on terrestrial ecosystems. *Water, Air and Soil Pollution* **85**:123-130.

Farrell, E.P. 1995b. Sustainability of the forest resource. p. 132-135. *In* F. Convery and J. Feehan (eds), *Assessing Sustainability in Ireland.* The Environmental Institute, University College Dublin.

Farrell, E.P., G.M. Boyle, and T. Cummins, 1996. Intensive Monitoring Network - Ireland, FOREM2 project, Final Report, Forest Ecosystem Research Group Report Number 18. Department of Environmental Resource Management, University College, Dublin.

Farrell, E.P., T. Cummins, G.M. Boyle, G.W. Smillie, and J.F. Collins, 1993. Intensive monitoring of forest ecosystems. *Ir. For.* **50**:53-69.

Foss, P.J. 1991. *Irish Peatlands, The Critical Decade*. Irish Peatland Conservation Council, Dublin, 164pp.

Franzen, l.G. 1994. Are wetlands the key to the ice-age cycle enigma? *Ambio* **23**:300-308.

Freeman, C., M.A. Lock, and B. Reynolds. 1993. Fluxes of CO_2, CH_4 and N_2O from a Welsh peatland following simulation of water table draw-down: Potential feedback from climatic change. *Biogeochemistry* **19**:51-60.

Grieve, I.C. 1990. Seasonal, hydrological and land management factors controlling dissolved organic carbon concentrations in the Loch Fleet catchments, southwest Scotland. *Hyd. Proc.* **4**:231-239.

Gorham, E. 1991. Northern peatlands: role in the carbon cycle and probable responses to climatic warming. *Ecol. Appl.* **1**:182-195.

Hammond, R.F. 1981. *The Peatlands of Ireland.* Soil Survey Bulletin No. 35. An Foras Talúntais, Dublin, 60pp.

Harriman, R. 1978. Nutrient leaching from fertilised watersheds in Scotland. *J. Appl. Ecol.* **15**:993-942.

Harriman, R. and Morrison, B.R.S. 1982. Ecology of streams draining forested and non-forested catchments in an area of Central Scotland subject to acid precipitation. *Hydrobiologia* **88**:251-263.

Henkens, C.H. 1972. Fertilisers and quality of surface water. *Stikstof* **10**:28-40.

Hickie, D. 1990. *Forestry in Ireland, Policy and Practice*. An Taisce, Dublin, 32pp.

Hornung, M., J.K. Adamson, B. Reynolds, and P.A. Stevens. 1988. Impact of forest management practices in plantation forests. Commission of the European Communities Air Pollution Research Report **13**:91-106.

Hultberg, H., L.Ying-Hua, U.Nyström, and S.I. Nilsson 1990. Chemical effects on surface-, ground-, and soil-water of adding acid and neutral sulphate to catchments in southwest Sweden. p. 167-182. *In* B.J. Mason (ed.), *The Surface Waters Acidification Programme*. Cambridge University Press.

Kaufmann, P.R., A.T. Herlihy, and L.A. Baker. 1992. Sources of acidity in lakes and streams of the United States. *Env. Poll.* **77**:115-122.

Kelly, A.M. 1993. The impact of coniferous afforestation on the physical and chemical properties of blanket peat. M. Agr. Sc. thesis. National University of Ireland, 184pp.

Kenttamies, K. 1980. Effects on water quality of forest drainage and fertilisation of peatlands. In: The influence of man on the hydrological regime with special reference to representatives and experimental regimes. IAHS-AISH Pub. no. 130.

Kilmer, V.J., J.W. Gillan, J.F. Lutz, R.T.Joyce, and C.D. Eklund. 1974. Nutrient losses from fertilised, grassed watersheds in western north Carolina. *J. Env. Qual.* **3**:214-219.

Kortelainen, P., J. Mannio, M. Forsius, J. Kämäri, and M. Verta. 1989. Finnish lake survey: the role of organic and anthropogenic acidity. *Water, Air and Soil Pollution* **46**:235-249.

Laine, J. and R. Laiho. 1992. Effect of forest drainage on the carbon balance and nutrient stores of peatland ecosystems. In: M. Kanninen, and P. Anttila, P. (eds), *The Finnnish research programme on climate change. Progress report. Publ. Acad. Finland,* **3:**205-210.

Laine, J., K. Minkkinen, K. Tolonen, J. Turunen, P.J. Martikainen, H. Nykanen, J. Sinisalo, and I. Savolainen. 1996. Greenhouse impact of Finnish peatlands 1900-2100. p. 230-235. *In* R. Laiho, J. Laine, and H. Vasander (eds), *Northern Peatlands in Global Climate Change*. Publ. Acad. of Finland, 1/96.

Laine, J. and J. Päivänen. 1992. Carbon balance of peatlands and global climate change: summary. *In* M. Kanninen and P. Anttila (eds), The Finnnish research programme on climate change. Progress report. *Publ. Acad. Finland* **3**:189-193.

Laine, J., J. Silvola, K. Tolonen, J. Alm, H. Nykänen, H. Vasander, T. Sallantaus, I. Savolainen, J. Sinisalo, and P.J Martikainen,. 1996. Effects of water level drawdown in northern peatlands on the global climatic warming. *Ambio* 25:179-184.

Lien, T., P.J. Martikainen, H. Nykänen, and L. Bakken. 1992. Methane oxidation and methane fluxes in two drained peat soils. *Suo*. **43**:231-236.

Lydersen, E. 1994. Long-term Monitored Catchments in Norway - a Hydrologic and Chemical Evaluation. Norwegian Institute for Water Research, Oslo, report 34-A. 306pp.

Maltby, E. and P. Immirzi. 1993. Carbon dynamics in peatlands and other wetland soils - regional and global perspectives. *Chemosphere* **27**:999-1023.

Makarevskii, M.F. 1992. Stocks and balance of organic carbon in forest and marsh biogeocenoses in Karelia. *Sov. J. Ecol.* **22**:133-139.

Marmorek, D.R., D.P. Bernard,.C.H.R. Wedeles, G. Sutherland, J.A. Malanchuk, and W.E. Fallon. 1989. A protocol for determining lake acidification pathways. *Water, Air and Soil Pollution* **44**:235-257.

Martikainen, P.J., H. Nykänen, P. Crill, and JSilvola. 1992. The effect of changing water table on methane fluxes at two Finnish mire sites. *Suo* **43:**237-240.

Martikainen, P.J., H. Nykänen, P. Crill, and J. Silvola. 1993. Effect of a lowered water table on nitrous oxide fluxes from northern peatlands. *Nature*. **366**:51-53.

Martikainen, P.J., H.Nykänen, K. Lång. J Alm, and J. Silvola. 1994. Emissions of methane and nitrogen oxides from peatland ecosystems. In: M. Kanninen and P. Heikinheimo (eds.), The Finnish Research Programme on Climate Change. Second Progress Report. The Academy of Finland, Helsinki.

Matthews, E. and I. Fung. 1987. Methane emissions from natural wetlands: Global distribution, area, and environmental characteristics of sources. *Global Biogeochemical Cycles* **1**:61-86.

McDowell, W.H. and G.E. Likens. 1988. Origin, composition and flow of dissolved organic carbon in the Hubbard Brook Valley. *Ecol. Mon.* **58**:177-195.

Moore, T.R. and R. Knowles. 1987. Methane and carbon dioxide evolution from subarctic fens. *Can. J. Soil Sci.* **67**:77-81

Moore, T.R. and Knowles, R. 1989. The influence of water table levels on methane and carbon dioxide emissions from peatland soils. *Can. J. Soil Sci.* **69**:33-38.

Nesterenko, I.M. 1976. Subsidence and wearing out of soils as a result of reclamation and agricultural utilisation of marshlands. *Proc. 5th. Int. Peat Congr.* (Poland). **3**:218-232.

Newson, M. 1980. The erosion of drainage ditches and its effect on bed-load yields in mid-Wales: Reconnaissance case studies. *Earth Surface Processes and Landforms* **5**:275-290.

Nykänen, H., K. Regina, . J. Silvola, J. Alm, and P.J. Martikainen. 1996. Fluxes of CH_4, N_20 and CO_2 on virgin and farmed peatlands in Finland. p. 136-139. *In* R. Laiho, J. Laine, and H. Vasander, H. (eds), *Northern Peatlands in Global Climate Change.*Publ. Acad. of Finland, 1/96.

Ormerod, S., A.P. Donald, and S.J. Brown. 1989. The influence of plantation forestry on the pH and aluminum of upland Welsh streams: a re-examination. *Env. Poll.* **62**:47-62.

Ormerod, S.J. and A.S. Gee. 1990. Chemical and ecological evidence on the acidification of Welsh lakes and streams. p. 203-221. In: R.W. Edwards, A.S. Gee, and J.H. Stoner (eds), *Acid Waters in Wales.*. Kluwer Academis Publishers, Dordrecht, Netherlands.

Ormerod, S.J., G.P. Rutt, N.S.Weatherly, and K.R. Wade. 1991. Detecting and managing the influence of forestry on river systems in Wales: results from surveys experiments and models. p. 163-184. *In* M.W. Steer (ed.), *Irish Rivers: Biology and Management.* Royal Irish Academy,

Paavalainen, E. and J. Päivänen. 1995. *Peatland Forestry - Ecology and Principles.* Ecological Studies Vol. 111. Springer.

Post, W.M., W.R. Emanuel, P.J. Zinke, and Stangenberger. 1982. Soil carbon pools and world life zones. *Nature* **298:**156-159.

Pyatt, D.G. 1976. Classification and improvement of upland soils: deep peats. Report on Forest Research 1976. *Forestry Commission:* **25** HMSO, London.

Pyatt, D.G. 1993. Multi-purpose forests on peatland. *Biodiversity and Conservation* **2**:548-555.

Regina, K., H. Nykänen, J. Silvola, and P.J. Martikainen. 1996. Nitrous oxide production in boreal peatlands of different hydrology and nutrient status. p. 158-166. *In* R. Laiho, J. Laine, and H. Vasander (eds), *Northern Peatlands in Global Climate Change.* Publ. Acad. of Finland, 1/96,

Robinson, M. 1980. The effect of pre-afforestation drainage on the streamflow and water quality of a small upland catchment. Report no. 73, Wallingford Institute of Hydrology.

Robinson, M. and K. Blyth. 1982. The effect of forestry drainage operations on upland sediment yields: a case study. *Earth Surface Processes and Landforms* **7**:85-90.

Roulet, N., T.R. Moore, J. Bubier, and P. Lafleur. 1992. Northern fens: methane flux and climate change. *Tellus* **44B**:100-105.

Ryan, J.B. and J.R. Cross. 1984. The conservation of peatlands in Ireland. *Proc. 7th Int. Peat Congr.* **1**:388-406.

Sakovets, V.V. and N. Germanova, 1992. Changes in the carbon balance of forested mires in Karelia due to drainage. *Suo* **43:**249-252.

Schothorst, C.J. 1976. Subsidence of low moor peat soils in the western Netherlands. *Proc. 5th. Int. Peat Congr.* (Poland) **3**:206-217.

Schothorst, C.J. 1979. Subsidence of low moor peat soils in the western Netherlands. *Geoderma* **17**:265-291.

Silvola, J. 1986. Carbon dioxide dynamics in mires reclaimed for forestry in eastern Finland. *Ann. Bot. Fennici* **23**:59-67.

Silvola, J. and J. Alm.1992. Dynamics of greenhouse gases in virgin and managed peatlands. In: M. Kanninen and P. Anttila (eds), *The Finnish Research Programme on Climate Change.* Progress Report. The Academy of Finland, Helsinki.

Silvola, J., Alm, J. and Ahlholm, U. 1992. The effect of plant roots on CO2 release from peat soil. *Suo.* **43**:259-269.

Silvola, J., J. Välijoki, and H. Aaltonen, 1985. Effect of draining and fertilization on soil respiration and three ameliorated peatland sites. *Acta For. Fenn.* **191**:1-32.

Sjörs, H. 1981. The zonation of northern peatlands and their importance for the carbon balance of the atmosphere. *Int. J.. Environ. Sci.* **7**:11-14.

Stephens, J.C. 1969. Peat and muck drainage problems. *J. Irrig. Drainage Div. Proc. Amer. Soc. Civil Eng.* **95**:285-305.

Stoner, J.H. and A.S. Gee. 1985. Effects of forestry on water quality and fish in Welsh rivers and lakes. *J. Inst. Water Engineers and Scientists* **39:**27-45.

Stretton, C. 1984. Water supply and forestry - a conflict of interests: Cray reservoir, a case study. *J. Inst. Water Eng. Sci.* **38**:323-330.

Svensson, B.H. and T. Rosswall. 1984. In situ methane production from acid peat in plant communities with different moisture regimes in a subarctic mire. *Oikos* **43**:341-350.

Tamm, C.O., H. Holmen, B. Popovic, and G. Wiklander. 1974 Leaching of plant nutrients from forest soils as a consequence of forestry operations. *Ambio* **3**:211-221.

Tate, R.L. 1980. Microbial oxidation of the organic matter of histosols. *Adv. Microb. Ecol.* **4**: 169-201.

Terry, R.E. 1980. Nitrogen mineralisation in Florida histosols. *Soil Sci. Soc. Amer. J.* **44:**747-755.

Vasander, H., A. Leivo, and T. Tanninen. 1992. Rehabilitation of a drained peatland area in the Seitseminen National Park in Southern Finland. p. 381-387. *In* O.M. Bragg, P.D. Hulme, H.A.P. Ingram, and R.A. Robertson (eds), *Peatland Ecosystems and Man: An Impact Assessment.* International Peat Society/ Department of Biological Sciences, University of Dundee.

Vasander, H. and M. Starr. 1992. Carbon Cycling in Boreal Peatlands and Climate Change. Proc. Int. Workshop, Hyytiälä Forestry Station, Finland, 28 September-1 October 1992. *Suo 43*:4-5.

Velthorst, E.J. and N. Van Breemen. 1989. Changes in the composition of rainwater upon passage through the canopies of trees and of ground vegetation in a Dutch oak-birch forest. *Plant and Soil* **119**:81-85.

Vompersky, S.E. and M.V. Smagina. 1984. The impact of hydroreclamation of forests on peat accumulation. *Proc. 7th Int. Peat Congr. Irish Nat. Peat Comm.*, Dublin, **4**:86-95.

Watts, W.A. 1990. Conservation of peatland ecosystems. p. 199-208. *In* G.J. Doyle (ed.), Ecology and conservation of Irish Peatlands. *Royal Irish Academy,* Dublin,

Wigington Jr, P.J., Davies, T.D., Tranter, M. and Eshleman, K.N. 1992. Comparison of episodic acidification in Canada, Europe and the United States. *Env. Poll.* **78**:29-35.

Wiklander, L. 1975. The role of neutral sea salts in the ion exchange between acid precipitation and soil. *Geoderma* **14**:93-105.

Williams, B.L., J.M. Cooper, and D.G. Pyatt. 1978. Effects of afforestation with *Pinus contorta* on nutrient content, acidity and exchangeable cations in peat. *Forestry* **51:**29-35.

Wright, R.F., S.A. Norton, .D.F. Brakke, and T. Frogner. 1988. Experimental verification of episodic acidification of freshwater by sea salts. *Nature* **334**:422-424.

Yevdokinova, N.V., M.N. Mostovyy, and Y.Z. Malyy. 1976. Subsidence and biochemical destruction of peat in the Ukranian Polesys. *Sov. Soil Sci.* **8**:345-347.

Is Peat Chemistry Useful in Assessing Acid Deposition Effects on Mineral Soils?

Catherine C. White, Abdulkadir M. Dawod, and Malcolm S. Cresser

UNIVERSITY OF ABERDEEN, DEPARTMENT of PLANT & SOIL SCIENCE, CRUICKSHANK BUILDING, OLD ABERDEEN AB24 3UU, SCOTLAND.

Abstract

Extensive regional surveys have been carried out on Scottish *Calluna* moorland peats and podzols to investigate the relative adverse effects of acid deposition on soil chemistry. The mineral soils were derived from quartzites or Devonian and Torridonian sandstones, and from granites. These Scottish soils are all reputed to be highly sensitive to acidification by acid deposition, and, moreover, over much of Scotland, the Critical Loads of these soils are exceeded. Peats have negligible reserves of base cations from mineral weathering, and tend to behave as simple organic cation exchangers, with soil pH depending primarily upon competition between atmospheric inputs of H^+ and Ca^{2+} for the exchange sites. Therefore, the deposition ratio of $[H+]:\sqrt{[Ca^{2+}]}$ has been tested as a predictor of acid deposition soil damage as reflected by soil pH. The pH values of mineral soils evolved from base-poor parent materials, such as quartzites or sandstones and, to a lesser extent, granites, are shown to relate to this atmospheric deposition parameter in a similar way to the pH values of peats. For these acidification-vulnerable soils, the relationships between soil pH and $[H^+]:\sqrt{[Ca^{2+}]}$ for the upper highly organic surface horizons are very similar to those for peats. As might be expected, the relationship is closest to that for peats for the organic surface horizons of the quartzites and sandstones group. Therefore, measurements of acid deposition effects on peat chemistry could be used as indicators of likely acidification damage to adjacent, sensitive mineral soils.

1 Introduction

The impacts of acid deposition on both terrestrial and aquatic ecosystems are of international concern, especially in the industrially developed areas of Europe and North America. Efforts to decrease these impacts have led to the lowering of emissions of acidifying pollutants, particularly for sulphur. In an attempt to assess the effectiveness of these abatement strategies, the critical loads approach was devised (Bull, 1992). Critical load is defined as "A quantitative estimate of an exposure to one or more pollutants below which significant harmful effects on specified elements of the environment do not occur according to present knowledge" (Nilsson and Grennfelt, 1988). These have been calculated for mineral soils primarily on the basis of the weatherable parent material, with soils derived from granites, quartzites, and sandstones quoted as being in the "most sensitive class" (Nilsson and Grennfelt, 1988). Critical loads for peats, in which biogeochemical weathering is negligible, have been calculated on the basis of the H^+ deposition load that causes a decrease in pH of 0.2 units from the pH value under pristine conditions (Smith *et al.*, 1993). Throughout much of Scotland, these critical loads are being exceeded, and soils are acidifying (Langan and Wilson, 1994).

The susceptibility of a soil to acidification is controlled by its rate of chemical weathering. If the rate of base cation production cannot compete with the rate of base cation depletion, the soil will acidify and become "damaged". In the context of critical loads, it is generally assumed that this damage will take the form of adverse changes in base cation:aluminium ratios, or in soil pH. The latter criterion is the most relevant one for peats. Ombrotrophic peats and soils derived from base-poor parent materials, such as granites, sandstones and quartzites, are particularly vulnerable, to varying degrees, to the adverse effects of acid deposition, unless, for the mineral soils, the parent materials contain significant occlusions of base-rich materials. Declines in soil pH of moorland podzol B horizon soils evolved from acid sensitive parent materials (Barton *et al.*, 1994), moorland peats (Skiba *et al.*, 1989; Smith *et al.*, 1993), and associated drainage waters (Skiba and Cresser, 1989) have been attributed to acid deposition effects.

In sensitive Scottish forest soils, sustained N deposition results in substantial increases in soil N content (Billett *et al.*, 1990). White *et al.* (1996b) reported a similar observation for the surface horizons of *Calluna* moorland soils on sensitive parent materials. Both these studies showed that the high N accumulation rate is accompanied by a far greater increase in C accumulation, so that soil C:N ratio in the L/F/H horizons actually increases with increasing N deposition. This was true up to an N deposition rate of about 9 kg ha^{-1} y^{-1} in the moorland soils studied. However, the C:N ratio then declined sharply with further increases in N deposition.

The evidence described above suggests causal links between acid deposition and mineral soil damage. However, it does not prove these unequivocally, because the pollution gradients exploited in regional surveys invariably are associated with climatic gradients and with parent material variability. A similar difficulty occurs when comparing pairs or small groups of sites. For deep ombrotrophic peats, mineral weathering variability at least is eliminated, although climate gradient effects could still cause systematic trends in peat properties. The lack of mineral weathering in peats generally allows reliable interpretation of long term trends from shorter-term simulation experiments, although it is important to take base cation inputs properly into account when assessing acid deposition effects on peats (Skiba and Cresser, 1989; Smith *et al.*, 1993).

Recently, White and Cresser (1995) have shown strong positive correlations between the pH values of soil pastes in $CaCl_2$ for all the major horizons of granite-derived moorland podzol soil profiles and wet deposited fluxes of the sea salt-derived ions, Na^+, Mg^{2+} and Cl^-. This prompted White *et al.* (1996a) to compare the relative size of base cation deposition fluxes from the atmosphere with those from biogeochemical weathering, as predicted by the PROFILE model. The latter were relatively much smaller than the former, which explained the relationship found and described briefly above.

For peats, it has been shown that the pH depends upon equilibrium between atmospheric inputs of Ca^{2+} and H^+, whereas Mg^{2+} and Na^+ deposition have negligible effects, at least in short-term simulation experiments (Smith *et al.*, 1993). However, there is also a positive correlation between Na^+ deposition and peat pH in regional survey data (Dawod and Cresser, unpublished), which reflects some longer-term influence. This suggests that peats and podzols derived from sensitive parent material may behave in a similar fashion. No correlations were found between climatic factors, such as rainfall amount and peat pH. Therefore, for the current study emphasis was placed upon Ca^{2+} and H^+ deposition only. If, as discussed above, base cation inputs from weathering are virtually negligible for soils derived from granites, these should be completely negligible for the surface horizons of soils derived from quartzites and Devonian and Torridonian sandstones (hereafter referred to as sandstones). It was therefore decided to sample moorland podzol profiles from this parent material group from all over Scotland, and nearby ombrotrophic peat soils. These soils would be used to test the hypothesis that chemical characteristics of both of these soil groups, which might indicate "damage", would behave in the same way in response to a pollution gradient. In other words, the hypothesis tested was that the surface horizons of podzols derived from quartzites and sandstones would function as peats containing a chemically inert diluent. If the hypothesis was sound, this would allow interpolation from peats to mineral soils when attempting to predict some "damage" effects when critical loads are exceeded.

2 Methodology

Sampling Techniques and Sites

The national soil map (1:250 000) of Scotland was used with the Ordinance Survey 20-km square grid system (UKRGAR, 1988) to select suitable soils from granite or granitic tills, quartzite or sandstones, and peats. The dominant geological parent materials for all these grid squares have been shown elsewhere (Barton *et al.*, 1994). For the mineral soils, two or three freely draining peaty podzol sites with a 5 - 45° slope, and at elevations of 200 - 500 m were selected from each grid square with granite or granitic till, or quartzite or sandstone as the dominant soil parent materials. All sites selected (Figure 1) also had *Calluna vulgaris* as the dominant moorland vegetation. Soil samples (each of approximately 1 kg) were taken from each horizon or horizon combination. Composite samples were taken for the L/F/H and A/E horizons. Occasionally, the C horizon could not be sampled because of excessive stoniness. Peats were sampled in triplicate (0-25 cm). The surface vegetation was removed with secateurs, and obvious root fragments and stones were discarded at all sites prior to sampling. The locations of the three soil types chosen for the study are shown in Figure 1.

Soil Preparation and Analyses

Samples were sieved and the < 2-mm fractions of fresh field-moist subsamples were analysed for pH and percent moisture. Soil pH was measured in both 0.01M $CaCl_2$ and water, after 1 hr equilibration of soil:solution pastes at a ratio of 1:5, using a low ionic strength glass calomel electrode. Total soil mineralogy for the soils derived from granites was determined by X-ray diffractrometry.

The atmospheric deposition data that are relevant to the present study have been described in other works involving one or more of the authors (White and Cresser, 1995; Barton *et al.*, 1994).

Figure 1 *The sites chosen for the regional survey*

3 Results and Discussion

Base cation contributions from biogeochemical weathering are effectively absent in peats, and, it is hypothesized, negligible for the soils derived from quartzites or sandstones. Soils derived from granites do receive varying quantities of base cations from weathering (Figure 2), though the amounts are relatively small. Therefore, negative correlations are to be expected between soil pH and the H^+ deposition flux for these soil groups. Figure 3 (v - ix) shows that some such relationships are indeed observed when soil pH values are measured using water pastes. Moreover, for the peat and for all the podzol horizons in the quartzite/sandstone group, the slopes and intercepts are similar (Table 1). For the granite-derived podzols the relationships between wet H^+ and soil pH_{water} were found not to be significant, and are therefore not shown here. However, significant but different relationships, compared to those for the peat and sandstone/quartzite-derived soils, were observed with pH_{CaCl_2} [Figures. 3 (i - iv)]; they are similar up to *ca.* 0.4 kg ha^{-1} y^{-1} deposition, in so far as negative trends are observed over this range for the granite-derived soils as for the peats, and for the quartzites and sandstones group of mineral soils (Figure 3). The data display considerable variation about the best-fit line, which is to be expected if atmospheric base cation inputs, especially inputs of total wet Ca^{2+}, have a compensatory influence *via* competitive cation exchange. This has been clearly shown to

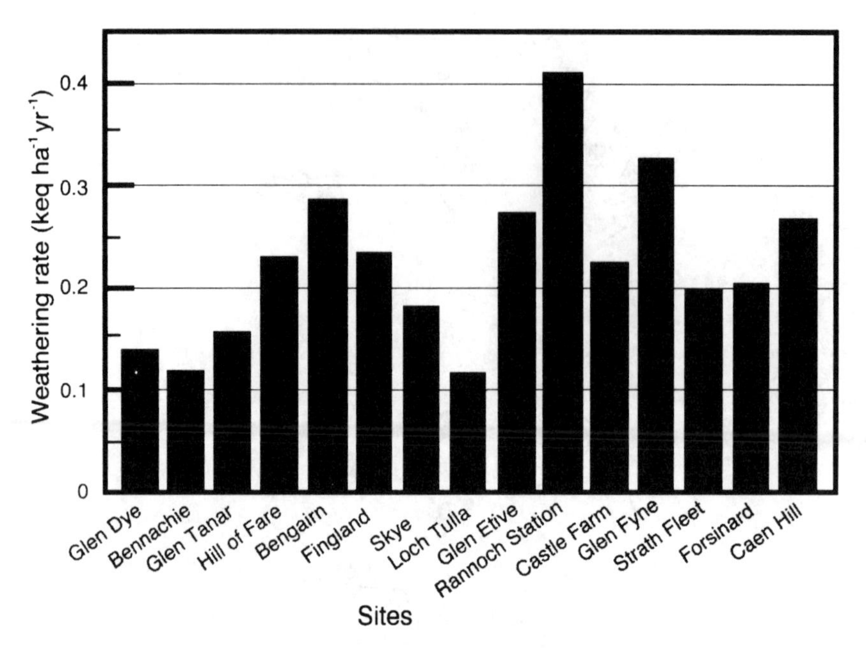

Figure 2 *The weathering rates of the podzols derived from granitic sites, as quantified by the PROFILE model, as described by White et al. (1996a)*

be the case for peat soils in earlier studies (Skiba and Cresser, 1989; Smith *et al.*, 1993). Data are plotted for the relationships between total estimated wet plus dry H^+- deposition and soil pH values. The relationships were less significant than those for the wet flux.

To establish that Ca^{2+} does have the expected compensatory influence upon soil pH, the relationships between soil pH and wet Ca^{2+} deposition fluxes were plotted for peat and the mineral soil horizons (Table 1 and Figure 4). This is to be expected, of course, since the plots for the granite-derived podzol horizons are for $pH_{CaCl_2,}$ not for pH_{water}. For the peats and the podzol L/F/H horizon soils derived from the quartzite/sandstone group, the intercepts were quite similar, considering the variability (Table 1 and Figure 4). On progressing from the L/F/H through the A/E to the C horizon soils, acidification, as reflected in pH at zero Ca^{2+} input, is progressively less, presumably because the soils are exhibiting some buffering at depth. The relationship between soil pH and wet Ca^{2+} deposition was not significant for the B horizon soils, possibly because of the influence of SO_4^{2-} <=> OH^- anion exchange on soil pH. Slopes were variable, as might be expected from the overall variability, which reflects the greater influence of H^+ deposition than Ca^{2+} deposition on soil pH. For the podzols evolved from granites, the trends were not significant for the pH in water values, although significant relationships were found between soil pH in calcium chloride and the calcium deposition flux (Table 1 and Figure 4).

The above results suggest that the pH of *Calluna* moorland podzols evolved from quartzites or sandstones may be regulated by competition between Ca^{2+} and H^+ for cation exchange sites, although the statistical relationships in Figure 4 are not particularly strong. The effect of other base cations, especially Na^+, may be small compared to that of Ca^{2+}, because of the relatively greater affinity of the latter for organic cation exchange sites. Thus a relationship might be expected between exchangeable H^+ saturation (and hence pH) and

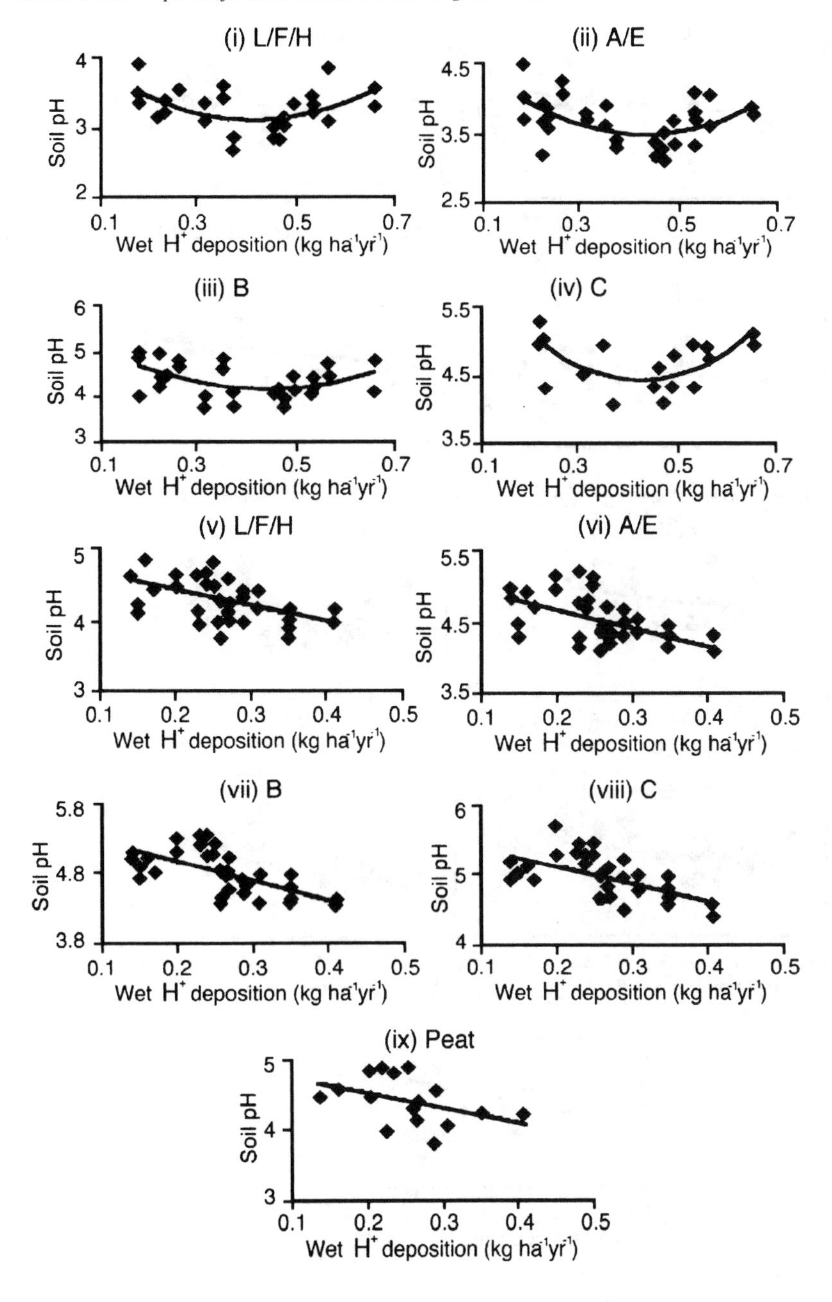

Figure 3 *Relationships between soil pH_{CaCl_2} and wet H^+ deposition flux for the major horizons of podzols derived from granites or granitic tills (i - iv), and soil pH_{water} and wet H^+ deposition flux for the major horizons of podzols derived from quartzites or sandstones (v - viii), and for peat (ix)*

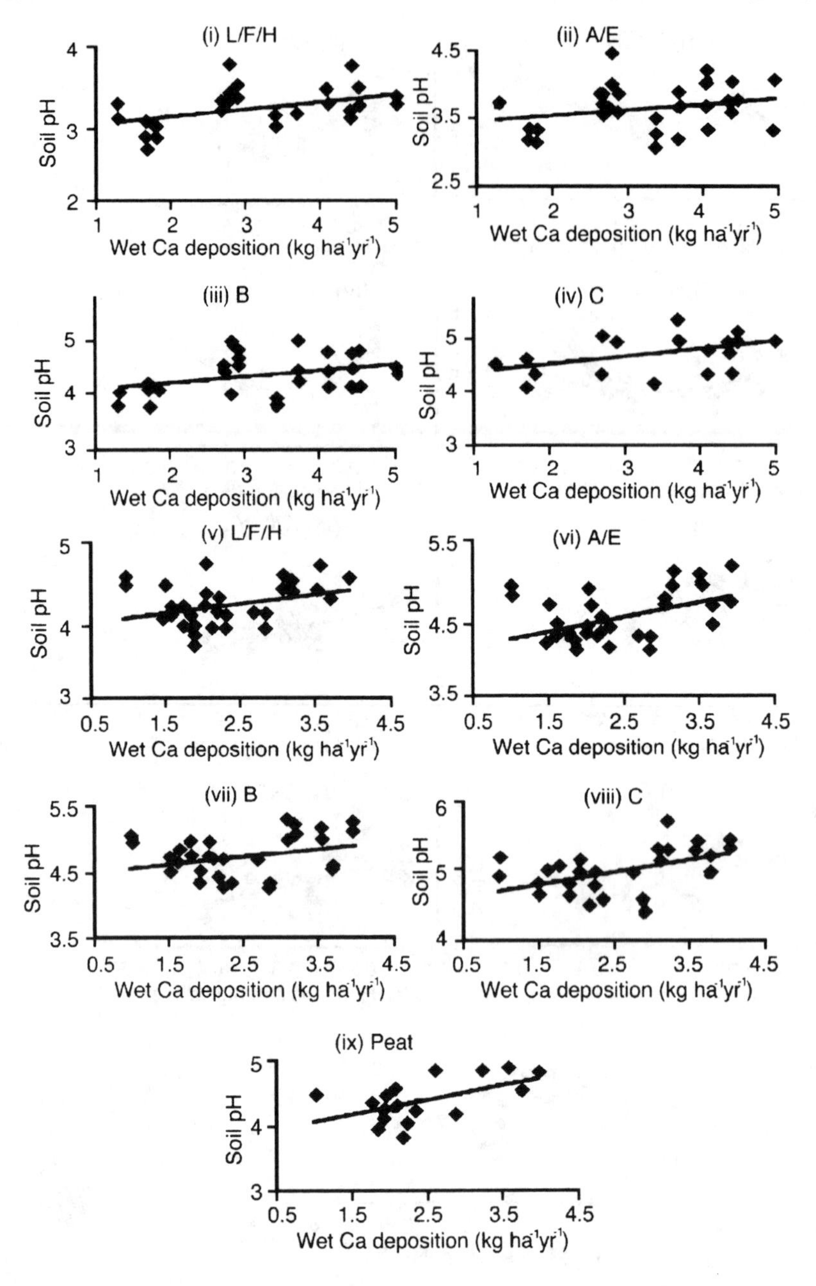

Figure 4 *Relationships between soil pH_{CaCl_2} and wet Ca^{2+} deposition flux for the major horizons of podzols derived from granites or granitic tills (i - iv), and soil pH_{water} and wet Ca^{2+} deposition flux for the major horizons of podzols derived from quartzites or sandstones (v - viii), and for peat (ix)*

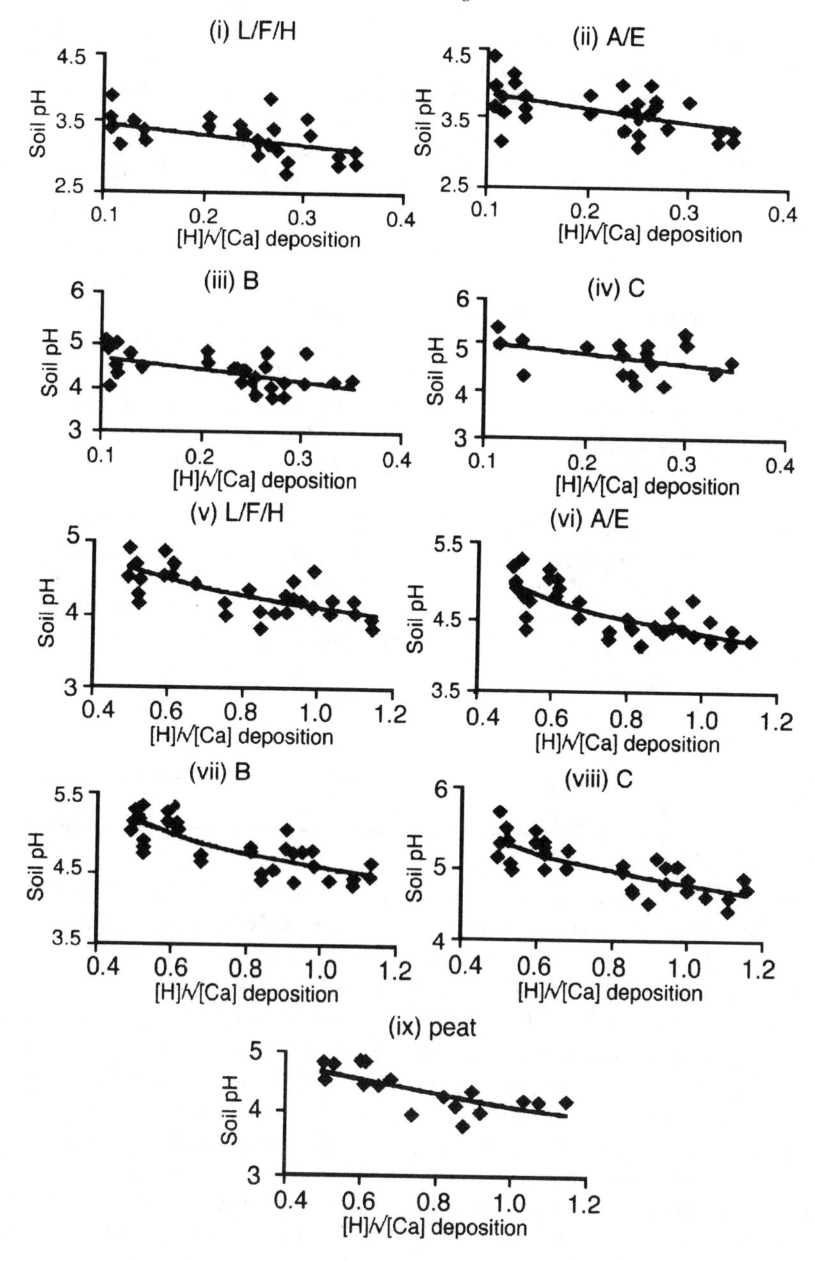

Figure 5 *Relationships between soil pH_{CaCl_2} and $[H^+]:\sqrt{[Ca^{2+}]}$ deposition flux for the major horizons of podzols derived from granites or granitic tills (i - iv), and soil pH_{water} and $[H^+]:\sqrt{[Ca^{2+}]}$ deposition flux for the major horizons of podzols derived from quartzites or sandstones (v - viii), and for peat (ix)*

Table 1 *Relationship between soil pH_{water} (peats and sandstone/quartzite derived soils) and pH_{CaCl2} (granite derived soils) and wet H^+, wet Ca^{2+}, and $[H^+]/\sqrt{[Ca^{2+}]}$ depositions and the correlation coefficient of that line. n = the number of sites*

Soil	n	Equations relating soil pH to		
		Wet H^+ deposition	Wet Ca^{2+} deposition	$[H^+]/\sqrt{[Ca^{2+}]}$ deposition
Granite L/F/H	34	$y = 6.744x^2-5.53x+4.27$ (r = -0.492)	y = 0.104x+2.94 (r = 0.453)	y = -1.52x+3.61 (r = -0.474)
A/E	35	$y = 8.5x^2 - 7.32x+ 5.03$ (r = -0.549)	y = 0.066x+3.36 (r = 0.295)	y = -2.11x+4.11 (r = -0.515)
B	35	$y = 8.52x^2-7.43x+ 5.80$ (r = -0.500)	y = 0.116x+3.99 (r = 0.348)	y = -2.54x+4.93 (r = -0.539)
C	19	$y = 13.81x^2 -11.71x+ 6.92$ (r = -0.66)	y = 0.15x+4.23 (r = 0.427)	y = -2.19x+5.23 (r = -0.427)
Quartzite/ Sandstone L/F/H	36	y = -2.25x+4.92 (r = -0.532)	y = 0.14x+4.00 (r = 0.377)	$y = 4.11x^{-0.178}$ (r = -0.701)
A/E	36	y = -2.56x+5.22 (r = -0.540)	y = 0.186x+4.11 (r = 0.445)	$y = 4.29x^{-0.20}$ (r = -0.755)
B	34	y = -2.85x+5.52 (r = -0.645)	y = 0.129x+4.46 (r = 0.329)	$y = 4.52x^{-0.18}$ (r = -0.766)
C	34	y = -2.456x+5.61 (r = -0.581)	y = 0.173x+455 (r = 0.460)	$y = 4.12x^{-0.17}$ (r = -0.774)
Peat 0-25cm	17	y = -2.19x+4.96 (r = -0.461)	y = 0.23x+3.83 (r = 0.548)	$y = 5.35x^{-0.26}$ (r = -0.706)

the ratio of $[H^+]:\sqrt{[Ca^{2+}]}$. The square brackets indicate concentrations in the soil solution, which depend upon the corresponding precipitation concentrations in the longer-term, according to the ratio law. This ratio, based for convenience on wet deposition fluxes (since evapotranspiration effects cancel out in the present context) was used to test the relationship strength. Figure 5 (v - viii) shows that the relationship between soil pH and $[H^+]:\sqrt{[Ca^{2+}]}$ is retained in the quartzite/sandstone profiles down to the B and C horizons, although the curves shift towards progressively higher soil pH values. Most probably this reflects the increasing importance of weathering, and of aluminium cationic species with depth. The similarity between these plots and that for peat [Figures 5 (ix)] is obvious. For completeness, the results for granite-derived *Calluna* moorland podzol surface horizons, but based upon soil pH_{CaCl_2} are shown in Figures 5 (i - iv). The relationships for these soils are weaker and show different trends from that of peats, probably reflecting the greater contribution of base cations from weathering . Direct comparison of Figures 4 (i - iv) and (v - ix) is not strictly valid because of different procedures used to measure pH.

5 Conclusions

This study confirms the hypothesis that the upper horizons of quartzite/sandstone-derived podzol profiles under *Calluna* moorland in Scotland respond to acid deposition and atmospheric inputs of Ca^{2+} in precipitation in a similar way to peat soils. Thus pH is governed primarily by cation exchange competition between Ca^{2+} and H^{+}. The ratio $[H^{+}]:\sqrt{Ca^{2+}}]$ is a useful indicator of soil pH, and could be used to predict effects on soil pH of increasing H^{+} deposition, or recovery effects as a consequence of pollution abatement.

Acknowledgements

The authors are indebted to the UK Department of the Environment and to NERC for financial support for this research. We also gratefully acknowledge Shimna Gammack for processing and supplying deposition data.

References

Barton, D., D. Hope, M.F. Billett, and M.S. Cresser. 1994. Sulphate adsorption capacity and pH of upland podzolic soils in Scotland: Effects of parent material, texture and precipitation chemistry. *Applied Geochemistry* **9**:128-139.

Billett, M.F., E.A. Fitzpatrick, and M.S. Cresser. 1990. Changes in carbon and nitrogen status of forest soils organic horizons between 1949/50 and 1987. *Environmental Pollution* **66**:67-79.

Bull, K. 1992. An introduction to critical loads. *Environmental Pollution*. **77**:173-176.

CLAG. 1994. *Critical Loads of Acidity in the United Kingdom. Summary Report of the Critical Loads Advisory Group*. Institute of Terrestrial Ecology, Edinburgh.

Langan, S.J. and M.J. Wilson. 1994. Critical loads of acid deposition on Scottish soils. *Water, Air and Soil Pollution*. **75**:177-191.

Nilsson, J. and P. Grennfelt (eds). 1988. p. 8-79. In *Critical Loads for Sulphur and Nitrogen*. Report from a Workshop Held at Skökloster, Sweden. 19-24 March 1988. Nordic Council of Ministers, Copenhagen.

Skiba, U. and M.S. Cresser. 1989. Prediction of long-term effects of rainwater acidity on peat and associated drainage water in upland areas. *Water Research* **23**:1477-1482.

Skiba, U., M.S. Cresser, R.G. Derwent, and D.W. Futty. 1989. Peat acidification in Scotland. *Nature* **337**:68-69.

Smith, C.M.S., M.S. Cresser, and R.D.J. Mitchell. 1993. Sensitivity to acid deposition of dystrophic peat in Great Britain. *Ambio* **22**:22-26.

UKRGAR. 1988. *Acid Deposition in the United Kingdom 1986-1988*. Department of the Environment, Warren Spring Laboratory, Stevenage, Herts., 1990.

White, C.C. and M.S. Cresser. 1995. A critical appraisal of field evidence from a regional survey for acid deposition effects on Scottish moorland podzols. *Chemistry and Ecology* **11**:117-129.

White, C.C., M.S. Cresser, and S.J. Langan. 1996a. The importance of marine derived base cations and sulphur in estimating critical loads in Scotland. *The Science of the Total Environment* **177**:225-236.

White, C.C., A. Dawod, and M.S. Cresser. 1996b. Nitrogen accumulation in surface horizons of moorland podzols: evidence from a Scottish survey. *The Science of the Total Environment* **184**:229-237.

The Direct Relationship between Base Cation Ratios in Precipitation and River Water Draining from Peats

Abdulkadir M. Dawod and Malcolm S. Cresser

UNIVERSITY OF ABERDEEN, DEPARTMENT OF PLANT & SOIL SCIENCE, MESTON BUILDING, OLD ABERDEEN AB9 2UE, SCOTLAND

ABSTRACT

The similarity between the relative proportions, on a mol_c basis, of Ca^{2+}, Mg^{2+},and Na^+ in precipitation and in river water draining from peats is demonstrated by plotting the proportions on triangular diagrams. Compared with these plots, plots of the relative proportions of the exchangeable base cations in peat show substantial differences. This reflects the differences in binding affinities of the three base cations by organic matter cation exchange sites, the divalent cations Ca^{2+} and Mg^{2+} being more strongly bound. It is hypothesized that the base cation proportions in stream water draining from peats should, however, closely reflect the relative amounts of cations in precipitation. These proportions in drainage water depend on the composition of ions on the exchange sites, which is itself directly dependent upon the composition of the precipitation. To test this hypothesis, data for stream water chemistry from three catchments with peats and peaty podzol soils in Glen Dye in northeast Scotland are plotted and compared with precipitation composition. The relative cation proportions during storms, i.e. periods of high flow when water drains predominantly from the surface organic horizon, are very close to those for precipitation. In the Small Burn catchment, which has negligible amounts of mineral soil underlying the peat, the relative cation proportions of the river water do not differ significantly, and are similar to the relative cation proportions in precipitation under all flow regimes.

1 Introduction

Stream water chemistry shows characteristics associated with a mixture of water derived from precipitation and specific soil types or soil horizons, and/or parent materials (Billett and Cresser, 1996). For base cations such as Ca^{2+}, the proportion of the cation exchange capacity (CEC) occupied by that particular base cation governs its solubility in soil drainage water, and hence in the associated river water (Billett and Cresser, 1992). However, the composition of the cations occupying the cation exchange sites of peat is itself regulated by atmospheric deposition, and not by biogeochemical mineral weathering (Skiba *et al.*, 1989; Skiba and Cresser, 1989). A direct link might be expected, therefore, between precipitation and river water compositions where the river water drains predominantly from peat soils.

In a study of catchments in northeast Scotland, which included a diverse range of soil types, Billett and Cresser (1996) found that a major difference between soil types and the most consistent change with depth was seen in the relative proportions of exchangeable Ca^{2+}, Mg^{2+}, and Na^{+}. These relative concentrations in the upper and lower horizons were compared with the relative amounts in stream water, in the parent material, and in precipitation, by plotting on triangular diagrams. Exchangeable cation concentrations are higher generally in the upper soil horizons of UK upland soils because of the high cation exchange capacity of the organic matter. It has been reported by several authors that Ca^{2+} and Mg^{2+} are the dominant exchangeable base cations in the surface soil horizons, whereas Na^{+} becomes increasingly important on the exchange complex with depth down the soil profile (Creasey, 1984; Reynolds *et al.*, 1988; Sanger *et al.*, 1994; Soulsby and Reynolds, 1994; Billett and Cresser, 1996).

In the study by Billett and Cresser (1996), stream water data were divided into values for low, medium, and high discharge. Changes were observed in the relative proportions of Ca^{2+}, Mg^{2+}, and Na^{+} stream water during storms (Billett and Cresser, 1996), and the results suggested that the chemical composition of the precipitation had more effect in modifying stream water chemistry during periods of high flow. River water in UK upland catchments is increasingly dominated by near-surface or surface flow rather than by deeper sub-surface flow during periods of high water discharge (Cresser and Edwards, 1987; Dewalle and Swistock, 1994). Soil water reaching streams at peak discharge is characterized by low Ca, Mg, Na, and Si and high dissolved organic carbon (DOC), Fe, and Al, and is thought to be strongly influenced by the ion-exchange chemistry of the upper (O/A) soil horizons (Cresser and Edwards, 1988). At lower discharge, water reaching rivers in UK upland catchments is characterized by the chemistry of the less acidic, lower (B/C) horizons, and/or water derived from groundwater sources.

This paper examines the variation between the relative amounts, on a $mol_c\ kg^{-1}$ basis, of Ca^{2+}, Mg^{2+} and Na^{+} for peats from 17 sites around Scotland compared with the corresponding relative concentrations of base cations for precipitation at the same sites. The hypothesis that the base cation chemistry of the streamwater draining peat soils closely echoes precipitation composition is then tested using data for three adjacent streams. The peats and peaty podzol soil distributions, and the topography of the three catchments used have been discussed in a previous paper in relation to river water pH differences between the streams (Rees *et al.*, 1989). It was hoped that stormflow water especially would have relative amounts of base cations very close to the ratio for precipitation at the site.

2 Sites, Materials and Methods

In a separate study of the relationship between peat chemistry and the levels of pollutants in the precipitation, 17 sites were chosen throughout Scotland, from Skye in the Inner Hebrides to Strichen in Grampian region, and from Benhorn in Sutherland to Langhom in Dumfriesshire, where peat was found overlying quartzite or sandstone parent materials. (See also p. 278 - 287, this Volume.) During the summer of 1994, three sub-samples, each of approximately 3 kg in weight, were collected from the upper 15 cm of the peat.

Once in the laboratory the peats were analysed for a range of physical and chemical properties. Of interest for this paper is the determination of exchangeable Ca^{2+}, Mg^{2+}, and Na^{+}. The extraction of exchangeable cations was carried out using 1 mol L^{-1} CH_3COONH_4, adjusted to peat pH. Triplicate 5-g subsamples of fresh, field-moist peat, which had been passed through a 3 mm mesh sieve, were leached with 250 mL portions of 1 mol L^{-1} CH_3COONH_4. The leachates were collected in 250-mL flasks and diluted to the mark with 1 mol L^{-1} CH_3COONH_4. The leachates were stored at 4 °C, and analysed within 3 days for Ca, Mg, and Na concentrations. The amounts of the three cations in the peat were then summed and the relative proportions of Ca, Mg, and Na calculated. These were then plotted on the triangular diagram, together with the same ratios from the precipitation data.

Precipitation composition data for Scottish sites for 1986-88 was obtained from the literature (UKRGAR, 1990), and values for the actual sites used estimated by krigging (Gammack, personal communication).

Three streams, Small Burn, Warm Burn and Brocky Burn, in the Glen Dye area of northeastern Scotland, were selected for stream water chemical analysis. These sites are described in Rees *et al.* (1989), as are the methods used to analyse the stream water. Rees *et al.* (1989) compared the stream water chemical data at baseflow along the length of each river, and during a sequence of three storm events in which the soils became progressively wetter, and near surface flow through organic surface horizons became increasingly dominant. Rees *et al.* commented upon water chemistry, including pH and cation and sulphate concentrations, in the light of the physical and chemical properties of the soil profiles adjacent to the streams.

The three streams at Glen Dye all were underlain by granitic parent material, and all had varying amounts of peat along the stream courses. The source of the Warm Burn was in relatively deep peats (1 - 2 m thick), but these quickly became thinner peats, and thin peaty podzolic soils in the lower parts of the catchment. Soils along the Brocky Burn were more mixed. The burn originated in deep peats, which on lower slopes became mixed with boulder deposits of glacial origin, and then gave way to peaty podzols and thin podzolic soils at the lower end of the catchment. The Small Burn was different, being dominated by deep peats throughout the stream length, with mineral soils found only at the very lowest sampling sites. Where mineral soils occurred at all on higher slopes, the mineral horizons were rarely more than a few cm deep, because of the convex topography of the catchment slope (Rees *et al.*, 1989).

3 Results and Discussion

The relative proportions of Ca^{2+}, Mg^{2+}, and Na^{+} on the peat sample exchange sites and in the precipitation for the 17 sites around Scotland are shown in Figure 1. The relative proportions in the precipitation all plot close to the 100% Na^{+} apex of the triangle, with between 70% and 78% Na^{+}, 4 and 13% Ca^{2+} and 16 and 18% Mg^{2+}. There is little variation between the ratios in the precipitation from the various sites, and especially between Na^{+} and Mg^{2+}. This is not surprising since both are of marine origin (Reid *et al.*, 1981a). Calcium deposition, on the other hand, is derived predominantly from terrestrial sources, and therefore its relative contribution is much more variable.

The peat samples show more variation, and the ratios are different from those in the precipitation. In the peat, Na^{+} varied between 10 and 41%, Ca^{2+} between 19 and 50%, and Mg^{2+} between 24 and 57%. Undoubtedly the variability primarily reflects that in

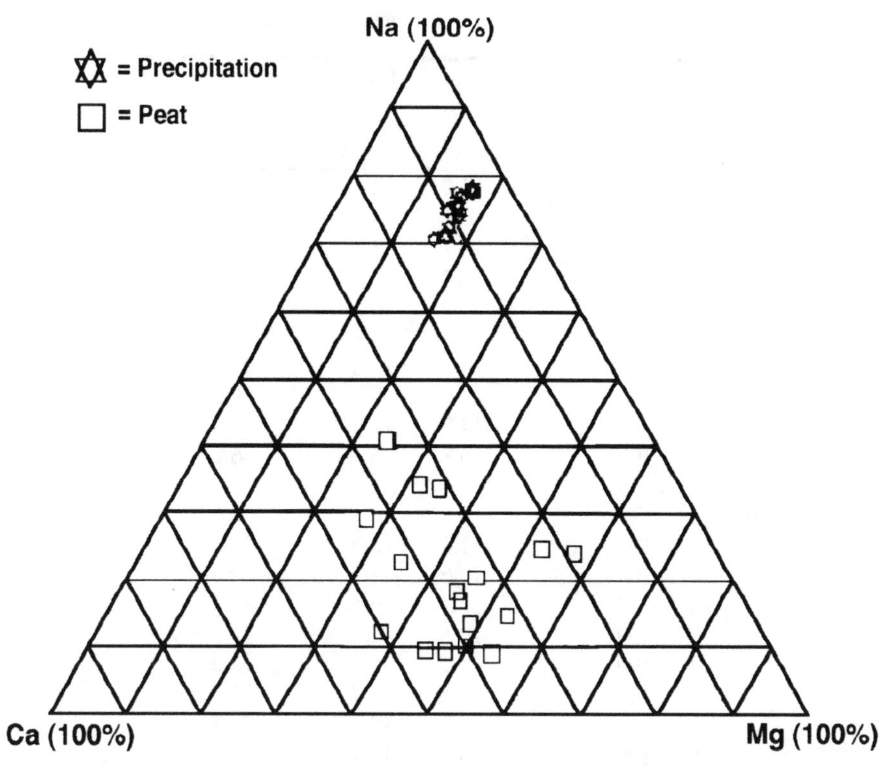

Figure 1 *Relative proportions (on a mol_c basis) of Ca^{2+}, Mg^{2+}, and Na^+ to the sum of the three cation concentrations in precipitation, and of exchangeable (mol_c kg^{-1} basis) Ca, Mg, and Na concentration to the sum of the three exchangeable cation concentrations for the peat from 17 sites in Scotland*

Ca^{2+} deposition, which is amplified because of the strong affinity of organic matter for binding Ca^{2+} on organic exchange sites.

The relative proportions of Ca^{2+}, Mg^{2+}, and Na^+ for the three streams at Glen Dye, measured at baseflow, are shown in Table 1. Baseflow water samples were collected at the end of July 1986, after a 4-week period that had been unusually dry (Rees *et al.*, 1989). Figures 2-4 show the relative amounts of Ca^{2+}, Mg^{2+} and Na^+ for the three streams at baseflow and stormflow, and in the precipitation, plotted on triangular diagrams. Analysis of variance was carried out on the baseflow and stormflow data for the three streams to look for significant differences between the relative proportions of Ca^{2+}, Mg^{2+} and Na^+ in the different streamflow conditions. For all three streams the chemistry of the stream water at baseflow and stormflow plot relatively closely to the precipitation chemistry. Data in Billett and Cresser (1996) for three other streams in northeast Scotland show stream water chemistry plotting well away from the precipitation, towards the centre or Ca^{2+}-rich area of the triangle, because of a strong influence from mineral weathering in the catchments which they studied. For Small Burn (Figure 2), there was no clear separation between baseflow stream water chemistry and precipitation chemistry.

Table 1 *Baseflow chemistry for the three streams at Glen Dye*

Stream Name	Distance downstream (m)	Na	Ca	Mg
		as % of total of Ca + Mg + Na		
Small Burn	0	13.0	12.8	74.2
	100	16.1	18.9	65.0
	200	19.1	23.8	57.1
	300	16.7	19.4	64.0
	400	16.9	16.8	66.3
	500	16.8	15.3	67.9
	600	14.7	15.2	70.1
	700	14.6	16.8	68.6
	800	14.0	15.6	70.4
	900	16.8	16.9	66.2
Warm Burn	0	11.5	16.1	72.5
	100	11.1	15.7	73.2
	200	10.6	15.6	73.8
	300	9.1	16.4	74.5
	400	10.5	18.9	70.6
	500	9.6	18.1	72.3
	600	10.0	18.7	71.3
	700	9.2	18.7	72.1
	800	8.4	17.7	73.9
	900	9.2	16.1	74.7
	1000	9.4	17.0	73.6
	1100	8.6	16.0	75.4
	1200	8.2	14.1	77.7
	1300	8.8	14.0	77.2
Brocky Burn	0	14.3	27.3	58.5
	100	13.9	29.3	56.8
	200	14.4	30.6	55.1
	300	14.0	30.5	55.4
	400	13.6	28.6	57.8
	500	14.1	27.3	58.6
	600	14.1	27.0	59.0
	700	13.2	24.7	62.1
	800	13.4	27.0	59.6
	900	15.2	28.2	56.6
	1000	14.5	24.6	60.9
	1100	14.7	23.7	61.6
	1200	12.5	33.7	53.8

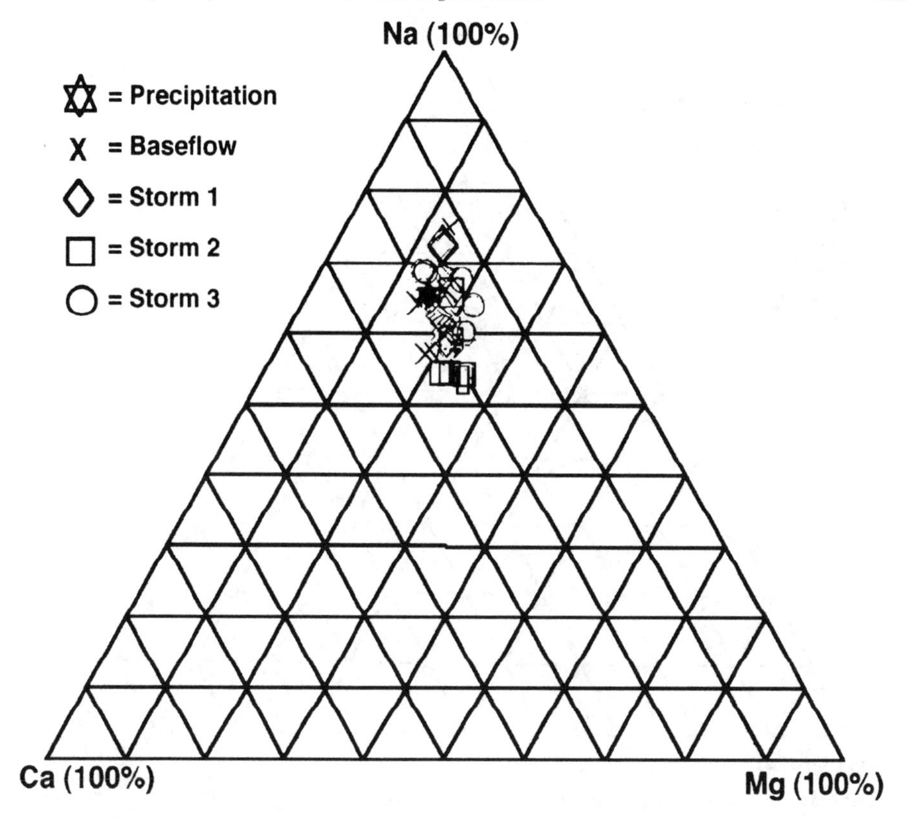

Figure 2 *Relative proportions of Ca^{2+}, Mg^{2+}, and Na^{+} (mol_c) to the sum of the three cation concentrations in precipitation, and in river water at base flow (along the length of the stream), and during three sequential storm events for the Small Burn sampled at the bottom of the catchment*

The stream water chemistry for storms 1 and 2 showed slight separations from both baseflow and precipitation chemistry and between each other, but for storm 3, although there is variation in stream water chemistry, the relative proportions are not clearly distinct from either those for baseflow or from that for precipitation. The analysis of variance showed that, for both Na% and Ca%, there were no significant differences between the relative proportions for baseflow and the stormflow, but the proportion of Mg^{2+} varied between discharge types with a significance of more than 0.1%. In storm 1, stream water contained relatively less Na^{+} and more Mg^{2+} than baseflow river water or precipitation. During storm 2, this same trend was more pronounced, with relatively even less Na^{+} and more Mg^{2+} and, in this instance, Ca^{2+}. By storm 3 in the sequence, the river water was heavily dominated by near-surface peat drainage water, and therefore in this storm precipitation and river water concentration ratios are closest together.

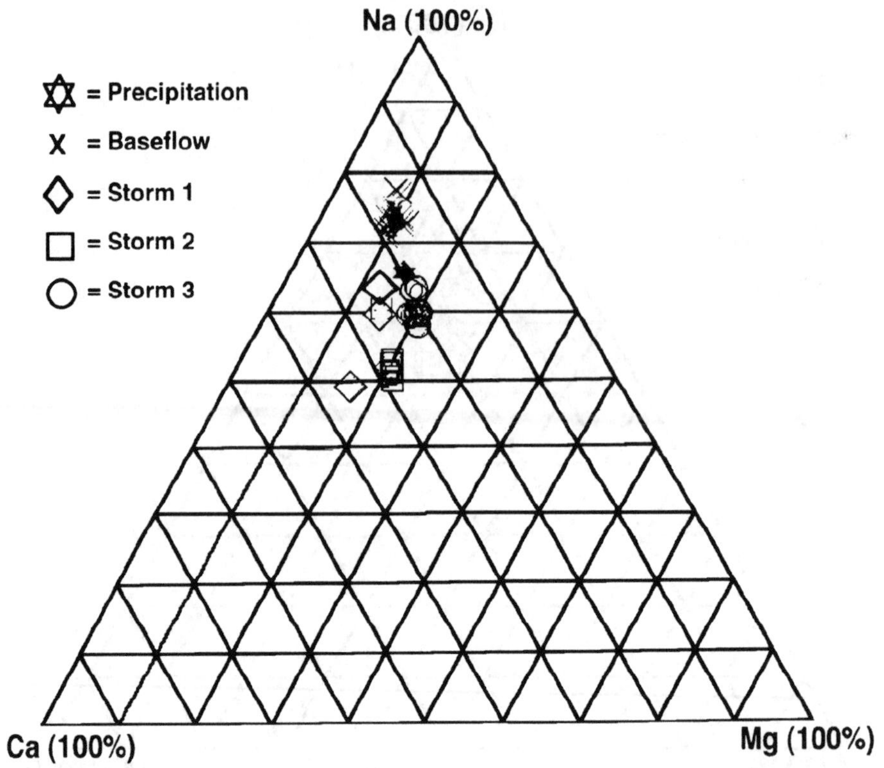

Figure 3 *Relative proportions of Ca^{2+}, Mg^{2+} and Na^{+} (mol_c) to the sum of the three cation concentrations in precipitation, and in river water at base flow (along the length of the stream) and during three sequential storm events for the Warm Burn sampled at the bottom of the catchment*

For Warm Burn (Figure 3), the relative proportions of all three components varied very significantly (at more than 0.1%) between baseflow and the three stormflows. Whereas in the Small Burn catchment, the mineral soil effect is minimal because of the very thin mineral soils, it is more important in the Warm Burn and Brocky Burn catchments. There was a clear separation between baseflow stream water chemistry and precipitation chemistry in that the stream water contained a greater proportion of Na^{+} and less Mg^{2+} in the Warm Burn. Storms 2 and 3 were clearly distinct from each other and from the precipitation, but storm 1 showed more variable stream water chemistry, which was only distinct from the other two storms in having slightly less Mg^{2+}. The stream water for storm 3 contained relatively slightly less Na^{+} and more Mg^{2+} than the precipitation, and for storm 2 this trend, especially for Na^{+}, was more pronounced. The stream water at storm 2 also contained relatively more Ca^{2+} than both the precipitation and the stream water at storm 3. For all three storms, the stream water contained relatively less Na^{+} and more Ca^{2+} and Mg^{2+} than at baseflow conditions. By the third storm, as water draining the surface organic horizons becomes more important, the relative proportions are again very similar for precipitation and for river water.

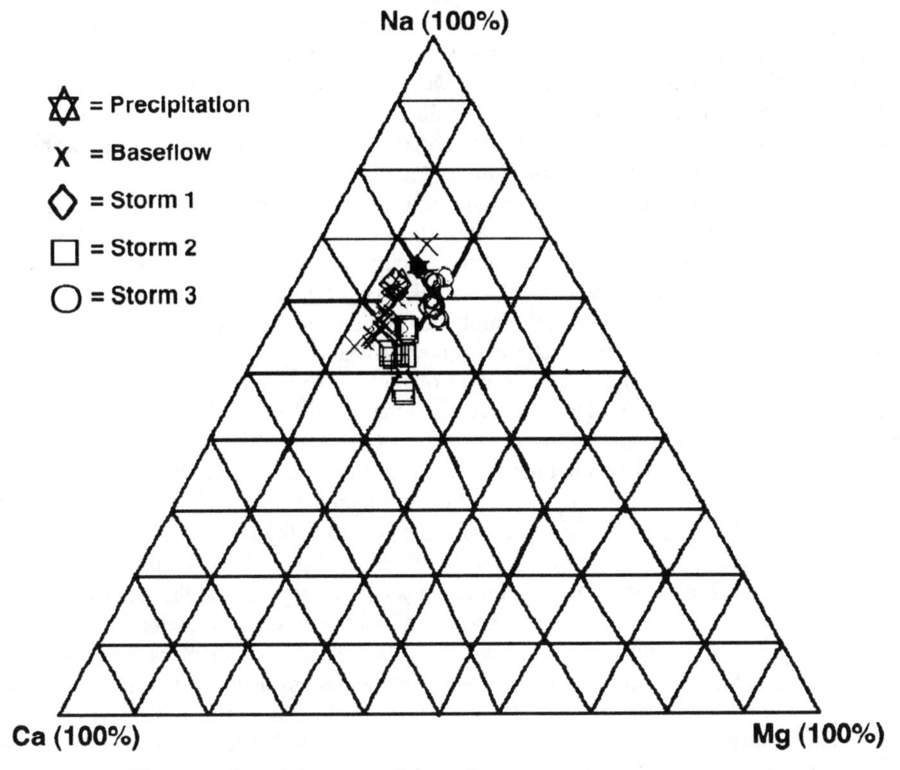

Figure 4 *Relative proportions of Ca^{2+}, Mg^{2+}, and Na^{+} (mol_c) to the sum of the three cation concentrations in precipitation, and in river water at base flow (along the length of the stream) and during three sequential storm events for the Brocky Burn sampled at the bottom of the catchment*

For Brocky Burn (Figure 4), the relative proportions of Ca^{2+}, Mg^{2+}, and Na^{+} in the streamwater vary very significantly (at more than 0.1%) between the baseflow and stormflows, although the p-values were slightly less than for the Warm Burn. There are clear separations in the relative proportions of Ca^{2+}, Mg^{2+}, and Na^{+} in the stream water between the different storm flows, between the second and third stormflows and the baseflow, and between all three stormflows and the precipitation. Most importantly, once again by the third storm, when water draining the surface organic horizons makes a major contribution to discharge, the relative proportions of stream water and precipitation base cation are most similar.

Na^{+} was found to be the dominant cation in all three streams at Glen Dye (Rees *et al.*, 1989) and so the baseflow relative proportions, although differing from that for the precipitation to some extent in Warm Burn and in Brocky Burn, were closer to that for precipitation (heavily Na^{+} dominated) than to what might be expected from mineral soil weathering (with Ca^{2+} more dominant, Reid *et al.*, 1981b). In their study of upland catchments with mineral soils in northeast Scotland, Billett and Cresser (1996) found that, although Ca^{2+} and Mg^{2+} were the dominant exchangeable base cations throughout the profile in peats and podzols, reflecting the selective retention on the soil solid phase surfaces of the divalent cations relative to Na^+. Na^+ became relatively more important with depth. Na^+ originating in the soils would therefore be expected to contribute even more to baseflow waters than to stormflow waters. This effect is especially noticeable in the Small Burn and in the Warm Burn (Figures 2 and 3).

The baseflow in the Small Burn did not in fact differ significantly from the precipitation (Figure 2). At its source, Small Burn originates in deep peats which remain permanently wet, and so the water in the upper part of this stream probably never reacts significantly with the underlying mineral material (Rees *et al.*, 1989). Unless there are large cracks or natural pipes in deep peats, drainage is very slow (McCaig, 1983). Another consideration in the Small Burn catchment is the fact that the peat is generally underlain by bedrock rather than by a significant amount of mineral soil, and so the input of cations from weathering will be less. This is reflected in the Small Burn having the lowest stream water Ca^{2+} concentrations and a low Na^+:Cl^- ratio (Rees *et al.*, 1989).

The baseflow stream waters in both the Warm Burn and Brocky Burn differed from the precipitation. In the Warm Burn, the baseflow contained a notably higher proportion of Na^+, as mentioned already, and less Mg^{2+} (Figure 3), and in the Brocky Burn the baseflow contained a notably larger proportion of Ca^{2+} and less Na^+ (Figure 4). At the source of the Warm Burn the 1-2 m of peat develops large cracks in dry periods and these allow the rain water to reach the lower soil horizons quickly. At baseflow there is, therefore, a high input of base cations from weathering, especially for Na^+. Reid *et al.* (1981b) found that the element output from weathering in the Glen Dye catchment decreased in the following order: Si > Ca > Na > Mg > Fe > K > Al. Water flowing laterally through the soil becomes increasingly base-cation enriched and, therefore, water entering the stream on the lower slopes will have even higher concentrations of base cations. This is reflected in the increasing Na^+:Cl^- ratio downstream (Rees *et al.*, 1989).

At the source of the Brocky Burn the peat again develops large cracks during dry weather and these allow the rainwater to drain quickly to the deep mineral soils below (often > 2 m in depth). These deep mineral soils provide a substantial input of Ca^{2+} into the stream water (Rees *et al.*, 1989), and so the relative proportion of Ca^{2+} at baseflow is increased.

During storms there is an increase in water flowing laterally through the organic surface horizons, and sometimes an increase in overland flow, especially during prolonged heavy storms. This is reflected in the Glen Dye streams by the increasing concentration of dissolved organic matter in the stream waters (Rees *et al.*, 1989). Also, stormflow waters generally contain smaller concentrations of base cations. Creasey *et al.* (1986) also argued that storm water in Glen Dye was dominated by drainage from acid, organic-rich soil horizons. In all three streams at Glen Dye there are small differences in the relative proportions of Ca^{2+}, Mg^{2+}, and Na^+ between the baseflow and stormflows. Rainfall chemistry is known to vary between and within storms (e.g. Edwards *et al.*, 1984), but the chemistry of the associated stream water often varies far more. This occurs as discharge switches from baseflow through the lower soil horizons to flow through the upper soil horizons, or even to overland flow in prolonged heavy storms (Cresser and Edwards, 1987). It is expected that stormflow water chemistry would more closely resemble precipitation chemistry than would baseflow water chemistry, due to the greater influence of precipitation upon organic surface soils with a low weathering mineral content.

It has been reported elsewhere (e.g. Glover and Johnson, 1974; Giusti and Neal, 1993) that many upland catchments show a dilution of weathering-derived cations during storms, and hence a negative correlation of Ca^{2+} and Mg^{2+} with discharge. With Na^{+}, which is also derived from weathering, there was also a negative correlation with discharge, but it was less significant than for Ca^{2+} and Mg^{2+}. Billett and Cresser (1996) considered this to be due to the large amounts of Na^{+} in the precipitation in northeast Scotland. In the three burns studied in this work, the very strong influence of sea salt inputs on river water composition is very clearly apparent in the Na^{+}-richness of the water.

4 Conclusions

Differences in the relative importance of Ca^{2+} concentration in atmospheric deposition between sites have a marked effect upon the relative distributions of base cations on the cation exchange complex of peat soils. However, the water draining from those soils will, like the deposition in the UK, be dominated by Na^{+} and, to a much lesser extent, Mg^{2+} in a relatively fixed ratio.

Whenever stream water is dominated by water draining from surface organic horizons of peats and peaty podzols, there is a direct link between precipitation chemistry and river water chemistry. This can be clearly seen in comparisons of the relative proportions of Ca^{2+}, Mg^{2+} and Na^{+}. The relationship is very strong if the catchment soil is dominated by deep peat, as for example in the Small Burn catchment, or when heavy precipitation falls upon already very wet soils, favouring near surface or overland flow, as in the third of a sequence of three storms at the Warm Burn and Brocky Burn catchments.

The result implies that peat samples could potentially be used, with an appropriate aqueous extraction, for prediction of mean base cation deposition ratios at the site. Thus, it should be feasible to estimate Ca^{2+} and Mg^{2+} depositions at a site in close proximity to peat if, for example, Na^{+} deposition is known for the site. The results also have implications in the development of hydrochemical models based upon soil and catchment characteristics.

References

Billett, M.F. and M. S. Cresser. 1992. Predicting river water quality using catchment and soil chemical characteristics. *Environ. Pollut.* 77:263-268.

Billett, M.F. and M.S. Cresser. 1996. Evaluation of the use of soil ion exchange chemistry for identifying the origins of streamwaters in catchments. *J. Hydrol.* In press.

Creasey, J. 1984. *The Geochemistry of a Small Upland Catchment in North-East Scotland.* Unpubl. Ph.D. thesis, Aberdeen University.

Creasey, J., A. C. Edwards, J. M. Reid, D. A. Macleod, and M. S. Cresser. 1986. The use of catchment studies for assessing chemical weathering rates in two contrasting upland areas in northeast Scotland. p. 468-499. *In* S. M. Colman and D. P. Dethier (eds), *Rates of Chemical Weathering of Rocks and Minerals.* Academic Press,. London.

Cresser, M. S. and A. C. Edwards. 1987. *Acidification of Freshwaters.* Cambridge University Press, Cambridge.

Cresser, M. S. and A. C. Edwards. 1988. Natural processes in freshwater acidification. *Endeavour* **12**:16-20.

Dewalle, D. R. and B. R. Swistock, 1994. Causes of episodic acidification in five Pennsylvania streams on the northern Appalachian Plateau. *Water Resour. Res.* **30**:1955-1963.

Edwards, A. C., J. Creasey, and M. S. Cresser. 1984. The conditions and frequency of sampling for elucidation of transport mechanisms and element budgets in upland drainage basins. p. 187-202. *Proceedings of the International Symposium on Hydrochemical Balances of Freshwater Systems*. Uppsala, Sweden.

Glover, B. J. and P. Johnson. 1974. Variations in the natural chemical composition of river water during flood flows, and the lag effect. *J. Hydrol.* **22**:303-316.

Giusti, L. and C. Neal. 1993. Hydrological pathways and solute chemistry of storm runoff at Dargall Lane, southwest Scotland. *J. Hydrol.* **142**:1-27.

McCaig, M. 1983. Contributions to storm quickflow in a small headwater catchment - the role of natural pipes and soil macropores. *Earth Surf. Process. Landforms* **8**:239-252.

Rees, R. M., F. Parker-Jervis, and M. S. Cresser. 1989. Soil effects on water chemistry in three adjacent upland streams at Glen dye in northeast Scotland. *Water. Res.* **23**:511-517.

Reid, J.M., D.A. MacLeod, and M.S. Cresser. 1981a. Factors affecting the chemistry of precipitation and river water in an upland catchment. *J. Hydrol.* **50**:129-145.

Reid, J. M., D. A. MacLeod, and M. S. Cresser, 1981b. The assessment of chemical weathering rates within an upland catchment in northeast Scotland. *Earth Surf. Proc. Landforms* **6**:447-457.

Reynolds, B., C. Neal, M. Hornung, S. Hughes, and P. A. Stevens. 1988. Impact of afforestation on the soil solution chemistry of stagnopodzols in mid-Wales. *Water, Air Soil Pollut.* **38**:55-70.

Sanger, L. J. M. F. Billet, and M. S. Cresser. 1994. Changes in sulphate retention, soil chemistry and drainage water quality along an upland soil transect. *Environ. Pollut.* **86**:119-128.

Skiba, U. and M.S. Cresser. 1989. Prediction of long-term effects of rainwater acidity on peat and associated drainage water in upland areas. *Water Res.* **23**:1477-1482.

Skiba, U., M.S. Cresser, R.G. Derwent, and D.W. Futty. 1989. Peat acidification in Scotland. *Nature* **337**:68-69.

Soulsby, C. and B. Reynolds. 1994. The chemistry of throughfall, stremflow and soil water beneath oak woodland and moorland vegetation in upland Wales. *Chem. and Ecol.* **9**:115-134.

UKRGAR 1990. *Acid Deposition in the United Kingdom 1986-1988*. Department of the Environment, Warren Spring Laboratory, Stevenage, Herts.

The Effects of Acidification on the Carbohydrate Contents of Humic Substances from the Watershed Soils and Lake Waters of Lake Skjervatjern

M.H.B. Hayes, I.P. Kenworthy, M.T. Quane, and B.E. Watt

THE UNIVERSITY OF BIRMINGHAM, SCHOOL OF CHEMISTRY, EDGBASTON, BIRMINGHAM B15 2TT, ENGLAND

Abstract

The major purposes of the humic lake acidification experiment (HUMEX-project) was to study the qualitative and quantitative impacts of additions of sulphuric acid (H_2SO_4) and ammonium nitrate (NH_4NO_3) on the humic substances (HS) and biota in the soil and the lake waters of Lake Skjervatjern in western Norway. After a two year settling period following the division of the lake using a plastic curtain in 1988, one part of the lake basin and its watershed were subjected to annual inputs of artificial acid rain similar to those that fall naturally in southern Norway. No acid was supplied to the *Control* part. The neutral sugar (NS) contents of the HS isolated from soil drainage waters and from the lake waters between 1992 and 1995 indicate that acidification caused considerable changes to the compositions, abundances, and certain ratio values of the NS. The acidification process was in operation for five years when the last of the samples studied in this communication were taken. The information obtained can be of value for predictions of changes resulting from inputs of acid rain, but ideally a more extended study of the system is desirable.

1 Introduction

Carbohydrates in Soils and in Waters

It is estimated that carbohydrates contribute between 5 to 25 per cent to the composition of soil organic matter (Stevenson 1994). These carbohydrates are derived from the remains of plants and biota, and from the metabolic processes of bacteria, actinomycetes, and fungi.

There are four major types of carbohydrate material in the soil. These are:

(i) free sugars in the soil solution;
(ii) complex polysaccharides which are separable from all other organic constituents;
(iii) polymeric molecules which are strongly bound to clay colloids; and

(iv) polymeric molecules which are strongly attached to humic substances.

All of these types of organic materials can be present in the drainage waters from soils, and types (i), (ii), and (iv) can be in solution.

Carbohydrates bound to humic substances (HS) are considered to account for about one-fifth of the total carbohydrate materials present in the soil (Thurman, 1985). The remainder, or the loosely bound carbohydrate, is separated using the XAD resin isolation technique (Malcolm and MacCarthy, 1992). The amounts of carbohydrate material in the HS in solution in natural waters can be as little as one tenth of those in the HS of the soils of the watershed (Hayes, 1996). Carbohydrates of type (iii) which are associated with the soil inorganic colloids and suspended in drainage waters, are removed in the filtration process which preceeds the use of the XAD resin processing procedures. Materials of type (iv) provided the basis for the present study.

Several studies have shown that environmental factors, including pollution, can lead to physiological stresses which markedly affect the pool of free amino acids (Hubberten *et al.*, 1994, 1995; Hedges *et al.*, 1994). However, the effects of pollution on the carbohydrate pool are not well understood. Chemical degradations of carbohydrates, particularly under acidic conditions, produce reductones, furan derivatives, pyruvaldehyde, etc., and these can condense with each other, or with amino compounds (e.g. the Maillard reaction) to produce dark coloured amorphous products similar to humic substances (Scheffer and Ulrich, 1960). Acidification processes could therefore be expected to have effects on the HS and the carbohydrates associated with these, whether in soils or in waters.

In this communication we report on the influence which simulated acidic rainfall has on the compositions of the carbohydrates associated with HS. The study set out to establish:

1, whether prolonged simulated acid rain could alter either the distribution or the concentration of carbohydrates bound to HS;
2, whether such rains could give rise to significant changes in the ratios of sugars with origins in plants and in microorganisms; and
3, whether any changes observed could prove useful as an early warning indicator of the effects of acid rain.

2 Materials and Methods

The Study Site

The study site is at Lake Skjervatjern, a dystrophic lake (pH 4.6) and its catchment, occupying an area of 8.9 ha, and located near Førde in western Norway. This lake was divided in 1988 by a plastic curtain imbedded in the silt of the lake bottom. Two years later applications of artificial acid rain were initiated to one part of the lake (0.9 ha) and its catchment (1.8 ha), and the remainder of the lake (1.5 ha) and its catchment (4.7 ha) was used as a control. The artificial acidity was supplied as additions of H_2SO_4 and NH_4NO_3 to irrigation waters applied with sprinklers, and the amounts added annually were the same as the acidity supplied in natural precipitation in southern Norway (Gjessing, 1992, 1994; Lydersen *et al.*, 1996). The total catchment (8.9 ha) lies on granitic bed-rock, and is covered by a mat of peat of varying thickness and extents of humification.

Sample Collection

The soil drainage water was taken from drainage pits dug to below the water table levels on the Acidified and Control sides of the catchment. Samples were taken from freshly dug pits in the same locations in late September 1992, 1993, 1994, and 1995. The lake water was taken from outlet pipes that drained the Lake. One tonne of drainage water was taken each year from the control and from the acidified cathments. In 1992 and in 1994, 50 L samples of water were taken from the outlets of the control and acidified parts of the lake, and in 1995 lake water samples were taken from the acidified part only. The waters were contained in aluminium kegs.

Processing of Samples

In order to avoid the proliferation of biological activity in the samples, the water was processed soon after arrival at the laboratory in Birmingham. Particulate matter was first removed by centrifugation (12 000 g, 45 min), and the supernatants were pressure filtered (10 psi, dinitrogen gas) through 0.2 µm pore size (142 mm diameter) Sartorius cellulose acetate membrane filters, using a Millipore stainless steel filtration set up.

Isolation of the Humic Substances

HS were isolated using the XAD-8 [(poly) methylmethacrylate] and XAD-4 (styrene divinylbenzene) resins in tandem (Malcolm and MacCarthy, 1992). The pH of the material back eluted from the XAD-8 resin was adjusted to 1 and the humic acid (HA) and fulvic acid (FA) fractions were recovered (see Hayes *et al.*, 1997; this issue) from that eluate. The XAD-4 acids were recovered from the back eluate in base (Hayes *et al.*, p. 107 - 120, this Volume).

Carbohydrate Analyses

Monosaccharides were determined as their alditol acetates, using a procedure based on that of Blakeneny *et al.* (1983). After the addition of an internal standard (*myo*-inositol), HS samples (ca 15 mg) were hydrolysed, using 2.0 mL of 2.0M trifluoroacetic acid (TFA). The solutions were mixed thoroughly and then heated at 120 °C for 2 h. Excess TFA was removed in a vacuum desiccator. Residual acid was neutralized by ammonia solution (0.1 mL, 1.0M). Monosaccharides were reduced by the addition of a sodium borohydride solution (1.0 mL; 2 g dissolved in 100 mL of DMSO heated to 100 °C). These mixtures were heated to 40 °C for 90 min. The solutions were allowed to cool and acidified by adding glacial acetic acid (0.1 mL). That destroys any excess borohydride. Next, 1-methylimidazole (0.2 mL) was added along with the derivatizing agent, acetic anhydride (2.0 mL). This was mixed thoroughly and allowed to stand at room temperature for 10 min. Excess acetic anhydride was destroyed by the addition of distilled water (5.0 mL), and the alditol acetates were extracted into dichloromethane (0.75 mL x 3 times). The extracts were washed with distilled water (3 x 1.0 mL) and evaporated to dryness. The residues were taken up into chloroform (0.2 mL). Analyses of the alditol acetates were carried out using FID detection on a Perkin Elmer AutoSystem XL gas chromatograph equipped with a BPX70 [SGE (UK Ltd.)] capillary column (25 m x 0.25 mm i.d.; 0.25 μm film thickness). Injection was made in the split mode (30:1). The oven temperature program was: 180 °C to 225 °C at 8 °C min^{-1}, held for 8 min, then raised to 240 °C at 3 °C min^{-1}, and held for 8 min. The injection temperature was 260 °C, and the FID temperature was 270 °C. The alditol acetates were identified according to their respective retention times by comparison with a mixture of standard monosaccharides.

3 Results and Discussion

Soil Drainage Waters

Neutral Sugar Contents, Abundances, and Mass Ratios Values. The neutral sugar (NS) contents of the samples, detected following hydrolysis, are given in Tables 1-4, and these range from ca 1% to 10% in the cases of soil drainage waters, and from ca 1% to 18% for the lake waters. In all cases the relative abundances of the NS in the humic fractions decreased in the order XAD-4 acids > HAs > FAs. The order is the same as that given by Watt *et al.* (1996) for humic fractions in the waters of rivers, streams, and reservoirs in Britain and Ireland. However, when the relative abundances of the different fractions are taken into account, the overall contents of NS in the various fractions decreased in the order XAD-4 acids > FAs > HAs, and the average distributions of the NS in the three major fractions for the soil drainage water HS was XAD-4 acids (6%) > HAs (3.7%) > FAs (1.4%). [The average distribution for the lake water fractions was XAD-4 acids (14%) >> HAs (3.0%) ≥ FAs (1.3%)].

Because crystalline cellulose is not digested by TFA (Botcher, 1984), the NS contents can be considered to be derived from hemicelluloses and saccharides of microbial origins. The presence of small amounts of ribose (Rib) could indicate microbial rather than plant origins for some of the saccharide materials (Forsyth, 1950). Some indications of the origins of the component NS may be obtained from the ratio values [galactose (Gal) + mannose (Man)]/[arabinose (Ara) + xylose (Xyl)] and [rhamnose (Rha) + fucose (Fuc)]/(Ara + Xyl). Xyl is found almost entirely in the saccharides of plants, and Ara is also found predominantly in plants. Man and Gal, and to a lesser extent Rha and Fuc are considered to be predominantly components of soil polysaccharides with origins in microorganisms. Ratio values [for (Man + Gal)/(Ara + Xyl)] < 0.5 would suggest that plants were the major sources of the sugars, and the more the values are > 0.5 the greater is the likelihood that the sugars were predominantly derived from microbial sources (Oades, 1984). Keefer and Mortensen (1963) proposed that soil polysaccharides are products of microbial activity and that constituent sugars undergo continual degradation and resynthesis (see also Hayes *et al.* p. 107 - 120, this Volume).

The contents of NS in the HA fractions in the Control for 1992, 1993, and 1994 range from 2.9% - 3.6%, and the range was similar for the fraction from the Acidified catchment (2.5% - 3.5%). The order of abundances of the NS in the HAs from the Control was broadly similar for the three years with Xyl = glucose (Glu) > Rha > Ara ≥ Gal > Man >> Fuc. Ribose (Rib) was not determined for these samples. The order was different for the Acidified catchment for the same three years (Glu >> Rha > Xyl ≥ Gal > Man >> Ara >> Fuc). Rib was not determined. The ratio values for (Man + Gal)/(Ara + Xyl) and

(Rha + Fuc)/(Ara + Xyl) were each of the order of two to three times greater for the Acidified than for the Control catchment.

Thus it would seem that the carbohydrate components of the HAs from the Control catchment were primarily from plant sources whereas those from the Acidified catchment had significantly greater inputs from microorganisms.

The NS contents of the FAs for the Control and Acidified catchments were similar for the years 1992 - 1994, inclusive, and accounted for 1.1% - 1.2% of the masses of the FAs of the Control and Acidified catchments. The relative abundances of the sugars (Rha ≥ Glu ≥ Xyl > Gal ≥ Ara > Man >> Fuc) were also similar for the two catchments. With the exception of the (Man + Gal)/(Ara + Xyl) values for 1993, the ratio values were also similar, and reflected the order of abundances of the component NS by indicating that there was a greater input by microbial processes to the FAs than to the HAs.

The total NS contents of the XAD-4 acids for the three years were greater in the drainage waters from the Acidified catchment than in those from the Control, and ranged from 4.2% - 5.1%, and from 4.9% - 7.2% of the contents of the fractions from the Control and Acidified catchments, respectively. The order of the abundances of the NS were similar for samples from both catchments (Glu > Rha ≥ Gal ≥ Man > Xyl >> Ara > Fuc >> Rib), and reflected enrichments in sugars that can be considered to be derived from microorganisms. This was also mirrored in the ratio values.

The NS data for the fractions from the drainage water samples taken from the two catchments in 1995 are different from those for the three previous years. In the case of the samples from the Control catchment, the sugar contents of the fractions are significantly greater in every case than for the previous years (vide infra).

The NS content of the HA from the Control side (4.1%) for 1995 was of the same order as that for the previous years, but that for the Acidified catchment (7.0%) was more than twice the contents for the previous three years. The order of abundances (Glu >> Rha > Man ≥ Gal ≥ Xyl >> Ara >> Fuc > Rib) for the HAs from the Control and the Acidified catchments were similar and indicative of greater contributions to the genesis from microbial sources than for the previous years. The ratio values for the Control for the HAs were in line with those for the Acidified catchment for the previous three years (vide infra).

The contribution of NS to the compositions of the FAs from the two catchments were significantly greater in 1995 than for the previous years, and that contribution was especially enhanced in the case of the FA from the Control catchment. There were differences in the relative abundances of the NS in the Control and Acidified catchments, and especially in the relative abundances of Man in the two fractions. The (Man + Gal)/(Ara + Xyl) ratio value for the FAs from the Control highlights the contribution to the genesis by microorganisms, but that ratio value is in line with those for the previous years in the case of the sample from the Acidified watershed.

The contribution of NS to the composition of the XAD-4 acids for the 1995 sample from the Control was significantly greater (10%) than those for the previous years. The order of abundances of the sugars in the samples from the Control and Acidified catchments for that year were the same (Glu > Man > Gal ≥ Rha > Xyl > Ara > Fuc > Rib), and only slightly different from the orders for the previous years, but that distribution, and the (Man + Gal)/(Ara + Xyl) ratio values highlight the contributions of microorganisms to the genesis of the saccharide components in the fractions. That trend is not mirrored in the data for the (Rha + Fuc)/(Ara + Xyl) ratio values.

Table 1 *Neutral sugar contents (NS, as relative molar percentages), the ratio values of (Man + Gal)/(Ara + Xyl) and of (Rha + Fuc)/(Arab + Xyl), and total NS contents (μg/mg) for the humic acid (HAs), fulvic acid (FAs), and XAD-4 acids isolated from drainage waters taken from the Control catchment of Lake Skjervatjern in September 1992-1995, inclusive*

Sample	Rha (R)	Fuc (F)	Rib	Ara (A)	Xyl (X)	Man (M)	Gal (G)	Glu	M+G / A+X	R+F / A+X	Total μg/mg
					1992						
HAs	14.6	2.5	n.d*.	14.3	24.9	9.3	12.1	22.1	0.5	0.4	32.1
FAs	19.8	1.7	n.d.	12.9	19.8	11.2	12.9	21.6	1.3	0.7	11.6
XAD-4	17.3	5.3	1.2	7.6	12.3	19.7	15.8	20.9	1.8	1.1	51.3
					1993						
HAs	16.1	1.7	n.d.	11.9	22.0	10.9	13.6	23.8	0.7	0.5	28.6
FAs	23.1	2.8	n.d.	11.1	19.4	10.2	12.0	21.3	0.7	0.8	10.8
XAD-4	17.4	6.7	1.6	8.4	15.8	14.9	15.9	15.3	1.0	1.3	43.1
					1994						
HAs	14.8	2.0	n.d.	12.8	24.3	9.8	12.0	24.3	0.5	0.6	35.8
FAs	21.9	1.9	n.d.	12.4	18.1	11.4	12.4	21.9	0.8	0.8	10.5
XAD-4	17.6	5.2	1.2	8.1	14.0	16.4	16.2	21.2	1.0	1.5	42.0
					1995						
HAs	18.2	1.0	0.7	5.4	14.3	15.5	14.8	30.5	1.6	1.0	40.6
FAs	10.8	2.7	1.2	8.8	11.5	23.0	13.5	28.5	1.8	0.7	26.0
XAD-4	10.8	4.1	1.5	6.8	9.0	21.6	19.7	26.5	2.6	0.9	99.5

*n.d = not determined

Figure 1 shows the total sugar concentrations of the HS for the Acidified minus the Control samples for the 1992-1995 period. In that it is evident that the FAs show only minor changes in the first three years following acidification. In that period the Acidified minus Control values were close to zero, and then dropped to ca -8 μg mg^{-1} in 1995. The HAs show a similar trend for 1993 and 1994, although the negative value for 1992 is ca 7.

The largest negative values for the three year period are for the XAD-4 acids. These differences would suggest that acidification gave rise to increases in the NS contents of the XAD-4 acids, and the (Man + Gal)/(Ara + Xyl) ratios, especially, would indicate that these sugars had their origins in microbial synthesis processes. However, the trends for the samples taken in 1995 were very different. The data show a large plus (ca 30) value for the HAs, and significant negative values for the FAs and XAD-4 acids (ca -8 and -36, respectively). The relative abundances of the sugars were similar for the HAs and XAD-4 acids from the Acidified and Control catchments. So too were the ratio values, and the increased (Man + Gal)/(Ara + Xyl) ratios for the Control catchment (compared with the values for the three previous years) would suggest that there was an input to that catchment (that was not evident from the analyses of the samples from the previous years) which had stimulated microbial activity during the period September, 1994 to September, 1995. Though NS contents for Acidified minus Control sides were negative

Table 2 *Neutral sugar contents (NS, as relative molar percentages), the mass ratio values of (Man + Gal)/(Ara + Xyl) and of (Rha + Fuc)/(Ara + Xyl), and total NS contents (µg/mg) for the humic acid (HA), fulvic acid (FA), and XAD-4 acids isolated from drainage waters taken from the Acidified catchment of Lake Skjervatjern in September 1992-1995, inclusive.*

Sample	Rham (R)	Fuc (F)	Rib	Arab (A)	Xyl (X)	Man (M)	Gal (G)	Glu	M+G / A+X	R+F / A+X	Total µg/mg
					1992						
HA	20.1	1.2.	n.d*.	7.2	16.5	13.7	15.7	25.7	1.2	0.9	24.9
FA	17.8	0.8	n.d.	11.9	16.9	15.3	16.1	21.2	1.1	0.6	11.8
XAD-4	16.6	6.6	1.4	7.0	12.8	18.4	15.7	21.5	1.2	1.7	72.4
					1993						
HA	20.9	1.7	n.d.	6.4	15.5	12.8	14.9	27.7	1.3	1.0	29.6
FA	20.6	0.9	n.d.	8.4	16.8	14.0	16.8	22.4	1.2	0.9	10.7
XAD-4	15.8	4.9	1.2	7.5	14.0	15.8	17.4	23.1	1.0	1.5	49.4
					1994						
HA	17.8	1.5	n.d.	6.2	14.2	12.5	15.7	32.0	1.4	0.9	33.7
FA	23.5	1.0	n.d.	10.8	16.7	13.7	14.7	19.6	1.0	0.9	10.2
XAD-4	17.5	5.1	1.2	6.9	12.6	16.8	17.2	22.7	1.2	1.8	56.5
					1995						
HA	18.0	1.4	0.7	5.8	14.5	13.2	14.7	31.4	1.4	1.0	70.3
FA	16.9	1.7	1.1	9.0	15.8	7.9	16.4	32.2	1.0	0.8	17.7
XAD-4	12.9	4.6	1.4	7.2	10.0	20.9	19.0	24.0	2.3	1.0	62.8

*n.d. = not determined

Table 3 *Neutral sugars contents (NS, as relative molar percentages), the mass ratio values of (Man + Gal)/ (Arab + Xyl) and of (Rham + Fuc)/(Arab + Xyl), and the total NS contents (µg/mg) for the humic acid (HA), fulvic acid (FA), and XAD-4 acids isolated from the lake waters of the Control basin of Lake Skjervatjern in September 1992 and 1994*

Sample	Rham (R)	Fuc (F)	Rib	Arab (A)	Xyl (X)	Man (M)	Gal (G)	Glu	M+G / A+X	R+F / A+X	Total µg/mg
					1992						
HA	12.6	1.9	1.5	15.0	12.8	9.7	16.3	30.1	0.9	0.5	27.8
FA	19.3	2.2	0.5	11.6	12.3	11.7	12.9	29.3	1.0	0.9	17.5
XAD-4	18.3	5.8	0.7	6.5	11.0	1.4	28.6	27.7	1.7	1.4	92.1
					1994						
HA	9.6	2.8	2.0	9.7	14.9	17.5	14.0	29.4	1.3	0.5	21.1
FA	12.8	3.2	1.8	9.8	14.0	15.6	13.1	29.8	1.2	0.7	10.9
XAD-4	11.9	5.6	0.6	5.4	21.4	11.6	14.8	28.7	1.0	0.6	142

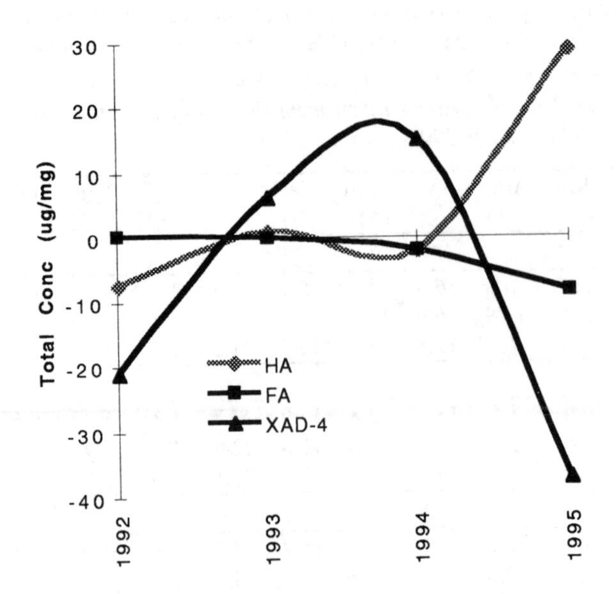

Figure 1 *The differences in total neutral sugar contents (in µg mg^{-1}) in the humic fractions from soil drainage waters isolated in the 1992-1995 period from the Acidified and Control catchments of Lake Skjervatjern*

Table 4 *Neutral sugar contents (NS, as relative molar percentages), the mass ratio values of (Man + Gal)/(Ara + Xyl) and of (Rha + Fuc)/(Ara + Xyl), and the total NS contents (µg/mg) for the humic acid (HAs), fulvic acid (FAs), and XAD-4 acids isolated from drainage waters taken from the Acidified basin of Lake Skjervatjern in September 1992, 1994, and 1995*

Sample	Rham (R)	Fuc (F)	Rib	Arab (A)	Xyl (X)	Man (M)	Gal (G)	Glu	M+G / A+X	R+F / A+X	Total µg/mg
					1992						
HAs	15.8	2.5	0.9	10.8	12.6	13.2	16.9	27.3	1.3	0.8	32.2
FAs	19.6	2.4	0.7	11.9	13.1	11.3	12.5	28.0	1.0	0.9	16.8
XAD-4	13.8	6.1	0.4	5.7	12.3	13.6	25.1	22.8	2.2	1.1	179
					1994						
HAs	11.3	4.2	1.4	7.1	12.1	19.8	19.4	24.7	2.0	0.8	37.9
FAs	14.2	3.1	1.8	13.5	14.4	15.1	13.5	24.4	1.0	0.6	10.3
XAD-4	14.5	6.7	0.8	6.3	13.1	14.3	18.5	25.7	1.6	1.1	119
					1995						
HAs	11.9	2.7	2.2	8.8	16.0	17.5	12.5	28.6	1.2	0.6	35.5
FAs	13.5	0.9	1.1	12.3	19.1	10.5	13.4	20.1	0.8	0.8	10.0
XAD-4	14.5	6.8	0.8	6.4	13.1	14.4	18.4	25.6	1.7	0.4	118

for the FAs and XAD-4 acids for 1995, the total NS contents in these fractions in these sides were greater than for the previous years. That was especially true for the Control catchment.

The differences for 1995 might be attributed to a climatic event that altered the continuing trend. Very high concentrations of NH_4^+ were recorded in the Acidified basin, from May 1994 to May 1995 (Lydersen *et al.*, 1996), and that may have had a significant effect on the HS sampled in October 1995. Lydersen has also speculated that an increase in NH_4^+ might have stimulated soil organisms, such as fungi, to exude more yellow coloured organic complexes. This phenomena is well known (Midwood and Felbeck, 1968; Felbeck, 1971).

Lake Waters

Neutral Sugar Contents, Abundances, and Mass Ratios Values. The relevant NS data for the humic fractions from the Control and Acidified sides of Lake Skjervatjern for 1992 and 1994 are presented in Tables 3 and 4, and data for samples from the Acidified side for 1995 are in Table 4.

The NS contents of the HA fractions from the Control and Acidified sides of the lake for 1992 are similar to those from the watershed catchments. Although the contents in the Acidified side for 1994 and 1995 are similar to those from the Control in 1992, the amounts in the Control side of the lake for 1994 are significantly less than those in the same side for 1992. Glu dominated the NS contents of all the lake HA samples, and there were significant differences in the relative abundances of Ara and Man in the HAs from the Control and Acidified parts of the lake in 1992 and 1994. The relative abundances of the NS were different in the HAs from the lake water and from the drainage waters. The relative abundances of the NS in the HAs from the Acidified side of the lake for the samples taken in 1992, 1994, and 1995 decreased more or less in the order Glu >> Gal = Man ≥ Xyl ≥ Rha > Ara > Fuc > Rib, and those for the Control for 1992 were of the order Glu >> Gal ≥ Ara ≥ Xyl = Rha > Man >> Fuc > Rib. Man was second in abundance in the case of the Control sample for 1994. The ratio (Man + Gal)/(Ara + Xyl) values for the HAs in the Control side in 1992 and 1994 were significantly greater than those for the drainage waters from the Control watershed, and the values for the Acidified side of the lake in 1992 and in 1995 were of the same order as those from the drainage waters from the Acidified side. The value for the HAs from the Acidified side of the lake in 1994 was significantly greater.

The NS contents and the ratio values for the FAs from the waters from the Control and the Acidified basins were similar for each year for the 1992 and 1994 samplings, and, apart from the ratio values for the Control catchment, were in line with the values for the FAs from the drainage waters. Glucose was invariably, and by a significant amount, the major component of the NS, and the order of the relative abundances was the more or less the same (but with minor variations) for the FAs from the Control and from the Acidified side for each year. For the Acidified basin, for example, for 1992 that order was: Glu >> Rha > Xyl ≥ Gal ≥ Ara ≥ Man >> Fuc > Rib. The differences in the relative abundances of Man, Gal, Xyl, and Ara were small.

The total NS contents were significantly greater for the XAD-4 acids, and especially so in the case of the sample taken from the Acidified basin in 1992. There was more variation in the order of the relative abundances of the NS in this fraction than for the HAs and FAs. For example, the order for the Control in 1992 (Gal ≥ Glu >> Rha >> Xyl

>> Ara ≥ Fuc >> Man > Rib) was significantly different from that for the Control for 1994 (Glu > Xyl >> Gal > Rha = Man >> Fuc = Ara >> Rib). The ratio values for 1992 would indicate higher inputs that year from sugars with origins in microorganisms (rather than with origins in plants).

The order of the abundance of the NS in the XAD-4 acids from the Acidified basin for 1992 was similar to that for the Control, except that in the Acidified instance the abundance of Man was similar to that of Xyl and Rha. The order for 1994 was roughly similar for the samples from the two basins except for the fact that the content of Xyl in the Acidified catchment sample was in line with the abundances of of Man and Rha. The ratio values indicated higher inputs from microbial sources in the cases of those from the Acidified basin.

The contents of the NS, their order of abundance, and the (Man + Gal)/(Ara + Xyl) ratio values for the XAD-4 acids from the Acidified basin in 1995 were almost identical to the contents, abundance, and ratio values for the same basin in 1994.

4 Conclusions

The acidification process has had a major effect on the NS contents of the HS isolated from the lake, and from soil drainage waters of Lake Skjervatjern. The effects were not the same for the different fractions of the HS isolated, regardless of the sources (from drainage waters or from lake waters). The artificial acid rain did have some effects on the overall totals of sugars bound to the HS fractions. However, the major influences involved changes to the relative abundances of the component NS in the different humic fractions as the result of the acidification process.

It is appropriate to consider the residence times (Gjessing, 1994) of the waters in the Acidified (1.6 months) and in the Control (4.5 months) basins of the lake, and the fact that one third of the acidity applied falls directly on the Acidified basin, and two thirds on its watershed. HS that are dissolved in the soil solution enter the lake basins in the drainage waters, and some HS will be generated from the plant and microbial life in the lake basins. Comparisons of the data in Tables 2 and 4 indicate that the NS contents of the HAs, FAs, and XAD-4 acids in the acidified basin are greater than those entering in the drainage waters. The relative abundances of the NS and the ratio values are similar for the HAs and the FAs, but significantly different for the XAD-4 acids. The compositions of the acids indicate that these have the greatest inputs from microbial processes. The same trends apply, though not for the FAs, for 1994. The results for 1995 do not follow the expected trends in so far as the abundances of the sugars are concerned, and the data for the XAD-4 acids would suggest that there was a higher input from microorganisms to the composition of the HS [as evidenced by the (Man + Gal)/(Ara + Xyl) ratio values] in the drainage waters. That might reflect a longer residence time in the soil solution for the HS on that occasion.

The data for the NS in the HS in the Control basin and its watershed for 1992 and 1994 would also point to higher inputs from microbial sources to the NS contents of the humic fractions in the lake basin. That would further suggest that sugars associated with the HS entering the lake are acted upon by microorganisms, and significant transformations are achieved in a relatively short time.

During four years of observation (in a five year period of acidification), it was observed that important changes had occurred in the abundance and relative proportions of one set of sugars to another as the result of acidification. In comparison, the changes to the Control or non-Acidified lake basin and its watershed have been very much less in the 1992 and 1994 period. However, dramatic differences were noted in the cases of samples taken in 1995, as was observed for the Acidified basin and its watershed. The data indicate that changes to sugar abundances, compositions, and ratios can be valuable for determining some influences of acid rain. However, our involvement has included four annual samplings in a five year acidification period, and we have seen evidence for what might be considered anomolous results during that time. Thus, in order to confirm the trends observed, it would be appropriate to extend the study of acidification at least for five additional years. It would seem appropriate also to sample the waters three times during the year. The study of recovery after the acidification cycle is complete would also be important.

Acknowledgements

This work was carried out as part of the HUMOR/HUMEX international project based at Lake Skjervatjern and its watershed, and supported by European Community funding.

References

Bochter, R. 1984. Bestimmung Cellulosisch und Nicthcellulosisch Gebundener Neutralzucker in Bodenhydrolysaten mit Hilfe von Hochleistungsdunnschicht-chromatographie. *Z. Pflanzenernahr. Bodenk.* **147**:203-209.

Blakeney, A.B., P.J. Harris, R.J. Henery, and B.A. Stone. 1983. A Simple and rapid preparation of alditol acetates for monosaccaride analysis. *Carbohydrate Res.* **113**:291-299.

Felbeck, G.T. Jr. 1971. Structural hypotheses of soil humic acids. *Soil Sci.* **111**:42-48.

Forsyth, W.G.C. 1950. The more soluble complexes of soil organic matter II. Composition of the soluble polysaccharide fraction. *Biochem. J.* **46**:141-146.

Gjessing, E.T. 1992. The HUMEX Project: Experimental acidification of a catchment and its humic lake. *Environ. Int.* **18**:535-543.

Gjessing, E.T. 1994. HUMEX (Humic Lake Acidification Experiment): Chemistry, hydrology, and meteorology. *Environ. Int.* **20**:267-276.

Hayes, T.M. 1996. Isolation and characterization of humic substances from soil , and the soil solution, and their interaction with anthropogenic organic chemicals. Ph.D. thesis, The University of Birmingham, England.

Hedges J.T., G.L. Cowie, J.E. Richie, P.D. Quay, R Benner, M. Strom, and B. Forsberg. 1994. Origins and processing of organic matter in the Amazon river as indicated by carbohydrate and amino acids. *Limnology and Oceanography* **39**:743-761.

Hubberten, U., R.J. Lara, and G. Kattner. 1994. Amino acid composition of seawater and dissolved humic substances in the Greenland sea. *Marine Chem.* **45**:121-128.

Hubberten, U., R.J. Lara, and G. Kattner. 1995. Refractory organic compounds in polar waters - Relationships between humic substances and amino acids in the Arctic and Antarctic. *Journal of Marine Res.* **53**:137-149.

Keefer, R.F. and J.L. Mortensen 1963. Biosynthesis of soil polysaccharides. I.-Glucose and alfalfa tissue substrates. *Soil Sci. Soc. Am. Proc.* **27**:156-162.

Lydersen, E., E. Fjeld and E.T. Gjessing. 1996. The humic lake acidification experiment (HUMEX): Main physico-chemical results after five years of artificial acidification. *Environ. Int.* **22**:591-604.

Malcolm, R.L. and P. MacCarthy. 1992. Quantitative evaluation of XAD-8 and XAD-4 resins in tandem for removing organic solutes from water. *Environ. Intern.* **18**:597-607.

Midwood, R.B. and G.T. Felbeck, Jr. 1968. Analysis of yellow organic matter from fresh water. *J. Am. Water Works Assoc.* **60**:357-366.

Oades, J.M. 1984. Soil organic matter and structural stability - mechanisms and implications for management. *Plant Soil* **76**:319-337.

Scheffer, F. and B. Ulrich. 1960. *Humus und Humusdungung*. Ferdinand Enke, Verlag, Stuttgart.

Stevenson, F.J. 1994. *Humus Chemistry-Genesis, Composition, Reactions*. Wiley, NY.

Thurman, E.M. 1985. *Organic Geochemistry of Natural Waters.* Martinus Nijhoff/Dr. Junk Publishers, Dordrecht, The Netherlands.

Watt, B.E., M.H.B. Hayes, T.M. Hayes, R.T. Price, R.L. Malcolm, and P. Jakeman. 1996. Sugars and amino acids in humic substances isolated from British and Irish waters. p.81-91. In C.E. Clapp, M.H.B. Hayes, N. Senesi, and S.M. Griffith (eds), *Humic Substances and Organic Matter in Soil and Water Environments: Characterization, Transformations and Interactions.* IHSS, Univ. Of Minnesota, St. Paul, Minn.

Decomposition in Soil of C4 and C3 Plant Material Grown at Ambient and at Elevated Atmospheric CO_2 Concentrations

A.S. Ball and J. Bullimore

DEPARTMENT OF BIOLOGICAL AND CHEMICAL SCIENCES, JOHN TABOR LABORATORIES, UNIVERSITY OF ESSEX, WIVENHOE PARK, COLCHESTER CO4 3SQ, ENGLAND

Abstract

The effects of growing plants under elevated atmospheric CO_2 concentrations on the degradability of the plant litter was investigated using aerial plant material from a range of C3 and C4 plants. The compositions and C:N contents of C4 plants grown in ambient and in elevated CO_2 did not differ significantly, nor did the plant materials decompose at different rates in soil. However, in all the C3 plant species examined the degradation rates were significantly less in the litter obtained from plants exposed to elevated atmospheric CO_2, presumably because of a 10% decrease in the N content. Any lowering of decomposition rates brought about by increasing atmospheric CO_2 concentrations will increase the storage of C in the soil. Long-term experiments are necessary to establish the effects of elevated CO_2 on C partitioning in terrestrial ecosystems.

1 Introduction

Changes in atmospheric CO_2 concentrations are likely to have significant effects on plant growth and ecosystem structure (Anderson, 1991). Both C3 and C4 plants respond to elevated atmospheric CO_2 concentrations by increased water use efficiency, while only C3 species show an instantaneous increase in net photosynthesis (Curtis and Whigham 1989). The overall effect of these responses is an increased production of plant biomass (Eamus and Jarvis 1989; Long and Drake, 1990).

The effects of elevated atmospheric CO_2 concentrations on soil C cycling has yet to be fully elucidated despite the fact that the amount of carbon and nitrogen in soils may be two to three times that found in plant tissues (Bouwman, 1990). It is thought that there would be no direct effects of increased atmospheric CO_2 on soil microbial processes due to the relatively high CO_2 concentrations present in most soils (Van de Geijn and Van Veen, 1993). However, to date this has yet to be confirmed experimentally. Nevertheless, it is likely that the main effects of elevated atmospheric CO_2 concentrations would be indirect, resulting from changes in plant productivity and composition (Van Veen *et al.*, 1991).

From our laboratory-based studies on the indirect effects of elevated atmospheric CO_2 concentrations on soil respiratory activity, we observed an increase in plant litter biodegradability for sorghum, a C4 plant, as assessed by the increase in soil respiratory activity (Taylor and Ball, 1994). Our previous study had shown that wheat, a C3 plant, showed an opposite effect because the plant material grown under elevated CO_2 was more difficult to degrade (Ball, 1991). We suggest that these two annual crops may have shown opposite responses according to their photosynthetic type. Although the reasons for the observed differences have not been elucidated, it is likely that the increase in C:N ratio of plant material from wheat may have decreased its biodegradability. The C:N ratio of plant material has been shown to be important in determining decomposition rates (Berg, 1984). The C:N ratio of plant material from sorghum grown under elevated atmospheric CO_2 remains unchanged (Taylor and Ball, 1994). Such a response, if repeated in perennial vegetations of natural ecosystems, will have important implications for future carbon and nitrogen cycling.

We examine in this paper the indirect effects of elevated atmospheric CO_2 on soil respiratory activity, using plant material from a range of C3 and C4 plants grown both in natural ecosystems and in controlled environmental chambers under elevated and ambient CO_2 conditions. Exposure of all the C3 plants used in this study to elevated CO_2 led to an increased photosynthetic rate, a decrease in plant respiration, and an increase in carbon accumulation (Arp and Drake 1991; Drake, 1992 ; Ziska *et al.*, 1990), and the C4 plants exhibited no significant change in productivity when exposed to elevated CO_2.

2 Materials and Methods

Plant Material and Soil Samples

Freshly fallen aerial plant materials from all the plants used in this study (see Table 1) were gathered from plants exposed to both elevated and ambient CO_2 and cut into uniform size (approx 2 cm in length) pieces before use. The soil was a sandy loam (pH 6.5), containing 1.3% C, 0.1% N, and 15% moisture (w/w; corresponding to an approximate moisture potential of -0.8 MPa) (Norby *et al.*, 1986). It was sampled (0-15 cm depth) from an undisturbed grassland site on the University campus. The soil was passed through a 2 mm mesh in order to remove plant material and provide a relatively homogeneous soil for measurement of respiration.

Experimental Details

Soil respiratory activity was measured using an infra red gas analyser (IRGA, ADC 225) attached to a twelve point multi-channel selector (ADC WA161). The degradation studies were carried out using water-jacketed respiration chambers through which sterile humidified air (containing ambient concentrations of CO_2) was passed. Carbon dioxide in the replaced air was assessed by IRGA. Both reference and sample air were maintained at a flow rate of 400 mL min^{-1}. Carbon dioxide flux was recorded for a dwell time of five minutes. The chambers were kept in the dark at 20 ^oC, and the moisture content was maintained at 15% (w/w) by the daily addition of sterile water. Each chamber contained 80 g of soil. Following the stabilization of respiratory activities in unamended soils, triplicate chambers were amended with 0.8 g (representing approximately 30% of the initial soil C) of aerial plant material from plants grown under either ambient (340 ppm) or elevated (650 ppm) atmospheric CO_2, and then incubated for 30 days. Soil respiratory measurements were recorded constantly using a chart recorder (Kipp and Zonen BD 111).

The carbon:nitrogen ratio of soil and leaf material used in the study was measured on an automated Perkin-Elmer CHNS/O analyser (Series ii, 2400). Soil (20 mg) and plant (5 mg) samples were ground to small particle sizes and placed in foil capsules for analysis by combustion and spectral characterization. The carbon content of the sample (expressed as a percentage) was calculated from a standard containing a known amount of carbon.

The analysis of the polymeric components of plant material used in this study were determined by sequential extraction of triplicate samples of plant material followed by gravimetric analysis as previously described (Harper and Lynch, 1981).

3 Results

Soil Respiratory Activity

Soil respiration measurements were taken daily over a period of thirty days for each set of plant materials. Table 1 shows the average mean daily total soil respiration rates over the 30 days for the unamended soils (control), and for soils amended with ambient and elevated CO_2-grown plant material. At the time of addition, the soil respiration rate in the chambers was found to vary from 60-78 μg CO_2-C g^{-1} oven dried soil day^{-1}. Table 1 shows the mean respiration rates for the soils from each of the different treatments, measured throughout the experiment. In each experiment, the addition of plant material (either ambient or elevated CO_2- grown) led to an increase (two-three fold) in soil respiratory activity over the length of the experiment. For plant material derived from the C4 plants, *Spartina patens* and Sorghum, soil respiration was largely unaffected by the original CO_2 concentration in which the plant was grown (Table 1). For chambers to which plant material from the C3 plants, *Scirpus olneyi, Triticum aestivum, and Lolium perenne* had been added, the respiration rates were highest in incubations to which plant material grown under ambient CO_2 had been added. Plant material grown in elevated CO_2 evolved less CO_2 than ambient CO_2 grown plant litter, with an average decrease of approximately 20% over the duration of the incubation period (Table 1).

Soil respiration did not vary in the unamended control chambers during the 30 day incubation (Figure 1). In contrast, the respiratory activity of chambers containing soil amended with plant material from ambient CO_2- grown wheat increased approximately ten fold after 72 h, compared to a five fold difference in soils amended with elevated CO_2 grown wheat. The respiration rates of soils amended with straw from wheat grown under ambient rather than elevated CO_2 remained significantly higher until the last three days of the experiment (Figure 1) by which time the rates had returned to the levels measured in the control (unamended) soil chambers.

Analysis of Plant Litter

Analysis of the carbon and nitrogen present in the plant litter used in this study, grown under elevated and ambient CO_2, indicated that while the C:N ratio of the C4 plants

Table 1 *Mean soil respiratory activity (µg CO_2-C g soil^{-1} day^{-1}) in soils incubated under ambient and elevated CO_2 and amended (1 g per 100 g dry weight soil) with plant material grown under elevated or ambient CO_2. The results presented are the mean values of 90 measurements*

Chamber containing soil amended with plant material from	Soil Respiration (µg CO_2-C g soil^{-1} day^{-1}) in chambers amended with material from plants grown under:	
	ambient CO_2	elevated CO_2
Soil only (control)	75	
C_3 Plants		
Scirpus olneyi	183	147[1]
Triticum aestivum	180	129[1]
Lolium perenne	216	162[1]
C_4 Plants		
Spartina patens	165	150
Sorghum bicolor	156	171

[1] Significant differences (T-test; $p < 0.05$) between chambers containing plant material grown under elevated (650 ppm) and ambient (340 ppm) CO_2

Spartina patens and Sorghum remained unchanged (no significant difference; unpaired t-test, $p > 0.86$) if grown under elevated atmospheric CO_2 concentrations (Table 2). A significant increase in the C:N ratio was observed in plant litter from all three C3 plants, *Triticum aestivum, Lolium perenne and Scirpus olneyi* (Table 2: unpaired t-test , $p < 0.05$). Gravimetric analysis of the main polymeric components of plant litter revealed no significant differences in the polymeric composition of plant material grown under elevated and ambient atmospheric CO_2 (Table 2). Likewise, lignin content remained unchanged in both the C3 and the C4 plant litter.

4 Discussion

Soil Respiratory Activity in Chambers

Despite experimental conditions that were far from natural (constant moisture and temperature, relatively large additions of substrate), our results indicate a significant difference between the respiratory rates of soils amended with plant material from C3 plants grown under elevated CO_2 and those grown under ambient CO_2. In contrast, no significant difference could be detected between respiratory rates in soils amended with C4 plant material from plants grown under elevated CO_2 and C4 material grown under ambient CO_2. This study necessarily examined the effects of elevated CO_2 on the initial stages of biodegradation. It is likely that the soil respiration rate would decrease once the labile components of the fresh litter had all decomposed.

The C:N ratio of plant material is an important determinant of decomposition rates. Analysis of the plant litter used in this study show that no significant differences were observed in the C:N ratios of C4 plant material grown under ambient or elevated CO_2

Table 2 *Elemental and gravimetric analysis of plant litter from C3 and C4 plants grown under elevated (650 ppm) and ambient (340 ppm) CO_2 concentrations. Results are the means of three replicates, with all standard deviations within 5% of the values shown*

Plant Litter from:	Carbon	Nitrogen	C:N Ratio	Lignin	Cellulose	Hemicellulose
	In oven-dry plant material (%)			In oven-dry plant material (%)		
C3 PLANTS						
Scirpus olneyi-ambient CO_2	38.0	0.44%	86:1	20.5%	42.2%	27.3%
Scirpus olneyi-elevated CO_2	42.1	0.40	105:1	20.1	44.4	24.8
Triticum aestivum-ambient CO_2	38.0	0.97	42:1	12.0	52.5	35.5
Triticim aestivum-elevated CO_2	41.8	0.52	79:1	11.4	51.1	37.5
Lolium perenne-ambient CO_2	40.2	1.90	22:1	9.0	58.3	32.7
Lolium perenne-elevated CO_2	41.1	1.60	26:1	10.0	59.4	30.6
C4 PLANTS						
Spartina patens-ambient CO_2	41.1	0.51	81:1	14.2	48.3	26.4
Spartina patens-elevated CO_2	40.2	0.53	77:1	14.4	49.1	25.3
Sorghum-ambient CO_2	42.2	3.0	14:1	19.2	48.8	30.0
Sorghum-elevated CO_2	44.2	3.1	14:1	18.8	49.6	31.6

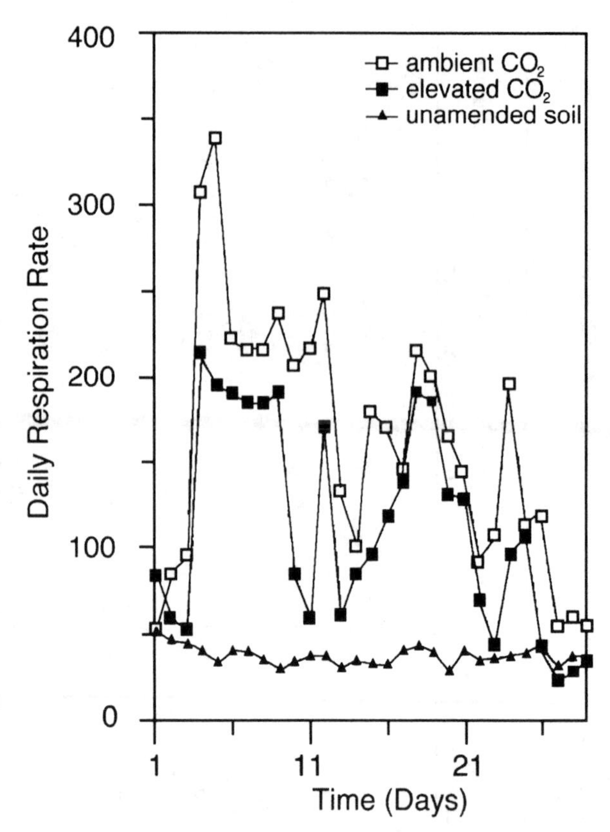

Figure 1 *Daily recorded respiration rates (g CO_2 - C g soil^{-1} day^{-1}) for unamended soil, and for soil containing straw from wheat grown at ambient and elevated CO_2 concentrations. Values represent the mean of three replicates, with standard errors within 10% of the mean values in all cases*

(Table 2). However, for the C3 plants, a significant increase was observed (unpaired t-test $p < 0.05$) in the C:N ratio of litter from plants grown under elevated CO_2. There was a decrease of around 10% in the N concentration of plant litter from C3 plants grown under elevated CO_2. Plants grown under elevated CO_2 commonly show changes in nitrogen content. The N content of plant litter is an important rate-regulating factor in the preliminary stages of decomposition (Berg, 1984). Lignin concentration has also been shown to be an important determinant in the decomposition of plant litter (Melillo *et al.*, 1982) and increased lignin concentrations have been reported in plants exposed to elevated CO_2 (Cipollini *et al.*, 1993), although our results show lignin concentrations remained unaffected by the atmospheric CO_2 concentration (Table 2).

In conclusion, C4 plants grown in ambient and elevated CO_2 did not differ significantly in composition or in C:N content, nor did the plants decompose at different rates in soil. In contrast, for all the C3 plant species examined, degradation rates were significantly less in litter obtained from plants exposed to elevated atmospheric CO_2, presumably because of a 10% decrease in the N content. Any lowering in decomposition rates brought about by increasing atmospheric CO_2 concentrations will increase the storage of C in soil. Long-term experiments are necessary to establish the effects of elevated CO_2 on C partitioning in terrestrial ecosystems.

References

Anderson, J.M. 1991. The effects of climate change on decomposition processes in grassland and coniferous forests. *Ecological Applications* **1**:326-347.

Anon. 1990. Operators Manual Version 1.8. PP Systems, Hitchin, UK.

Arp, W.J. and B.G. Drake. 1991. Increased photosynthetic capacity of *Scirpus olneyi* after 4 years exposure to elevated CO_2. *Plant, Cell and Environment* **14**:1003-1006.

Ball, A.S. 1991. Degradation by *Streptomyces viridosporus* T7A of plant material grown under elevated CO_2 conditions. *FEMS Microbiology Letters* **84**:139-142.

Berg, B. 1984. Decomposition of root litter and some factors regulating the process : long-term root litter decomposition in a Scots pine forest. *Soil Biology and Biochemistry* **16**:609-617.

Bouwman, A.F. 1990. Exchange of greenhouse gases between terrestrial ecosystems and the atmosphere. p 61-192. *In* A.F. Bouwman (ed.), *Soils and the Greenhouse Effect.* Wiley, Chichester.

Cipollini, M.L., B.G. Drake, and D. Whigham. 1993. Effects of elevated CO_2 on growth and carbon/nutrient balance in the deciduous woody shrub *Lindera benzoin* (L.) Blume (Lauracea). *Oceologia* **96**:339-346.

Couteaux, M.M., M. Mousseau, M.L. Celerier, and P. Bottner. 1991. Increased atmospheric CO_2 and litter quality : decomposition of sweet chestnut leaf litter with aminal feed webs of different complexities. *Oikos* **61**:54-64.

Curtis, P.S. and D.F. Whigham. 1989. Nitrogen and carbon dynamics in C_3 and C_4 estuarine marsh plants grown under elevated CO_2 *in situ*. *Oecologia* **78**:297-301.

Drake, B.G. 1992. A field study of the effects of elevated CO_2 on ecosystem processes in a Chesapeake Bay wetland. *Australian Journal of Botany* **40**:579-595.

Eamus, D. and P.G. Jarvis. 1989. The direct effcts of increase in the global atmospheric CO_2 concentration on natural and commercial trees and forests. *Advances in Ecological Research* **19**:1-55.

Harper, H.T. and J.M. Lynch. 1981. The chemical components and decomposition of wheat straw, leaves, internodes and nodes. *Journal of Science of Food and Agriculture* **32**:1057-1062.

Leadley, P.W. and B.G. Drake. 1992. Open top chambers for exposing plant canopies to elevated CO_2 concentration and for measuring net gas exchange. *Vegetatio* **104/105**:3-15.

Long, S.P. and B.G. Drake. 1992. Photosynthetic CO_2 assimilation and rising atmospheric CO_2 concentrations. p 69-107. *In* N.R. Baker and H. Thomas (eds), *Crop Photosynthesis : Spatial and Temporal Determinants*. Elsevier, Amsterdam.

Melillo, J.M., J.D. Aber, and J.F. Muratore. 1982. Nitrogen and lignin control of hardwood leaf litter decomposition dynamics. *Ecology* **63**:621-626.

Norby, R.J., J. Pastor, and J.M. Mellilo. 1986. Carbon-nitrogen interactions in CO_2-enriched white oak : physiological and long term perspectives. *Tree Physiology* **2**:233-241.

Sharabi, N.E. and R. Bartha. 1993. The use of soil respiration as a measurement of decomposition. *Applied and Environmental Microbiology* **9**:1201-1205.

Taylor, J. and A.S. Ball. 1994. The effect of plant material grown under elevated CO_2 on soil respiratory activity. *Plant and Soil* **162**:315-318.

Van de Geijn, S.C. and J.A. van Veen. 1993. Implications of increased carbon dioxide levels for carbon input and turnover in soils. *Vegetatio* **104/105**:283-292.

Van Veen, J.A., E. Liljeroth, L.J.A. Lekkerkerk, and S.C. van de Geijn. 1991. Carbon fluxes in plant soil systems at elevated atmospheric CO_2 levels. *Ecological Applications* **1**:175-181.

Ziska, L.H., B.G. Drake, and S. Chamberlain. 1990. Long-term photosynthetic response in single leaves of a C3 and C4 salt marsh species grown at elevated atmospheric CO_2 *in situ*. *Oecologia* **83**:469-472.

Humic Substances Research in Polar Waters

R. J. Lara, U. Hubberten, D. N. Thomas[1] and G. Kattner.

ALFRED-WEGENER-INSTITUT für POLAR- und MEERESFORSCHUNG, Postfach 120161, D-27515 BREMERHAVEN, GERMANY
[1]Present address: INSTITUT für CHEMIE und BIOLOGIE des MEERES, UNIVERSITÄT OLDENBURG, D-26111 OLDENBURG, GERMANY

Abstract

A summary is presented of field and laboratory research on marine humic substances (HS) from polar regions. The contribution and the role of organic nitrogenous compounds in humic fractions of dissolved organic matter (DOM) have been investigated in seawater samples from Arctic and Antarctic regions. On average the HS contributed 40% to the pool of dissolved organic nitrogen (DON). Total dissolved amino acids (TDAA) accounted for about 35% and 7% of the nitrogen in the hydrophobic neutral and hydrophobic acid fractions of HS, respectively. Differences in amino acid distribution and composition patterns in dissolved and suspended particulate material suggest the selective preservation of certain amino acids, in particular glycine. In surface waters ca 60%, and at depths > 500 m almost 100% of TDAA are found in the "humic" fractions. A background value of TDAA of around 200 nM, mostly contained in the hydrophobic neutral fraction, is present throughout the water column, probably forming part of the refractory molecules. Culture experiments with Antarctic diatoms showed that relatively fresh dissolved compounds of algal origins were actively taken up by bacteria, regardless of their hydrophobic ("humic") or hydrophilic ("non-humic") nature as determined by sorption onto XAD-2 resin. Part of the particulate amino acid pool was transformed in hydrophilic DON, which was not adequate for supporting sustained bacterial growth. This suggests that humification may begin in the hydrophilic fraction of DOM. Further studies with Antarctic algae were carried out to test the efficiency of combinations of the XAD-2, -4 and -7 resins for extracting ^{14}C-labelled algal derived DOM. The serial combinations XAD-7/-2/-4 and XAD-7/-4/-2 allowed the isolation of a very high percentage of DOM (65%) from the seawater matrix, whereas a mixed bed of the same resins showed a significantly lower recovery (39%).

1 Introduction

Humic substance (HS) research has traditionally focused on the characterization and dynamics of these compounds in soil organic matter. Although this has great importance for understanding the global carbon and nitrogen cycles in the biosphere, similar investigations in the marine ecosystem are comparatively scarce. However, the total mass of organic carbon in dissolved form (DOC) in the oceans is comparable with that in atmospheric carbon dioxide (Benner *et al.*, 1993).

In open ocean regions, the primary source of organic material is phytoplankton, which through the trophic network gives rise to the pools of particulate (POM) and dissolved organic matter (DOM). Marine humus is produced by the transformation of constituents of these pools into slowly degradable, so-called refractory substances. This process is of particular relevance, since it can remove organic matter from the active cycles in the ocean, thus acting as a functional sink of atmospheric carbon dioxide.

The amount of carbon in dissolved form in the world's oceans is about ten times higher than in particulate form (Cauwet, 1981). Therefore, models about the evolution and fate of atmospheric carbon dioxide should consider the role of oceanic DOM. Although theoretical considerations have been given to this aspect (Toggweiler, 1989), field work has been focused mainly on the downward vertical transport of POM in the water column.

For various reasons the Arctic and Antarctic are relevant in discussions of these processes. For example, downward transport of DOM in areas of deep-water formation may be an important process in both regions, as well as the transformation of organic material by increased UV- radiation. Furthermore, rivers flowing into the Arctic basins are sources of huge amounts of organic matter (OM). This may represent an additional, enlarged input of land-derived material to the Arctic Ocean should global warming cause an increase in permafrost melting, These factors could influence the budget and quality of OM in the ocean, and in particular the proportions of labile and refractory substances.

Clearly, in order to assess the relevance to the global carbon and nitrogen budgets of humification in the ocean, more information is needed about the mechanisms and dynamics of the formation of marine HS. It has been proposed (Harvey *et al.*, 1983) that marine HS originate by transformation of triglycerides with a high percentage of polyunsaturated fatty acids. That model considers incorporated nitrogen functions to be incidental to the basic structure. Another theory (Ishiwatari, 1992) explains the formation of HS by the synthesis of biopolymers on the basis of modifications of proteins and carbohydrates during the bacterial degradation of aggregates and senescent phytoplankton cells. This pathway includes the direct participation of amino acids in the genesis of HS. Amino acids are present in all organisms as components of proteins and are found ubiquitously in seawater. Although amino acids are usually included in the fraction of labile compounds of DOM, there is evidence (Hubberten *et al.*, 1994 and references therein) for an association between amino acids and compounds commonly considered refractory, such as HS.

Several investigations have been carried out in our laboratories in order to gain a deeper insight into the mechanisms of DOM and POM cycling in the polar oceans, and to obtain an appraisal of the bioavailability of the different OM fractions. Field studies on organic nitrogen compounds in dissolved HS were carried out in Arctic and Antarctic waters. Laboratory studies aimed to gain a deeper insight into the dynamics of dissolved amino acids and their role in the humification of algal derived DOM, as well as a better characterization of the nature of its XAD-fractions. A long-term experiment, using a batch unialgal diatom culture, was carried out in order to study the changes of these components in relation to algal and bacterial biomass. Related laboratory investigations were directed to the comparison of different types of XAD resins for studies on dissolved marine HS. Since oceanic DOM is mainly phytoplankton derived, it seemed appropriate to improve our understanding about the sorptive behaviour of algal produced organic OM towards these resins. There follows a summary of our research on these topics (Hubberten *et al.*, 1994, 1995; Lara and Thomas 1994a, b; Lara *et al.*, 1993, 1994).

2 Experimental

Field studies were carried out using samples collected in the Arctic and the Antarctic during the R. V. *Polarstern* expeditions ARK VIII/1 and ANT X/1b. During both cruises, areas with water depths covering a range of about 150-3000 m were sampled. Five samples were taken at each station, and these were distributed among the euphotic zone, at intermediate depths, and near the bottom. Filtered seawater and DOM fractionated into "humic" and "nonhumic" fractions were analysed for dissolved organic nitrogen (DON) and amino acids. Seawater (1000 mL) was filtered (Whatman GF/C, precombusted for 4 h at 450 °C), and the filtrate (500 mL) was adjusted to pH 2 with HCl, and then passed through XAD-2 resin. The adsorbed "humic" material was eluted with 0.2M NaOH (hydrophobic acid fraction, HbA), and thereafter with methanol (hydrophobic neutral fraction, HbN) (Kukkonen *et al.*, 1990). The untreated samples and the DOM fractions were analyzed for DON, DIN (nitrate + nitrite + ammonium) (Lara *et al.*, 1993), and total dissolved amino acids (TDAA) (Hubberten *et al.*, 1994). The filters with the particulate material were hydrolysed and analyzed for amino acids (Hubberten *et al.*, 1995).

Amino acid and DON dynamics were studied during a three month laboratory experiment using a batch culture of the Antarctic marine diatom *Thalassiosira tumida*. The diatoms were cultured at 0 °C for 83 days. The algae grew exponentially during the first 12 days and thereafter entered the stationary phase. On day 19 the culture was placed in the dark to induce the degradation phase. DOM was fractionated using XAD-2 resin, as described. Amino acid contents of unfractionated filtered seawater, of the XAD-fractions, and of particulate material were determined during the growth, stationary, and degradation phases of the culture. Results were related to changes in the DON in DOM fractions, to inorganic nutrients, and to algal and bacterial biomass (Lara *et al.*, 1995). To facilitate comparison with the data for HS in polar waters, the experiment was carried out in conditions of temperature and light that resembled those in the regions studied.

Comparisons of the performance of different types of XAD resins for the extraction of algal-produced DOM were carried out using ^{14}C-radiolabelled DOM produced by Antarctic diatoms grown in Antarctic seawater. The XAD types -2, -4, and -7 (particle size: 0.3 - 1.0 mm) resins were tested separately and in various combinations. Four combinations of columns connected in series were used: XAD-2/-4/-7, XAD-4/-2/-7, XAD-7/-2/-4, and XAD-7/-4/-2. These were compared with a mixed bed of all three resins (XAD-2+-4+-7). Seawater containing DO^{14}C was filtered through cellulose nitrate filters (0.2 μm, Sartorius), acidified to pH 2 with HCl, and passed through the different columns. Thereafter, the resin was washed with 50 mL of 0.01M HCl, and the retained DOM was fractionated by eluting with different solvents. The radioactivity of each fraction was measured by liquid scintillation counting, and the results were expressed as percentages of the total activity applied to the column (Lara and Thomas, 1994a, b).

3 Results and Discussion

Average DON concentrations in seawater (3 - 4 μM N). were similar in the Arctic and for

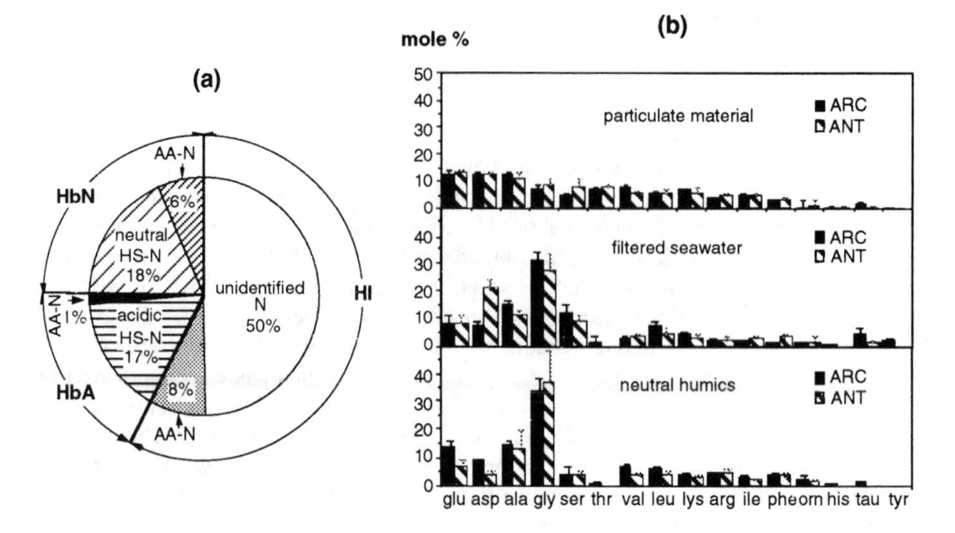

Figure 1(a) *Average contribution of humic substances (HS) to dissolved organic nitrogen in Arctic seawater. Dissolved organic matter fractions: HbA = hydrophobic acid; HbN = hydrophobic neutral; Hl = hydrophilic; AA-N = amino acid nitrogen: HS-N = non amino acid nitrogen in a humic fraction.*

(b) *Percentage amino acid composition of particulate material, filtered seawater and a humic fraction (HbN) in the Arctic (ARC) and the Antarctic (ANT).*

the Antarctic. Also the contribution of the organic nitrogen in the HbA to the DON was, within ca 20%, very similar in both regions. The HbN fraction was about 15% of the DON in the Antarctic and about 25% of that in the Arctic. The "non-humic" fraction contributed ca 60% to the DON. Therefore, it can be estimated that about 40% of the DON has a refractory character. This, and the contribution of TDAA to the different DON fractions, is illustrated in Figire 1a for samples from the Greenland Sea. Similar results were found for the Antarctic. In surface water about 60% of the amino acid content was bound to HS, while at depths > 500 m this percentage was almost 100%.

In filtered seawater TDAA represented about 15% of DON. In general the amino acid composition of both hydrophobic fractions was very similar, with TDAA concentrations in HbN about 4-5 times higher than in HbA. Significant differences were found in the amino acid composition of filtered seawater, humic fractions, and particulate organic material, as shown in Figure 1b. Glycine, alanine, glutamic acid, aspartic acid, and serine made up ca 70-80% of the TDAA in the seawater and in the humic fractions. In particulate organic material (POM), however, the same amino acids represented only ca 50% of the total. Glycine, especially, was distributed in a very different way. The mole percentage of glycine increased from particulate material (8%) to bulk filtered seawater (27%), and reached 33 - 45% in the humic fractions. The mole percentage of glycine in the POM at the surface was 7%; it ncreased progressively to about 11% at depths > 500 m.

The individual amino acids in the humic fractions were homogeneously distributed throughout the water column, and without any obvious relationship to plankton biomass parameters, such as particulate amino acids and chorophyll *a*. This suggests that these amino acids formed parts of recalcitrant molecules. Our data indicate that a significant amount of the amino acids (ca 200 nM) may be present as refractory compounds in the water column of the world's oceans. The high amount of glycine in the HS fractions, and the increase with depth of the relative amount of glycine in the suspended POM indicate that glycine is selectively preserved from bacterial attack on POM and on DOM.

The laboratory investigations on the characterization of exudates and degradation products in an Antarctic diatom culture attempted to improve the differentiation of the concepts of "refractory" DOM on a chemical and biological basis and of "humic substances" as an operational definition.

The results from the growth and from the stationary phase showed that relatively "fresh" substances of algal origin were actively taken up by bacteria, regardless of their hydrophobic ("humic") or hydrophilic ("non-humic") nature, as determined by sorption onto XAD-2. Bacteria were able to take up organic nitrogen as free and as bound amino acids in the growth phase, and also as hydrophobic and hydrophilic non-amino acid nitrogen in the growth and in the stationary phases.

The data from the degradation phase indicated that in short time scales, part of the particulate amino acid pool can be transformed in hydrophilic DON. This, in contrast to the stationary phase, did not seem to be adequate for supporting sustained bacterial growth. If substances in the hydrophobic fractions only are considered to be "humic", then the slight increase of the hydrophobic acid fraction after two months of degradation would indicate only first symptoms of humification. That might suggest that a significant number of cells were still alive. However, if in a wider sense humification is considered to be the formation of slowly degradable matter, then the behaviour of hydrophilic DOM suggests that this process may begin in this fraction.

Comparisons of the sorptive behaviour of the XAD-2, -4, and -7 resins for DOC derived from marine phytoplankton gave the following results. The sequence of the different serial combinations proved to be critical for determining the efficiency of DOC isolation. The sequences XAD-7/-2/-4 and XAD-7/-4/-2 allowed the isolation of a very high percentage of DOC (65%) from the seawater matrix, whereas a mixed bed of the same resins gave a significantly lower recovery (39%). For the combinations XAD-4/-2/-7 and XAD-2/-4/-7 the recoveries were 49% and 36%, respectively.

Varying amounts of the material adsorbed by the individual resins could not be recovered by NaOH/methanol elution. The percentage of DOC firmly retained by each resin was 32%, 21%, and 2% in XAD-2, XAD-4, and XAD-7, respectively. In the case of XAD-2, after grinding and (hot) hydrolysis with H_2SO_4/MeOH and KOH/MeOH, it was possible to recover most (ca 90%) of the radioactive material tightly held by the resin. This suggests a lipid character for this fraction. However, it is also possible that part of this fraction might be trapped in the pores of the macroreticular resin, since grinding and drastic hydrolysis probably disrupted resin structure, liberating adsorbed/absorbed DOC. Previous work with XAD-2 gave excellent recoveries of organic nitrogenous compounds. Therefore, we can assume that this fraction is not proteinaceous in nature.

4 Conclusions

Our field investigations showed that a very high percentage of the DON (ca 60%) in seawater from polar regions is present in operationally defined humic fractions of DOM.

Should the reaction of amino acids with other molecules be mainly responsible for the incorporation of nitrogen moieties to marine HS, then the resulting product must have been intensively transformed, since only a relatively low percentage of its nitrogen is accounted for by amino acids. The nature of most of the nitrogenous compounds in the hydrophobic "humic" fractions of seawater remains unidentified. The selective preservation of glycine in the HS fractions, and in particulate material makes this amino acid an interesting model substance for improving the understanding of the formation of refractory amino acid-containing macromolecules in the ocean.

The combination of XAD resins in the sequences XAD-7/-2/ and -4 and -7/-4/ and -2 showed an interesting potential for the extraction of algal derived DOM. Our work is directed to the evaluation of the suitability of these combinations for field extraction of mixed marine DOM pools. The use of XAD-2 in diatom culture experiments indicated that, far from collecting only an "amorphous" fraction of post-mortal components, this resin is also able to adsorb substances from pools that are undergoing rapid changes, thereby reflecting the changing diatom and bacterial population dynamics. The fact that organic substances in algal derived DOM were adsorbed by this resin does not necessarily mean that they are "humic", and hence refractory, since these were readily taken up by bacteria. Thus, the ecological function of HS may not necessarily coincide with the properties of the "humic" material isolated using a specific method. Thus, in field studies on aquatic HS, data from surface waters should be compared with those from deep layers in order to account for the possible inclusion of "fresh" DOM in fractions assumed to be "humic","old" or "refractory". On the basis of the field data obtained in polar waters, the homogeneous depth distribution of humic nitrogen as well as of the amino acids incorporated in HS, suggests a refractory character of this material.

An intensification of humic substance research in the marine environment is urgently needed. For that purpose, the cooperation with other related disciplines is necessary. Ongoing studies within our group, in colaboration with marine biologists, deal with the processes leading to the humification of DOM arising from decomposition of phytoplankton and "marine snow". Our work in the Arctic is being extended to the determination of terrigenic components tracing the input of DOM from the Siberian rivers in the Greenland Sea via Polar Drift. At this point there is a growing interface connecting our work in oceanic areas with the humic research in soil material.

Contribution Nr. 831 of the Alfred Wegener Institute for Polar and Marine Research.

References

Benner, R., B. von Bodungen, J. Farrington, J. Hedges, C. Lee, R.F.C. Mantoura, Y. Suzuki, and P.M. Williams. 1993. Measurement of dissolved organic carbon and nitrogen in natural waters: Workshop report. *Mar. Chem.* **41**:5-10.

Cauwet, C. Non-living particulate matter. 1981. p.71-86. *In* E.K. Duursma and R. Dawson, (eds), *Marine Organic Chemistry*. Elsevier, Amsterdam.

Harvey, R.H., D.A. Boran, L.A. Chesal, and J.M. Tokar. 1983. The structure of marine fulvic and humic acids. *Mar. Chem.* **12**:119-132.

Hubberten, U., R.J. Lara, G. Kattner. 1994. Amino acid composition of seawater and dissolved humic substances in the Greenland Sea. *Mar. Chem.* **45**:121-128.

Hubberten, U., R.J. Lara, and G. Kattner. 1995. Refractory organic compounds in polar waters: Relationship between humic substances and amino acids in the Arctic and Antarctic. *J. Mar. Res.* **53**:1-13.

Ishiwatari, R. 1992. Macromolecular material (humic substance) in the water column and sediments. *Mar. Chem.* **39**:151-166.

Kukkonen, J., J.F. McCarthy, and A. Oikari. 1990. Effects of XAD-8 fractions of dissolved organic carbon on the sorption and bioavailability of organic micropollutants. *Arch. Environ. Contam. Toxicol.* **19**:551-557.

Lara, R.J., U. Hubberten, and G. Kattner. 1993. Contribution of humic substances to the dissolved nitrogen pool in the Greenland Sea. *Mar. Chem.* **41**:327-336.

Lara, R.J., and D.N. Thomas. 1994a. XAD-fractionation of "new" dissolved organic matter: is the hydrophobic fraction seriously underestimated? *Mar. Chem.* **47**:93-96.

Lara R.J., and D.N. Thomas 1994b. Isolation of marine dissolved organic matter: Evaluation of sequential combinations of XAD resins 2, 4 and 7. *Anal.Chem.* **66**:2417-2419.

Lara, R.J., U. Hubberten, D.N. Thomas, M.E.M. Baumann, and G. Kattner. 1994. Amino acids and dissolved organic nitrogen in an unialgal Antarctic diatom culture: Turnover and humification. Limnol. Oceanogr. (submitted).

Toggweiler, J.R. 1989. Is the downward dissolved organic matter (DOM) flux important in carbon transport? p. 65-83. *In* W.H. Berger, V.S. Smetacek, and G. Wefer . (eds.), *Productivity of the ocean: Present and past.* S. Bernhard, Dahlem Konferenzen. Wiley, New York.

The Effects of Humic Substances on Enhanced Chemiluminescence Produced by the HRP-Catalyzed Oxidation of Luminol

B.E. Watt, E.E. Robinson[1], K.E. Sawcer[1], G.H.G. Thorpe[1], and M.H.B. Hayes

THE UNIVERSITY OF BIRMINGHAM, SCHOOL OF CHEMISTRY, EDGBASTON, BIRMINGHAM, B15 2TT, ENGLAND
[1]WOLFSON APPLIED TECHNOLOGY LABORATORY, QUEEN ELIZABETH MEDICAL CENTRE, EDGBASTON, BIRMINGHAM, B15 2TH, ENGLAND

Abstract

An assay for the detection of the inhibition of enhanced chemiluminescence (ECL) was applied to humic substances (HS) from soils and waters, and to a sample of leonardite. The results showed that the different fractions of the HS had varying capabilities to inhibit ECL. When the data were related to the ^{13}C NMR spectra of the samples, it appeared that, for the most part, the fractions which had compositions/functionalities with possible origins in lignin-/tannin-type structures had the greatest ECL inhibitory effects. The study suggests that the amounts and the origins of the HS in waters should be taken into account when using the inhibition of ECL for the detection of pollutants in waters.

1 Introduction

Considerable research is focused on analytical techniques to assess levels of contamination which decrease the availability of oxygen in natural waters. Radical scavengers, such as those in sewage, silage liquors, and farmyard and dairy washings can decrease the levels of oxygen in watercourses and thereby suppress the biota in the waters. Measurements of biological oxygen demand (BOD) and of chemical oxygen demand (COD) are used to monitor the presence of oxygen scavenging materials in waters. BOD determinations require five days to complete and there may be a considerable delay between the occurrence of a pollution incident and the ability to trace its source. The costs of BOD and COD determinations are relatively expensive, and new legislation in Europe requires that the levels of water quality monitoring are increased.

There is a need for a screening approach for water samples whereby a simple test is employed at the point of sampling to determine whether or not it would be appropriate to take a sample for testing in the laboratory. Whitehead *et al.* (1993) have developed a simple and rapid assay for this purpose, based on the enhanced chemiluminescence (ECL) reaction of luminol (5-amino-2,3-dihydro-1,4-phthalazinedione) using an enhancer such as 4-iodophenol. ECL is used increasingly as a detection system in various fields of analytical biochemistry. Assays based on this system have the advantage of being more sensitive and rapid than conventional techniques (Egorov *et al.*, 1993).

Luminol is conventionally oxidized by hydrogen peroxide, in a complex reaction that is catalysed by horseradish peroxidase (HRP), to give low level chemiluminescent light emission which decays rapidly. More intense (2500 fold), prolonged, and stable light emission can be achieved by the addition of the enhancer 4-iodophenol (Thorpe *et al.*, 1985). Continuous light output is dependant on the constant production of free radical intermediates derived from 4-iodophenol, luminol, and oxygen. The presence of radical scavenging (chain breaking) antioxidants results in the suppression or delay of light emission, in a manner directly related to the amount of antioxidant present. Any substance capable of directly or indirectly inhibiting the HRP-catalyzed reaction will also cause a decrease, or a complete inhibition of the light output. Such substances include phenols, amines, heavy metals, and cyanide.

The assay developed by Whitehead *et al.* (1993) has been used to assess the quality of water in rivers, particularly with respect to the levels of sewage, heavy metals, and phenols. The results have been shown to compare favourably with those obtained from more conventional BOD and COD tests (Whitehead *et al.*, 1993). This communication describes some influences of humic substances (HS), and some model compounds on the ECL based assay.

2 Experimental

Sources of Samples

Eight humic samples were obtained from the International Humic Substances Society (IHSS). These included two from the Suwannee River, [the aquatic humic acid (HA) and fulvic acid (FA) standards], five soil-derived samples [the peat HA and FA references (from the Everglades, Florida), the soil Mollisol (from Illinois) HA and FA standards from the initial (Soil I) extraction (1984), and the new soil FA standard (Soil II) taken from the same soil in 1995], and one coal-derived sample (the Leonardite HA standard). The Yorkshire HAs, FAs, and XAD-4 acids were isolated from waters taken from Upper Bardon Reservoir in the Bardon Moors, north of Skipton, Yorkshire, England, and the watershed soils were mainly upland peats. The Yorkshire HS samples were isolated by adsorption on to XAD-8 [poly(methylmethacrylate)] and XAD-4 (styrene divinylbenzene) resins, using the procedure described by Malcolm and MacCarthy (1992). Lignin (alkali) and lignin (organosolv) were obtained from Aldrich Chemical Co., Gillingham, UK. Carbon contents are expressed on a dry, ash free (d.a.f.) basis.

Enhanced Chemiluminescent Assay

Instrumentation. A BioOrbit 1250 luminometer (BioOrbit, Uckfield, UK) was used to measure light emission. Kinetics of light emission were recorded using a chart recorder and a PC. A 0.3 neutral density wedge (Kodak, UK) was placed between the sample and detector, to reduce light to within measurable limits.

Reagents. ECLOX (enhanced chemiluminescence reagents) were obtained from Aztec Environmental Control Ltd., Didcot, Oxon, UK.

Procedure. HS were dissolved in 0.1M NaOH (10 mL) and the pH of each solution was adjusted to approximately 7 using 0.5M HNO_3. The final concentration of each of these stock solutions was ~250 mg L^{-1} carbon. A control sample was prepared by addition of sufficient 0.5M HNO_3 to 0.1 M NaOH (10 mL) to give a solution pH of 7. For model compounds, any solvent used was incorporated in the control.

The procedure used for the ECL assay was based on that developed by Billings *et al.* (1994), and Whitehead *et al.* (1993). Samples were tested over a range of concentrations by serial dilutions of the humic stock solutions. The following were added in sequence to a cuvette: 1000 μL sample (or control), 100 μL reagent 1 (luminol plus enhancer), 100 μL reagent 2 (perborate), and 50 μL reagent 3 (HRP conjugate). The time between the addition of reagents 1 and 2 did not exceed one minute. The reaction was timed from the point of addition of the HRP. The cuvette was shaken thoroughly and introduced into the luminometer.

Per cent ECL inhibition was measured between 0 and 4 min after initialisation, using the following relationship:

$$\% \text{ ECL inhibition (0-4 min)} = (A \times 100)/(A + B)$$

where:

A = area between the curve for the deionised water blank and the curve for the sample (0-4 min); and

B = area under curve for the sample (0-4 min).

Limits of detection (LODs) were determined as the concentration of sample causing a 10% inhibition of ECL. IC_{50} values were determined as the concentration causing a 50% inhibition of ECL.

3 Results and Discussion

HS were found to cause a concentration dependent inhibition of ECL light emission in this assay (Figures 1 to 3). Figure 1 shows the per cent inhibition of ECL caused by varying concentrations of the Yorkshire HAs, FAs, and XAD-4 acids. HAs gave the greatest inhibition (essentially complete suppression of ECL was obtained at the 20 mg C L^{-1} level), and the XAD-4 acids gave the least. Based on this study, at 2 mg C L^{-1}, which is representative of the concentrations of total HS found in many natural waters, the per cent inhibition attributable to the Yorkshire HS (using 1 mL sample) ranged from 18% and 30%.

The ratio of FAs:HAs in such waters is generally of the order of 9:1 (Malcolm, 1991). Thus, in the application of this assay to the Yorkshire waters, the majority of the inhibition caused by HS would be attributable to FAs. The per cent ECL inhibition caused by 2 mg C L^{-1} Yorkshire FAs was approximately 18% (Figure 1). At a carbon concentration of 4 mg L^{-1}, there was approximately 30% inhibition of ECL. It is likely, therefore, that effects due to HS may occur when this assay is applied to waters which have high levels of dissolved organic carbon (DOC).

The inhibition of ECL caused by the IHSS Standard (or Reference) HAs and FAs are shown in Figures 2 and 3, respectively. In general, the per cent inhibition for the Standard/Reference HAs (Figure 2) followed the order: Suwannee River = Leonardite >

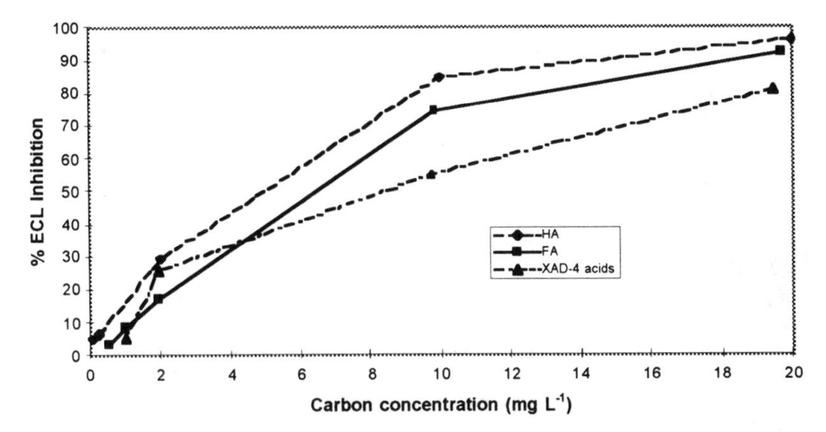

Figure 1 *Plot of per cent ECL inhibition versus carbon (d.a.f) concentration for solutions of HA, FA and XAD-4 acids isolated from a Yorkshire reservoir*

Peat >> Soil, whereas that for the FAs followed the order Suwannee River > Soil II > Soil I >> Peat.

The data in Table 1 indicate the concentrations of the Standard or Reference materials that give rise to 10% inhibition (LOD), and to 50% inhibition (IC_{50}) of the ECL. The combination of the data for the LOD and for the IC_{50} would suggest that the abilities of the Standard and Reference materials to inhibit ECL decreased in the order Suwannee River HA Standard > Leonardite HA Standard > Peat HA Reference = Suwannee River FA Standard >> Soil HA Standard = Soil I FA Standard = Soil II FA Standard >> Peat FA Reference.

Table 2 lists the model compounds tested in the ECL assay, and their calculated LOD values. On a mg L^{-1} basis, commercially obtained lignin preparations caused the greatest ECL inhibition. Tannic acid and catechin also had significant inhibiting effects. The simple chemicals tested in the study were products that could be related to oxidative degradation products of lignins and of humic substances. Those with the greatest inhibitory effects had two hydroxyl groups *ortho*, *meta*, or *para* to each other, or carboxyl groups, or another deactivating substituent *ortho* or *para* to the phenolic hydroxyl, and with an activating substituent, such as -OH or $-OCH_3$ *meta* to the deactivating group. 1,4-Benzoquinone, the non-aromatic oxidation product of quinol was also a strong inhibitor.

The CPMAS ^{13}C NMR spectra for the samples taken from the Yorkshire Reservoir (Watt *et al.*, 1996) provide clear evidence for phenolic functionality in the HAs (chemical shift 140 - 150 ppm). Such evidence was less clear in the spectra for the FAs. However, there was abundant evidence for resonances in the 110 - 140 ppm chemical shift range for both the HA and FA samples, and such resonances are characteristic of lignin- and tannin-type structures (Wilson, 1987). The NMR spectrum for the XAD-4 acids indicated that aromaticity was low, and resonances characteristic of lignin- and tannin-type structures were absent. However, inhibition at the 2 mg L^{-1} concentration was similar to that for the HAs, but the inhibitory effect was less at the higher concentrations (Figure 1). The compositions of the XAD-4 acids were different from those of the HAs and FAs, and it would appear to be likely that the inhibiting species that operated were different. We have not carried out the experiments that would indicate whether or not the inhibition effect is additive when the fractions are mixed.

The data in Figures 2 and 3 indicate that at the 10 mg L^{-1} concentration, the Suwannee

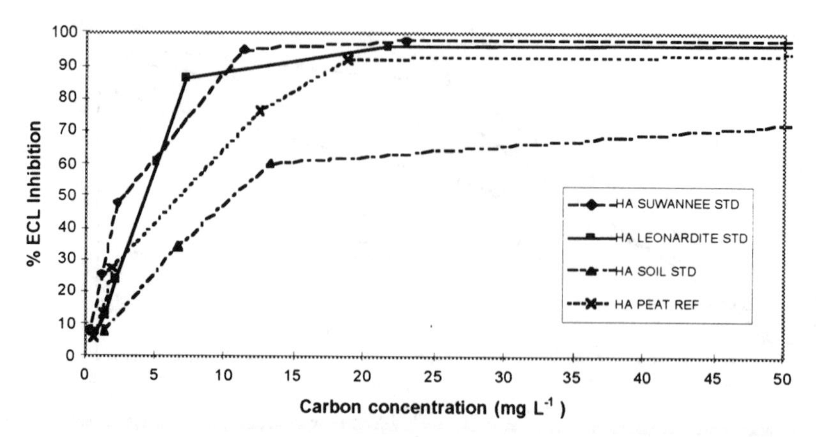

Figure 2 *Plot of per cent ECL inhibition versus carbon (d.a.f.) concentration for solutions of IHSS Standard humic acids*

River HA is a slightly stronger inhibitor of ECL than the Yorkshire HA, but inhibitory effects of the Suwannee and Yorkshire FAs are similar at that concentration. The NMR spectrum of the Suwannee River HA (available from IHSS, and awaiting publication) provides evidence for functionalities associated with the compositions of lignin- and tannin-type structures, and it has a strong resonance (140 - 150 ppm) characteristic of -OR substitution in the aromatic nucleus. Likewise, the NMR spectrum for the Suwannee FA is similar to that for the Yorkshire FA. The NMR spectrum for the Everglades Peat Reference HA showed evidence for -OAr substitution, and only weak evidence for lignin-/tannin-type constituents, and there was no convincing evidence for the presence of such constituents in the FA samples. ECL inhibition responses were in line with expectations from the NMR spectra, although in a comparative sense, the inhibition by the peat FA was

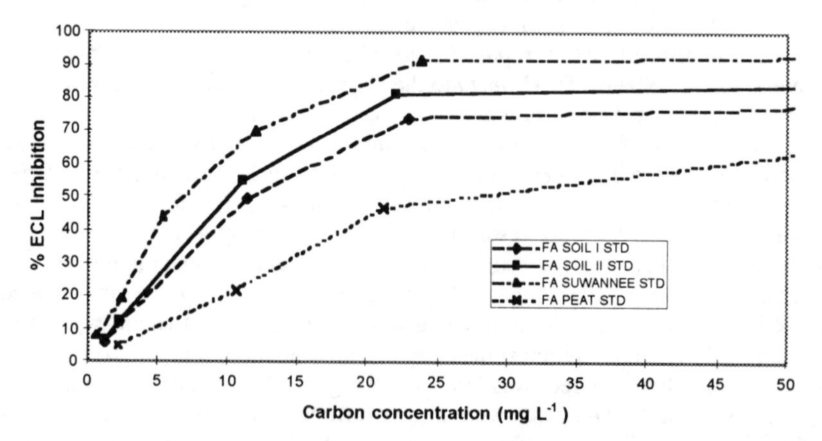

Figure 3 *Plot of per cent ECL inhibition versus carbon (d.a.f.) concentration for solutions of IHSS Standard fulvic acids*

Table 1 *Limits of detection (LOD) for IHSS Standard and Reference HS in the enhanced chemiluminescence assay*

Sample	LOD (mg L^{-1})	IC_{50} (mg L^{-1})
Suwannee River HA Standard	0.4	7
Suwannee River FA Standard	0.9	10
Leonardite HA Standard	0.9	6
Mollisol Soil HA Standard	2.0	20
Mollisol Soil I FA Standard	2.0	20
Mollisol Soil II FA Standard	2.0	20
Everglades Peat HA Reference	0.8	10
Everglades Peat FA Reference	4.0	40

surprisingly low.

Comparison of the ECL inhibition by the Mollisol soil HA and FA Standards indicates that the inhibition by these fractions was similar in the 10 - 15 mg L^{-1} concentration range, but the FA fractions were more inhibitory at the higher concentrations. There was no evidence in the NMR spectra for -OAr substitution or for lignin-/tannin -type structures in

Table 2 *Limits of detection for model compounds in the enhanced chemiluminescence assay*

Compound	Limit of Detection mol L^{-1} ($\times 10^{-7}$)	Limit of Detection mg L^{-1}
tannic acid	0.05	0.007
lignin (alkali)*	n/a	0.0001
1,2-dihydroxybenzene (catechol)	0.1	0.0009
lignin (organosolv.)*	n/a	0.0003
2,5-dihydroxybenzoic acid (gentisic acid)	0.1	0.0012
4-hydroxy-3-methoxybenzoic acid (vanillic acid)	0.1	0.0013
3,5-dimethoxy-4-hydrocinnamic acid (sinapinic acid)	0.1	0.002
1,3-dihydroxybenzene (resorcinol)	0.25	0.003
1,4-dihydroxybenzene (quinol)	0.5	0.004
catechin	0.5	0.012
1,4-benzoquinone	1	0.009
phenol	5	0.04
3-hydroxybenzoic acid	10	0.11
4-hydroxybenzoic acid	100	1.1
1,3,5-benzenetricarboxylic acid	100	1.7
3,4-dimethoxybenzaldehyde	250	3.3
4-hydroxybenzophenone	250	4.0
2-hydroxybenzoic acid (salicylic acid)	1000	11.0

*Approximate value (compound partially insoluble in ethanol); n/a = not applicable

the case of the HA, but there was some such evidence in the cases of the FA fractions.

The ECL inhibition pattern for the Leonardite HA Standard does not fit into the patterns observed for the other HAs studied. The aromaticity resonance in the ^{13}C NMR spectrum is symmetrical, with no evidence for lignin-/tannin-type functionalities, and only a very small shoulder would indicate some -OAr functionality. The carboxyl functionality is small. Leonardite was, however, a strong inhibitor of ECL on the scale of HAs studied.

4 Conclusions

The method used for the isolation and fractionation of the HS would free the fractions from contamination by small organic chemicals that were not components of the macromolecular structures. Thus, the ECL inhibitions observed arose from the macromolecular HS.

The model studies used lignins and tannic acid, and these were relatively strong inhibitors of ECL. Other model compounds were hydroxy- and methoxybenzenes with various other substituents, and some of these compounds could be derived from the oxidation products of lignins and tannins.

The evidence which we have clearly indicates that HS are inhibitors of ECL, and that the different fractions can inhibit to different degrees. There would appear to be a link between the abilities to inhibit ECL and the extents to which the compositions/functionalities of the HS are related to altered lignin-/tannin- type structures.

It is clear that the assay used in this study is capable of detecting ECL inhibitors at low concentrations in water. HS are ubiquitous in waters, and the properties of these will depend on the nature of the soils and the vegetation of the watershed, and on the extents to which anthropogenic inputs from, for example, sewage treatment works, contribute to the genesis of the aquatic HS. The aquatic HS used in the present study were from pristine sources, and did not have inputs to their genesis from anthropogenic sources. The presence of HS in waters should be taken into account when assaying for polluting substances using techniques based on the inhibition of ECL.

Acknowledgements

The contribution of B. E. W. to this study was supported by a grant from NERC for studies of the chemistry of humic substances in soils and waters

References

Billings, C., M. Lane, A. Watson, T. P. Whitehead, and G. Thorpe. 1994. A rapid and simple chemiluminescent assay for water quality monitoring. *Analusis* **22**:27-30.

Egorov, A.M., B.B. Kim, V.V. Pisarev, Y.L. Kapeliuch, and I.G. Gazarian. 1993. Fundamental and applied aspects. Reaction of enhanced chemiluminescence. p. 286-290. *In* A. Z. Szalay, L. J. Kricka and P. E. Stanley (eds), *Bioluminescence and. Chemiluminescence-Status report*. Wiley, New York.

Malcolm, R.L. 1991. Factors to be considered in the isolation and characterization of aquatic humic substances. p. 369-391. *In* H. Boren and B. Allard (eds), *Humic Substances in the Aquatic and Terrestrial Environment*. Wiley, Chichester.

Malcolm, R. L., and P. MacCarthy. 1992. Quantitative evaluation of XAD-8 and XAD-4 resins used in tandem for removing organic solutes from water. *Environ. Internat.* **18**:597-607.

Thorpe, G.H.G., L.J. Kricka, S.B. Moseley, and T.P. Whitehead. 1985. Phenols as enhancers of the chemiluminescent horseradish peroxidase-luminol-hydrogen peroxide reaction: Application in luminescence-monitored enzyme immunoassays. *Clinical Chem.* **31**:1335-1341.

Watt, B.E., R.L. Malcolm, M.H.B. Hayes, N.W.E. Clark, and J.K. Chipman. 1996. Chemistry and potential mutagenicity of humic substances in waters from different watersheds in Britain and Ireland. *Water Res.* **30**:1502-1516.

Whitehead, T.P., G. Thorpe, M. Lane, A. Watson and C. Billings. 1993. A rapid and simple chemiluminescent assay for water quality monitoring. p. 425-429. *In* A.Z. Szalay, L.J. Kricka and P.E. Stanley (eds), *Bioluminescence and Chemiluminescence-Status report.* Wiley, New York.

Wilson, M.A. (ed). 1987. *NMR Techniques and Applications in Geochemistry and Soil Chemistry.* Pergamon Press, New York.

Section 4

Biological Impacts of Humic Substances

There has always been a mystique about the influences of humic substances (HS) on health, and throughout the ages there have been claims that HS can benefit health. Unfortunately, the evidence for many of the claims in the past of the benefits to health of HS has been anecdotal, and rigorous assessments of the claims are only now being introduced. There is, however, no acceptable evidence for deleterious influences to health of HS when the substances are used as they occur in nature.

Because HS are organic molecules they will, like all organic substances, give rise to artefact degradation products which depend on the degradative agents used and the conditions of degradation. This theme impacts on human health when waters with significant amounts of HS are chlorinated. There is epidemiological evidence from the USA, from Scandanavia, and to a lesser extent from Britain which indicates that persons who consume waters that are chlorinated and rich in HS are more susceptible to certain cancers (especially those of the digestive and urinary tracts) than those who do not consume such waters, or use chlorinated waters low in HS. There is abundant evidence to show that 3-chloro-4-(dichloromethyl)-5-hydroxy-2(5H)-furanone, or MX, is formed when humic waters and HS are chlorinated. MX is found in significant abundance as a product of the chlorination of lignin, and has been known for some time to be a potent mutagen, and more recently it has been shown to be carcinogenic in animals.

Chlorination of HS will give rise to numerous organic molecules, many of which are organochlorine compounds, and many of these are, to some extent, mutagenic. A compound that is shown to be mutagenic in bacterial assay systems is not necessarily carcinogenic. The effects of MX *in vivo*, for instance, would appear to be restricted because cellular defensive mechanisms restrict the extents to which it can cross cell membranes.

The contribution by MacDonald and Chipman (p. 337 - 345) asks "Does the genetic toxicity of water chlorination products pose a risk to public health?" Any subject area that asks questions relating to public health is, of course, emotive. Chemical and oxidative treatments of organic matter will give rise to compounds other than carbon dioxide and water. Thus it is a matter of deciding which treatment will provide appropriate disinfection while producing the least toxic byproducts. Chlorination does provide effective disinfection, and its residual effects are sufficiently long to provide protection during the residence time of water in the supply system. It does, however, as indicated, give rise to mutagenic, and possibly to carcinogenic substances. The extents to which such substances are formed will depend on the nature and the amounts of organic substances present in the water, and humic substances are among the organic materials that give rise to undesirable products. The dangers from chlorination are lessened when the HS are removed prior to chlorination (see p. 219 - 236). There is a search for oxidative alternatives to chlorine, and interest is focusing on the uses of ozone. The organic artefacts of ozonolysis would appear to be less threatening than those from chlorination, but the residual disinfection effects of chlorine appear to be longer than those of ozone.

The contribution by MacDonald and Chipman has highlighted evidence for the potential of MX to damage DNA and to induce DNA repair in mammalian cells, but it has stressed that the positive results with mammalian cells were often obtained by employing concentrations of MX which approached cytotoxic concentrations. These concentrations are many orders of magnitude higher than the levels established in chlorinated drinking water. They stress that, "from *in vitro* findings in mammalian cells there emerges the hypothesis that, at sub-toxic concentrations, the DNA strand breaks observed arise (to some extent) as a consequence of DNA base damage repair (as in bacterial mutagenicity assays), and not through a direct-acting clastogenic activity of MX". MacDonald and Chipman conclude that it is justified to propose that the risk associated with low levels of exposure to MX in drinking water is substantially lower than the risk that might be predicted from studies with higher doses in animals, and from high concentration genotoxicity studies. Because disinfection by chlorination has been a major success in preventing the spread of water borne diseases from microbiological contaminations, the risks versus benefits should be always weighed in decision making processes. However, because DNA base modification by MX is a feasible event, though somewhat remote from low level exposure in vivo, the possibility exists that exposure throughout a lifetime can be a health hazard.

The contribution by Flaig (p. 346 - 356) has focused attention on aspects of the biochemistry of the healing effects of humic substances from peat. As indicated above, most of the evidence for healing effects of HS is anecdotal. Thus it is encouraging to see from this article reference to more rigorous studies which indicate that HS and humic-type substances (H-TS) can have beneficial physiological effects. The article focuses on low molecular weight phenolic compounds that can be isolated from peats. Such compounds are readily oxidized to quinonoid structures, and these can interact with amino acids and proteins. Flaig concludes that it is possible that healing effects of HS can be biochemically explained by interactions of phenolic or quinonoid monomeric or polymeric precursors of HS (or derived from HS), or humic acids per se, with proteinaceous materials.

A viable industry in Germany is based on the reported health benefits of peat baths. Peats suspended/dissolved in bath waters allow the bather to withstand bath temperatures significantly greater than those that can be tolerated in the absence of the peat. Claims have been made for beneficial effects, on skin especially, and there have been claims for other benefits that are less well substantiated. Inevitably, dissolved HS, and low molecular weight phenolic and quinonoid compounds will be in the bathing waters.

It might be appropriate to consider the colour in upland waters as a desirable property. However, should it be possible to overcome the aesthetic objections to humic-coloured waters, it would be necessary to introduce an effective filtration mechanism that would remove biological carriers of diseases before such waters could be introduced to water supplies, and it would be necessary to alter European Community recommendations with regard to minimum colour levels in potable waters.

Does the Genetic Toxicity of Water Chlorination Products Pose a Risk to Public Health?

D. L. MacDonald and J. K. Chipman

THE UNIVERSITY OF BIRMINGHAM, SCHOOL OF BIOCHEMISTRY, EDGBASTON, BIRMINGHAM, B15 2TT, ENGLAND.

Abstract

Several reports from epidemiological studies have indicated a weak but significant correlation between the consumption of chlorinated water and cancers of the bladder and colon. Experimental studies have concentrated on the effects of various genotoxic components of chlorinated water (particularly the chloro-hydroxy-furanone, MX, which can be derived from humic substances). Despite evidence for the potential of MX to damage DNA and induce DNA repair in mammalian cells, the effects of this agent *in vivo* would appear to be very restricted because of the limited extent to which it appears to be absorbed as reactive MX, and because of the presence of effective cellular defence systems. Even though the concentrations of MX in potable water are very low, the recent finding that MX is carcinogenic in animals highlights the need for a better mechanistic understanding.

1 Introduction

The use of chlorine as a disinfectant in the production of potable water devoid of microbiological contamination has been crucial for the prevention of the transmission of waterborne infectious diseases. Though highly valuable in this respect, chlorine has been implicated in contributing potentially hazardous organic micropollutants to drinking water. Currently, more than 1100 organic compounds have been identified in drinking water, and most of these are at or below the $\mu g\ L^{-1}$ level. Many of these compounds are thought to arise as a consequence of the interaction between chlorine and naturally occurring humic substances (HS) in water (Bull *et al.*, 1982). These ubiquitous substances are a mixture of complex, high molecular weight constituents which occur largely as the products of chemical and microbiological degradation of plant residues. For more than two decades, it has been recognized that the chlorination of water leads to the formation of by-products derived from HS which are highly reactive and exhibit an ability to cause mutation in bacterial cells (Meier *et al.*, 1987a).

Epidemiological studies have also given rise to concern that water chlorination is associated with an elevated risk of certain cancers. A meta-analysis from a study in the

USA has suggested that the risks of bladder and of colon cancers are increased by 21% and 38%, respectively, because of the presence of chlorination by-products (Morris *et al.*, 1992). However, all epidemiological studies have their inherent limitations and none of the existing studies has inferred a causal relationship, and so such data are inconclusive.

It is important that the real risk, if any, from the chlorination of water needs to be carefully assessed and balanced against the benefits. Epidemiological studies, though a very useful tool in risk assessment, are inevitably fraught with uncertainties. Thus, experimental studies provide a major source of information upon which critical and meaningful risk assessments can be made.

Many chlorinated by-products are produced when water is chlorinated (e.g. trihalomethanes and chloroacetic acids) and several of these have provided evidence of carcinogenicity in rodents (Jorgenson *et al.*, 1985; Dunnick and Melnick, 1993; Tumasonis *et al.*, 1985; Angelo and McMillan, 1994). However, the majority of these products are not genotoxic and may be carcinogenic via threshold-dose mechanisms (Purchase, 1994). Particular attention has been given to the presence of genotoxic compounds in extracts of tap waters disinfected with chlorine (Loper, 1980; Fawell *et al.*, 1986; Vartiainen and Liimatainen, 1986). The discovery of 3-chloro-4-(dichloromethyl)-5-hydroxy-2(5H)-furanone (compound **I**), otherwise referred to as MX, has aroused considerable concern. It

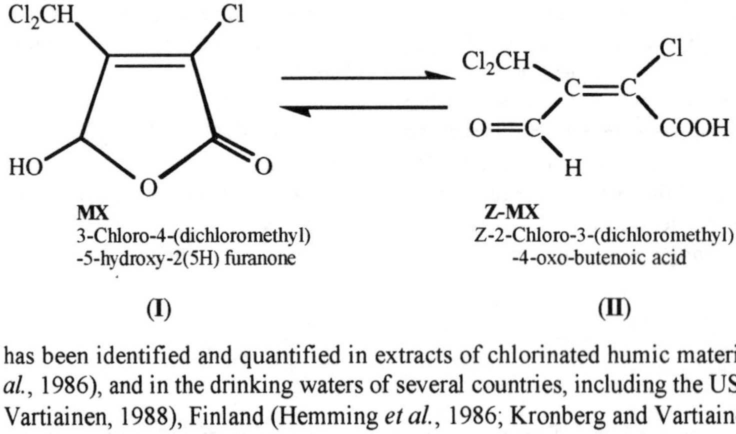

Cl_2CH Cl HO O O

MX
3-Chloro-4-(dichloromethyl)
-5-hydroxy-2(5H) furanone

(I)

Cl_2CH Cl C=C O=C COOH H

Z-MX
Z-2-Chloro-3-(dichloromethyl)
-4-oxo-butenoic acid

(II)

has been identified and quantified in extracts of chlorinated humic materials (Hemming *et al.*, 1986), and in the drinking waters of several countries, including the USA (Kronberg and Vartiainen, 1988), Finland (Hemming *et al.*, 1986; Kronberg and Vartiainen, 1988) and the UK (Horth *et al.*, 1989). In water, MX is in equilibrium with its non-cyclic structural isomer, compound **II** (Z-2-chloro-3-(dichloromethyl)-4-oxo-butenoic acid). Although present in low concentrations only (ranging from 2 - 67 ng L^{-1}) in chlorinated waters, MX is considered to account for approximately 5-60% of the total bacterial mutagenicity in such waters (Meier *et al.*, 1990). The mutagenicity of MX has been reported to be in the range of 10^3-10^4 revertants $nmol^{-1}$ in *Salmonella typhimurium* strain TA100 (Holmbom *et al.*, 1984; Ishiguro *et al.*, 1987; Meier *et al.*, 1987b), and this ranks the compound amongst the most potent of the base-pair mutagens that give a positive response in this strain. Thus, it is not surprising that the potential genotoxicity of MX in mammalian cells has been studied in order to ascertain the possible deleterious effects on public health which could arise from the consumption of chlorinated water supplies containing this compound.

2 Experimental Findings With MX

In Vitro. MX has consistently demonstrated an ability to induce mutagenicity in bacterial assay systems (Holmbom *et al.*, 1984; Ishiguro *et al.*, 1987; Meier *et al.*, 1987b). This ability has been observed in the absence of a metabolic activation system in all strains (Meier *et al.*, 1987c), indicating that MX does not require metabolism to convert the compound to a mutagenic form. Indeed, the cytosolic thiol, glutathione, offers protection against bacterial mutagenicity (Meier *et al.*, 1987c; Ubom *et al.*, 1994) (see below). The potential for MX to exert genotoxicity has also been studied in a variety of cultured mammalian cells. These studies have yielded positive results with respect to (MX) concentration-related increases in structural chromosome aberrations (Meier *et al.*, 1987; Mäki-Paakkanen *et al.*, 1994), sister chromatid exchange (Meier *et al.*, 1987c; Mäki-Paakkanen *et al.*, 1994; Jansson *et al.* 1993; Brunborg *et al.*, 1991), gene mutation (Mäki-Paakkanen *et al.*, 1994; Jansson and Hyttinen, 1994), DNA strand breaks (Meier *et al.*, 1990; Daniel *et al.*, 1985; Chang *et al.*, 1991), and unscheduled DNA synthesis (Nunn *et al.*, 1997) in a variety of mammalian cell types. Again, the experimental evidence supports the fact that MX does not require metabolic activation to exert an effect (Brunborg *et al.*, 1991). MX has been shown to cause the inhibition of metabolic cooperation between 6-thioguanine resistant and 6-thioguanine sensitive cells in contact with each other, and this indicates an ability to inhibit intercellular communication (Matsumura *et al.*, 1994). The results suggested that MX produced a weak inhibitory effect, indicating that in addition to genotoxicity, the compound might have potential tumour promoting activity. It should be borne in mind that a number of these positive findings were obtained from experiments in which $\mu g\ mL^{-1}$ concentrations of MX were employed. Such concentrations approach toxic limits, and are orders of magnitude higher than the concentrations in water ($ng\ L^{-1}$).

In Vivo. Data are scarce on the effects of MX *in vivo*. Generally, results on genotoxicity from rodent studies are negative where MX is administered orally at doses approaching the LD_{50} value. Negative results have also been obtained from the mouse bone micronucleus assay (Meier *et al.*, 1990), from DNA strand break assays in a range of rat organs (Brunborg *et al.*, 1991), and from *ex vivo* unscheduled DNA synthesis studies in mice following oral dosing of MX (Nunn *et al.*, 1997).

Genetic damage induced by MX *in vivo* has been reported in a few studies only. One of these studies showed nuclear anomalies in the intestines of rats (Daniel *et al.*, 1991), and another reported sister chromatid exchange in rat peripheral lymphocytes (Jansson *et al.*, 1993). Genetic damage was also reported in rat glandular stomach (Furihata *et al.*, 1992). However, these effects were observed only for doses at near-toxic levels, which were many times higher than the concentrations likely to be found in water.

3 How Can the Discrepancies Between the *In Vitro* and *In Vivo* Findings be Interpreted?

The *in vivo* findings, which are generally negative (despite the *in vitro* genotoxicity), may suggest that there is limited absorption of MX (or of MX in its intact genotoxic form), and/or that there is adequate cellular protection against the high reactivity of MX with DNA.

Absorption of MX

Disposition studies of ^{14}C- labelled MX, following oral administration (Ringhand *et al.*, 1989), have indicated that > 40% of MX-related radioactivity was absorbed into the systemic circulation, where it was distributed to tissues and eliminated in the urine. This study was unable to clarify whether the radiolabel in the faeces (> 44%) arose from incomplete absorption, or from biliary excretion. Neither could it distinguish between unchanged (potentially genotoxic) MX and derivative products. A recent study examined MX absorption through isolated rat small intestine. The compound was detected indirectly by measuring bacterial mutagenesis. The results showed that measurable but limited uptake of MX took place through the gut wall, even in the absence of intestinal contents (Clark and Chipman, 1995). Restrictions to absorption may involve factors such as MX hydrolysis, and its interaction with thiols or other cellular and extracellular macromolecules (vide infra) as the result of the high reactivity of the compound.

Cellular Protection Against MX

Various studies have indicated that glutathione and cysteine provide significant protection against mutagenic compounds, and especially MX, recovered when HS isolated by resin techniques were chlorinated. Studies have illustrated that the presence of glutathione and glutathione-S-transferases in liver cytosol are responsible for a marked (30-40%) reduction of MX-induced mutagenicity when rat liver 9000g supernatant (S9) was added to assays (Meier *et al.*, 1990; Meier *et al.*, 1987b; Meier *et al.*, 1987c; Ubom and Chipman, 1994). The direct reactivity of MX with nucleophiles such as glutathione is likely to limit dramatically the genotoxic effects of MX *in vivo*. However, it was surprising that *Salmonella typhimurium* strains deficient in glutathione gave an approximately 10-fold lower mutagenic response than that of the parent glutathione-rich strain (Ubom *et al.*, 1993). Furthermore, other studies employing mammalian testicular cells (Brunborg *et al.*, 1991), primary rat hepatocytes (Chang *et al.*, 1991), and rodent hepatocytes (Nunn *et al.*, 1997) reported that depletion of intracellular glutathione does not enhance DNA damage by MX. These anomalous findings suggest that nucleophiles do not offer protection against some form(s) of genetic damage (Ubom and Chipman, 1994). Collectively, these findings have demonstrated effectively the protective role of exogenous glutathione and other thiol compounds against the genotoxicity of MX, but the role of intracellular glutathione in cellular defence is unclear. There is a tendency to imply that the access of MX to cellular DNA is restricted by the high reactivity towards thiol-containing and other compounds as reflected by the lack of positive findings from *in vivo* data. If there is both limited absorption of MX and adequate cellular protection, as suggested from the data, does this mean that our DNA can effectively be protected? There is the emergence of a strong foundation upon which a meaningful and critical risk assessment can be built with respect to human consumption of MX in chlorinated drinking water. A critical question may relate to the precise nature of DNA damage and the concentration-dependency of the damage.

4 Is DNA Strand Breakage Induced by MX Linked to Cytotoxicity Rather than to Genotoxicity?

A basic knowledge of the possible mechanism(s) underlying the genetic damage induced by MX is essential before a critical risk assessment can be made with respect to the compound. Findings of DNA strand breakages resulting from treatments with MX in some *in vitro* assays have suggested that these effects might actually be linked to

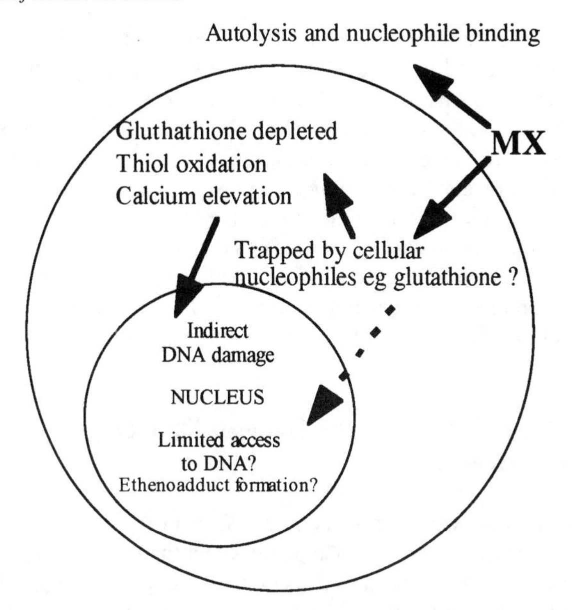

Figure 1 *Potential mechanisms of MX genotoxicity*

cytotoxicity rather than to genotoxicity. This is especially important in the light of the limited information from *in vivo* genotoxicity data relating to DNA strand breakage and not mutations. The hypothesis that DNA strand breakage in human white blood cells is secondary to the depletion of glutathione was examined by Nunn and Chipman (1994a), (Figure 1). This study postulated that such DNA damage may also be mediated through secondary thiol oxidation and disruption of calcium homeostasis leading to the activation of calcium-dependent endonucleases and/or topoisomerases. The results suggested that, although a relationship between MX and induced strand breakage exists, the mirrored depletion of glutathione was coincidental and not causal. Moreover, the strand breakage induced by MX did not appear to be mediated via an increase in intracellular cytosolic calcium which can arise following thiol depletion and inhibition of Ca^{2+} ATPase. The resultant stand breakage could not thus be explained by activation of calcium- dependent endonucleases, though it was proposed that calcium ions may be required for strand breakages as inhibition occurred in the presence of the calcium chelating agent 'Quin 2'. In particular, calcium deprivation may restrict the activity of DNA repair enzymes that mediate strand breakage (Nunn and Chipman, 1994a). The possible involvement of DNA repair enzymes that mediate strand scissions remains the question of critical importance.

More specifically, it may be asked if DNA strand breaks produced by MX occur as the result of excision repair of DNA adducts rather than as secondary events linked to cell toxicity.

5 Is there Evidence for DNA Adduct Formation?

The formation of a small etheno adduct (1, N'-ethenodeoxyadenosine) by MX or related compounds has been reported (Nunn and Chipman, 1994b; Kronberg *et al.*, 1992, 1993). Such ethenobases are repaired by glycosylase, albeit not rapidly (Bartsch *et al.*, 1994). The level of etheno adduct formation from MX is, however, extremely low, and is unlikely to provide assistance as an *in vivo* biomarker of DNA damage. Not surprisingly, bulky hydrophobic adducts were not evident from studies with ^{32}P-postlabelling (Nunn and Chipman, 1994b). One recent study obtained a positive result in an unscheduled DNA synthesis (UDS) assay employing rat and mouse hepatocytes *in vitro* (but not *in vivo*). That would implicate nucleotide excision repair of DNA damage induced by MX *in vitro*. The nature of the damage responsible for UDS is not known, but this result is suggestive of DNA base modification *in vitro*, requiring long-patch DNA repair. This damage was not exacerbated by depletion of the potential protectant glutathione (Nunn *et al.*, 1997).

6 The Ultimate Question: Does the Genetic Toxicology of MX Pose a Risk to Human Health?

It is not surprising that the positive results obtained from *in vitro* studies have aroused concern among scientists from different disciplines about the possible detrimental effect of MX to public health. It must be emphasised that the positive results obtained for mammalian cells were obtained often by employing concentrations of MX approaching cytotoxic concentrations ($\mu g\ mL^{-1}$) many orders of magnitude higher than the levels established in chlorinated drinking water ($ng\ L^{-1}$). From the *in vitro* findings in mammalian cells there emerges the hypothesis that, at sub-toxic concentrations, the DNA strand breaks observed arise (to some extent) as a consequence of DNA base damage repair (as in bacterial mutagenicity assays), and not through a direct-acting clastogenic activity of MX. At relatively high concentrations of MX, it is feasible that the occurrence of DNA strand breaks may be associated also with cell toxicity.

Most of the studies of genotoxicity which have involved dosing animals with MX have proved negative, even when doses approaching the LD_{50} were used. The negative UDS result *ex vivo* is especially important in contrast to the positive result in the analogous *in vitro* assay. Positive *in vivo* effects were limited to clastogenicity-related endpoints at high dosages. This suggests evidence for limited absorption, for restraints on MX-tissue availability, and/or for adequate cellular defence. Absorption of mutagenic MX across the GI tract appears to be limited. Due to the highly reactive nature of MX, it is possible that the presence of gastrointestinal contents may contribute to inactivation and to decreased absorption. The role of glutathione in cellular defence against MX is strongly implicated, although the role of intracellular glutathione is still unclear.

There is a strong argument from the experimental findings for the existence of conceptual thresholds at each level of the process, from the ingestion of water through to the possible access of MX to cellular DNA. In view of this discussion, we feel that it is now justified to propose that the risk associated with low level exposure to MX in chlorinated drinking water is substantially lower than the risk that might be predicted from studies with high doses on animals, and from high concentration genotoxicity studies. In particular, these observations need to be considered when interpreting the two year carcinogenicity study employing MX (Tuomisto et al., 1995) in which an increased incidence of liver and thyroid tumours in female mice has been provisionally reported.

The risk versus benefit equation associated with the use of chlorination as a means of water disinfection should be built into the decision-making process. Disinfection by chlorination has been a major success this century in disease prevention world-wide. It has been crucial in producing water devoid of microbiological contamination. In coming to this conclusion, however, it has to be recognized that DNA base modification by MX is nevertheless a feasible event, albeit of remote likelihood from low level exposures *in vivo*. The development of techniques to remove MX and related mutagenic components from drinking water (or to avoid their formation) are therefore to be encouraged.

Acknowledgements

JKC acknowledges financial support from the NERC and BBSRC for various aspects of laboratory work referred to. DLM was the recipient of a BBSRC MSc Toxicology studentship.

References

Angelo, A.B. and L.P. McMillan. 1994. In: Jolley, *et al.* (eds), *Water Chlorination Chemistry, Environmental Impact and Health Effects.* Vol.6 Lewis Publishers, Chapter 16, 193.

Bartsch, H., A. Barbin, M.J. Marion, J. Nair, and Y. Guichard. 1994. Formation, detection, and role in carcinogenesis of ethenobases in DNA. *Drug Metabolism Reviews* **26**:349-371.

Brunborg, G., J.A. Holme, E.J. Søderland, J.K. Hongslo, T. Vartiainen, S. Lötjonen, and G. Becher. 1991. Genotoxic effects of 3-chloro-4-(dichloromethyl)-5-hydroxy-2[5H]-furanone in mammalian cells in vitro and in rats in vivo. *Mutation Res.* **260**:55-64.

Bull, R.J., M. Robinson, J.R. Meier, and J. Strober. 1982. Use of biological assay systems to assess the relative carcinogenic hazards of disinfection by-products. *Environ Health Perspect.* **46**:215-227.

Chang, L.W., F.B. Daniel, and A.B. DeAngelo. 1991. DNA strand breaks induced in cultured human and rodent cells by chlorohydroxyfuranones - mutagens isolated from drinking water. *Terat. Carc. Mut.* **11**:103-114.

Clark, N.W.E. and J.K. Chipman. 1995. Absorption of 3-chloro-4-(dichloromethyl)-5-hydroxy-2[5H]-furanone (MX) through rat small intestine in vitro. *Toxicol. Lett.* **81**:33-38.

Daniel, F.B., D.L. Haas, and S.M. Pyle. 1985. Quantitation of chemically-induced DNA strand breaks in human-cells via an alkaline unwinding assay. *Anal. Biochem. 144:390-*402.

Daniel, F.B., G.R. Olson, and J.A. Stober. 1991. Induction of gastrointestinal tract nuclear anomalies in B6C3F1 mice by 3-chloro-4-(dichloromethyl)-5-hydroxy-2[5H]-furanone and 3-chloro-3,4-(dichloro)-5-hydroxy-2[5H]-furanone, mutagenic by-products of chlorine disinfection. *Environ. Mol. Mut.* **17**:32-39.

Dunnick, J.K. and R.L. Melnick. 1993. Assessment of the carcinogenic potential of chlorinated water - experimental studies of chlorine, chloramine, and trihalomethanes *J. Natl. Cancer Inst.* **85**:817-822.

Fawell, J.K., M. Fielding, H. Horth, H. James, R.F. Lacey, J.W. Ridgway, P. Wilcox, and I Wilson. 1986. Health aspects of organics in drinking water. *Publication no. TR 231.* Water Research Centre, Marlow, Bucks, U.K.

Furihata, C., M. Yamashita, N. Kinae, and T. Matsushima. 1992. Genotoxicity and cell proliferative activity of 3-chloro-4-(dichloromethyl)-5-hydroxy-2[5H]-furanone [MX] in rat glandular stomach. *Water Sci. Tech.* **25**:341.

Hemming, J., B. Holmbom, M. Reunanen, and L. Kronberg. 1986. Determination of the strong mutagen 3-chloro-4-(dichloromethyl)-5-hydroxy-2[5H]-furanone in chlorinated drinking and humic waters. *Chemosphere* **15**:549-556.

Holmbom, B., R.H. Vass, R.D. Motimer, and A. Wong. 1984. Fractionation, isolation and characterization of Ames mutagenic compounds in kraft chlorination effluents. *Environ. Sci. Technol.* **18**: 333-337.

Horth, H., M. Fielding, T. Gibson, H.A. James, and H. Russ. 1989. Identification of mutagens in drinking water. *Aqua* **38**:80-100.

Ishiguro, Y., R.T. Lalonde, C.W. Dence, and J. Santodonato. 1987. Mutagenicity of chlorine-substituted furanones and their inactivation by reaction with nucleophiles. *Environ. Toxicol. Chem.* **6**: 935-946.

Jansson, K., J. Mäki-Paakkanen, S. Vaittinen, T. Vartiainen, H. Komulainen, and J. Tuomisto. 1993. Cytogenic effects of 3-chloro-4-(dichloromethyl)-5-hydroxy-2[5H]-furanone (MX) in rat peripheral lymphocytes in vitro and in vivo. *Mutation. Res.* **229**:25-28.

Jansson, K. and J.M.T. Hyttinen. 1994. Induction of gene mutation in mammalian-cells by 3-chloro-4-(dichloromethyl)-5-hydroxy-2(5H)-furanone (MX), a chlorine disinfection by-product in drinking-water. *Mutation. Res.* **322**:129-132.

Jorgenson, T.A., E.F. Meierhenry, C.J. Rushbrook, R.J. Bull, M. Robinson, and C.E. Whitmore. 1985. Carcinogenicity of chloroform in drinking-water to male Osborne-Mendel rats and female B6C3F1 mice. *Fundam. Appl. Toxicol.* **5**:760-769.

Kronberg, L., S. Karlsson, and R. Sjöholm. 1993. Formation of ethenocarbaldehyde derivatives of adenosine and cytodine interactions with mucocholic acid. *Chem. Res. Toxicol.* **6**:495-499.

Kronberg, L., R. Sjöholm, and S. Karlsson. 1992. Formation of 3,N^4-ethenocytidine, 1,N^6-ethenoadenosine, and 1,N^2-ethenoguanosine in reaction of mucochloric acid with nucleosides. *Chem. Res. Toxicol.* **5**:852-855.

Kronberg, L. and T. Vartiainen. 1988. Ames mutagenicity and concentration of the strong mutagen 3-chloro-4-(dichloromethyl)-5-hydroxy-2[5H]-furanone and of its geometric isomer E-2-chloro-3-dichloromethyl)-4-oxobutanoic acid in chlorine-treated tap waters. *Mutation. Res.* 206:177-182.

Loper, J.C. 1980. Mutagenic effects of organic compounds in drinking water. *Mutation. Res.* **76**:241-268.

Mäki-Paakkanen, J., K. Jansson, and T. Vartiainen. 1994. Induction of mutation, sister-chromatid exchanges, and chromosome-aberrations by 3-chloro-4-(dichloromethyl)-5-hydroxy-2(5H)-furanone in Chinese-hamster ovary cells. *Mutation. Res.* **310**:117-123.

Matsumura, H., M. Watanabe, K. Matsumoto, T. Ohta. 1994. 3-Chloro-4-(dichloromethyl)-5-hydroxy-2(5H)-furanone (MX) induces gene-mutations and inhibits metabolic cooperation in cultured Chinese-hamster cells. *J. Tox. Environ. Health* **43**:65-72.

Meier, J.R., W.F. Blazak, and R.B. Knohl. 1987a. Mutagenic and clastogenic properties of 3-chloro-4-(dichloromethyl)-5-hydroxy-2[5H]-furanone: a potent bacterial mutagen in drinking water. *Environ. Mol. Mut.* **10**:411-424.

Meier, J.R., W.F. Blazak, and R.B. Knohl. 1987b. mutagenic and clastogenic properties of 3-chloro-4-(dichloromethyl)-5-hydroxy-2(5H)-furanone - a potent bacterial mutagen in drinking-water *Environ. Mol. Mut.* **10**:414-424.

Meier, J.R , A.B. DeAngelo, F.B. Daniel, K.M. Schenk, J.W. Doerger, L.W. Chang, F.C. Kopler, M. Robinson, and H.P. Ringhand. 1990. Genotoxic and carcinogenic properties of chlorinated furanones: important byproducts of water chlorination. p. 185-195. *In* M.D. Waters, F.B. Daniel, J. Lewtas, M.M. Moore, and S. Nesnow (eds), *Genetic Toxicology of Complex Mixtures*. Plenum, New York.

Meier, J.R., P.B. Knohl, W.E Coleman, H.P. Ringhand, J.W. Munch, W.H. Kaylor, R.P. Streicher, and F.C. Kopfler. 1987c. Studies on the potent bacterial mutagen 3-chloro-4-(dichloromethyl)-5-hydroxy-2[5H]-furanone: aqueous stability, XAD recovery and analytical determination in drinking water and in chlorinated humic acid solutions. *Mutation Res.* **189**:363-373.

Morris, R.D., A.M. Audet, I.F. Angelillo, T.C. Chalmers, and F. Morsteller. 1992. Chlorination, chlorination by-products, and cancer: a meta-analysis. *Am. J. Public Health.* **82**:955-963.

Nunn, J.W. and J.K. Chipman. 1994a. Induction of DNA strand breaks by 3-chloro-4-(dichloromethyl)-5-hydroxy-2[5H]-furanone and humic substances in relation to glutathione and calcium status in human white blood cells. *Mutation Res.* **341**:133-140.

Nunn, J.W. and J.K. Chipman. 1994b. Genotoxicity of 3-chloro-4-(dichloromethyl)-5-hydroxy-2[5H]-furanone (MX). *Human Exptl. Toxicol.* **13**: 635.

Nunn J.W., J.E. Davies, and J.K. Chipman. 1997. Production of unscheduled DNA synthesis in rodent hepatocytes in vitro, but not in vivo, by 3-chloro-4-(dichloromethyl)-5-hydroxy-2[5H]-furanone (MX). *Mutation Res.* **373**:67-73.

Purchase, I.F.H. 1994. Current knowledge of mechanisms of carcinogenicity - genotoxins versus non-genotoxins. *Human Exptl. Toxicol.* **13**:17-28.

Ringhand, H.P., W.H. Kaylor, R.G. Miller, and F.C. Kopfler. 1989. Synthesis of 3-C-14-3-chloro-4-(dichloromethyl)-5-hydroxy-2(5H)-furanone and its use in a tissue distribution study in the rat. *Chemosphere* **18**:2229-2236.

Tuomisto, J., T. Vartiainen, H. Komulainen, M. Koivusalo, K. Jansson, V.M. Kosma, J. Mäki-Paakkanen, Vaittinen, S. Lotjonen, and T. Hakulinene. 1995. Cancer risk of 3-chloro-4-(dichloromethyl)-5-hydroxy-2-(5H)-furanone (MX) and related non-volatile chlorination products. *Presented at the Society of Toxicology (SOT) 34th meeting,* Baltimore, USA, March 5-9.

Tumasonis, C.F., D.N. McMartin, and B. Bush. 1985. Lifetime toxicity of chloroform and bromodichloromethane when administered over a lifetime in rats *Ecotoxicol. and Environ. Safety* **9**:233-240.

Ubom, G.A., J.K. Chipman, and M.H.B. Hayes. 1994. Glutathione deficiency does not elevate susceptibility of bacteria to the mutagenicity of chlorinated humic acids. *Human Exptl. Tox.* **13**:558-562.

Ubom, G. and J.K. Chipman. 1994. Absence of unscheduled DNA-synthesis in rat hepatocytes treated with mutagenic and cytotoxic chlorinated humic substances. *Mutation. Res.* **321**:57-63.

Vartiainen, T. and A. Liimatainen. 1986. High-levels of mutagenic activity in chlorinated drinking-water in Finland. *Mutation. Res.* **169**:29-34.

Aspects of the Biochemistry of the Healing Effects of Humic Substances from Peat

W. Flaig

OTTO-HAHN-STRASSE 132, 97 218 GERBRUNN, GERMANY

Abstract

The interactions of di- and of polyhydroxybenzene compounds, and their derivatives from lignin-type precursors give rise to humic-type substances. These, and their quinonoid precursors, as well as naturally occurring humic substances can interact with amino acid, peptide, and proteinaceous materials to give products which may have significant healing effects.

1 Introduction

Physiologically-active substances in peat have been characterized as belonging to the so-called 'bitumen-fraction' that is derived mostly from cyclopentane-phenanthrene compounds, which also include hormones. Other physiologically-active compounds can be derived from the degradation of plant lignins.

It has been known for some time that a variety of low molecular weight phenolic compounds can be isolated from peat (Flaig, 1993). The transformations of these compounds to quinonoid derivatives have been investigated extensively. For example, Quecke and Loschen (1989) have described the uses of lignin degradation products, such as vanillic and gallic acids, as inhibitiors of the inflammatory effects caused by products of the arachidonic acid cycle. The effects of synthetic, higher molecular weight humic-type substances (H-TS) are summarised by Klöcking and Helbig (1993).

Tolpa (1992) has reported positive results with regard to the application of products derived from peat for the treatment of different types of cancers.

Some Transformations of Vanillic Acid and of Gallic Acid

There are questions which arise with regard to the effect of vanillic acid (4-hydroxy-3-methoxybenzoic acid) on the inhibition of inflammation caused by compounds released in the arachidonic cycle. Firstly: 'Is it vanillic acid itself, or one of its transformation products, which gives rise to the inhibition under the prevailing physiological conditions?' One of the first steps in the transformation process in plants has been determined using vanillic acid with a ^{14}C-labelled methoxyl group. The results show that vanillic acid is transformed to 2-methoxy(^{14}C)-1,4-benzoquinone.

The transformations of vanillic acid have been studied under a variety of physiological conditions. The reactions of interest are summarised in Figure 1. These indicate that:

1, vanillic acid (Compound 1) dimerises (to Compound 3), and then undergoes oxidative decarboxylation to give a methoxybenzoquinone dimer (Compound 4);

2, direct oxidative decarboxylation of vanillic acid leads to the formation of 2-methoxy-1,4-benzoquinone (Compound 5);

3, demethylation of vanillic acid produces the more reactive protocatechuic acid (3,4-dihydroxybenzoic acid, Compound 2); and

4, compounds with methoxyl groups, such as vanillic acid, form light coloured macromolecules (H-TS) under oxidative conditions. However, when demethylation takes place, the subsequent dimerization and macromolecularization sequence can lead to the formation dark coloured H-TS.

Gallic acid (3,4,5-trihydroxybenzoic acid; Compound 9) is formed by demethylation of the lignin degradation product, syringic acid (3,5-dimethoxy-4-hydroxybenzoic acid; Compound 8), as shown in Figure 2. Enzymatic oxidation of gallic acid gives rise to Compound 10 by 1:1 addition or, under stronger oxidative conditions, decarboxylation can take place to give brown substances with properties similar to those of low molecular

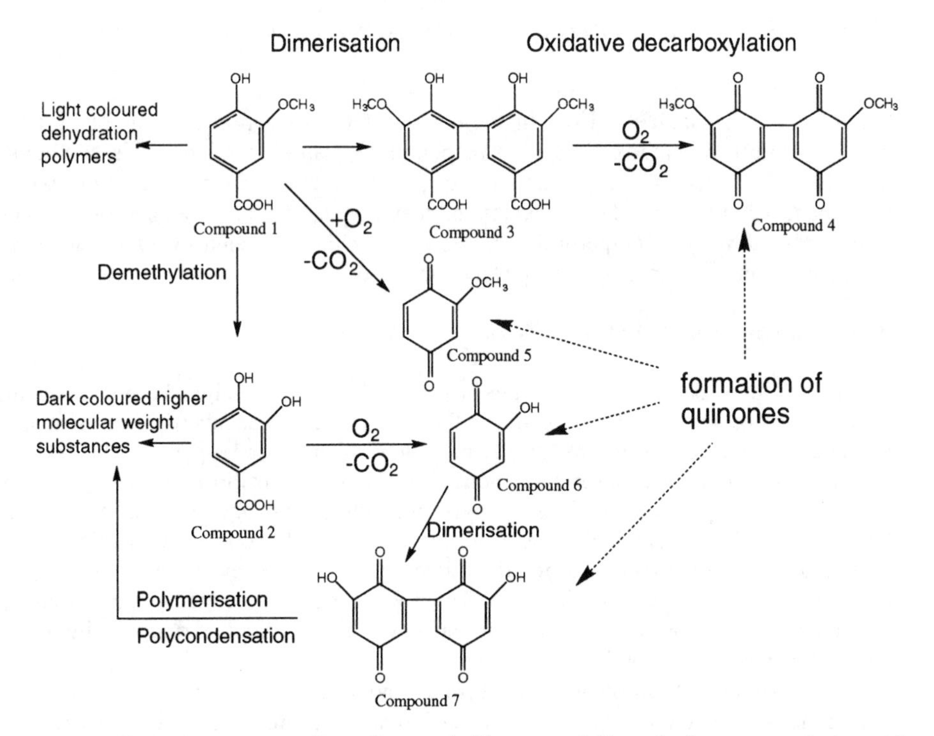

Figure 1 *Transformations of vanillic acid (Compound 1) and of protocatechuic acid (Compound 2)*

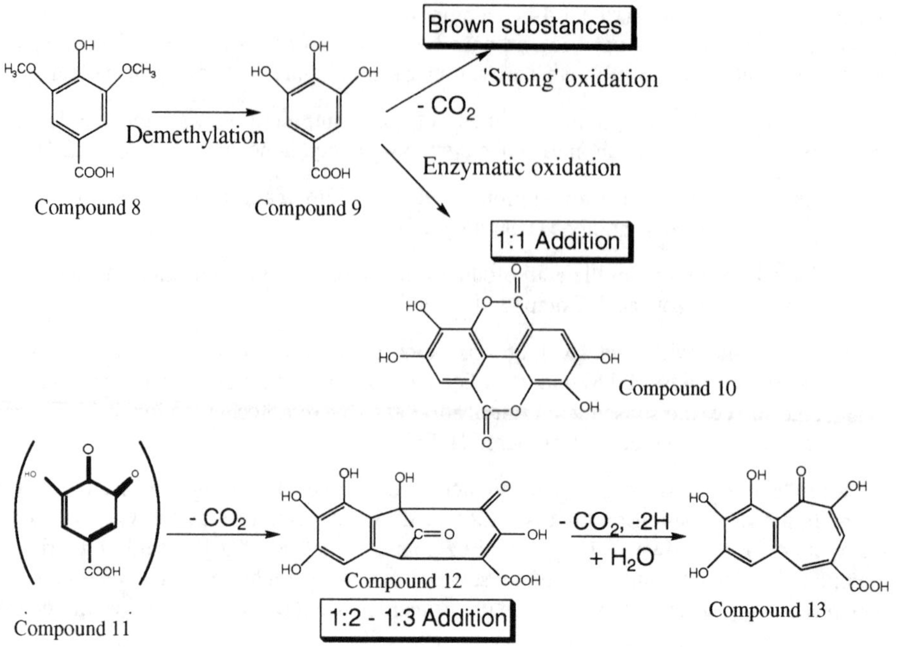

Figure 2 *Products formed that involve the dimerization of gallic acid*

weight humic substances (HS). In the course of the oxidation to Compound 10, a hydroxy-*o*-quinone-carboxylic acid (Compound 11) can be formed. This compound dimerises at the 1,2- and 1,3- positions to give an intermediate product, from which Compound 12 is formed. Decarboxylation, dehydrogenation, and the addition of water leads to the formation of purpurogallin-8-carboxylic acid (Compound 13). The role of this compound in metabolism is not completely understood.

o-, *p*- Tautomerism of Hydroxyquinones

In terms of physiological activity, the healing effects of HS might be related to the compositions of hydroxylated quinones in the medium. Experiments with plants suggest that hydroxyquinones are physiologically more active than quinones *per se*. The basis for this is considered to be the *o*-, *p*- tautomerism involved. This is illustrated in Figure 3 for derivatives of thymoquinone (2-isopropyl-5-methyl-1,4-benzoquinone; Compound I), hydroxylated in the 6- and in the 3- positions (Compounds II and III, respectively).

The *o*-, *p*-tautomerism can be illustrated by using the example of differently hydroxylated thymoquinones (Figure 3). The ring positions of the methyl and *iso*-propyl substituents can be considered to be representative of the positions of ring substitution on the degradation products of lignin.

The maximum UV absorbance of the *p*-quinonoid configuration is at ca 320 nm, and that of the *o*-quinonoid configuration is at ca 260 nm. According to the UV spectra of 3- and 6-hydroxythymoquinone (Figure 3), relatively high proportions of these compounds are present in the *o*-form, whereas for 2-hydroxy-l,4-naphthoquinone, only 0.2% is in the

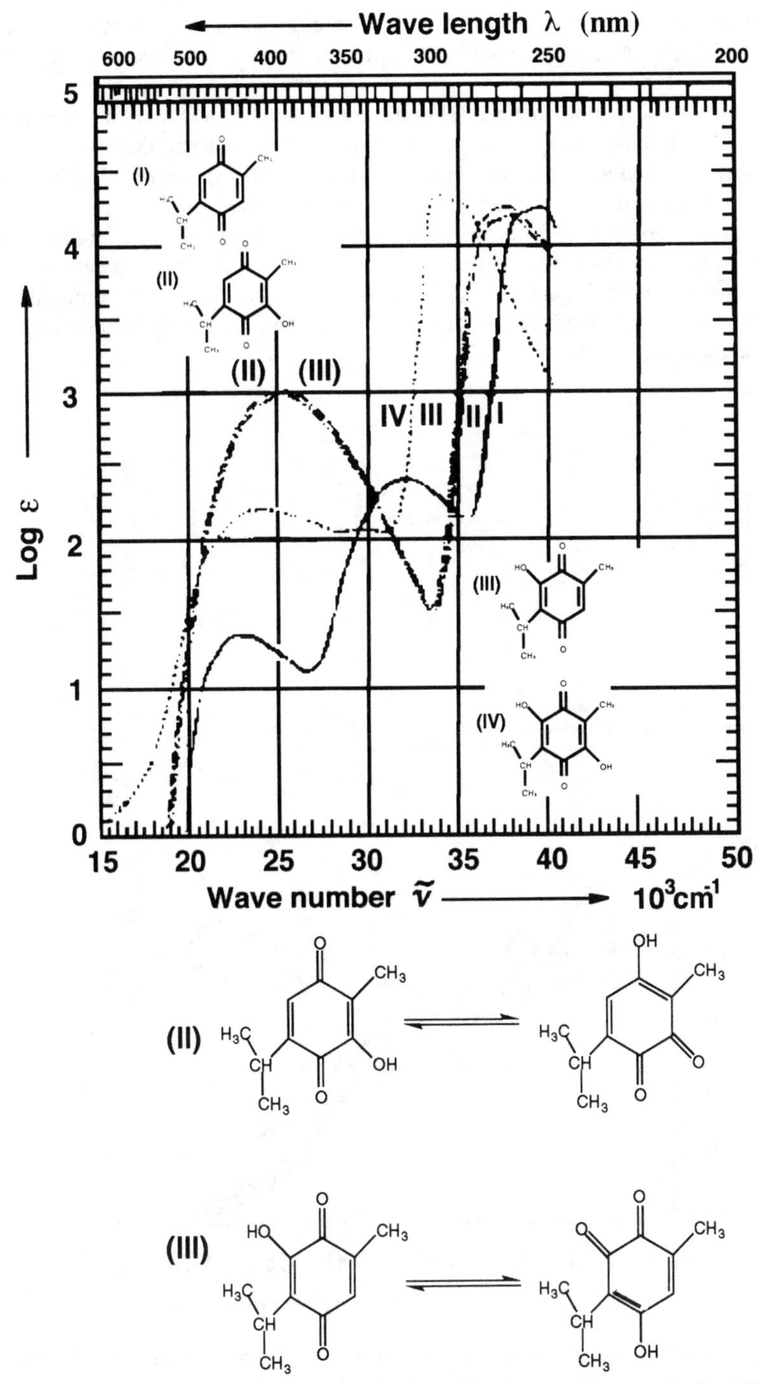

Figure 3 *o-, p-Tautomerism of hydroxythymoquinones and the UV spectra of thymoquinone and hydroxythymoquinone compounds*

o-form (Fieser and Fieser, 1954).

It is thought that in natural and synthetic HAs, aromatic rings are partially linked as conjugated ring systems. As an example of this type of system, di-quinonylbenzene was synthesised (Ploetz, 1955).

The UV spectrum of di-quinonylbenzene (Figure 4) suggests that a radical intermediate with an *o*-quinonoid structure also exists. After the addition of water in the presence of oxygen, the spectrum changes more or less to the characteristic spectrum of naturally occurring HAs, in which a mixture of *o*- and *p*- configurations can be assumed.

The maximum UV absorbance of the *p*-quinonoid configuration is at ca 320 nm, and that of the *o*-quinonoid configuration is at ca 260 nm. According to the UV spectra of 3- and 6-hydroxythymoquinone (Figure 3), relatively high proportions of these compounds are present in the *o*-form, whereas for 2-hydroxy-1,4-naphthoquinone, only 0.2% is in the *o*-form (Fieser and Fieser, 1954).

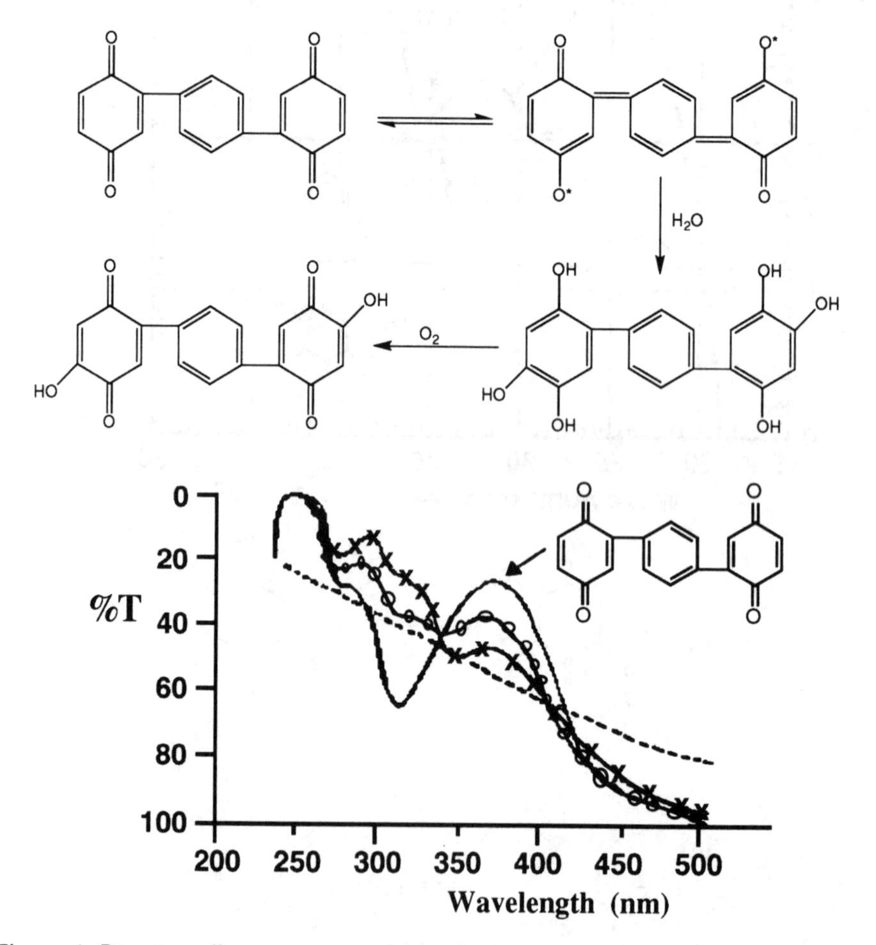

Figure 4 *Di-quinonylbenzene as a model for the formation of humic substances, showing the change over time in the UV spectrum of the transformation products of 1,4-diquinonylbenzene in the presence of water (Time — = 0 h, ɵ = 1h, x = 2h)*

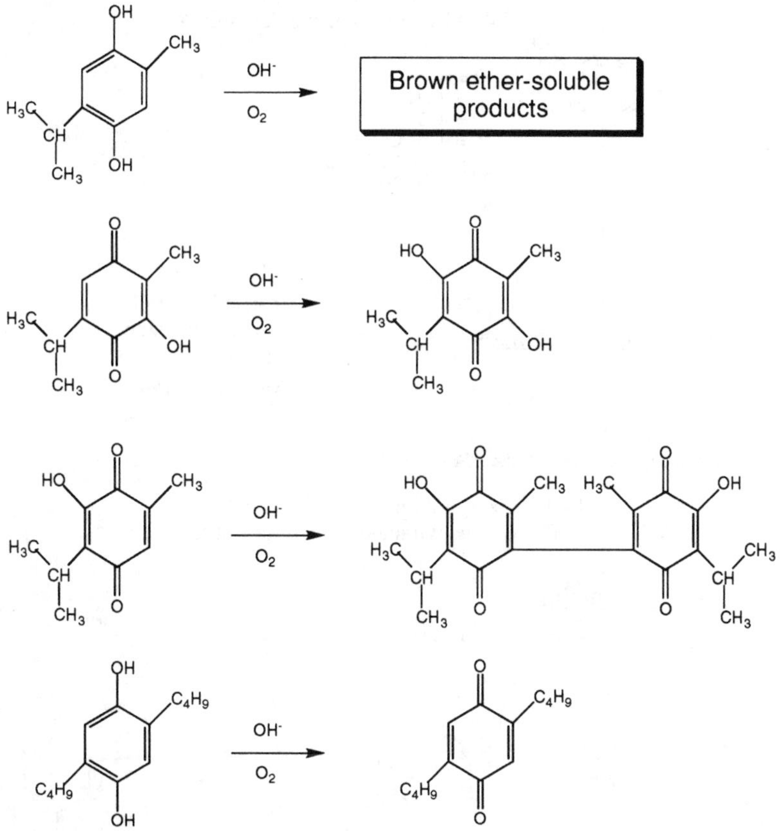

Figure 5 *Substitution and oxidation of hydroxythymoquinones*

Model Experiments with Differently Substituted Benzoquinones

Model compounds, such as substituted hydroxythymoquinones, were used for studies of the formation of natural and/or synthetic HAs. This is relevant to HS with origins in lignins because lignin transformation products have a variety of substituents on the aromatic nucleus, including hydroxyl and methoxyl groups.

Oxidation of unsubstituted hydroxythymoquinone in alkaline media leads predominantly to the formation of brown ether-soluble products. These products may in some ways be comparable to some of the components of HS. In the cases of the substituted thymoquinones, 3-hydroxythymoquinone is dimerised in the 6-position, and 6-hydroxythymoquinone is hydroxylated in the 3-position (Figure 5).

If the same experiment is carried out after substitution of the methyl- and the *iso*-propyl groups by two *tert*-butyl groups, which are more bulky, the oxidation stops at the corresponding *p*-quinone (Figure 5). Such experiments demonstrate the dependence of oxidation reactions upon the nature of the ring substitution, both in the case of lignin degradation and in the formation of synthetic humic substances from low molecular weight phenolic compounds.

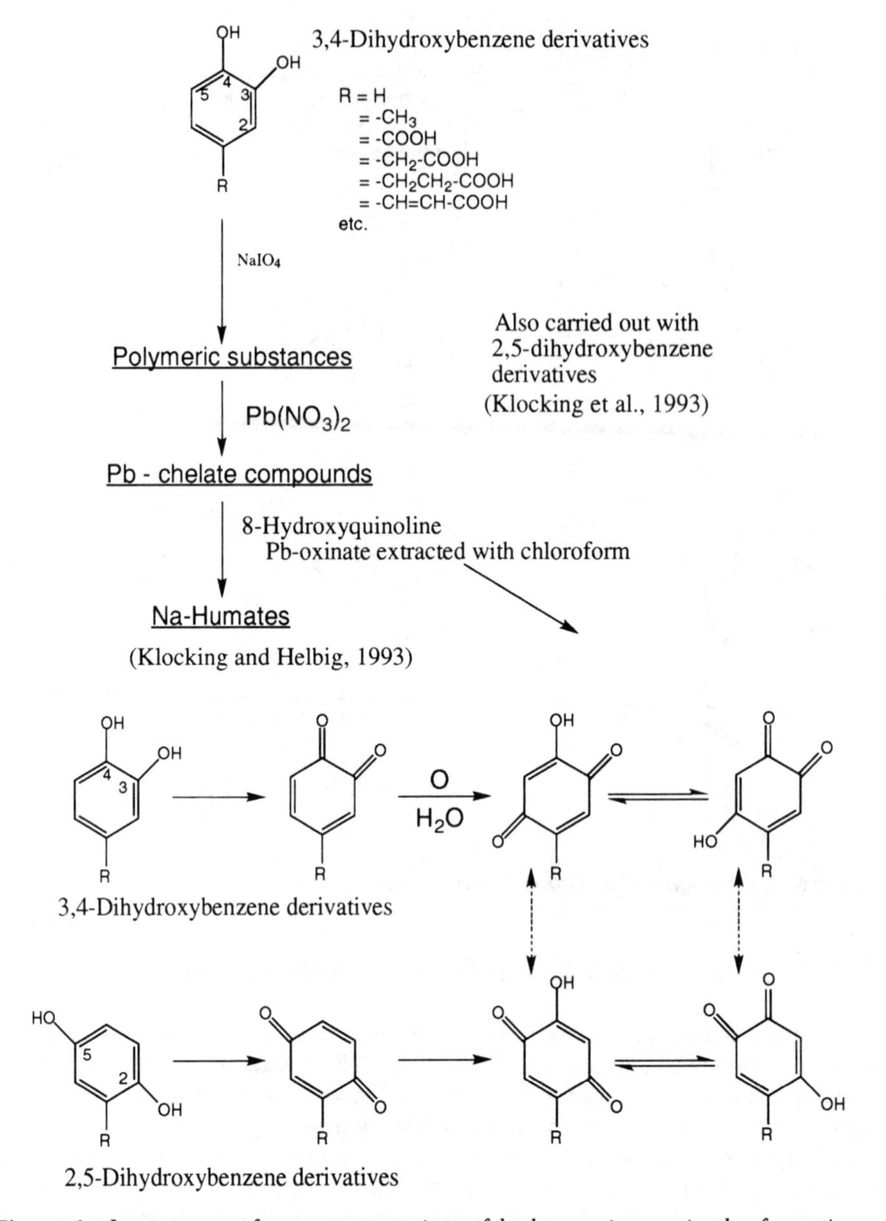

Figure 6 *Importance of o-, p-tautomerism of hydroxyquinones in the formation of synthetic humic acids*

In general, it can be concluded that during the formation of natural and synthetic humic substances different types of oxidation reactions occur and, in most cases, a quinonoid structure is ultimately formed.

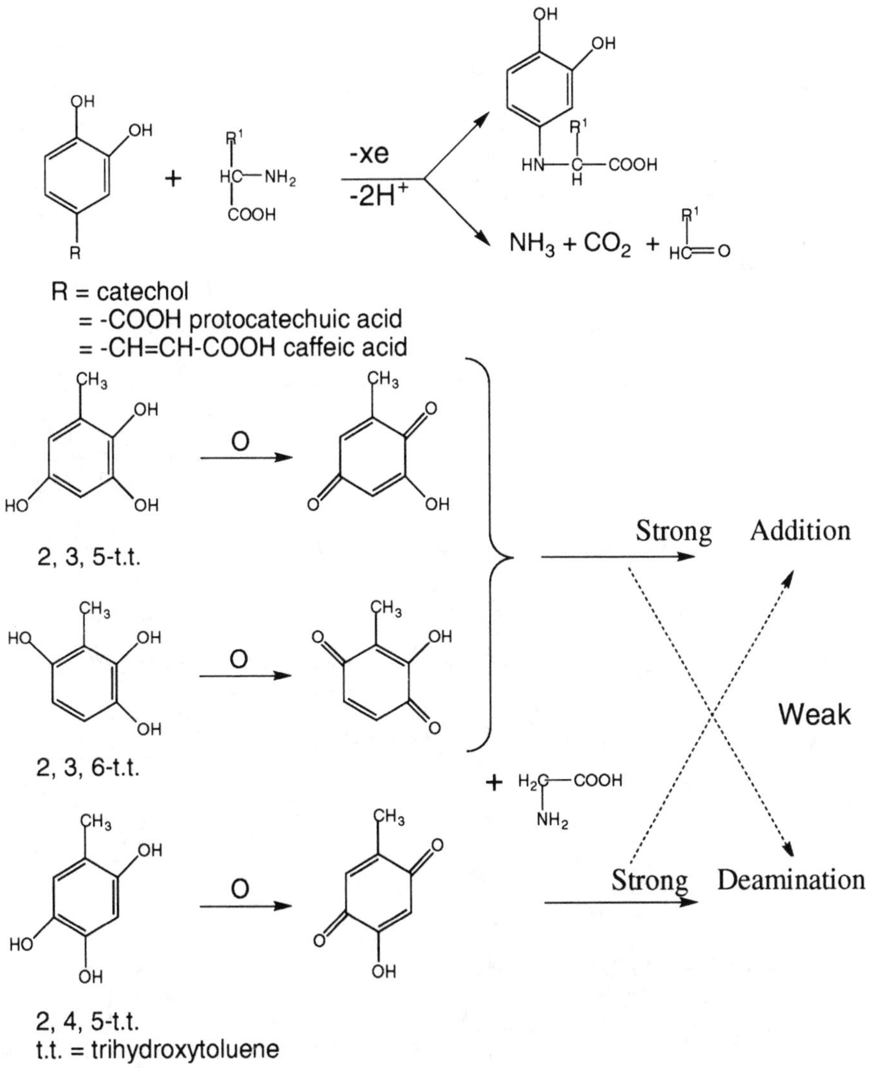

R = catechol
= -COOH protocatechuic acid
= -CH=CH-COOH caffeic acid

-xe
$-2H^+$

$NH_3 + CO_2 +$

2, 3, 5-t.t.

2, 3, 6-t.t.

2, 4, 5-t.t.
t.t. = trihydroxytoluene

Strong Addition

Weak

Strong Deamination

Figure 7 *Addition and/or oxidative deamination of amino acids by phenolic or quinonoid compounds in oxidizing media*

Principal Reactions in the Formation of Synthetic Humic Acids

As indicated in Figure 6, Klöcking and Helbig (1993) synthesized H-TAs from a series of substituted 3,4-dihydroxybenzene compounds, and Klöcking *et al.* (1994) used 2,5-dihydroxybenzene compounds with the same substituents. The H-TAs synthesized from the 3,4- and the 2,5-dihydroxybenzene derivatives do not have different properties, because *o*-, *p*- tautomerism of the starting materials leads to the same configuration (Figure 6; dashed lines).

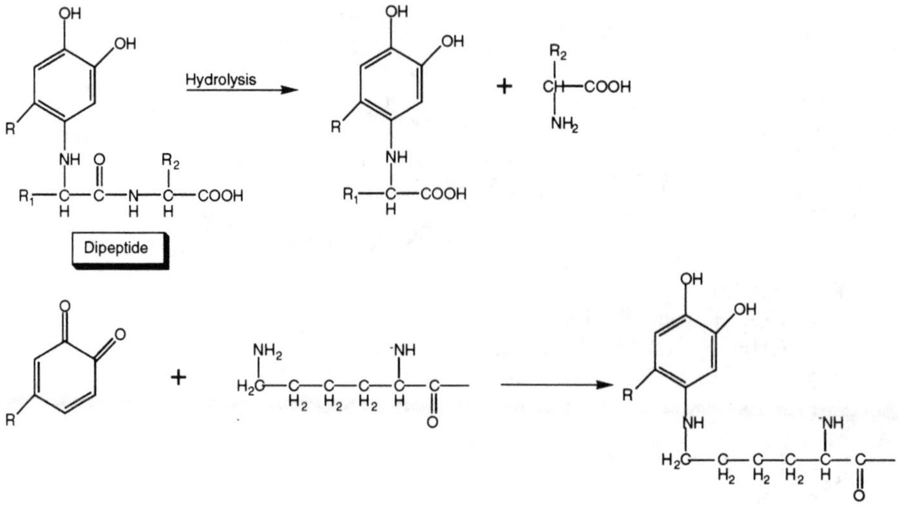

Figure 8 *Hydrolysis of addition products of quinones and peptides*

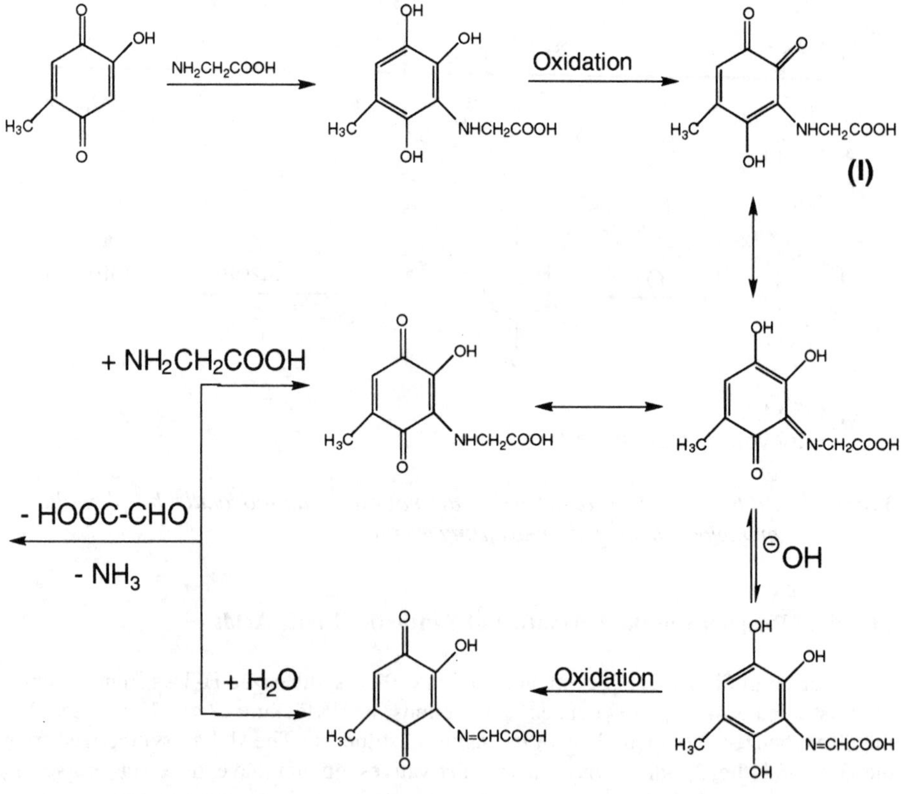

Figure 9 *Mechanism of oxidative deamination*

Catechol, protocatechuic acid, and caffeic acid (which have also been used for the synthesis of H-TAs) add to and/or deaminate amino acids in oxidizing media. The extent of addition or deamination depends on the nature of the substitution of the phenolic or of the quinonoid compounds.

Differently substituted trihydroxytoluenes are given as examples to demonstrate the principles of these two types of reactions (Figure 7). In the cases of 2,3,5- and 2,3,6-trihydroxytoluene, strong addition and weak deamination reactions occur; in the case of 2,4,5-trihydroxytoluene strong deamination and weak addition is observed (Flaig, 1988).

Healing Effects and Interactions of Hydroxyquinones with Amino Acids and Proteins

According to the results of Klöcking *et al.* (1994), and of Klöcking and Helbig (1993), the nitrogen in naturally occurring HS does not play a role in their healing effects. The healing effects which have been described for natural or synthetic HS occur mainly in physiological processes where amino acids, peptides, or proteins play a role. It is of interest, therefore, to study the interactions of synthetic H-TS or their precursors with amino acids and their polymers.

Free amino acids can be liberated on hydrolysis of the addition products, leaving a mixture of amino acids, quinone peptide adjuncts, and eventually on prolonged hydrolysis, quinone structures (Figure 8). The freed quinone could react with the ε-amino group of the peptide-bound lysine.

The mechanism of oxidative deamination was elucidated mainly by polarographic methods (Flaig and Riemer, 1971).

When glycine is added to hydroxyquinones the phenolic addition product is oxidised to the corresponding quinone, as shown in Figure 9. By rearrangements of the double bonds and further oxidation, glyoxalic acid and ammonia are released, and the original quinone is formed again. After addition of a further molecule of glycine, the reaction cycle is continued. For the elucidation of this mechanism it was important that the *o*-quinone compound (Figure 9; Compound I) could be identified.

The transformation of tryptophan to ß-indoleacetic acid and other derivatives was also investigated using ^{14}C-labelled compounds. The proposed reaction mechanism of Trautner and Roberts (1950) is not relevant, because the reaction cycle does not occur.

2 Conclusions

It is possible that the healing effects of humic substances in the illnesses mentioned by Klöcking and Helbig (1993) and by others can be biochemically explained by the interactions of phenolic or quinonoid monomeric or polymeric precursors of humic acids, or of humic acids *per se*, with proteinaceous materials.

References

Fieser, L.T. and M. Fieser. 1954. Lehrbuch der Organischen Chemie. Weinheim: Verlag Chemie.

Flaig, W. 1993. Verkommen von Ligninabbauprodukten, stickstoff haltigen Verbindungen aus dem Proteinabbau und von polyzyklischen Aromaten in Torf, in Moortherapie, Grundlagen und Anwendungen. p. 78-91. *In* W. Flaig, C. Goecke und W. Kauffels, (eds), *Uberreuter Wissenschaftsverlag*, Ges.m.b.H. , Berlin.

Flaig, W. 1988. Generation of model chemical precursors. p. 75- 92. *In* F.H. Frimmel and R.F. Christman (eds), *Humic Substances and Their Role in the Environment.* Wiley S. Bernhard, Dahlem Konferenzen, 1988.

Flaig, W. and H. Riemer. 1971. Polarographische Untersuchungen zum Verhalten von Trihydroxytoluolen bei der Reaktion mit Glycin unter oxidierenden Bedingungen. *Lieb. Ann. Chem.* **746**:81-85.

Klöcking, H.P., R. Klöcking and B. Helbig. 1994. Anti-Factor-IIa activity of humic acid-like polymers derived from *p*-diphenolic compounds. p. 411-415. *In* C.E. Clapp, M.H.B. Hayes, N. Senesi, and S.M. Griffith (eds), *Humic Substances and Organic Matter in Soil and Water Environments.* Proc. 7th Intern. Meeting, IHSS (St. Augustine, Trinidad and Tobago, 1994). IHSS, St. Paul Minn.

Klöcking, R. and B. Helbig. 1993. Physiologische Wirkung von Huminstoffen als Grundlage für ihre medizinische Anwendung p. 173-189. *In* DFG, Deutsche Forschungsgemeinschaft, Mitteil. XII der Senatskommission für Wasserforschung.

Ploetz, T. 1955. Polymere Chinone als Huminsäuremodelle. *Z. Pflanzenern, Bodenkde*, **69**, 50 - 58.

Quecke, K. and C. Loschen. 1989. Beeinflussung der Arachidonsäure Kaskade durch pflanzliche Inhaltsstoffe und Torfbestandteile. p. 166 - 174. *In* M. Hornig (ed), *"Gynäkologische Balneotherpie"* 25. - 27 Mai, 1989, Bad Waldsee.

Tolpa, S. 1992. p. 1 - 149. *In* A. Danysz (ed.), *Preclinical Investigations on Tolpa Torf Preparation. Collective Works.* Torf Corporation, Wroclaw.

Trautner, E.M. and E.A.H. Roberts. 1950. T he chemical mechanism of the oxidative deamination by catechol and polyphenol-oxidase. *Aust. J. Sci. Res. **Ser.*** **B3**, 356 - 380.

Section 5

Composts, Peats, and Sludges

The emphasis on the recycling of organic wastes, and the ban from the end of 1998 on the disposal of sludges at sea has led in recent years to considerable research activity on composting processes and on the utilization of sewage sludge on land. There have been extensive investigations world wide of the uses of sludges as soil amenders, and especial attention has been given to the deleterious influences which the heavy metal contents of municipal sludges can have on plant and animal health. Many of these studies have followed the line "apply and see", and too few have employed a fundamental approach in which the properties of the soils and of the amendments were carefully studied, and the changes resulting from the applications explained on the basis of the science involved.

Traditional agriculturalists attribute to farmyard manure (FYM) growth promotion properties that cannot be explained by its manurial value based on its elemental composition. Gardeners have always extolled the value of composts.

The contribution by Hoitink and Grebus (p. 359 - 366) provides references to appropriate procedures for the production of composts of consistent quality and maturity, and it lists a number of soil borne pathogens whose activities are suppressed by composts of appropriate compositions and quality. The temperatures experienced during composting can kill biocontrol agents as well as pathogens, and so it is necessary for the biocontrol agents to recolonize the composts in order to achieve appropriate suppression of disease. Emphasis is placed on the fact that, in order to produce composts of consistent quality, it is necessary to use the same starting materials, and to apply rigidly the same composting procedures because the levels of compost maturity are important in order to realise disease suppression effects. Immature composts can provide substrates for pathogens, and excessively matured composts do not support the biological activities of biocontrol agents. Highly humified soil organic matter, for example, is a poor substrate for biocontrol agents.

The contribution by Johnston *et al.* (p. 367 - 383) has provided evidence to indicate that amendments of peat to sandy loam soils with low organic matter (OM) contents (1.0 to 1.4%) provided 'points of weakness' in the surface cap or crust which can develop after heavy rainfall. These 'points' improved seedling emergence. Peat was shown always to have had a beneficial effect on crumb stabilization, even though the build up of organic matter from site to site varied, probably because of differences in soil compaction.

Vaidyanathan (p. 384 - 409) has provided an extensive review of the influences of sewage sludge on the growth of crops, based on observations of research carried out by the Agricultural Development and Advisory Service (ADAS). His review has addressed experiments in several sites where heavy loadings of metals (especially zinc, copper, nickel, and chromium) in sludges were added to different soils supporting different vegetable and grass/clover crops. The responses were different for the different metals, for the different soils, and for the different crops.

The contribution by Luo and Christie (p. 410 - 424) provides results for the growth of barley plants (*Hordeum vulgare* L.) in pots in greenhouse experiments in which two different soils were amended with increasing rates of lime stabilized sewage sludges. The influences of the sludges on dissolved organic substances (DOS) and the heavy metal concentrations in the soil solution were measured. Increasing amendment rates gave increases of DOS, and (with the exception of zinc) increased concentrations of heavy metals (Cu, Ni, and Cr) in the soil solution; concentrations of Cd and of Pb were too low to be detected. The effects on the compositions of the soil solution were related to soil type, and the data indicated that DOS from sludge amendments could have affected the concentrations of Cu, Ni, and Cr in solution, and the formation of metal-organic complexes with the DOS may have influenced the mobilities of these metals.

Aitken and Cummins (p. 425 - 437) investigated the effects of different application rates of sewage sludges on the behaviour of heavy metals in three soils in Scotland, and on the uptake of the metals by grass. After four years the high application rates increased the soil concentrations of heavy metals to a depth of 100 - 150 mm. The highest concentration was at 0 - 25 mm, and the distribution was not affected by soil pH. The uptake by plants of Cd, Ni, Cr, and Pb was not affected by the sludge applications, but plant contents of Cu and Zn were increased. The concentrations in plants did not exceed the UK regulations for herbage. It was concluded that the sludge appeared to provide a sustainable soil conditioner, although its long term use in grassland could lead to a build up of metals in the topsoil (0 - 25 mm).

O'Riordan and McGrath (p. 438 - 445) also studied the effects on soil and on herbage of repeated applications sewage sludge at two sites in Ireland. To one site relatively high rates of Cu, Zn, and Pb were applied in the sludges, whereas the application rates to the second site were very much lower. Although increased soil levels of total and of EDTA-extractable metals were evident for the two application rates, the increases were not significant at depths below 52 mm and 13 mm in the cases of the heavy and light applications, respectively. Spring applications of metal-rich sludges usually increased herbage metal concentrations, but the effect was carried over to the second year only in the case of Zn. The low level applications had little or no effects on the metal contents of the herbage. The authors conclude that sludge applications that conform to Irish law are unlikely to give rise to undesirable effects.

In their investigations of the effects of cattle slurry and FYM on spring herbage production from grass/white clover, Humphreys *et al.* (p. 446 - 461) applied slurry in mid-August or in mid-March, and FYM only in mid-March. The spring applications gave significantly improved yields of grass compared with the summer applications. The manure treatments decreased white clover stolon development, and the white clover content. Spring-applied slurry favoured stolon survival during the winter, and the clover content in the spring. The manure treatments maintained soil P, K, and Mg levels.

Disposal of cheese whey and other products of the dairy industry can present problems. Ross and Mullen (p. 462 - 474) have shown from soil column experiments that repeated and high application rates of cheese whey can lead to groundwater contamination, especially of nitrate and orthophosphate, but the applications did give rise to low though acceptable levels of herbage production.

The study of Lyons and Reidy (p. 475 - 484) investigated the potential of unmodified virgin peats from high moor and low moor sources for the treatment of leachates from landfill. Their results are encouraging, and suggest that virgin peat would be suitable as a biological filter. However, peat clogging presents a problem. Decreases in the leachate concentrations of the order of 70% - 80% were obserbed for metals and for P.

Composts and the Control of Plant Diseases

Harry A.J. Hoitink and Marcella E. Grebus

DEPARTMENT OF PLANT PATHOLOGY AND GRADUATE ENVIRONMENTAL STUDIES PROGRAM,
OHIO AGRICULTURAL RESEARCH AND DEVELOPMENT CENTER, THE OHIO STATE UNIVERSITY, WOOSTER, OHIO 44691, USA

Abstract

Composts have the potential to provide biological control of plant diseases. Foliar as well as root pathogens may be affected. Many factors control these effects. Heat exposure during composting kills or inactivates pathogens if the process is monitored properly. Unfortunately, most biocontrol agents are also killed by this heat treatment. Thus, biocontrol agents must recolonize composts after peak heating. The raw feedstock, the environment in which the compost is produced, as well as conditions during curing and utilization determine the potential for recolonization by this microflora and disease suppression. Controlled inoculation of compost with biocontrol agents has proved necessary to induce consistent levels of suppression on a commercial scale. The decomposition level (stability) of composts also affects suppressiveness. Immature composts serve as food for pathogens and increase disease even when biocontrol agents are present. On the other hand, excessively stabilized organic matter does not support the activity of biocontrol agents. Composts also may induce detrimental effects. For example, chemical factors may negate suppressiveness. Salinity, and the concentration of nitrogen in composts are important chemical factors to be considered.

1 Introduction

During the 1960s, nurserymen across the United States explored the possibility of using wood industry wastes as peat substitutes to lower production costs. Since that time, procedures for the composting of tree bark have been developed that avoid nitrogen immobilization and allelopathy toxin problems associated with fresh bark (Hoitink and Kuter 1986). In the past decade, similar procedures have been developed for composts produced from the food industry, agricultural wastes, and from municipal wastes. At this time there is emphasis on composting of yard trimmings and municipal solid wastes.

Improved plant growth and decreased losses caused by Phytophthora root rots were soon observed as side benefits from the utilization of composts by the nursery industry in the U.S. In practice, control of such diseases with composts is at least as effective as that obtained with fungicides. In many parts of the world, therefore, agriculture relies heavily on composts for the control of diseases caused by these soil-borne plant pathogens.

Composts must be of consistent quality and maturity to be used successfully for the biological control of diseases of horticultural crops, particularly in container media (Inbar *et al.*,1993). Variability in this quality parameter is one of the principal factors limiting compost utilization for this purpose. In ground bed or field agriculture, maturity is less important as long as composts are applied well ahead of planting to allow for additional stabilization. Lack in maturity frequently causes problems here as well, however.

Predicting maturity of composts (Chen and Inbar, 1993), as related to the potential for improved plant health, is now possible. In addition, as a result of increased research on biological control in the 1980s, information is now available that facilitates the predictable formulation of composts for the suppression of soilborne diseases caused by *Fusarium* spp., *Phytophthora* spp., *Pythium* spp., *Rhizoctonia solani* and other pathogens (Hoitink *et al.*, 1993). To maintain quality related to both plant growth and disease control, compost producers must develop a better understanding of the processes involved in this method of biological control. A brief review of the principles involved is presented here.

2 Fate of Biocontrol Agents During Composting

The composting process is often divided into three phases. The first phase occurs during the initial 24-48 h of composting as temperatures gradually rise to 40-50 °C, and when sugars and other easily biodegradable substances are lost. During the second phase, when temperatures of 40-65 °C prevail, cellulosic substances that are less biodegradable are destroyed. Lignins break down even more slowly, but the rate varies among plant species. Thermophilic micro-organsims predominate during this part of the process. Plant pathogens and seeds are killed by the heat generated during this high temperature phase (Bollen 1993; Farrell 1993). Compost piles must be turned frequently to expose all parts to high temperature to produce a homogeneous product free of pathogens and weed seeds.

The third, or the curing phase of composting begins as the concentrations of readily biodegradable components in wastes decline. As a result, rates of decomposition, heat outputs, and temperatures decline. At this time, mesophilic microorganisms recolonize the compost moving into the pile from the outer low temperature layer. Humic substances accumulate in increasing quantities at this time. Mature composts consist largely of lignins, humic substances, and biomass, and have a dark colour.

Beneficial as well as detrimental micro-organisms are killed during the high temperature phase of composting. Therefore, suppression of pathogens and/or disease is largely induced during curing as biocontrol agents recolonize composts after peak heating. *Bacillus* spp., *Enterobacter* spp., *Flavobacterium balustinum*, *Pseudomonas* spp., other bacterial genera, and *Streptomyces* spp., as well as *Trichoderma* spp., *Gliocladium virens*, and other fungi have been identified as biocontrol agents in composts (Chung and Hoitink 1990, Hardy and Sivasithamparam 1991, Hoitink and Fahy 1986, Phae *et al.*,1990).

3 Mechanisms of Biological Control in Composts

Two mechanisms of biological control, based on competition, antibiosis, hyperparasitism, and possibly also induced systemic resistance in plants, have been described for compost-amended substrates. Propagules of plant pathogens, including *Pythium* and *Phytophthora* spp., are suppressed through a mechanism known as "general suppression" (Chen *et al.*, 1988, Cook and Baker, 1983, Hardy and Sivasithamparam, 1991, Mandelbaum and Hadar, 1990). Many different kinds of microorganisms present in compost-amended container media function as biocontrol agents for diseases caused by *Phytophthora* and *Pythium* spp. (Chen *et al.*,1988, Hardy and Sivasithamparam, 1991). Coliforms and other faecal pathogens are also suppressed through this mechanism. Propagules of these pathogens, if inadvertently introduced into composts, do not germinate in response to nutrients released in the form of seed or root exudates. The high microbial activity and biomass in composts caused by the "general soil microflora", prevent germination of spores of the pathogen and infection of the host, presumably through microbiostasis (Chen *et al.*, 1988, Mandelbaum and Hadar, 1990). The spores of these pathogens remain dormant and are not killed if introduced in compost-amended soil. An enzyme assay that determines microbial activity, based on the rate of hydrolysis of fluorescein diacetate, predicts suppressiveness of potting mixes to Pythium diseases (Chen *et al.*, 1988). There is similar information for "organic farms" where soil-borne diseases are less prevalent (Workneh *et al.*, 1993).

The mechanism of biological control for a plant pathogen such as *Rhizoctonia solani*, a common damping-off pathogen, is entirely different from that of *Pythium* and *Phytophthora* spp. *R. solani* produces large progagules known as sclerotia. These sclerotia do not rely on root or seed exudates as sources of energy for germination and infection of plants, and respond to volatile substances given off by the host. A much more narrow group of microorganisms is responsible for supression of *Rhizoctonia*. This type of suppression is referred to as "specific suppression" (Hoitink *et al.*, 1993).

Variability in suppression of Rhizoctonia damping-off encountered in soils amended with mature composts is due in part to random recolonization after peak heating of compost by a microflora that varies in efficacy against *R. solani.* Composts produced in the open near a forest, an environment that is high in microbial species diversity, are more consistently suppressive to Rhizoctonia diseases than those from the same materials in facilities that are partially enclosed, where few of these microbial species survive (Kuter *et al.*, 1983). Composts prepared from municipal sewage sludges and MSW are consistently conducive to *R. solani* because care is taken to kill faecal pathogens and parasites with heat exposure. As mentioned, this process kills most beneficial microorganisms as well. These composts, although naturally suppressive to Pythium diseases, have to be incubated for a month or more before becoming colonized by chance by the right-specific microflora and thus also become naturally suppressive to *R. solani* (Kuter *et al.*, 1988). In field soil, several months may pass before suppression is induced (Lumsden *et al.*, 1983).

To solve the problem of variability in suppressiveness of compost to *R. solani*, specific fungal and bacterial inoculants have been developed that, when introduced into compost after peak heating, but before significant levels of recolonization have occurred, induce consistent levels of suppression. Patents have been issued to the Ohio State University for this process (Hoitink 1990). In Japan, Phae *et al.* (1990) have isolated a *Bacillus* strain that induces predictable biological control in composts. In the U.S., efforts are now underway to commercialize controlled inoculation of composts with biocontrol agents in cooperation with Earthgro Inc., Lebanon, CT, and Chris Hansen's Biosystems, Milwaukee, WI. The disease suppressive composts must be registered with US-EPA.

4 Role of Organic Matter Decomposition Level

The decomposition level of organic matter (OM) in compost has a major impact an disease suppression. *R. solani* is highly competitive as a saprophyte. It can utilize cellulose and colonize fresh wastes. It cannot colonize the low-in-cellulose, mature compost. On the other hand, *Trichoderma* isolates that function as biocontrol agents of *R. solani* are capable of colonizing mature compost. In fresh, undecomposed organic matter, biological control does not occur because both fungi grow as saprophytes and *R. solani* remains capable of causing disease. In mature compost, on the other hand, sclerotia of *R. solani* are killed by the hyperparasite, and biological control prevails.

Because the OM decomposition level is so important, composts must be stabilized adequately to reach that decomposition level where biological control is possible. In practice, this occurs in composts (e.g. tree barks and yard wastes) that have been stabilized far enough not to induce phytotoxicity due to lack of stability, nor immobilize nitrogen during plant growth, but are colonized by the appropriate specific microflora. Practical guidelines that define this critical stage of decomposition are not yet available. At the present time the industry controls the decomposition level by maintaining constant conditions during the entire process, and by adhering to a given time schedule.

Excessively stabilized OM does not support adequate activity of biocontrol agents. In a highly mineralized soil, where humic substances (HS) are the predominant forms of OM, suppression is lacking and soilborne diseases are severe. It is not yet known how long a compost incorporated in soil supports adequate levels of activity for biocontrol agents. Presumably, the time varies with soil temperature, and with the type of OM from which the compost was prepared. Loading rates and farming practices also play a role.

In order to bring a solution to this problem we have studied the "carrying capacity" of soil OM in potting mixes prepared with sphagnum peat. Sphagnum peat typically competes with compost as a source of organic amendments for horticulture. Both the microflora and the OM in peat can affect suppression of soil-borne diseases. The literature on that effect is reviewed briefly here. OM in sphagnum peat generally does not support high microbial activity because of its stability (resistance to decomposition). Sphagnum peat, harvested from depths of 30 cm and deeper in peat bogs, is darker and more highly decomposed than the surface peats, and is low in activity and consistently conducive to Pythium and Phytophthora root rots. On the other hand, light sources of sphagnum peat, harvested from the surface of peat bogs, are less decomposed and have a greater microbial activity. The suppressive effect of light peat to Pythium root rots is of short duration, however (Boehm and Hoitink, 1992; Tahvonen, 1982; Wolffhechel, 1988). Light peats, therefore, can be used most effectively for short production cycles, such as in plug and flat mixes used in the ornamentals industry. Composts have longer lasting effects (Boehm and Hoitink, 1992; Boehm *et al.*, 1993).

The "slow release" nature of the organic nutrients tied up as "carbohydrates" in mature composts and light peat support the activity of the microflora and thus sustain biocontrol. This is thought to be the essence of biological control associated with "organic farming." Non-destructive, direct spectroscopic (NMR and FT-IR) procedures are now being used to predict the OM decomposition level (carbohydrate content) and compost maturity. These techniques also are used to determine the amount of undecomposed readily biodegradable organic matter present as carbohydrates in compost (Inbar *et al.*, 1989; Chen and Inbar, 1993). This approach to the analysis of maturity promises to yield a technology that can assess the "carrying capacity", or the potential for a soil to support sustained microbial activity, and thus "general suppression" of these soil borne plant pathogens (Hoitink *et al.*, 1993). This field of science is still in its early stages, however.

5 Effects of Chemical Properties of Composts on Disease Suppression

The beneficial effects induced by composts can be temporarily destroyed by unfavorable chemical properties of composts. We showed some years ago that salinity in composts affects the potential for the induction of biological control of Phytophthora and Pythium diseases. For example, composted municipal sludge with a high salinity value (> 10 mS in the saturated paste extract) applied to soybean just before planting increased Phytophthora root rot of soybean over that in the control. However, the same compost applied several months ahead of planting suppressed the disease. Addition of salt (NaCl) to this soil just before planting, in an amount equivalent to that applied in the sludge compost earlier, destroyed the suppressive effect (reviewed in Hoitink and Fahy, 1986). Composted manures may be high in salinity. A procedure has been developed in Israel for the preparation of compost from separated cow manure solids (Chen and Inbar, 1993). The liquid fraction separated from the manure before composting contains the salt and is applied as fertilizer in field agriculture. The low-in-salinity compost has a broad spectrum suppressive effect. It is incorporated in potting mixes at a volumetric ratio of 40%, rather than the maximum level of 15% typically recommended for manure compost high in salts. The low-in-salinity, composted, separated cattle manure effectively controls soil borne diseases, including those caused by *Pythium* spp.

Some composts, such as municipal sewage sludge and poultry manure composts, release high levels of nitrogen after application in field or container culture. When used as a peat substitute in potting mixes, composted sludges often are used at 15% on a volume basis. This results in excessive nitrogen availability to some crops. Fireblight and Phytophthora blights in the foliage and stems of plants and Fusarium wilts are increased by such excessive fertility practices. These responses in compost-treated plants can be explained, based on interactions between fertility, plants, and plant pathogens. Composts that do not release high levels of N (bark composts) do not enhance the severity of these diseases, and are used widely as peat substitutes and for the control of Fusarium wilts.

Technology for the prediction of mineralization rates of essential plant nutrients from composts is still in the early stages of development. Therefore, composts can only be used on a predictable basis for disease control, or even for plant growth, if the raw organic matter from which it was prepared is kept constant, the composting process is controlled and kept constant, and finally, curing is carried out in a such a way that a uniform product is made available. Pine bark is one of the few organic resources for which a total process has been described (Hoitink *et al.*, 1991). Much research needs to be carried out to extend this procedure into practice for yard waste, MSW, or other composts.

6 Compost Steepages for Disease Control

During the past decade, Dr. H. C. Weltzhien from the Rheinische Friedrich- Wilhelms Universitat in Bonn, Germany, has directed a series of research projects on the control of plant diseases of above ground plant parts using compost water extracts (Weltzhien, 1992). The steepages are prepared by soaking mature manure composts (not pure chicken manure) in water (still culture; 1:1, w/w) for 7-10 days. The steepage is filtered and then sprayed on plants. Plants must be sprayed frequently, and the efficacy varies with crops and with the disease under question.

Sackenheim (1993), from the same laboratory, has shown that aerobic microorganisms predominate in these steepages. The microflora include strains of bacteria and isolates of fungi already known as biocontrol agents. Enrichment strategies, which include nutrients as well as microorganisms, have been developed to improve the efficacies of steepages.

Disease control induced by compost steepages in the past has been attributed in part to systemic aquired resistance induced in plants by microbes present in the extracts. The recent work on grape by Sackenheim (1993) does not support that assumption. Plants differ widely, however, in their mechanisms of resistance to plant pathogens. Lack of activity of extracts under some conditions could possibly be due to soil factors.

A factor that has been entirely overlooked, but could play a role in efficacy of steepages, is the condition of the soil on which treated plants are produced. Soils that are naturally suppressive to soil-borne plant pathogens (e.g. compost-amended soils), harbour active populations of biocontrol agents. Some of these microorganisms, while colonizing roots can, in the leaves of plants, induce protection to foliar pathogens (Wei *et al.*, 1991; Maurhofer *et al.*, 1994). We have learned recently at The Ohio State University that anthracnose of cucumber is suppressed in the foliage of the plants produced in suppressive bark composts as compared with to those grown in a 'conducive-to-.Pythium root rot' peat mix (Tseng, W., W.A. Dick and Hoitink, H.A.J., unpublished information). Thus, further research may show that plants produced in suppressive soils not only have less root rot but also could be less prone to attack by vascular wilt and foliar pathogens.

7 Disease Suppression - Future Outlook

Success in biological control of diseases caused by *Fusarium, Pythium* and *Phytophthora* spp. is possible only if all factors involved in the production of composts are defined and kept constant. Most composts cannot meet these criteria because emphasis on quality control in general is still inadequate. Composted bark and sphagnum peat, therefore, remain the principal organic materials used for the preparation of potting mixes that have suppressive effects on soilborne plant pathogens. Uses of composted yard wastes are steadily increasing, but mostly in field agriculture and in the consumer markets .

Natural suppression, based on present concepts, at best covers those diseases caused by pathogens suppressed through microbiostasis. Microbial activity in a mix, based on the rate of hydrolysis of fluorescein diacetate, is one procedure that can now be used effectively to determine the suppressiveness to Pythium root rot. By itself this procedure does not predict how long the effect will last. The potential for biodegradable carbon in a mix to support an active biomass determines that phenomenon. CPMAS ^{13}C NMR spectroscopy predicts this property for peat. The same basic information will have to be developed for composts. Successful utilization of composts will remain an art unless fundamental, as well as applied research and demonstration programmes are funded,

Controlled inoculation of composts with biocontrol agents is a procedure that must be developed on a commercial scale to induce consistent levels of suppression to pathogens such as *R. solani*. Much development is needed in this new field of biotechnology, however. Major research and development efforts must be directed into this approach to disease control. Recycling through composting is being chosen increasingly as the preferred strategy for waste treatment. For that reason, composts are becoming available in greater quantities. Peat, on the other hand, is a limited resource that cannot be recycled. Future opportunities for both natural and controlled-induced suppression of soil-borne plant pathogens, using composts as the food stuff for biocontrol agents, therefore, appear bright. Unfortunately, the waste generating industry does not yet fully recognize that disease suppression often is the most valuable beneficial effect obtained after application of composts.

References

Boehm, M.J. and H.A.J. Hoitink. 1992. Sustenance of microbial activity and severity of Pythium root rot of poinsettia. *Phytopathology* **82**:259-264.

Boehm, M.J.L.V. Madden, and H.A.J. Hoitink. 1993 Effect of organic matter decomposition level on bacterial species diversity and composition in relationship to Pythium damping-off sverity. *Applied Environ. Microbial.* **59**:4171-4179.

Bollen, G.J. 1993. Factors involved in inactivation of plant pathogens during composting of crop residues. p. 301-318. *In* H.A.J. Hoitink and H.M. Keener (eds), *Science and Engineering of Composting: Design, Environmental, Microbiological and Utilization Aspects.* Renaissance Publications, Worthington, OH.

Chen, Y., H.A.J. Hoitink, and L.V. Madden. 1988. Microbial activity and biomass in container media predicting suppressiveness to damping-off caused by *Pythium ultimum. Phytopathology* **78**: 1447-1450.

Chen, Y. and Y. Inbar. 1993. Chemical and spectroscopical analyses of organic matter transformations during composting in relation to compost maturity. p. 551-600. *In* H.A.J. Hoitink and H.M. Keener (eds), *Science and Engineering of Composting: Design, Environmental, Microbiological and Utilization Aspects.* Renaissance Publications, Worthington, OH.

Chung, Y.R. and H.A.J. Hoitink. 1990. Interactions between thermophilic fungi and *Trichoderma hamatum* in suppression of Rhizoctonia damping-off in a bark compost-amended container medium. *Phytopathology* **80**:73-77.

Cook, R.J. and K.F. Baker. 1983. *The Nature and Practice of Biological Control of Plant Pathogens.* Am. Phytopathol. Soc., St. Paul, MN.

Farrell, J.B. 1993. Fecal pathogen control during composting. p. 282-300. *In* H.A.J. Hoitink and H.M. Keener (eds), *Science and Engineering of Composting: Design, Environmental, Microbiological and Utilization Aspects.* Renaissance Publications, Worthington, OH.

Hardy, G.E.St.J. and K. Sivasithamparam. 1991. Effects of sterile and non-sterile leachates extracted from composted eucalyptus bark and pine-bark container media on *Phytophthora* spp. *Soil Biol. Biochem.* **23**:25-30.

Hoitink, H.A.J. 1990. Production of disease suppressive compost and container media, and microorganism culture for use therein. *US Patent 4960348.* Feb. 13, 1990.

Hoitink, H.A.J., M.J. Boehm, and Y. Hadar. 1993. Mechanisms of suppression of soilborne plant pathogens in compost-amended substrates. p. 601-621. *In* H.A.J. Hoitink and H.M. Keener (eds), *Science and Engineering of Composting: Design, Environmental, Microbiological and Utilization Aspects.* Renaissance Publications, Worthington, OH.

Hoitink, H.A.J. and P.C. Fahy. 1986. Basis for the control of soilborne plant pathogens with composts. *Ann. Rev. Phytopathol.* **24**:93-114.

Hoitink, H.A.J., Y. Inhbar, and M.J.. Boehm. 1991. Status of composted-amended potting mixes naturally suppressive to soilborne diseases of floricultural crops. *Plant Disease* **75**:869-873.

Hoitink, H.A.J. and G.A.Kuter. 1986. Effects of composts in growth media on soilborne plant pathogens. p. 289-306. *In* Y. Chen and Y. Avnimelich (eds). *The Role of Organic Matter in Modern Agriculture.* Martinus Nyhoff Publishers, Dordrecht and The Netherlands.

Inbar, Y., Y. Chen, and Y. Hadar. 1989. Solid-state carbon-13 nuclear magnetic resonance and infrared spectroscopy of composted organic matter. *Soil Sci. Soc. Am. J.* **53**:1695-1701.

Inbar, Y., Y. Chen, and H.A.J. Hoitink. 1993. Properties for establishing standards for utilization of composts in container media. p. 668-694. *In* H.A.J. Hoitink and H.M. Keener (eds), *Science and Engineering of Composting: Design, Environmental, Microbiological and Utilization Aspects.* Renaissance Publications, Worthington, OH.

Kuter, G.A., H.A.J. Hoitink, and W. Chen. 1988. Effects of municipal sludge compost curing time on suppression of Pythium and Rhizoctonia diseases of ornamental plants. *Plant Disease* **72**:751-756.

Kuter, G.A., E.B. Nelson, H.A.J. Hoitink, and L.V. Madden. 1983. Fungal populations in container media amended with composted hardwood bark suppressive and conducive to Rhizoctonia damping-off. *Phytopathology* **73**:1450-1456.

Lumsden, R.D., J.A. Lewis, and P.D. Millner. 1983. Effect of composted sewage sludge on several soilborne pathogens and diseases. *Phytopathology* **73:**1543-1548.

Mandelbaum, R. and Y. Hadar. 1990. Effects of available carbon source on microbial activity and suppression of *Pythium aphanidermatum* in compost and peat container media. *Phytopathology* **80**:794-804.

Maurhofer, M., Hase, C., Meuwly P., Metraux, J.-P., and Defago, G. 1994. Induction of systemic resistance of tobacco to tobacco necrosis virus by the root-colonizing *Pseudomonas fluorescens* strain CHAO: Influence of the *gacA* Gene and of Pyoverdine Production. *Phytopathology* **84**:139-146.

Phae, C.G., M. Saski, M. Shoda, and H. Kubota. 1990. Characteristics of *Bacillus subtilis* isolated from composts suppressing phytopathogenic microorganisms. *Soil Sci. Plant Nutr.* **36**:575-586.

Sackenheim, R. 1993. Untersuchungen uber Wirkungen von wasserigen, mikrobiologisch aktiven Extracten aus kompostierten Substraten auf den Befall der Weinrebe (*Vitis viniferal* mit *Plasmopora viticola, Uncinula necator, Botrytis cenerea* und *Pseudopezicula tracheiphila.* PhD. Thesis. Rheinische Friedrich-Wilhelms Universitat, Bonn, Germany.

Tahvonen, R. 1982. The suppressiveness of Finnish light coloured Sphagnum peat. *J. Sci. Agric. Soc. Finl.* **54**:345-356.

Wei, G., Kloepper, J.W., and Tuzun, S. 1991. Induction of systemic resistance of cucumber to *Colletotrichum orbiculare* by select strains of plant growth-promoting rhizobacteria. *Phytopathology* **81**:1508-1512.

Weltzhien, H. C. 1992. Biocontrol of foliar fungal diseases with compost extracts. p 430-450. *In* J.H. Andrews and S. Hirano (eds), *Microbial Ecology of Leaves.* Brock Springer Series in Contemporary Bioscience. BSBN 0387-97579-9.

Wolffhechel, H. 1988. The suppressiveness of Sphagnum peat to Pythium spp. *Acta Hortic.* **221**:217-222.

Workneh, F., A.H.C. Van Bruggen, L.E. Drinkwater, and C. Shernnan. 1993. Variables associated with a reduction in corky root and Phytophthora root rot of tomatoes in organic compared to conventional farms. *Phytopathology* **83**:581-589.

Peat - A Valuable Resource

A.E. Johnston, M.V. Hewitt[1], P.R. Poulton[2] and P.W. Lane[3]

LAWES TRUST SENIOR FELLOW
[1]CROP AND DISEASE MANAGEMENT DEPARTMENT
[2]SOIL SCIENCE DEPARTMENT
[3]STATISTICS DEPARTMENT, IACR - ROTHAMSTED, HARPENDEN, HERTS AL5 2JQ

Abstract

Beneficial effects on crop growth and yield from applying a range of organic amendments to a sandy loam soil suggest that such amendments had the largest effect when incorporated into the seed bed. The benefits varied between seasons, depending on rainfall, and on the stability of soil crumbs in the seed bed. Peat appeared to provide points of weakness in the surface cap or crust that readily develop on sandy loams after heavy rainfall. Such points of weakness improved seedling emergence. Stabilizing soil crumbs with an inorganic soil amendment was deleterious when little rainfall fell after drilling in the spring, probably because the surface soil dried too rapidly and water was removed from around the germinating seed. Often the benefits, and occasionally the adverse effects were statistically significant. Peat had a consistent beneficial effect.

The build up of soil organic matter from the addition of peat in one of these experiments was much less than in another experiment on the same soil type. The reason was probably related to differences in soil compaction, which may be another factor to be considered when the turnover of carbon is being compared.

The rate of decline in soil organic carbon once the application of peat ceased was faster than the rate of decline in another experiment where a range of different organic amendments had been tested. In the latter experiment the added materials had small C:N ratios, and because these had been applied for periods of up to 25 years most had been subjected to microbial decomposition to produce humus which further decomposed at a uniform decay rate. Peat had only been added for six years and some, perhaps 30%, still existed in the soil as peat.

1 Introduction

Peat, often metres deep, has accumulated where climate and topography have conspired against nature's more usual way of dealing with plant residues, i.e. by microbial transformations. Such transformation processes prevent the build up of organic residues and recycle nutrients, and are especially important in natural ecosystems where nutrient availability, amongst other factors, often controls the type of vegetative cover. In agricultural systems, a shallow surface mulch of undecomposed plant residues can help control soil erosion where this is a problem, but in many temperate climates microbial decomposition of unwanted crop residues makes many aspects of crop husbandry easier.

In the early decades of this century it was accepted that peat was an asset of considerable value and that its removal was sound in principle (Russell, 1927). Peat is widely used for a variety of purposes including industrial and domestic fuel, peat based composts in amenity horticulture, and perhaps in future in effluent treatment (Lyons and Reidy, p. 475 - 484, this Volume). More importantly, in an agronomic sense many very productive agricultural soils have been developed on peat deposits. For example, in England the fen peats were brought into cultivation in East Anglia in the 17th century following the installation of extensive drainage systems to remove surface water. These soils are still very productive, although much of the peat has been lost. It was the observation that the rate of decomposition of peat increased following drainage which led Russell (1927) to consider the peats to be a wasting asset, and if these could not be conserved for future generations, their use was justified. However, there are those nowadays who hold diametrically opposite views to those of Russell. The oxidation of peat releases carbon dioxide, CO_2, a greenhouse gas. Thus any unnecessary use of peat should be avoided. It has even been suggested that, in appropriate situations, peat formation should be encouraged to remove CO_2 from the atmosphere. There is also the view that existing peat deposits should be preserved at all costs because peat, which has developed in both upland and lowland situations from contrasted plant communities, now supports very specific communities of plants and animals of considerable scientific and amenity interest.

The case for or against such views is not to be argued here. Peat, however, can be a valuable resource for applications in agriculture, and especially in horticulture. This paper discusses data from field experiments in which peat was used at Rothamsted, and where the aim was to try to separate the nutrient and other, non-specified effects on crop productivity of soil organic matter (SOM) .

Background

For a very long time farmers, especially in temperate climates, have considered that there was a strong link between the organic matter (OM) content of their soil and its productivity (Russell, 1977). However, summarising much experimental data from temperate agriculture, Cooke (1967) considered that there was little evidence to support such a view. Certainly in recent times this relationship has received less prominence, with one notable exception, probably because it has been possible to grow large crops with inputs of inorganic fertilizers and well-timed cultivations. The exception was the comments in the Strutt Report (MAFF, 1970) which recommended that arable soils should contain not less than 3% organic matter, or 1.75% organic carbon. For a short period this suggestion aroused much debate, and later Russell (1977) summarised and discussed some aspects of the role of humus in soil fertility and noted:

> "Thus a major problem facing the agricultural research community is to quantify the effects of soil organic matter on the complex of properties subsumed under the phrase soil fertility, so that it can help farmers develop systems which will minimise any harmful effect this lowering brings about".

However, it has been difficult to get the reliable data needed to satisfy Russell's exhortation. In temperate climates it takes a long time to set up experiments with different levels of organic matter, and especially with differences which can be reliably measured. This is because soil organic matter changes only very slowly in temperate regions under normal husbandry practices (Johnston, 1986, 1991). Interest at Rothamsted in soil organic matter has been stimulated in recent decades for two reasons:

1, because of its role in the nitrogen nutrition of crops, and because the mineralization of organic nitrogen, especially in the autumn, produces nitrate which is at risk to loss by leaching; and
2, because for 100 years and more the yields for crops grown in the Classical Experiments at Rothamsted were the same on farmyard manure (FYM) and on fertilizer-treated plots despite the fact that SOM was increasing with several additions of farmyard manure.

Now, however, in the last 20-30 years, yields have been larger on soils with more OM derived from added farmyard manure (FYM), provided extra fertilizer nitrogen was applied. One example is given in Table 1; others were given by Johnston (1986, 1991), who suggested that these differences in yield did not arise because of differences in the OM levels in the two groups of soils. These levels have changed only little, if at all, in the last 30 years. The effect appears to be related to the introduction of higher yielding cultivars which require optimum soil conditions as well as nutrients, and freedom from weeds, pests and diseases to achieve their potential. As well as affecting soil structure, SOM has other important functions in soil that need to be defined and quantified.

Beneficial effects from extra SOM have been measured in Rothamsted experiments from the late 1950s, but the magnitude of the effect was not the same for all crops and it was not consistent between experiments. On the sandy loam at Woburn, larger yields of vegetable crops on soils with higher OM contents appeared to be associated with earlier and more uniform germination of the crop. In other experiments, more SOM was associated with larger concentrations of readily soluble P and K, both in the topsoil and in the subsoil. Thus it was impossible to separate the effects on yield of the availability of these and other nutrients, as well as other attributes assigned to SOM, such as the effects on soil structure and on the water holding capacity. To resolve some of these uncertainties, a programme of small plot field experiments was initiated in 1960. The background and rationale are discussed elsewhere (Warren and Johnston, 1961). The common theme was to quantify, if possible, some of the attributes of SOM. The effects of subsoil enrichment with P and K have been discussed previously (Johnston and McEwen, 1984). In the experiments discussed here a range of soil amendments was tested. These included peat, FYM, a synthetic soil conditioner, and sodium alginate. Only peat was tested in all experiments because, compared to other readily available organic additives, it contains very little P and K (Table 2). One disadvantage is that the organic matter tends to be in distinct pieces whilst SOM is usually intimately mixed within the soil mineral matrix and is often closely associated with the clay sized fraction.

2 Materials and Methods

All the experiments were carried out on sandy loam soils at Stackyard Field, Woburn.

Table 1 *Yields of winter wheat (grain t ha^{-1}), grown on soils with different amounts of organic matter, in Broadbalk, Rothamsted*

Treatment and % organic matter	Cultivar grown			
	Flanders 1979-84		Brimstone 1985-90	
	Continuously	In rotation	Continuously	In rotation
NPK[a] 1.8	6.93	8.09	6.69	8.61
FYM[b] 4.5	6.40	7.20	6.17	7.89
FYM[c] + N[d] 3.8	8.13	8.52	7.92	9.36

a average values for "continuous" and "rotation" plots. Best yields of cv Flanders were given by 192 kg N ha^{-1} and of cv Brimstone by 228 kg N ha^{-1}

b FYM 35 t ha^{-1} since 1843

c FYM 35 t ha^{-1} since 1885

d 96 kg ha^{-1} fertilizer N

The soils had between 1.0 and 1.4% organic matter, and the soil pH (water) was about 6.5. For some experiments a new site was chosen each year to test the short-term effects of the different amendments, whilst in other experiments the cumulative effects were estimated by repeated applications on the same plots. The treatments were usually replicated three or four times in randomised blocks separated by narrow paths to aid access. The plots were small, usually 3.04 x 2.12 m. The width was chosen to accommodate sugarbeet grown in rows 53 cm apart and other vegetable crops grown in 30.5 cm rows. For all cultivations, including the primary cultivation, digging to 25 cm was done by hand. The primary cultivation was usually done about eight weeks before the intended date of sowing. Digging left the soil surface reasonably level, and so a minimum of secondary cultivations, usually with a hand rake, were needed to level the soil and to prepare a seed bed.

Sedge peats only were used, and the analyses of these were given by Johnston and Brookes (1979). The FYM, a mixture of pig and cattle manure produced by overwintering these animals on straw in covered yards at Woburn, contained a large amount of potassium and, at the rates applied, supplied the equivalent of about 400 kg K ha^{-1}. The synthetic soil conditioner, coded CRD 189 [a hydrolyzed poly(acrylonitrile)], had been supplied for experimental use at Rothamsted by Monsanto Chemicals in March 1952.

Table 2 *Average annual amount of carbon, nitrogen, phosphorus, and potassium (kg ha^{-1}), and the C:N ratio of the three organic amendments used in the Organic Manuring experiment, Woburn, 1965-1970*

Treatment	C	N	C:N ratio	P	K
Peat	3472	76	45:1	1	1
Straw	4205	48	87:1	6	86
FYM	3508	264	13:1	98	375

Annual application of peat and straw to supply 7.5 t ha^{-1} dry matter, FYM, 50 t ha^{-1} fresh weight.
Adapted from Mattingly, 1974

Its primary aim was to bind soil particles together. This, or a similar compound, was available commercially under the brand name 'Krilium'. Sodium alginate was the laboratory grade chemical. It was thought that it might simulate the binding effects of some of the organic components in humus. Shredded coir fibre was obtained from a factory making door mats in East London. It was considered that this fibrous material, similar in appearance to the larger roots of cereals and grasses, could help prevent individual soil particles compacting together.

Nitrogen (N), phosphorus (P), potassium (K), and magnesium (Mg) were applied in large amounts (see individual experiments). Except in the first year of the Fertilizer-FYM experiments, P, K, and Mg were applied before digging, and so were well incorporated into the cultivated soil layer. Nitrogen was usually part dug in and part applied to the seedbed. The seed (drilled one row at a time) was uniformly drilled along the row. The seedlings were then thinned by hand to the required plant spacing: 25 cm for sugar beet, 10 cm for globe beet, 30.5 cm for lettuce, and 5 cm for carrots. All the singlings were collected, and either weighed or counted to assess the effect of treatment on germination and early growth. All rows were harvested and top and root weights recorded for each row. Yields of the outer two rows were not usually included in the yields reported. Samples of tops and roots were taken for dry matter determination and another sample, dried at 80 °C, was kept for chemical analyses.

3 Results

Fertilizer-FYM Experiments

The first in this series of experiments in 1960 sought to answer queries arising in the Woburn Ley Arable experiment started in 1938 (Boyd *et al.*, 1968). By the mid 1950s potassium deficiency, especially on plots growing large crops of lucerne [*Medicago sativa* (L)] for conservation, had been identified as the cause of the poor potato yields which followed lucerne (Johnston, 1975). Potassium deficiency did not occur where 37.5 t ha^{-1} FYM was ploughed in before planting potatoes because it supplied about 400 kg K ha^{-1}, but it was considered that to apply such a large amount of K as fertilizer would affect germination. Thus, FYM was compared with the same amount of K and a large amount of N, both as fertilizers either to the seedbed or dug in. The dug-in fertilizer treatment also compared with and without peat dug in to supply about the same amount of organic matter as in the FYM. All plots received a small basal application of NPK fertilizer incorporated into the top 5 cm prior to drilling sugar beet. The results, given in more detail elsewhere (Warren and Johnston, 1961), showed a benefit from applying extra NK fertilizer with peat. Sugar yields were 6.09, 5.88 and 5.50 t ha^{-1} with FYM, peat + NK dug in, and NK dug in alone, respectively. This benefit from a single application of peat led to its being tested in subsequent experiments. Digging in became the standard practice in these experiments following the demonstration of the need to incorporate thoroughly the large applications of P and K required to ensure that these elements were not limiting.

Soil Amendments Experiments

In 1962 and 1963 peat, CRD 189, sodium alginate, coir fibre, and other amendments were tested, and yields were compared with those on unamended soils. The response to the four main treatments was very different in the two years. This highlighted the problem of assessing the effects of soil structure, especially when rainfall and structure interact to affect plant population.

In 1962 all the treatments were incorporated by digging and sugar beet was the test crop. The treatments had little effect on the weight of seedlings, because of the incorporation into the cultivated soil layer. Yields ranged from 282 to 332 g plot^{-1} at the first singling to 12 cm spacing. Yields of sugar were good and, although largest with peat and least with coir fibre (Table 3), the differences were not statistically significant.

To avoid bringing the coir fibre to the surface the plots were not dug in 1963. Instead CRD 189, sodium alginate, and peat (at three rates) were applied to the surface and worked into the top 7.5 cm. The residual effects of the coir applied in 1962 were measured also. The crop was globe beet singled to 10 cm spacing. Numbers of plants at both singling and harvest were recorded. The theoretical plant population, 323 000 ha^{-1}, was not achieved with any of the treatments, although numbers for treatments with CRD 189, sodium alginate, and the largest amounts of peat applied were 272 000, 242 000 and 232 000 ha^{-1}, respectively. CRD 189 almost doubled the yield of bulbs compared to the unamended control (18.3 and 10.9 t ha^{-1}, respectively). The largest amount of peat gave the second largest yield (16.4 t ha^{-1}) of bulbs. The benefit from each of the soil amendments was statistically significant with the exception of the coir fibre which slightly decreased the yield.

The lack of benefit from the soil amendments in 1962, and the substantial benefits in 1963 are of considerable interest. These indicate a possible but complicated relation between crop and soil structure. In 1962 the comparatively small amounts of the soil amendments tested were considerably diluted when incorporated into the top 25 cm of soil, and these had no effect on either plant population or on the yield of sugar beet. In 1963, however, the amendments were incorporated only into the top 7.5 cm. These had an appreciable effect on germination, and hence on plant number at harvest and on the yield of globe beet. But equally of interest, the amendments affected soil conditions at the surface in very different ways.

Secondary cultivation, by hand raking, produced in 1963 a range of soil crumb sizes at the surface, and many of these "mechanically produced" crumbs were stabilized by CRD 189 against the impact of rain which fell shortly after the seed was drilled. Rainfall caused the soil crumbs on the unamended plots to slump badly, and to form a cap or crust. When the cap dried it was difficult for the seedlings to emerge. Seedling numbers, in 000s ha^{-1}, were 198 and 689, respectively,.on the unamended and CRD 189 treated plots. Compared with CRD 189, there were somewhat fewer seedlings, 516, for the sodium alginate. This treatment appeared to impart some stability to the crumbs, but it was less than that for CRD 189. There were still fewer seedlings (range 264 to 450) for the peat treatments. Visually, the peat did not appear to stabilize the crumbs at all. The surface soil slumped as badly after the peat treatments as it did for the unamended plots. However, there was more than twice the number of seedlings in the soil treated with the largest amount of peat compared to the untreated soil. The discrete pieces of peat within the surface crust may have provided many points of weakness between the mineral particles which enabled the cotyledons to emerge. Lower yields for coir fibre applications might be attributable to its fibrous, springy nature. This made it difficult to compact the soil appropriately, and caused a greater loss of water by evaporation throughout the growing season. The residues of coir fibre incorporated the previous year had little effect

Table 3 *Effect of four soil amendments on the yields (Y) of sugar, from sugar beet and globe beet. Soil Amendments experiment, Woburn.*

[a]Treatment and rate of application (t ha^{-1})	1962 Sugar (Y t ha^{-1})	1963 Bulbs (Y t ha^{-1})
None	7.28	10.88
CRD 189 (1.26[b])	7.21	18.34
Sodium alginate (1.26)	7.34	15.40
Peat		
(3.7[c])	-	14.21
(7.5)		15.12
(10)	7.75	-
(11.3)	-	16.41
Coir fibre (10e)	6.85	9.1[e]
SED[d]	0.295	1.094

[a] All treatments were incorporated into the top 25 cm soil by hand digging in 1962 and into the top 7.5 cm only in 1963
[b] 1.26 t ha^{-1} active ingredient (5.02 t ha^{-1} as supplied)
[c] dry matter
[d] SED: 1962, 16 d.f; 1963, 15 d.f; SEDs for difference between none and any amendment
[e] Residues of the 1962 application not included in the statistical analysis
Basal fertilizers, kg ha^{-1}, N, 176; P, 82; K, 377 to sugar beet.
N, 168; P, 84; K, 252 to globe beet

on seedling numbers or on the numbers of plants at harvest (compared to those on the unamended soil), and yields were less than on the unamended soil.

Peat - CRD 189 Experiment

Because the amendments had the largest effects when applied to the seed bed, and because peat and CRD 189 appeared to act in very different ways, treatments with peat, with CRD 189, and with peat plus CRD 189 were tested in 1964 and in 1965. In both years all three amendments were cultivated into the top 5 cm, and in 1965 these were applied after the plots had been dug (which incorporated the applications of the previous year). For both globe beet in 1964 and lettuce in 1965, the peat treatment gave the largest number of seedlings. The numbers for the CRD 189 treatment were least; less even than for the unamended soil. Addition of CRD 189 with peat also gave less seedlings than peat alone. In both years there was little rainfall after drilling the seed. The well stabilized crumbs with CRD left a rather open seedbed which probably led to a rapid loss of water from soil in contact with the seed. This adverse effect was opposite to the beneficial effects seen for CRD in the previous experiment (in 1963), and it highlights the difficulty of generalizing about the type of seedbed which is beneficial for plant establishment.

Although the treatments affected seedling numbers in the same way in the two years, the effects on yields were different. Yields of globe beet in 1964 were largest for applications of peat plus CRD 189, and second largest with CRD 189 alone (Table 4).

Table 4 *Effects of two soil amendments on the yields of globe beet and lettuce. Peat -CRD 189 experiment, Woburn*

	Globe beet, bulbs t ha^{-1}, 1964		Lettuce t ha^{-1}, 1965	
	0	Peat	0	Peat
0	31.74	32.74	30.73	32.49
CRD 189	34.15	40.48	25.28	27.52

SEDs 1. For vertical comparisons 1.029, 1964; 2.105, 1965
(6 d.f.) 2. For horizontal and interaction comparisons 1.456, 1964; 2.979, 1965
Basal fertilizers, kg ha^{-1}
for globe beet: N, 224; P, 84; K, 280; Mg, 56
for lettuce: N, 112; P, 84; K, 224; Mg, 56

The effects of both peat and CRD 189 alone were not statistically significant. However, these interacted to give a highly significant yield increase. This result may be attributable to the better stabilized surface structure which allowed improved gaseous exchange between the soil and the atmosphere in the latter part of the growing season when rain had caused the surface soil structure to collapse in the other plots. Conversely, there was much less rain during the growing season in 1965, and yields of lettuce were largest for amendments with peat alone, and were second largest for the unamended soil (Table 4). The influence of peat alone was not statistically significant, whereas CRD 189 gave a significant decrease in yield. Addition of peat with CRD 189 mitigated some of the adverse effects of the CRD 189. The very open surface structure maintained by CRD 189, both in the presence and absence of peat, gave the lowest yields. Again this was probably because of excessive water loss in a year when rainfall was low. In these experiments no attempt was made to consolidate the seedbed by rolling except that the seed drill had a press wheel which lightly compacted the soil for about 2.5 cm on either side of the slit down which the seed fell. The plots were not rolled because, when first the CRD 189 was used, it was observed that the soil slumped following heavy rain where there were footprints. Presumably the CRD 189 provided only a protective 'skin' around the artificially created crumbs, and once this was broken, even by foot pressure, there was no longer any additional binding strength other than that already in the soil.

Peat Experiment 1963-1983

After showing that the differences in the surface soil structure depended, in part, on rainfall, and that peat could be beneficial under both wet and dry conditions in the spring, it was decided to concentrate further work on peat and on its cumulative effects over a number of years. In this experiment the effects of peat at three rates were tested:

(i) 7.8 t ha^{-1} dry matter was applied to the seedbed;
(ii) 7.8 t ha^{-1} dry matter plus 7.8 or 15.6 t ha^{-1} dug in; and
(iii) a control without peat.

In the years when peat was applied, it was applied cumulatively. Yields and uptake of nutrients have been published elsewhere (Johnston and Brookes, 1979). Yields for the first three years are summarised here (Table 5) because peat was applied each year and globe beet, 1963 and 1965, and carrots, 1964, were grown. The effect of peat on the levels of soil carbon is discussed below.

Seedling numbers and yields were better on peat amended soils than on the control. In

Table 5 *Yields of globe beet and carrots grown on peat amended soils. Peat Experiment, Woburn, 1963-1965. Adapted from Johnston and Brookes (1979)*

Crop and yield t ha^{-1}	Treatment applied cumulatively				
	None	Peat to seedbed 7.8 t ha^{-1}			SED$^{(a)}$
		only	plus peat dug in		
			7.8 t ha^{-1}	15.6 t ha^{-1}	
1963 Globe beet					
Bulbs	14.16	18.58	17.93	18.88	0.934
1964 Carrots					
Roots	39.12	40.10	42.46	42.84	0.864
1965 Globe beet					
Bulbs	38.26	41.30	41.51	41.38	1.814

$^{(a)}$ d.f. 8

Basal fertilizers, kg ha^{-1}

to globe beet, both years: N, 224; P, 84; K, 280, Mg 56

to carrots: N, 112; P, 84; K, 224; Mg 56

the first year, applications of peat to the seedbed increased significantly the yields of the shallow rooted globe beet. No further benefits were obtained from the dug-in peat. In 1965, when globe beet was grown again, the numbers of bulbs were increased by all peat treatments. Although yields were consistently larger on peat-treated soils, the effect was not statistically significant. The deeper rooted carrots grown in 1964 did not benefit from surface applied peat, but yields were significantly increased where peat was incorporated. That year the May to August rainfall was least for the the six year period, 1963-1968.

'Market Garden' and 'Organic Manuring' Experiments

One aim of the small plot experiments discussed here was to decide on modifications to treatments in existing experiments, and also on treatments to be tested in a major new experiment, The Organic Manuring Experiment, to be started at Woburn.

'Market garden' experiment

This experiment was started in 1942 to test four organic amendments, and to compare the yields which these gave with those on an unamended soil (Johnston and Wedderburn, 1975). Each amendment was made at two rates. It was one of the first experiments at Rothamsted where benefits to yields were reported from additional SOM.

In 1965 a test of peat was introduced because consistent, and sometimes statistically significant benefits from peat were observed in the above experiments. It was decided to test whether 31.4 t ha^{-1} peat dry matter would cause yields on the fertilizer-treated plots (with about 1.8% OM) to approach those on FYM-treated plots (with about 3.9% OM). Four rates of application of nitrogen were also tested (Table 6). The peat treatment increased significantly yields for all rates of N, except for the 150 kg ha^{-1} application rate. Even for the largest amount of N applied, the yield (37.8 t ha^{-1}) of bulbs was not increased to that (39.2 t ha^{-1}) for FYM alone. It would not be reasonable to expect that a single large application of peat would make good a difference of just over 2% in the SOM content. The result indicates the need to give some attention to the OM content of soil.

'Organic manuring' experiment

A major experiment on the effects of organic amendments on SOM and on crop yields was started in 1965 (Mattingly, 1974) after consideration had been given to results from several previous experiments. For the first six years, annual treatments included:

(i), peat and straw, to supply 7.5 t ha^{-1} dry matter;
(ii), FYM, 50 t ha^{-1} fresh weight;
(iii), grass clover ley; and
(iv), all grass ley plus fertilizer N.

Yields from these treatments were compared with those for two plots given fertilizer treatments only. There were two fertilizer-only plots because the large application of FYM added much P, K, and Mg, and it was considered unrealistic to adjust the P, K, and Mg on all plots to match these large additions. Thus the P, K, and Mg balance on one fertilizer plot was made equal to that on the FYM plot, whereas the P, K, and Mg balance on the other fertilizer plot was made the same as that for the remaining organic amendment plots. At the end of six years each plot in the two groups had very similar concentrations of bicarbonate soluble P and exchangeable K and Mg. However, these nutrients were much larger in the FYM and fertilizers-equivalent-to-FYM plots than in the others (Mattingly, 1974). After this treatment phase, potatoes [*Solanum tuberosa* (L)], winter wheat [*Triticum arvense* (L)], sugar beet and spring barley [(*Hordeum vulgare* (L)], each given eight amounts of N fertilizer on sub plots, were grown in rotation. Yields on the two fertilizer plots were closely similar for each amount of N tested, suggesting that P, K, and Mg were not limiting. Because of the lack of any response to P, K, and Mg, it was considered that the differences in yield between fertilizer-treated and organic-amended

Table 6 *Yields (t ha^{-1}) of globe beet on a low organic matter soil with and without one application of peat compared to those on a soil with a higher organic matter content. Market Garden Experiment, Woburn*

Treatment[a]	N applied kg ha^{-1}			
	0	75	150	225
Fertilizers	12.60	26.76	33.57	33.27
Fertilizers + peat	16.47	32.29	35.20	37.79
FYM	39.17	42.51	44.67	45.98

[a] Fertilizers and FYM cumulative to appropriate plots since 1942; Peat, 31.4 t ha^{-1} dry matter applied in 1965 to one half the plots previously given fertilizers only. FYM, 50 t ha^{-1}; P, 160 kg ha^{-1}; K, 310 kg ha^{-1}.
SED for comparison between fertilizers and fertilizers plus peat only
12 d.f. Horizontal comparisons 2.720
Vertical comparisons 1.360

Table 7 *Yields of four arable crops grown in rotation after applying peat annually for six years. Organic Manuring Experiment, Woburn 1972-1976*

Treatment and % organic matter in soil		*N treatment to test crop				
		N0	N1	N2	N3	N4+
		Potatoes, tubers, t ha^{-1}				
None	1.2	22.4	29.7	34.4	37.7	43.5
Peat	2.4	20.8	31.2	34.4	39.6	47.9
		Sugar beet, sugar, t ha^{-1}				
None	1.2	0.73	1.76	2.60	2.86	2.90
Peat	2.4	0.95	2.22	2.72	3.08	3.08
		Winter wheat, grain, t ha^{-1}				
None	1.2	1.60	2.65	3.76	4.68	5.23
Peat	2.4	1.58	2.78	3.72	4.84	5.63
		Spring barley, grain, t ha^{-1}				
None	1.2	1.35	2.40	3.06	3.64	3.65
Peat	2.4	1.56	2.76	3.39	3.92	3.83

*N rates, kg ha^{-1}:	N0	N1	N2	N3	N4+
Potatoes	0	50	100	150	200-300
Sugar beet	0	40	80	120	160-280
Winter wheat and Spring barley	0	25	50	75	100-175

Yields for N4+ were the means given by the four largest amounts of nitrogen applied to each crop. The ranges of yields are shown under the N rates above.

Table 8 *Amounts of peat added each year and the increase in soil carbon after six 'Peat' and 'Organic manuring' experiments, Woburn*

	Experiment			
		Peat		Organic Manuring
Amount of peat added each year, dry matter, t ha^{-1}	7.84	15.69	23.52	7.5
%C in the top soil after 6[a]	1.21	1.59	2.26	1.39
Gain, t ha^{-1}, in soil carbon[b]	16.52	27.82	49.49	21.47
% retention of added carbon	63	53	62	89

[a] 30 cm in the 'Peat experiment, 23 cm in Organic Manuring experiment

[b] Soil carbon at the start of each experiment, 30.56 t ha^{-1} Peat experiment, 26.38 t ha^{-1} in the Organic Manuring experiment

soils were due to differences in SOM (unless there was some subtle interaction between nutrients and SOM). For each of the treatments where fertilizer N was applied, the yields of all four crops were greater for the peat-treated soils than for the unamended soil (Table 7). In that Table only the mean yield for the four largest amounts of N is given because there was little response for applications above the N3 rate.

Effects of Adding Peat on the Build Up of Soil Organic Matter

Peat was added for six years in both the 'Peat' and 'Organic manuring' experiments. In the former experiment the smallest annual rate of addition was almost the same as that in the latter. In the 'Peat experiment', the percentage retention of all three amounts of added carbon was similar, and averaged 60% (Table 8). This was much less than the value of 90% for the 'Organic manuring' experiment. This large difference suggests that there is a major factor, other than soil type and climate (which was the same for both of these experiments) that must be considered when comparing turnover rates of carbon in different experiments. The availability of nitrogen for microbial decay of the peat and soil aeration were two possible causes considered. Each year more nitrogen, 209 kg ha^{-1} on average, was given to the vegetable crops in the 'Peat experiment' compared to only 61 kg N ha^{-1} to the arable crops. However, the amount of N was almost the same (8.9 kg and 8.1 kg in the 'Peat' and Organic manuring' experiments, respectively) per tonne of peat applied. Thus the availability of fertilizer N to aid microbial decomposition of the peat does not appear to offer an explanation for the different rates of decomposition. This suggests that soil aeration could have been important, and was probably responsible for the more rapid microbial decomposition of the peat in the 'Peat experiment'. Certainly the soils of the 'Peat experiment' were never subjected to the compaction effects seen in the 'Organic manuring' experiment because of the machinery used in the latter for husbandry operations from seedbed preparation to harvest.

The Rate of Decline of Soil Organic Matter Derived from Applying Peat

The breakdown and turnover of plant derived soil organic matter is of considerable interest and importance. Except for the fact that peat dressings were repeated on the appropriate plots in 1972, application of peat in the 'Peat experiment' ceased after 1968. Also, after 1968 mainly arable crops were grown until 1983 when the final soil samples were taken and the experiment was stopped. Visual observation during hand cultivation (1972 to 1983) suggested that the peat was disappearing quite rapidly. This was confirmed by the decline in the soil carbon content (Table 9). The untreated soil had a fairly stable carbon content, average 30.0 t C ha^{-1} (s.e. 0.39). The decline in the carbon content with time on the soils which had received peat was not inconsistent with an exponential decay model. Two such models were fitted to the data for soils treated with peat.

First, a decay curve was fitted to each of the three sets of data with separate asymptotes, but with a common decay rate (Table 10, Model 1). The difference between each asymptote and that for the carbon content of the unamended soil was 37, 29, and 24% of the carbon added in the three treatments where 26.4, 52.7 and 79.1 t C ha^{-1}, respectively, was applied as peat. That there was a difference between each of the three asymptotes and that for the unamended soil would not be inconsistent with the concept of an inert carbon fraction in SOM, used in many models for carbon turnover in soil, (for

Table 9 *Percentage carbon in soils given different amounts of peat after peat applications ceased in 1972. Peat experiment, Woburn*

Year	Total amount of peat dry matter applied, t ha^{-1}			
	0	55	110	165
	% carbon in top 30 cm			
1972	0.675	1.276	1.924	2.573
1974	0.752	1.252	1.744	2.317
1977	0.738	1.207	1.672	2.224
1979	0.734	1.175	1.553	2.071
1980	0.734	1.186	1.549	1.974
1982	0.737	1.132	1.599	1.920
1983	0.714	1.108	1.476	1.838

example, Jenkinson *et al.*, 1994). However, the average increase (30%) between the estimated asymptotes for the peat-treated and the unamended soil suggests that there was still some readily mineralizable peat in each soil.

Second, a model with a common decay rate and common asymptote was fitted to the three sets of data combined (Table 10, Model 2). This gave an asymptote of 37.2 t C ha^{-1} (s.e. 3.3). This standard error was appreciably less than where separate asymptotes were fitted (Table 10) because there was now more information to estimate some of the model parameters. The fit was almost the same for each peat treatment using either model, and

Table 10 *Estimated asymptotes to which soil carbon would decline after ceasing to add peat and the shift in years needed to bring the individual decay curves into coincidence, together with half lives for decay of soil carbon using two different models. 'Peat experiment', Woburn*

Peat treatment number	and t C ha^{-1} applied	Asymptote t ha^{-1} carbon	Shift in years to bring decay curves for treatments 3 and 4 into coincidence with that for 2	Number of years for carbon to decline by half [a]
Model 1, which estimated a separate asymptote for each treatment				
2	26.4	39.7 (3.8) [b]		
3	52.7	45.1 (7.8)	10.3 (6.2)	
4	79.1	49.3 (13.1)	16.9 (8.9)	
Mean				8.5 (4.1)
Model 2 which assumed a common asymptote for the three peat treatments				
2	26.4			
3	52.7	37.2 (3.3)	16.6 (2.4)	
4	79.1		25.5 (2.6)	
Mean				12.4 (1.5)

[a] Derived from the unified decay curve produced by each model
[b] Standard errors in parenthesis

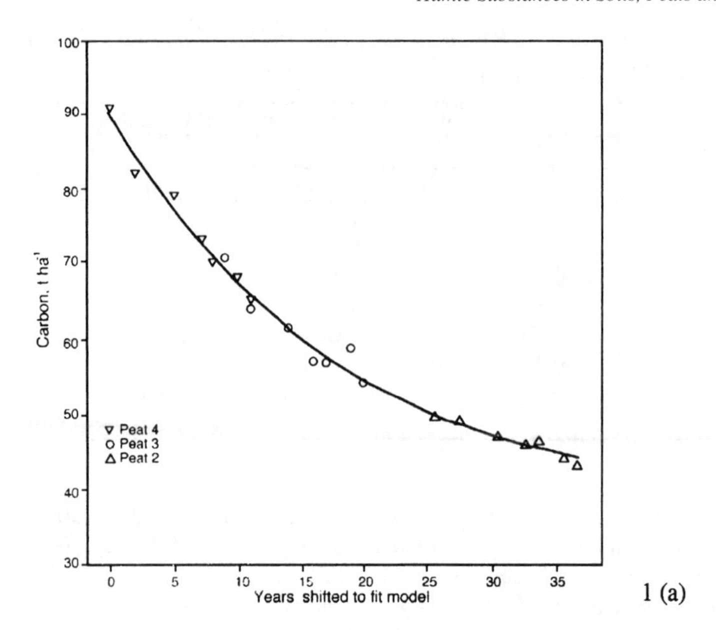

Figure 1*a*. *Unified organic carbon decay curve, half-life 12.4 (s.e. 1.5) y, asymptote 37.2 t C ha^{-1}, for the three amounts of peat added in the 'Peat experiment', Woburn ∇, O, Δ, 26.4, 52.7, 79.1 t C ha^{-1}, respectively, added between 1965-1972. Decay measured during 1972-1983 (see also Table 10)*

the difference was not statistically significant. However, there was a considerable difference in half life, i.e. the number of years to halve the difference in carbon content of a treated soil and the asymptote to which it is estimated to decline.

Using either model it was possible to produce a unified carbon decay curve for the three peat treatments, but the time shifts needed to bring the three individual curves into coincidence, i.e. to bring peat treatments 3 and 4 into coincidence with peat treatment 2, were different (Table 10). Figure 1a shows the unified curve for a common asymptote. Where separate asymptotes were used, the shift in years to bring the curves into coincidence was much shorter, but with much larger standard errors than where a common asymptote was used.

The rate of decay of organic matter in this 'Peat experiment', using data from the second model, can be compared with that for the organic matter in the 'Market garden' experiment at Woburn (Johnston *et al.*, 1989). In both cases the same methodology was used, and the half life of 12.4 y for the 'Peat experiment' was shorter than that of 20.1 (s.e. 1.4) y estimated for the decay of soil organic matter in the 'Market garden' experiment.

Although not on the same field as the 'Peat experiment', the 'Market garden' experiment was sited on the same soil type. Four organic materials, FYM, vegetable compost, sewage sludge, and sludge-straw compost were each added at annual rates of 37.5 and 75.0 t ha^{-1} fresh material from 1942 until 1961 for the materials with sludge, and until 1967 for the FYM and vegetable compost (Johnston and Wedderburn, 1975). The differences in dry matter content and C:N ratios, in the amounts applied, and the number of years of

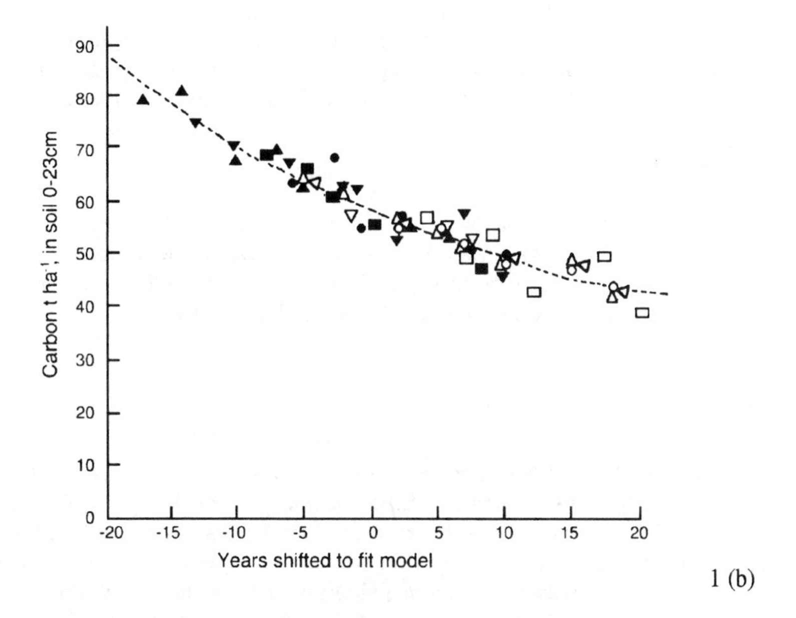

Figure 1b. *Unified organic carbon decay curve, half-life 20.1 (s.e. 1.4) years, asymptote 27.5 t C ha^{-1}, for four organic amendments added in the 'Market garden' experiment, Woburn.* □, *FYM;* Δ, *sewage sludge;* O, *FYM compost;* ▼, *sludge straw compost; open and closed symbols, 37.5 and 75 t ha^{-1} fresh material annually 1942-1967 except the two with sludge 1942-1961 only. Decay measured during 1967-1980 (Adapted from Johnston et al., 1989)*

application of these organic manures led to large differences in the quantity of SOM. No further organic manures were applied after either 1961 or 1967, but arable crops continued to be grown. The soils had been sampled periodically during the build up period, and sampling was continued to monitor the decline in SOM once the addition of organic manures ceased. Visual observation suggested that each of the individual decay curves was a segment of a single decay curve. When shifted horizontally, this was shown to be the case (Figure 1b) (Johnston *et al.*, 1989).

The fact that the decay curve for each experiment had a different half life is interesting. That derived from the 'Peat experiment', 12.4 y (s.e. 1.5), was significantly shorter than than that (20.1, s.e.1.4) derived from the 'Market garden' experiment. The difference may simply reflect the fact that in the 'Peat experiment' the added material had a large C:N ratio (45:1), and that it had been in the soil for a comparatively short period so that some of the peat still remained as discrete pieces liable to rapid microbial decomposition. In the 'Market garden' experiment, the organic amendments had much smaller C:N ratios, with the range of 9.5 for sewage sludge to 13.8 for FYM compost, and these had been added over a minimum period of 20 years (Johnston, 1975). The single decay curve for soil organic matter derived from different organic amendments in the Market garden experiment would be expected from the fact that such amendments are a food source for the soil microbial population, and that soil humus, the end product of that initial decomposition, could well be expected to decay at a constant rate. The shorter half life of soil carbon in the 'Peat experiment' would appear to relate to the fact that not all the added peat had been decomposed further by microbial activity to relatively stable soil humus.

Acknowledgements

We thank the many members of the then Chemistry Department, who helped in many ways with the field work, and Alan Todd (Statistics Department) for the statistical analysis of the yield data. IACR receives grant-aided support from the Biotechnology and Biological Sciences Research Council (BBSRC) of the United Kingdom.

References

Boyd, D.A. and The Ley Arable Sponsors. 1968. p. 316–331. Experiments with Leyand Arable Farming Systems. *Rothamsted Experimental Station Report for 1967.*

Cooke, G.W. 1967. *The Control of Soil Fertility*, Crosby Lockwood & Son Ltd, London.

Jenkinson, D.S., N.J. Bradbury, and K. Coleman. 1994. How the Rothamsted Classical Experiments have been used to develop and test models for the turnover of carbon and nitrogen in soil. p. 117-138. *In* R.A. Leigh and A.E. Johnston (eds), *Long-term Experiments in Agricultural and Ecological Sciences*, CAB International, Wallingford.

Johnston, A.E. 1975. The Woburn Market Garden experiment, 1942–69. II. Effects of the treatments on soil pH, soil carbon, nitrogen, phosphorus and potassium. p. 102–131. *Rothamsted Experimental Station Report for 1974*, Part 2.

Johnston, A. E. 1986. Soil organic matter, effects on soils and crops. *Soil Use and Management* **2**, 97–105.

Johnston, A. E. 1991. Soil fertility and soil organic matter. p. 299-313. *In* W.S. Wilson (ed),. *Advances in Soil Organic Matter Research: the Impact on Agriculture and the Environment.* Royal Society of Chemistry, Cambridge.

Johnston, A. E. 1995. The value of long-term field experiments in agricultural and environmental research. p. 31-50. *In* B.T. Christensen and U. Trentemøller (eds), *The Askov Long-term Experiments on Animal Manure and Mineral Fertilizers,* 100th Anniversary Workshop, Danish Institute of Plant and Soil Science, Tjele.

Johnston, A.E. and P.B. Brookes. 1979. Yields of, and P, K, Ca, Mg uptakes by, crops grown in an experiment testing the effects of adding peat to a sandy loam soil at Woburn, 1963–77. p. 83–98. *Rothamsted Experimental Station Report for 1978*, Part 2.

Johnston, A.E. and J. McEwan. 1984. The special value for crop production of reserves of nutrients in the subsoil and the use of special methods of deep placement in raising yields. p. 157-176. *In Nutrient balances and fertilizer needs in temperate agriculture.* International Potash Institute, Basel.

Johnston, A.E., S.P.McGrath, P.R. Poulton, and P.W. Lane. 1989. Accumulation and loss of nitrogen from manure, sludge and compost: long–term experiments at Rothamsted and Woburn. p. 126–139. *In* J.A.A. Hansen and K. Henriksen (eds), *Nitrogen in Organic Wastes Applied to Soils.* Academic Press, London.

Johnston, A.E. and R.W.M. Wedderburn. 1975. The Woburn Market Garden experiment,

1942–69. I. A history of the experiment, details of treatments and the yields of the crops. p. 79–101. *Rothamsted Experimental Station Report for 1974*, Part 2.

MAFF 1970. *Modern Farming and the Soil*. Report of the Agricultural Advisory Council on Soil Structure and Soil Fertility (The Strutt Report), HMSO, London.

Mattingly, G.E.G. 1974. The Woburn Organic Manuring experiment. I. Design, crop yields and nutrient balance, 1964-72. p. 98-133. *Rothamsted Experimental Station Report for 1963,* Part 2.

Russell, E.J. 1927. *Soil Conditions and Plant Growth. 5th Edition*. Longmans, Green & Co. Ltd., London.

Russell, E.W. 1977. *Philosophical Transactions of the Royal Society, London, B* **281**, 209-219.

Warren, R.G. and A. E. Johnston. 1961. Farmyard manure. p. 45-48. *Rothamsted Experimental Station Report for 1960*.

Growth of Crops in Land Treated with Sewage Sludge: Observations from Experiments made by the Agricultural Development and Advisory Service (ADAS) of England and Wales

L V Vaidyanathan*

SOIL SCIENCE DEPARTMENT, ADAS, MAFF, CAMBRIDGE CB2 2DR
*PRESENT ADDRESS: SCHOOL OF CHEMISTRY, THE UNIVERSITY OF BIRMINGHAM, EDGBASTON, BIRMINGHAM B15 2TT, ENGLAND

Abstract

Experiments on vegetable crops at ADAS Lee Valley and ADAS Luddington (1968-71), adding sludge cakes contaminated with zinc, copper, nickel or chromium to the soils, tested the "zinc equivalent" concept, based on metal contents of slurries and sludge cakes. Yields of red beet (*Beta vulgaris*) and lettuce (*Lactuca sativa*) were decreased by the large amounts of zinc, and to a lesser extent, of copper added; yield of celery (*Apiem graveolens*) was unaffected. Chromium had no adverse effects on any of the crops. The data did not conform to the "zinc equivalent" formula. In experiments at ADAS Wolverhampton (1971-76), zinc and nickel were much less toxic to different vegetable crops at neutral than at acid pH of the soil given metals as sludge cake, largely exceeding the "zinc equivalent". In a ploughed up grassland receiving metal contaminated sewage sludge from 1957 to 1974, increases in yields of several vegetable crops resulted from burying the contaminated topsoil, by increasing soil pH by liming, or by the addition of peat; yields decreased when soil pH was made acidic by incorporating sulphur. In the ADAS Lee Valley site, sown to grass:clover ley in 1973, yields of white clover (*Trifolium repens* L) were significantly decreased by the large amounts of zinc added in 1968; copper effect was less, while effects of nickel and chromium were confounded by the accompanying zinc in the sludge added. Contents of zinc, copper, and nickel in the topsoil and in the herbage were positively correlated with the amounts added. Yields of red clover (*Trifolium pratense*) at ADAS Luddington, grown in 1985, after sludge application in 1968, were <7 t ha^{-1} as the result of the large zinc additions and >10 t ha^{-1} in the control (without sludge amendments). The effects of copper and nickel were confounded by accompanying zinc. In an experiment started at ADAS Gleadthorpe in 1982, yields of grass, sugar beet (*Beta vulgaris*) and barley (*Hordeum vulgare*) were unaffected by zinc, copper and nickel added as sludge cake (1200, 1050 and 200 kg ha^{-1}). More zinc and copper applied in sludge cake to selected plots in 1986 increased substantially the metal contents in the soil. Yields of white clover (*Trifolium repens* L), sown in 1988, were greatly depressed by the large (1550 - 2760 kg ha^{-1}) zinc aditions. The effect of copper could not be separated from that of the accompanying zinc; nickel was not toxic. Contents of metals in soil dissolved by boiling strong mineral acid, and the contents in harvested crop parts reflected very closely the amounts of metals added as sludge cakes.

1 Introduction

Uptake by plants of elements that are either required in trace amounts, e.g., zinc, copper, manganese, molybdenum, and boron, or are non-essential e.g. lead, cadmium, and nickel, may lead to crop loss through phytotoxicity from larger than normal uptake of these metals, and/or present a health risk from produce consumed by animals (Mills *et al.*, 1980) and humans (Lindsay, 1980). Controlling metals in sewage sludge spread on farmland is, therefore, an essential component of soil protection. Such concerns had prompted greenhouse studies by the National Agricultural Advisory Service (NAAS, predecessor of ADAS) of the Ministry of Agriculture, Fisheries and Food, in the 1960's. These measured crop growth in soils to which inorganic metal salts were added, with or without relatively uncontaminated sewage sludge (Davies, 1980; Webber, 1980). Although it was known that there are differences between the action of the metal salts and metals contained in organic wastes, the data from these experiments were used to devise the ADAS "zinc equivalent" (Chumbley, 1971), which expressed the combined effect of metal contents in sludge according to the formula:

zinc equivalent = [zinc] + 4 x [copper] + 8 x [nickel]

This formed the basis of ADAS guidance on sewage sludge additions to farm fields in England and Wales until the Department of the Environment (DOE) issued guidelines in 1978. These were revised in 1981 (DOE, 1981). Subsequently, the 1989 Regulations (Anon, 1989) introduced the requirements of the 1985 EC Directive.

Brief Review of Published ADAS Results

Tests of the "zinc equivalent" concept began in 1968 with field experiments at ADAS Lee Valley and ADAS Luddington. Vegetable crops were grown for four seasons in plots that were amended with different amounts of sewage sludge cakes containing large contents of either zinc, copper, nickel or chromium. Yields of crops, and contents of the metals in soil and crops from this work have been published in detail (Webber, 1972; Williams, 1975; Marks *et al.*, 1980). At both sites, nickel was by far the most toxic metal for all crops. Zinc was also toxic to all crops, except celery (*Apiem graveolens*), which tolerated more than twice as much metal as red beet (*Beta vulgaris*) and lettuce (*Lactuca sativa*). The relative toxicity of copper varied between the two sites, being less than that of zinc for red beet at Lee Valley, but greater than that of zinc at Luddington. Celery growth was not decreased by copper additions in this experiment. Chromium had no adverse effect on crop yields at either of the sites. A summary of results is given in Table 1 a-d. The results did not conform to the 'zinc equivalent' formula (Table 1d; Vaidyanathan, 1976a; Marks *et al.*, 1980).

A microplot experiment at ADAS Wolverhampton tested the residual effect of additions to soil of sludge-borne zinc and nickel. Several vegetable crops were grown

Table 1a *Amounts of metals added* (kg ha^{-1}) *at the ADAS Lee Valley and Luddington experiments*

Metals added (kg ha^{-1}), in the single 125 t ha^{-1} dry solids application						
	Lee Valley			*Luddington*		
	Zn	*Cu*	*Ni*	*Zn*	*Cu*	*Ni*
Sludge alone	130	44	15	120	40	10
+ Low Zn	1100	60	15	1100	55	10
+ High Zn	2100	70	15	2090	65	10
+ Low Cu	310	430	20	310	515	20
+ High Cu	500	755	20	500	990	20
+ Low Ni	390	65	255	410	60	250
+ High Ni	720	75	555	600	70	450

Metals added (kg ha^{-1}) in each of the four annual 31.25 t ha^{-1} dry solids application									
	Up to 1969			*Up to 1970*			*Up to 1971-72*		
	Zn	*Cu*	*Ni*	*Zn*	*Cu*	*Ni*	*Zn*	*Cu*	*Ni*
Sludge alone									
Lee Valley	110	45	10	190	75	20	220	90	20
Luddington	60	20	5	90	30	5	112	40	10
Zinc									
Lee Valley	360	20	10	560	50	15	720	80	20
Luddington	1050	30	5	1570	50	10	2090	65	10
Copper									
Lee Valley	130	160	10	220	250	15	260	320	20
Luddington	250	500	10	370	740	15	500	990	20
Nickel									
Lee Valley	160	40	70	260	70	105	330	90	140
Luddington	300	35	225	450	50	335	600	70	445

during the period 1971 - 1976 in soil having a pH in the range of 5.8 to 7.2. Zinc and nickel toxicity was significantly decreased at near neutral pH despite the fact that metal doses were greatly in excess of the "zinc equivalent" (Williams, 1980). A summary of the results is given in Table 2.

Continuing investigations have followed the expected persistence and residual effects of sludge-borne metals added to land. A notable feature since the mid 1980s has been the likely long term effects of the metals on legume crop productivity and soil biomass activity. The clover component of a grass:clover ley established at the ADAS Lee Valley site in 1973 showed marked decrease in growth, measured in 1975, 1976 and 1977, associated with the sludge metals added in 1968 (Vaidyanthan, 1976b). Further results of these observations are presented below.

Observations from non-ADAS Site Investigations

In another study, metal contents were determined for a large number of crops and for the topsoils from fields where the crops were grown in England and Wales. The fields had

Table 1b *ADAS Lee Valley: Crop yields after the addition of metal contaminated sewage sludge cake*

Treatment	*Rate*	*Crop Yield (% of control*, average of 4 replicates)*						
		Red beet				*Celery*		
		1969	*1970*	*1971*	*1972*	*1970*	*1971*	*1972*
Sludge alone	S	115	135	88	85	90	101	95
+Low Zn	S	0	47	41	9	77	89	101
+High Zn	S	0	0	9	0	39	24	68
+Low Cu	S	92	93	60	79	103	92	148
+High Cu	S	42	89	16	45	106	101	107
+Low Ni	S	54	63	15	39	77	21	48
+High Ni	S	0	74	0	23	31	2	12
Sludge alone	M	84	159	99	108	148	118	119
+ Zn	M	57	106	86	52	136	111	101
+ Cu	M	130	91	83	91	95	100	102
+ Ni	M	117	93	72	87	100	105	112

S = 1 x 125 t ha^{-1} application of sludge cake in 1968. * = expressed as percentage of yield from no sludge application.
M = 4 x 31.25 t ha^{-1} annual applications of sludge cake commencing in 1968.

Table 1c *ADAS Luddington: Crop yields after the addition of metal contaminated sewage sludge cake*

Treatment	*Rate*	*Crop Yield (% of control, average of 4 replicates)**							
		Red beet				*Lettuce*			
		1969	*1970*	*1971*	*1972*	*1969*	*1970*	*1971*	*1972*
Sludge alone	S	93	106	94	132	107	101	120	90
+ low Zn	S	22	0	0	0	12	0	0	0
+ high Zn	S	18	0	0	0	7	38	0	0
+ low Cu	S	35	0	59	0	62	104	42	57
+ high Cu	S	3	0	0	0	63	0	28	2
+ low Ni	S	25	14	16	0	48	43	57	58
+ high Ni	S	0	0	0	0	28	0	0	0
+ low Cr	S	96	69	93	144	93	92	150	101
+ high Cr	S	119	58	127	125	119	101	121	159
Sludge alone	M	92	114	89	126	70	15	174	94
+ Zn	M	32	0	0	0	21	0	0	0
+ Cu	M	49	0	0	0	89	56	23	4
+ Ni	M	48	8	86	0	61	51	0	118
+ Cr	M	90	152	109	107	113	135	103	131

S = 1 x 125 t ha^{-1} application of sludge cake in 1968. * = expressed as percentage of yield from no sludge applicati
M = 4 x 31.25 t ha^{-1} annual applications of sludge cake commencing in 1968.

received several applications of sewage sludge. Examples of differences between treated and untreated fields, and of metal distribution within components of the crop are included in Table 3a-e (Richardson, 1980).

Although it is often difficult to identify appropriate control sites that were not

Table 1d *Equations for regression of crop yields as percentage of control receiving no sludge application (%Y), against metals added as sludge cake* (kg ha^{-1})

Lee Valley	Red beet	%Y = 115.41 - 0.057[Zn] - 0.043[Cu] - 0.100[Ni]
	Celery	%Y = 123.32 - 0.033[Zn] - 0.001[Cu] - 0.163[Ni]
Luddington	Redbeet	%Y = 114.85 - 0.069[Zn] - 0.090[Cu] - 0.148[Ni]
	Celery	%Y = 118.53 - 0.066[Zn] - 0.060[Cu] - 0.116[Ni]

Table 2 *ADAS Wolverhampton: Effects on yields of vegetable crops (as percentage of control receiving no sludge) of soil pH and applications of metal-contaminated sewage sludge*

Metals applied		*Zn/Ni* (kg ha^{-1}) *640/285*	*Zn/Ni* (kg ha^{-1}) *1280/570*	*Zn/Ni* (kg ha^{-1}) *2560/1140*
Crop	*pH*	*Crop yield (% of control receiving no sludge)*		
Carrots, 1971	5.8	5.5*	0.7*	0.0*
(*Dauca carota*)	6.6	64.0*	30.6*	6.8*
Red beet, 1972	6.2	13.9*	7.1*	3.3*
(*Beta vulgaris*)	7.0	100	97.4	47.2*
Onions, 1973	6.2	67.3	22.4*	6.1*
(*Allium cepa*)	7.0	103	74.6	61.9*
Swedes, 1974	6.7	87.8	29.7*	0.0*
(*Brassica napus* ssp. *rapifera*)	7.2	117	113	136
Lettuce, 1975	6.5	100	70.0*	3.7*
(*Lactuca sativa*)	7.0	87.7	76.3*	71.6*
Onions, 1976	6.3	76.7	70.0	1.0*
(*Allium cepa*)	6.8	75.0	70.0	63.6*

* Significantly different from control at P = 0.05; for horizontal comparisons only.

given sludge amendments, the results indicated in general that there were elevated contents of metals in soils, and in the plants grown on those where there was a history of sewage sludge applications, when compared to sites which had not received sludge. In addition, there were significant differences in metal contents between different parts of the sampled plants, and these differences did not show the same pattern for all metals.

Scientists at the IACR, Rothamsted Experimental Station (Brooks and McGrath, 1984; 1987; Brooks *et al.*, 1984; 1986a; 1986b) found in a soil given annual doses of sludge-borne metals during 1947-1961, decreases in soil microbial biomass, in soil ATP concentration, in dehydrogenase activity, in N fixation by heterotrophs and by blue green algae, and in nitrification of added nitrite and ammonium nitrogen. This historic Woburn Market Garden experiment, investigated for the residual effect of metals, some 20 years after cessation of sludge application, showed severe decreases of rhizobial symbiosis and nitrogen fixation in the sludge-treated soil (McGrath and Brooks, 1988). Compared with

Table 3a *Comparison of metal contents in grass from fields which had partial applications of metal contaminated sewage sludge*

Sewage sludge	*Metal contents in grass dry matter* (mg kg^{-1})				
	Zn	*Cu*	*Ni*	*Pb*	*Cd*
T	58	10.5	9.7	6.8	Trace
U	36	7.8	7.7	5.0	Trace
T	125	28	21	22	1.6
U	51	17	4.3	11	1.1
T	150	25	12	19	2.3
U	67	22	4	11	0.8
T	85	24	8	12	0.8
U	45	13	6	7	0.5
T	41	8.1	2.3	4	0.5
U	40	9.9	1.6	4	0.5
T	32	8.0	4.3	4	0.4
U	33	8.0	3.0	3	0.4
T	51	16	4.0	12	1.4
U	48	7.5	5.0	7	0.3

T = treated; U = untreated.

Table 3b *Comparison of metal contents in soil* (0-15 cm) *and in crops between untreated and sites treated for about 15 years with digested (metal contaminated) sewage sludge on a sandy loam soil*

		Metal contents						
		Zn	*Cu*	*Ni*	*Pb*	*Cd*	*Hg*	*Cr*
Soil (total)	T	310	91	28	297	11	-	-
(mg kg^{-1})	U	59	24	18	44	0.3	-	-
Soil (EDTA soluble)	T	263	64	11.2	-	-	-	-
(mg L^{-1})	U	28	11	0.4	-	-	-	-
Lettuce (mg kg^{-1} DM)	T	276	14	18	38	13	0.13	3.4
(*Lactuca sativa*)	U	50	12	17	14	0.8	0.12	6.2
Carrots (mg kg^{-1} DM)	T	107	6.2	15	5.3	5.3	0.03	5.2
(*Dauca carota*)	U	19	5.0	4.7	3.2	0.2	0.03	6.2
Parsnips (mg kg^{-1} DM)	T	104	7.5	9.5	3.2	5.2	0.01	1.8
(*Pastinaca sativa*)	U	18	4.0	2.7	2.2	0.3	0.02	1.6
Red beet (mg kg^{-1} DM)	T	181	16	35	2.2	3.5	0.01	2.1
(*Beta vulgaris*)	U	37	7.8	19	3.2	0.5	0.01	2.3
Leeks (mg kg^{-1} DM)	T	141	10	85	7.5	5.7	0.01	2.9
(*Allium ampeloprasum var. porrum*)	U	32	4.8	50	4.2	0.2	0.02	1.3
Cabbage (mg kg^{-1} DM)	T	124	5.5	20	5.3	1.2	0.01	2.6
(*Brassica oleoracea var. capita*)	U	33	4.8	5.5	3.2	0.1	0.01	0.8
Runner beans (mg/kg^{-1} DM)	T	48	4.5	24	2.2	0.2	0.01	0.8
(*Phaseolus vulgaris*)	U	35	4.3	6.3	2.2	0.1	0.01	2.3

T = Treated (pH = 6.4, Organic matter = 5.0%) - = not determined.
U = Untreated (pH = 6.6, organic matter = 2.2%), DM = dry matter.

Table 3c *Effect of long term applications of metal contaminated sewage sludge on metal contents in soils and plants*

Crop	Treatment	Metal contents									
		plant material (mg kg^{-1} DM)					*soil* (total, mg kg^{-1})				
		Zn	*Cu*	*Ni*	*Pb*	*Cd*	*Zn*	*Cu*	*Ni*	*Pb*	*Cd*
Barley grain	Nil	43	8.2	3.1	4	0.3	61	10	18	60	1.5
(*Hordeum*	Recent	34	6.8	2.3	6	0.5	75	17	18	60	1.0
vulgare)	Prolonged	116	9.4	3.8	4	0.3	420	70	31	285	4.5
Whole potatoes	Nil	16	7.2	3.1	3	0.3	64	12	20	40	2.0
(*Solanum*	Recent	10	8.0	3.1	6	0.5	99	20	25	70	1.5
tuberosum)	Prolonged	22	11.7	3.8	4	0.5	420	70	31	285	4.5
Barley grain	Nil	38	6.2	1.5	2	0.2	76	14	30	80	1.0
(*Hordeum*	Recent	29	10.1	2.1	2	0.3	112	20	31	90	1.5
vulgare)	Prolonged	25	14.9	2.1	2	0.3	200	37	50	140	2.5

Nil = no sludge applied. DM = dry matter. Recent = sludge applied only for past five years. Prolonged = sludge applied since the beginning of the century.

Table 3d *Metal contents in potatoes and peas grown in soil receiving metal contaminated sewage sludge*

	Metal content (mg kg^{-1})					
	Zn	*Cu*	*Ni*	*Pb*	*Cd*	*Hg*
Whole potatoes	18	4	2.5	5	0.8	-
Peel	32	23	6.3	12.5	1.1	-
Whole potatoes	31	10	17.5	1.3	0.4	-
Peel	54	38	17.5	7.5	1.3	-
Peeled potatoes	10	7.4	nil	nil	0.08	-
Peel	22	6.9	0.15	1.0	0.4	-
Shelled peas	82	8.0	13	1.0	0.2	0.03
Pea pods	106	7.8	50	4.2	1.2	0.03

- = not determined.

soil that had received FYM during the same period, clover yield was decreased by 40% and nitrogen fixation was halved.

Zinc and copper contents in sludge treated soil were similar to, and nickel content was half of, the current EC limits. However, the cadmium content was more than three times the EC limit. An unpublished internal report from WRc, Medmenham (Smith, 1991), suggests that the large cadmium content resulting from sludge application may be the major cause of the adverse effects.

Table 3e *Metal contents in foliage and storage organs of potatoes, carrots, and sugar beet grown in soil receiving metal contaminated sewage sludge*

Metal	Metal content (mg kg^{-1})*																	
	Potatoes			Carrots			Sugar beet			Sugar beet			Sugar beet			Sugar beet		
	Tubers	*Tops*	*Soil*	*Roots*	*Tops*	*Soil*	*Roots*	*Tops*	*Soil*	*Roots*	*Tops*	*Soil*	*Roots*	*Tops*	*Soil*	*Roots*	*Tops*	*Soil*
Zn	35	280	285	73	230	320	26	72	84	28	48	31	118	348	214	31	144	67
Cu	7.5	22.5	161	10	11	94	8.4	31	28	7.9	14.3	16	12	20	124	7	28	17
Ni	13.8	70	103	11	10	26	2.4	2.1	19	1.9	3.5	11	7	24	76	2.6	2.1	18
Pb	6.3	15	124	nil	3.5	350	1.7	2.4	57	3.2	2.0	20	4.3	3.0	101	3.5	5.8	26
Cd	0.38	3.4	1.4	2.5	5.5	8.0	0.5	4.0		0.1	0.5	-	0.3	1.5	1.1	0.6	1.8	-

* = in dry matter of plant tissues; in air dry soil, extracted by boiling strong mineral acid. - = not determined.

2 Unpublished and Current ADAS Experiments

Husbandry Measures to Alleviate the Effect of Sludge Metals in Treated Land. A range of vegetables was grown during 1976-1981 on a well drained sandy loam (Sutton series, pH 6.5) site that was in grass up to 1975 and had received frequent applications of liquid digested sewage sludge during the period 1957-1974 (Richardson and Chumbley, 1981). After ploughing, the effects on crop yields of double digging, of peat addition, of chalk addition, of sulphur addition, and of replacement of topsoil with uncontaminated topsoil of similar texture (not undertaken until 1977) were investigated. Double digging was carried out once only, at the start of the experiment. The top 15 cm of soil was placed at a depth of 15-30 cm, and the soil from 15-30 cm was placed on top in March 1976. The aim was to dilute the metals held in the topsoil with less contaminated soil from below. Peat was added to raise the soil organic matter level. A dressing of 6 t ha^{-1} peat was applied on seven occasions between May 1975 and March 1977, and on each occasion 1.5 t ha^{-1} of ground chalk was applied with the peat to prevent the peat making the soil acid. Ground chalk (10 t ha^{-1}) was applied to other plots in December 1975, and again in January 1977 to raise the soil pH to over 7.0. Sulphur powder (3 t ha^{-1}) was forked into the topsoil as an additioinal treatment in March 1976 to decrease the pH to less than 6.0. Topsoil replacement was carried out in June 1977 by bringing uncontaminated topsoil of similar texture from another uncontaminated site; the whole original topsoil was replaced to a depth of 15 cm. The first data for yields for this treatment were obtained in 1978.

The crops grown were turnips (*Brassica campestris*, 1976), beetroot (*Beta vulgaris*, 1977), lettuce (*Lactuca sativa*) followed by transplanted Brussels sprouts (*Brassica oleoracea* var. Gemmifera D.C., 1978), maincrop potatoes (*Solanum tuberosum*, 1979), transplanted cabbage (*Brassica oleoracea* var. Capitata, 1980), and transplanted cauliflower (*Brassica oleoracea* var. Botrytis, L., 1981).

Samples of the edible parts of all crops were analysed for zinc, copper, nickel, cadmium and lead. Beetroot leaves were also analysed in 1977. No special washing procedure was used to remove dust from leafy samples because these were generally free from obvious contamination. Soil samples (0-15 cm) were analysed for pH, for organic matter, for available zinc, copper, and nickel (extracted with EDTA) and for total zinc, copper, nickel, lead, and cadmium (digestion with boiling nitric and perchloric acids) using standard ADAS methods (MAFF,1986).

The yields (Table 4) indicate that the metals had a phytotoxic effect in most of the crops, especially when the soil pH was decreased by sulphur additions. Double digging and peat and chalk additions gave appreciable increases in the yields of beetroot and of cabbage. Sulphur addition resulted in the complete failure of beetroot. Only peat addition gave any appreciable yield improvement of lettuce or of cauliflower. All treatments had very little effect on yields of Brussels sprouts or of potatoes.

The results of the plant analyses are given in Table 5. Sulphur addition increased metal contents in the plant, whereas chalk addition, or replaced topsoil, decreased the metal contents. Both double digging and peat addition had variable effects on plant metal contents. These trends were most clearly expressed in the zinc content results; zinc contents were the largest for all plants.

The results for the soil analyses for 1975 and 1979 (Table 6) show that the site was initially contaminated with moderate contents of zinc, copper and nickel, and with relatively large contents of lead and cadmium. The contents did not change much from

Table 4 *Effects of husbandry measures to alleviate the influences on crop yields (*t ha^{-1} *fresh weight) of metals in sludge treated land*

Treatment	1976		1977		1978		1979	1980	1981
	Turnip		Beetroot		Lettuce	Brussel sprouts	Potato tubers	Cabbage	Cauliflower
	Roots	Tops	Roots	Tops					
	t ha^{-1} fresh weight								
Control	0.79	9.2	4.8	4.0	29.2	4.4	40.3	83.0	19.6
Double digging	0.66	10.3	17.0	8.1	30.6	3.4	31.9	90.5	18.4
Peat addition	0.87	10.2	12.2	7.1	41.7	4.6	41.8	99.5	22.1
Chalk "	0.91	8.6	20.3	9.6	23.8	4.1	35.7	90.5	16.0
Sulphur "	0.54	6.1	0.0	0.0	8.1	2.9	25.5	77.0	13.1
Replaced TS	-	-	-	-	29.1	5.3	41.1	91.5	18.0
Mean	0.74	8.9	13.6	7.2	27.1	4.1	36.1	88.7	17.9
SED	0.132	2.39	6.42	2.24	5.17	0.55	3.07	5.29	4.74
CV (%)	15.4	23.3	47.3	31.2	19.1	13.3	8.5	6.0	26.5

TS = topsoil; SED = standard error of the difference between means; CV = coefficient of variation.

1975 to 1979. Double digging decreased the amount of metals in the top 15 cm, but tended to decrease soil pH (Table 7).

Metal contents in the replaced topsoil treatment were significantly less than those of the control, whereas metal contents remained similar in the other treatments. The peat treatment raised the organic matter content to over nine percent. However, the indigenous organic matter content in the soil at this site was larger than is usual for arable land because of a management history of long term grass.

Overall, the results indicate that increasing the soil pH and/or exporting contaminated soil is the most reliable method for decreasing metal contents of plant tissue. However, different crops take up different amounts of various metals in the soil, and large contents of metals in the soil do not necessarily (or always) result in large uptakes by the plants grown in the contaminated soils.

Residual Effect of Sludge-Borne Metals on Clover: Experiments at ADAS Lee Valley and ADAS Luddington

The ADAS Lee Valley and ADAS Luddington sites are on a silty loam soil with a pH of 6.2 -7.0 and a sandy loam with a pH of 5.8, respectively. Initial applications of sludge were made in 1968. Yields of vegetable crops and metal contents in soil, associated with the metals added in sludge cake have been published (Marks *et al.*, 1980).

Experiments at ADAS Lee Valley. The Lee Valley plots were maintained in a grass/clover ley sown in April 1973. The ley was mown as necessary, and clippings were removed, though yields were not recorded until mid 1975. Regrowth in summer 1975 and growth in 1976 and in 1977 were measured. Water shortage presented a major constraint in the very dry seasons in 1975 and 1976.

Clover and grass were carefully separated and the yields of both were determined. The tissues from the 1975 harvest were analysed for metal contents. Yields of clover and grass in 1975 have been reported (Vaidyanathan, 1976b).

Table 5 *Effect on the metal contents in plants of husbandry measures to alleviate the influences of metals in sludge treated land*

Year		Metal Content (mg kg^{-1} dry matter)						
	Crop	*Element*	*Control*	*Double digging*	*Peat addition*	*Chalk addition*	*Sulphur addition*	*Replaced topsoil**
1976	Turnips (edible roots)	Zn	151	108	121	93	193	-
		Cu	7	7	7	7	7	-
		Ni	10	8	9	7	14	-
		Cd	1.4	1.4	1.7	0.8	1.8	-
		Pb	3	4	3	3	3	-
1977	Beetroot+ (leaves)	Zn	416	832	748	102	-	-
		Cu	18	18	17	12	-	-
		Ni	6	11	7	2	-	-
		Cd	8	15	14	2	-	-
		Pb	16	10	22	9	-	-
1977	Beetroot+ (edible roots)	Zn	167	242	250	82	-	-
		Cu	19	19	17	19	-	-
		Ni	6	8	8	3	-	-
		Cd	3	3	4	2	-	-
		Pb	2	4	3	3	-	-
1978	Lettuce (edible leaves)	Zn	189	192	240	125	303	114
		Cu	20	20	20	21	23	19
		Ni	15	14	16	11	22	12
		Cd	9	9	13	6	13	5
		Pb	13	11	13	16	34	16
1978	Brussels sprouts (edible sprouts)	Zn	91	86	86	69	114	48
		Cu	6	6	7	6	7	5
		Ni	9	10	7	5	15	6
		Cd	0.6	0.6	0.8	0.3	1.0	0.3
		Pb	2	2	2	2	2	2
1979	Potato (tubers)	Zn	23	23	21	15	34	19
		Cu	11	11	10	10	13	10
		Ni	2	2	1	2	3	2
		Cd	0.7	0.6	0.6	0.4	0.9	0.4
		Pb	0.2	0.2	0.2	0.2	0.2	0.2
1980	Cabbage (inner leaves)	Zn	95	150	108	41	139	66
		Cu	5	6	6	4	7	5
		Ni	9	14	7	4	16	8
		Cd	1.1	1.8	1.5	0.5		0.9
		Pb	2	2	1	2	3	2
1981	Cauli-flower (edible curd)	Zn	125	124	117	75	149	65
		Cu	6	6	6	6	7	6
		Ni	11	13	9	4	20	5
		Cd	0.7	0.7	0.8	0.4	1.1	0.2
		Pb	2	2	2	2	2	2

* = Topsoil was replaced only in 1977. + = Beetroot crops failed to grow.

Summaries of the treatments and data for the harvest of 1975, and for those of 1976 and 1977 are given in Table 8.

Variability between the four replicates was large. Plots given peat yielded the largest percentage of clover [mostly white clover, as red clover in the initial sowing mixture had

Figure 6 *Effects of husbandry measures used to alleviate the contents of metals in the 0 - 15 cm soil layer of sludge treated land*

Year and Treatment	*pH*	*OM* (%)*	EDTA-extractable (mg L^{-1}) *Zn*	*Cu*	*Ni*	Strong acid extractable (mg kg^{-1}) *Zn*	*Cu*	*Ni*	*Pb*	*Cd*
1975										
Control	6.4	4.5	158	55	8	341	80	27	365	7.2
Double digging	6.3	6.6	209	74	8	401	106	31	354	12.4
Peat addition	6.4	4.6	158	63	7	333	92	27	448	6.2
Chalk addition	6.5	4.0	161	63	10	328	82	28	410	9.2
Sulphur addition	6.4	5.3	172	45	6	362	99	32	484	12.8
Replaced topsoil	6.8	3.3	8	6	3	70	16	17	43	0.3
1979										
Control	6.7	4.6	221	63	12	388	111	31	454	11.3
Double digging	6.6	3.3	133	45	10	252	67	25	272	5.4
Peat addition	6.8	9.4	150	38	8	377	100	30	554	11.0
Chalk addition	7.7	5.0	217	62	10	375	98	30	438	10.7
Sulphur addition	6.2	4.7	133	56	10	298	107	27	544	9.8
Replaced topsoil	6.8	3.6	9	6	2	75	17	18	70	0.1

* OM = organic matter

Table 7 *Effects of husbandry measures used to alleviate the influence on soil pH of metals in sludge treated land*

Treatment	*pH value* *1975*	*1977*	*1979*	*1981*
Control	6.4	6.2	6.7	6.4
Double digging	6.3	6.3	6.6	6.1
Peat	6.4	6.0	6.8	6.4
Chalk	6.5	7.5	7.7	7.4
Sulphur	6.4	5.6	6.2	5.7
Replaced topsoil	-	-	6.8	6.5

- = Topsoil was replaced only in 1977.

died off after the first two seasons]. All sludge treatments had significantly decreased the clover component of the sward in the 1975 harvest. Grass yields were not affected significantly but the yield of total herbage from most metal amended sludges, except the 125 t ha^{-1} + Cr (low and high), + Ni (high) and the 31.25 t ha^{-1} annual + Cu treatments, were significantly decreased.

Contents of zinc, copper and nickel extractable from soil in boiling strong mineral acid and the amounts added as sludge cake in 1968 were strongly, positively correlated. Similar correlations apply for copper and for nickel. These data demonstrate that metal extractable

Table 8 *Residual effect on yields of grass and clover of sludge-borne metals at ADAS Lee Valley*

			Total metal added (kg ha^{-1})			*Dry matter, average of 4 replicates*								
						1975				*1976*				*1977*
						grass	*clover*	*total*	*clover*	*grass*	*clover*	*total*	*clover*	*total*
Plot nos	*Treatment*	*Rate*	*Zn*	*Cu*	*Ni*	(g m^{-2})			(%)	(g m^{-2})			(%)	(g m^{-2})
1,26,38,51	Peat		0	0	0	625	751	1376	55	1941	1860	3801	49	939
13,32,42,57	Peat		0	0	0	579	675	1254	54	1564	917	2481	37	870
10,27,45,58	Sludge alone	S	130	44	15	746	349	1095	32	1185	504	1689	30	1012
2,29,33,49	+ low Zn	S	1100	60	15	429	86	505	17	1443	213	1656	13	911
16,22,40,56	+ high Zn	S	2100	70	15	311	123	434	28	1264	45	1309	34	661
3,25,39,55	+ low Cu	S	310	430	20	499	184	683	27	1759	161	1920	8	680
5,20,46,62	+ high Cu	S	500	755	20	549	206	755	27	1095	174	1269	14	932
12,30,34,54	+ low Ni	S	390	65	255	495	361	856	42	1046	552	1598	35	853
11,21,35,52	+ high Ni	S	720	75	555	562	391	953	41	829	820	1649	50	840
8,17,48,53	+ low Cr	S	125	70	15	615	355	970	37	1315	206	1521	14	733
4,19,67,59	+ high Cr	S	120	95	15	691	348	1039	33	1932	467	2399	19	986
6,28,43,63	Sludge alone	M	110	45	10	819	511	1330	38	2104	378	2482	15	913
7,31,44,60	+ Zn	M	360	20	10	566	270	836	32	1046	482	1528	32	950
14,24,27,50	+ Cu	M	130	160	10	681	350	1031	34	1955	722	2677	27	892
1,18,36,61	+ Ni	M	160	40	70	611	274	885	31	1124	489	1613	30	1102
15,23,41,64	+ Cr	M	210	90	20	756	163	919	18	1470	363	1470	25	847

S = 1 x 125 t ha^{-1} application of sludge cake in 1968.

M = 4 x 31.25 t ha^{-1} annual applications of sludge cake commencing in 1968.

Table 9 *Residual effect of sludge-borne metals on red clover yield (1985) and metal concentration in dry matter at ADAS Luddington*

		Total metal added (kg ha^{-1})			Clover dry matter (t ha^{-1})				Metal contents												
									in clover (mg kg^{-1} dry matter)										in soil (mg kg^{-1})		
					Cut				*Cut* 1			*Cut* 2			*Cut* 3						
					1	*2*	*3*	*tot*													
Treatment	*Rate*	*Zn*	*Cu*	*Ni*					*Zn*	*Cu*	*Ni*	*Zn*	*Cu*	*Ni*	*Zn*	*Cu*	*Ni*	*Zn*	*Cu*	*Ni*
None		-	-	-	2.4	6.5	1.9	10.8	59	14	6	65	19	6	50	13	9	73	28	15
None		-	-	-	2.9	7.0	2.1	12.0	64	14	8	50	12	6	62	14	10	54	36	15
Sludge*	S	120	40	10	2.4	6.5	1.7	10.6	84	14	8	80	12	7	70	13	10	81	26	14
+ low Zn	S	1100	55	10	2.7	6.3	1.5	10.4	232	12	6	257	12	4	283	13	7	238	23	14
+ high Zn	S	2090	65	10	0.6	5.0	0.8	**6.3**	236	13	4	308	9	4	301	12	6	455	29	15
+ low Cu	S	310	515	20	2.0	7.4	1.5	10.9	111	17	9	126	13	8	137	14	13	184	148	23
+ high Cu	S	500	990	20	2.4	6.5	1.5	10.4	152	14	13	141	15	10	156	15	14	247	290	24
+ low Ni	S	410	60	250	0.9	6.2	1.7	10.5	133	17	25	115	8	27	104	10	32	176	34	84
+ high Ni	S	600	70	450	0.9	5.1	1.2	*7.2**	131	11	26	128	8	33	106	11	33	214	30	118
Sludge*	M	112	40	10	3.3	6.5	1.8	11.6	73	14	6	52	11	5	59	13	9	75	20	15
+ Zn	M	2090	65	10	0.1	4.8	0.7	**5.7**	261	6	13	227	9	4	269	12	11	511	25	19
+ Cu	M	500	990	20	2.2	6.3	1.7	10.2	139	17	9	185	15	9	184	15	14	235	232	18
+ Ni	M	600	70	445	1.9	6.0	1.5	9.4	110	10	23	104	8	20	101	9	28	192	27	92
SED					0.52	0.81	0.27	1.34	19.4	3.47	3.18	38.4	3.01	2.98	15.72	0.71	2.26	27.5	13.6	6.7
CV(%)					35.6	18.3	24.3	15.1	21.9	37.0	40.1	42.3	37.5	41.9	17.0	8.1	22.0	20.2	29.5	28.8

SED = standard error of the difference between means. CV = coefficient of variation.

S = 1 x 125 t ha^{-1} application of sludge cake in 1968. M = 4 x 31.25 t ha^{-1} annual applications of sludge cake commencing in 1968.

Sludge* = uncontaminated sludge only. **Treatments shown bold** = significant at P = 0.001. *Treatment shown in italics* and * = significant at P = 0.01.

Table 10 *ADAS Luddington: Effects on seed and haulm yields and zinc contents of leaves of peas (Pisum sativum) grown in soil receiving zinc as sewage sludge cake, added in 1968 (topsoil reinstated at ADAS Rosemaund in July 1991)*

Treatment	Rate	Yields of		Zinc contents of	
		Haulm (t ha^{-1} DM)§	Seeds (t ha^{-1} @ 85% DM)	Pea leaves (mg kg^{-1} in DM)	Topsoil 1993 (1985) (mg kg^{-1})
No sludge		4.55	4.36	23	86 (64)
Sludge alone	S†	4.77	4.56	17	125 (81)
+ low Zn	S	4.06	5.45	42*	177 (238)
+ high Zn	S	3.60*	4.30	68***	293 (455)
+low Cu	S	4.02	4.37	32	150 (184)
+ high Cu	S	5.10	4.99	39	208 (247)
+ low Ni	S	4.08	4.42	28	140 (176)
+ high Ni	S	4.66	4.93	43*	148 (214)
Sludge alone	M‡	5.26	4.97	22	75 (75)
+ Zn	M	3.47*	4.99	148***	457 (511)
+ Cu	M	5.29	5.13	39	198 (235)
+ Ni	M	4.46	4.87	27	196 (192)
SED¶		0.496		11.9	
CV≠ (%)		16	15	42	

† S = 1 x 125 t ha^{-1} application of sludge cake in 1968.
‡ M= 4 x 31.25 t ha^{-1} annual applications of sludge cake commencing in 1968.
§ DM=dry matter. *, *** = significant at P=0.05 and 0.001 levels respectively.
¶ SED=standard error of the difference between means. ≠CV=coefficient of variation

in boiling strong mineral acid is a good indicator of metal content in the soil. No measurements of chromium were made. Both grass and clover yields were negatively correlated with the amounts of zinc added as sludge cake, and with its contents in soil and plant dry matter. The relationship between yields and copper contents was negative, but was not readily distinguishable from the effect of the accompanying zinc in the sludges.

Yields in 1976 from two cuts were larger than those for the single late cut measured in 1975 (where the yield from an earlier first cut had not been measured). The dry weights yields in 1976 from two cuts were larger than those for the single late cut measured in 1975 (where the yield from an earlier first cut had not been measured). The pattern of the dry weights of the grass and clover components broadly followed that of the earlier data, but the proportion of clover had declined in several of the treatments. The severe drought in 1976 may have stressed the clover plants more than the grass.

The total harvest from two cuts in 1977 was considerably less than that for 1976, and for the single late cut of herbage of 1975.

Experiment at ADAS Luddington. Red clover (*Trifolium pratense*) was grown at the ADAS Luddington site as a single season crop in 1985. Dry weight yields for three cuts, and the contents of metals in the plant tissue were measured. These data, together with contents of metals in the topsoil samples collected in January 1975, are given in Table 9.

Again, zinc in large doses (exceeding early ADAS guide lines and current EU limits) applied 18 years previously in 1968, decreased yield by almost one third from > 10 t ha^{-1} (for no large zinc additions) to < 7 t ha^{-1}. These sludges also contained significant contents of zinc, and this is reflected in the large zinc contents of the tissue. There was strong positive correlation between the contents of metals in the herbage and those in soil as determined in the boiling strong mineral acid extracting solution. The zinc and nickel contents in the tissue increased much more rapidly than that for copper in response to the increased metal contents in the soil.

The site was maintained in grass after the 1985 tests, but no measurements were made. Topsoil of the different treatments, omitting those given chromium, were dug up in July 1991, transported to ADAS Rosemaund and re-established as separate 1.2.m x 1.2 m plots (consequent to closure of the ADAS Luddington centre by MAFF); further tests with legume and other crops continued at the new site (Chambers, 1995).

Growth and yields of runner beans (*Phaseolus vulgaris*) grown in 1992 were unaffected by any of the treatments given in 1968.

Seed yields of peas (*Pisum sativum*) grown in 1993 were also not affected by the sludge-borne metals, while haulm dry weights were decreased significantly (Table 10) where 2090 kg ha^{-1} zinc had been added as sludge cake in 1968. Zinc contents of pea leaves dry matter were strongly positively correlated with zinc contents in soils extracted by boiling strong mineral acid, measured after transfer from ADAS Luddington and reinstatement at ADAS Rosemaund (Chambers *et al.*, 1994; Chambers, 1995). Digging up the topsoil and re-establishment at the new site, involving more thorough mixing within each treatment, had altered boiling strong acid extractable metal contents without changing the original relative order. (The last coloumn in Table 10 shows, in parenthesis, the soil zinc contents measured in soil sampled in 1975.)

Grain yields of winter wheat grown in 1994 were decreased (by > 2 t ha^{-1}) significantly in the two treatments receiving large amounts (500 kg ha^{-1}) of copper as sludge cake in 1968. Straw yields were also decreased (by > 1.5 t ha^{-1}, $P < 0.01$) significantly in all the three treatments receiving copper as the major metal in sludge cake (Chambers, 1995), but these had large amounts of accompanying zinc as well, making it difficult to separate the individual metal effects.

3 Current Experiment at ADAS Gleadthorpe

An experiment was initiated in 1982, by the soil scientists of ADAS Wolverhampton, at ADAS Gleadthorpe using custom made sewage sludges to which zinc, copper, and nickel were added, either alone or in combination. Cakes produced at Coleshill and Coton Park (Severn Trent Water Region Sewage Treatment facilities) were used as the control. The cake was obtained mostly from sewage from domestic sources, and it had very small amounts of the metals listed. Metal salts were bled into the crude sewage stream entering the WRc Coleshill Experimental Sewage Treatment facility in order to manufacture raw pressed cakes containing the three individual metals. Uniform mixtures of these products enriched with different metals and the control (unspiked) cake were prepared to give sludge cakes with a range of contents of the three metals. Addition of 100 t ha^{-1} of each preparation provided the different treatments listed in Table 11 (ADAS Wolverhampton, 1983).

Table 11 *Metals applied as sewage sludge cake to a sandy loam site in 1982 at ADAS Gleadthorpe*

Treatment No.	Sludge type	Metal applied as sludge cake (kg ha^{-1})		
		Zn	*Cu*	*Ni*
0	No sludge	-	-	-
1	Coleshill Control	127	31	10.5
2	Coton Park Control	48	12	1.3
3	Zn	119	27	9.6
4		633	42	8.4
5		986	38	9.4
6		1191	36	7.9
7		158	69	10.4
8		287	187	7.5
9		436	289	5.5
10		639	444	7.1
11	Cu	125	129	9.2
12		115	424	3.3
13		112	606	3.3
14		130	1278	3.6
15		143	15	29.0
16		288	20	49.8
17		432	25	77.6
18		578	29	105.4
19	Ni	51	12	49.8
20		51	14	98.0
21		51	14	153.0
22		52	16	208.2

The soil is a sandy loam of the Newport series. It is maintained at pH 7 or above, and has two per cent organic matter. Treatments were applied in February 1982 to duplicate plots in a randomized block design.

A succession of spring barley (*Hordeum vulgare* in 1982), rye grass (*Lolium perenne* in 1983 and 1984) and sugar beet (*Beta vulgaris* in 1985) was grown to test the effects on the crops of the sludge metals. The yields for all these crops were not affected by any of the treatments. Analysis of soil, sampled in 15 cm layers to 90 cm in January 1985, showed that the contents of metals soluble in boiling strong mineral acid were not as large as expected, despite the addition as cake of intended doses. The soil pH, and much smaller metal contents in soil than intended meant that crops grew unaffected by the treatments made.

Almost all (97%) of the added zinc, and most (84%) of the added nickel were extracted from the soil to 90 cm depth by boiling strong mineral acid. In comparison, only 28% of the added copper was extracted in this treatment process. Contents of boiling strong mineral acid-soluble metals were strongly, and positively correlated with the amounts added; the linear relationship was better for zinc and nickel than for copper.

Although the sugar beet grown in 1985 showed no adverse effects on yields of crops, contents of metals in the foliage had increased, approximately reflecting the metal dosage. A selection of the plots originally treated in 1982 were given further doses of sludges in March 1986 in order to give rise to potentially more toxic zinc and copper contents in the soil. In addition, sulphur powder was incorporated as appropriate to plots during the winter of 1987 - 88 to decrease the soil pH to 6.2 for all treatments.

Spring barley, sown in March 1986, and winter barley harvested in 1987 showed that these increased metal additions had no effects on the yields of either of the crops.

White clover (*Trifolium repens.* L), a legume expected to be sensitive to metal toxicity, was sown in spring 1988, and its growth was followed through the following seasons up till 1990.

Yields of dry matter for the three seasons are given in Table 12 (Royle *et al.*, 1989; Chambers *et al.*, 1994).

In 1988, toxicity symptoms were seen as uneven or diminished growth, and there were some bare patches. Yellowing of foliage occurred where larger zinc and copper doses were applied. There were very few plants, and little mowable foliage growth in the plots that were given the two largest zinc doses. Average yields from two cuts are given in Table 12. Growth of clover within each plot varied considerably. Zinc was the most toxic metal. Copper had a greater effect in the first cut compared to the second, while nickel did not appear to be toxic.

In the second season (after the initial drilling), dry weight yields were consistently larger than in 1988. Remarkably, the two treatments with very large zinc doses (Treatments 5, 6) yielded 2.75 and 2.18 t ha^{-1}, respectively, although there was little mowable foliage growth in either of the two 1988 cuts. Copper, at the two largest doses, accompanied by large and small amounts of zinc (Treatments 9, 10 and 13, 14) also decreased yield, but less severely than the large zinc, small copper treatments. This effect was more noticeable in the first than in the second cuts, as for 1988.

The yields for 1990 were considerably smaller than for the previous two seasons. There was no mowable growth from the largest zinc dose (Treatment 6), but Treatment 5 (where less zinc was applied) gave a modest crop. The pattern of decreases in yields associated with zinc and copper were similar in all three seasons. Notably, the treatments with nickel as the main metal (Treatments 15 to 22) consistently improved clover growth even more than the 'unspiked' control sludges of Treatments 2 and 3.

Table 13 presents the metal contents in the harvested dry matter of 1988 and 1989. Increases of up to four or five fold, and roughly proportional to the amounts of metals added to soil as sludge cake (in 1982 and 1986) are evident for the metal contents in the dry matter. Contents of metals in the herbage from the second cut were generally smaller than for the first cut.

Contents of metals dissolved from the soil by boiling strong mineral acid were positively correlated with the quantities of metals added as sludge cake. Clover yields were negatively correlated with the amounts of metals added, and with the acid soluble contents of metals, but there was much scatter in the data. Influences on yield were less for the 1989 compared with the 1988 season, and also the differences between the toxicities of copper were small.

Drought in 1990 resulted in extensive death of the clover and the site was re-drilled in 1991 and maintained in white clover during 1992 (Chambers *et al.*, 1994)). Seed yields of peas (*Pisum sativum*) grown in 1993 were greatly decreased (by > 2 t ha^{-1}, $P < 0.01$) where large additiions of zinc had been made (Treatments 5 and 6); haulm dry weights were also

Table 12 *Effects of sewage sludge (after increasing zinc, copper and nickel contents in soil in March 1986) on yields of clover at ADAS Gleadthorpe (1988 and 1989)*

Treatment No	*Treatment*	*Total metal added* (kg ha^{-1})			*Clover dry matter yield* (t ha^{-1})							
					1988			*1989*			*1990*	*Grand*
		Zn	*Cu*	*Ni*	*Cut 1*	*Cut 2*	*Total*	*Cut 1*	*Cut 2*	*Total*	*Total*	*Total*
0	No sludge	-	-	-	3.09	0.69	3.78	2.51	2.07	4.58	0.77	9.13
1	Coleshill control	127	31	10	3.09	0.59	3.69	2.22	2.62	4.84	0.96	9.49
2	Coton Park control	48	12	1	2.71	0.52	3.23	2.56	2.17	4.74	0.72	8.69
3	Zinc	633	42	8	2.83	0.60	3.43	2.40	2.24	4.64	1.26	9.33
4	Zinc	1191	36	8	2.61	0.39	3.00	1.53	2.68	4.21	1.50	8.71
5	Zinc	1896	100	18	-	-	**0.16**	1.22	1.53	*2.75**	0.34	3.09
6	Zinc	2763	111	17	-	-	**0.70**	0.69	1.49	*2.18**	0.00	2.18
7	Zinc:copper	295	258	8	3.20	0.60	3.80	2.51	2.01	4.52	0.69	9.01
8	Zinc:copper	916	1244	17	2.38	0.51	2.89	2.21	1.81	4.01	0.51	8.94
9	Zinc:copper	1557	1087	17	0.99	0.49	**1.48**	1.36	1.82	*3.18**	0.70	5.99
10	Zinc:copper	2299	1617	22	2.04	0.45	2.49	1.70	2.28	3.98	0.94	6.62
11	Copper	242	1121	16	2.27	0.66	**2.93***	2.30	2.26	4.56	1.25	8.74
12	Copper	130	1278	4	3.63	0.62	4.25	2.14	2.49	4.63	1.99	8.76
13	Copper	310	2077	15	1.97	0.53	**2.49***	1.96	2.32	4.28	0.86	7.63
14	Copper	391	2968	21	0.96	0.40	**1.36**	1.92	2.15	4.07	0.47	5.90
15	Zinc:nickel	143	15	29	3.38	0.68	4.06	2.17	2.36	4.53	1.05	9.64
16	Zinc:nickel	288	20	50	3.60	0.59	4.19	2.51	2.07	4.58	0.95	9.72
17	Zinc:nickel	432	25	78	3.54	0.65	4.19	2.40	2.31	4.72	1.30	10.21
18	Zinc:nickel	578	29	105	2.95	0.59	3.54	2.40	2.27	4.67	1.08	9.29
19	Nickel	51	12	50	3.18	0.59	3.77	2.39	2.64	5.03	1.52	10.32
20	Nickel	51	14	98	3.91	0.61	4.52	2.26	2.57	4.83	1.33	10.68
21	Nickel	51	14	153	3.44	0.56	4.00	2.47	2.43	4.90	1.42	10.32
22	Nickel	52	16	208	3.17	0.65	3.82	2.54	2.20	4.74	0.85	9.41
SED for comparison with "no sludge"							0.444	0.27	0.35	0.49		
SED for all other treatments							0.513	0.31	0.41	0.56		
CV (%)							17.1	14.6	18.2	13.0		

SED = standard error of the difference between means. CV = coefficient of variation. **Treatments shown bold** = significant at P= 0.001. *Treatments shown in italics and* * = significant at P = 0.01. **Treatments shown bold with *** = significant at P = 0.05. - = total of two cuts measured after bulking.

Table 13 *Effects of sewage sludge on the contents of metals in white clover (Trifolium repens.L) at ADAS Gleadthorpe (1988 and 1989)*

Treatment*		Total metal added (kg ha^{-1})			Clover metal concentrations (mg kg^{-1} in dry matter)											
					1988						*1989*					
		Zn	*Cu*	*Ni*	*Zn*		*Cu*		*Ni*		*Zn*		*Cu*		*Ni*	
No					*Cut 1*	*Cut 2*	*Cut 1*	*Cut 2*	*Cut 1*	*Cut 2*	*Cut 1*	*Cut 2*	*Cut 1*	*Cut 2*	*Cut 1*	*Cut 2*
0	No sludge	-	-	-	39	53	13.0	10.1	1.8	1.9	33	41	11.0	11.3	0.8	4.9
1	Coleshill control	127	31	10	61	56	12.5	11.1	3.2	1.1	101	51	12.8	9.0	2.7	5.3
2	Coton Park control	48	12	1	44	39	13.0	11.0	1.7	0.2	40	46	10.2	12.0	0.6	1.0
3	Zinc	633	42	8	160	127	13.0	10.9	2.7	1.1	112	140	11.3	10.0	1.3	2.0
4	Zinc	1191	36	8	220	130	13.0	13.5	1.6	0.7	175	98	109	10.5	0.8	0.7
5	Zinc	1896	100	18	-	-	-	-	-	-	203	200	11.3	10.5	1.9	1.5
6	Zinc	2763	111	17	-	-	-	-	-	-	200	364	10.3	14.0	1.4	1.6
7	Zinc:copper	295	258	8	91	92	18.5	14.0	3.4	1.8	73	83	12.5	12.0	2.3	1.6
8	Zinc:copper	916	1244	17	195	130	15.5	10.0	3.7	1.8	160	131	12.7	14.5	2.7	1.9
9	Zinc:copper	1557	1087	17	155	170	11.5	13.0	2.8	3.9	135	152	12.1	12.0	2.4	1.6
10	Zinc:copper	2299	1617	22	175	195	16.5	15.0	2.6	1.7	118	151	12.9	12.5	2.3	1.3
11	Copper	242	1121	16	77	56	17.5	12.0	4.8	2.4	131	77	130	11.0	3.3	6.1
12	Copper	130	1278	4	52	48	15.0	31.5	2.2	1.2	55	71	11.1	11.5	4.4	4.6
13	Copper	310	2077	15	61	58	19.0	16.5	3.4	2.6	61	79	12.0	9.5	5.5	4.4
14	Copper	391	2968	21	62	54	22.0	15.5	3.5	2.4	49	57	15.4	13.5	2.3	1.8
15	Zinc:nickel	143	15	29	64	47	11.0	9.5	4.4	2.1	44	73	11.5	12.0	3.7	6.1
16	Zinc:nickel	288	20	50	72	80	14.0	10.7	6.1	5.0	55	71	11.1	11.5	4.4	4.6
17	Zinc:nickel	432	25	78	77	35	12.0	9.0	8.0	5.1	61	79	12.0	9.5	5.5	4.4
18	Zinc:nickel	578	29	105	97	110	14.0	10.2	12.5	9.1	83	64	10.9	9.5	8.1	10.7
19	Nickel	51	12	50	49	36	13.5	10.0	6.6	5.4	37	48	10.1	11.0	5.8	3.5
20	Nickel	51	14	98	42	28	12.5	9.4	10.7	8.4	33	37	10.1	12.0	6.3	5.9
21	Nickel	51	14	153	46	40	12.5	11.5	21.0	12.7	36	41	10.2	9.5	18.3	4.9
22	Nickkel	52	16	208	48	37	16.5	6.9	19.0	17.7	37	38	10.4	9.0	17.8	11.9

* The Treatment No and the treatments applied are the same as given in Table 8. - = not determined.

significantly decreased by these treatments. Zinc contents of pea seeds and zinc extracted from soil by boiling strong mineral acid were strongly, positively correlated. Seed yields were also decreased where large amounts of copper had been added (Treatment 14, > 1.5 t ha^{-1}, $P < 0.05$).

The evidence from the legumes grown at all the ADAS sites point to toxicity from excessive uptake of metals, particularly zinc, to be more likely than any effect on rhizobial symbiosis to cause adverse effects (if any) on crop performance.

4 Discussion

Data from all of the ADAS studies discussed here suggest that toxicity symptoms arise in all crops from the uptake of larger than normal amounts of zinc. Red beet and lettuce, among the vegetable crops tested, and clover suffered significant yield decreases associated with the uptake of large amounts of metals. Data for white clover from the ADAS Gleadthorpe experiment did not show clearly that the lack of or the inadequacy of rhizobial symbiosis was involved. The two treatments that received the large zinc doses (and modest doses of copper) gave measurable regrowth in the 1989 season, following very little mowable foliage in 1988 - the year the crop was drilled initially (Table 12). Sufficient numbers of clover plants may have established after sowing, but severe toxicity probably resulted in stunted growth. The habitually prostrate plants would not have leaves with petioles sufficiently long to produce herbage at and above mowing height and this, most probably, resulted in records of very small yields in 1988. [The decrease in pea haulm dry weight, but not pea seeds, found in 1993 in the ADAS Luddington soil (transferred to ADAS Rosemaund) supports the possibility of an impact on vegetative growth by large metal contents in tissue]. Growth of clover foliage in 1989 was more successful and both cuts in that season gave measurable, but significantly smaller yields than for unspiked or modestly zinc contaminated sludges. These two treatments (with large zinc doses), again showed evidence in 1990 of the large zinc contents in the tissues. Should inadequate N fixation, through impaired rhizobial activity, be the major cause of the lack of growth in 1988, then the rather impressive regrowth of the crop in 1989 would be hard to explain.

McGrath and Brooks have presented evidence (Figure 2b and Tables 3, 4 and 5 of McGrath and Brooks, 1988) of severe decreases in the efficiency of *Rhizobium* infecting roots of white clover (*Trifolium repens*. L), grown in a sandy loam soil that had received sewage sludge with large metal contents many years previously. There was no lack of nodulation of clover roots. Bacteria isolated from the nodules would not fix dinitrogen in the absence of heavy metals in *in vitro* tests (Giller *et al.*, 1989; McGrath and Hirsch, 1989) leading to the conclusion that the *Rhizobium* infecting the clover roots was an ineffective genotype. Direct toxicity from metal uptake by clover plants was not the cause of decreased crop growth.

However, because of the development of a microbial population tolerant to heavy metals, there is good evidence that dinitrogen fixation by non-symbiotic, heterotrophic, free living microorganisms was not seriously inhibited by heavy metals in mine spoils (Rother *et al.*, 1982; 1983). El Aziz *et al.*, (1991) found little evidence of damage to *Rhizobium mellitoti* in root nodules of alfalfa (*Medicago sativa*) plants growing in soils with large contents of zinc, accrued from zinc ore smelters, in Palmerston, Pennsylvania, USA. These investigators suggested that "it is possible that the macrosymbiont hosts differ in their symbiotic capacities when grown under stressed conditions. White clover may be more sensitive to the toxic effects of heavy metals, thereby failing to achieve an effective symbiotic relationship with *Rhizobium leguminosarum* bio *trifolii*. Secondly, the microsymbionts may differ in their sensitivity to the toxic effects of heavy metals". They also stated that "adaptations of isolates (of microsymbiont bacteria) to local conditions is possible and could result in divergent lines of the same species".

Adaptation of bacteria to overcome stress induced by metals is well recognized; an excess, or a scarcity of metallic elements in the substrate seems "to trigger the production of unique biochelates" (Appanna and Viswanathan, 1986; Anderson and Appanna, 1993) by bacteria. An arctic *Rhizobium*, in the presence of a large concentration of manganese in the culture medium, secreted a novel exopolysaccharide (Appanna and Preston, 1981) which immobilized the metal. This response was considered as a mechanism "to trap this metal and hence may be playing a role in manganese homeostasis".

Smith of WRc Medmenham, UK (Smith, 1991) made a "*Rhizobium* screening assay of historically sludge treated soils" that "show potentially detrimental effects, although the bacteria was (sic) never completely eliminated and no ineffective strains have so far been identified". Smith's inference was that large cadmium contents, considerably in excess of the 3 mg kg^{-1} limit, as in the Woburn site, may be the toxic component damaging effectiveness of *Rhizobia*. "However, because of improved sludge quality, the maximum soil limit of 3 mg kg^{-1} for cadmium is rarely approached and the main limiting potentially toxic elements to sludge recycling in agriculture now are copper and zinc. Consequently the model of Woburn and other sites supplied with sewage sludge of similar quality (to that applied at Woburn) represents a legacy of historical sludge application, and does not reflect the likely long-term effect of modern sludge application practices on soil microorganisms and (soil) fertility". But, very large cadmium contents in soil (mine spoil) - up to around 200 mg kg^{-1} cadmium accompanied by > 20 000 mg kg^{-1} zinc, as well as up to 30 000 mg kg^{-1} lead - did not significantly inhibit dinitrogen fixation by *Rhizobia* associated with white clover (*Trifolium repens*. L) in grass swards (Rother *et al*., 1982, 1983). Up to 80 kg ha^{-1} nitrogen was fixed during spring growth, and except for small decreases of plant size and nitrogenase activity, plants and nodules were normal and looked healthy. More recent tests (Obbard and Jones, 1993a) also showed that, where host plants of white clover (*Trifolium repens*. L) were indigenous to the sward, roots were nodulated. *Rhizobia* capable of effective symbiosis and dinitrogen fixation were present both in the root nodules and in the rhizosphere soil even when metal contents greatly exceeded UK/EC recommnded limits. However, there was some evidence that "metals may have had a quantitative effect on the free-living population of *Rhizobia* ...... where *Trifolium repens* was not indigenous to the contaminated soils". Thus, evidence for severe damage to symbiotic rhizobial activity - and activity of free living dinitrogen fixers - is not as compelling as might be inferred from the Woburn study of McGrath *et al*.. Smith's suggestion that cadmium may be the cause of any damage observed to nitrogenase efficiency is also unsupported unless cadmium and other heavy metals introduced via sewage sludges speciate differently (and in ways that affect the dinitrogen fixing ability of *Rhizobia*) compared with those derived from the transformations of metals in mine spoils.

Adverse effects of sludge-derived heavy metals on other soil microflora - the soil biomass - are also more likely to arise from a possible smaller density of their population than from a severe and permanent decrease in their activity. For example, cellulose degraders have been shown (Obbard and Jones, 1993b) to be active in the ADAS Luddington soils (with zinc and copper nearly 3.5 times, and nickel about two times larger than prescribed by UK/EC limits at pH 5.2-5.6). Obbard and Jones state: "although initial populations of the (relevant) microorganisms in the contaminated soils may be smaller, once full colonisation of the available substrate has taken place, the actual process of decomposition is not significantly limited. Heavy metal effects on the decomposition of cellulose in sludge-amended soils are short-term".

Farmers who regularly apply repeated doses of sewage slurry or sludge cake (carefully screened to avoid larger than prescribed limits of heavy metals being added, cumulatively), accept as a welcome benefit - for light textured soils where arable crops including root crops are grown - the nitrogen, the increases in available phosphate, and the improvement in soil physical condition through the addition of organic matter. Leys with clover receiving such sewage additions would appear to benefit also (Blake, 1991).

ADAS soil mineral nitrogen service (Vaidyanathan *et al.*, 1991), for identifying fields with large mineral (ammonium + nitrate) nitrogen supplies, found a range of 45 to 1740 kg ha^{-1} (N) in 33 fields (sampled in the 1989-1990 season) receiving sewage sludge. Of these, 24 fields had more than 100 kg ha^{-1} and 12 had in excess of 200 kg ha^{-1}; the latter fields would not require any fertilizer nitrogen for all the arable crops grown. The ammonium N in slurry/sludge - which could be > 30% of total N in digested sludge - gets quickly oxidised to nitrate N that is vulnerable to leaching (Misslebrook *et al.*, 1996), especially when applied to warm soil in the autumn, when there is little crop uptake and periods of excess rain occur subsequently. There can be substantial build up of labile phosphate in soil receiving repeated applications of sewage slurry/sludge; egress of phosphate out of such land through soil erosion and in the drainage water may also become a problem.

All the adverse effects observed in the ADAS studies were in treatments where sludge-borne metals were applied in amounts vastly greater than the limits prescribed. These ADAS studies, which emphasize the adverse impacts of deliberate and gross overloading of soils with sewage-borne heavy metals, contrasts sharply with the meticulously planned long term experimental strategy of the scientists and funding agencies in the USA, on the rational use of plant nutrients in sewage (and other organic wastes) during the past > 25 years (Clapp *et al.*, 1994). Sewage slurries or sludge cakes applied to farmland are strictly monitored and the soils regularly analysed for heavy metal contents to ensure compliance with recommended safe limits. Thus, results discussed here are useful only to confirm and emphasize the importance of avoiding addition of excessive amounts of sludge-borne metals to land. Also, the fact that a major proportion of the metals in sewage slurry and sludge may be as sulphides and/or hydroxy carbonates (Mathews, 1980) and may not be strongly chelated by the organic matter is of concern. However, where necessary, raw sewage slurry may be chlorinated, or treated with hydrogen peroxide as an initial odour abatement step (Dao *et al.*, 1994). Metal sulphides are oxidised facilitating some interaction with the organic constituents.

Research is needed to:

(i) distinguish between straightforward toxicity from excessive uptake of metals, particularly zinc, which cause possible decreases in the growth of crops, especially leguminous crops, and damage to soil biomass activity and/or efficiency of rhizobial symbiosis leading to lack of, or impaired dinitrogen fixation, and to a consequent nitrogen deficiency;

(ii) develop methods for predicting post-spring mineralization of nitrogen (and phosphorus) in the accumulating sludge-derived organic matter by identifying and characterizing the components, as humification progresses; and

(iii) determine more reliably the forms in which metals occur in sewage slurries and sludges, the association between the metals and the organic components in the sludges, and their interactions with indigenous soil organic matter as humification progresses.

Acknowledgements

The experiments discussed were the joint effort of many ADAS Soil Scientists and Analytical Chemists at the Cambridge, Wolverhampton, and other centres, funded by the Ministry of Agriculture, Fisheries and Food. Mr R.J. Unwin, ADAS Bristol and Dr R.Harrison, ADAS Cambridge advised on the initial draft. Dr T.M. Hayes, School of Chemistry, The University of Birmingham, provided word processing expertise. Help from WRc Medmenham, and members of the water industry, and cooperation from farmers is gratefully acknowledged.

References

ADAS Wolverhampton. 1983. Unpublished Internal Report of the Soil Science Department.

Anderson, S. and V.D. Appanna. 1993. Indium detoxification in *Pseudammonus fluorescens*. *Environmental Pollution* **82**:33-37.

Anon. 1989. The sludge (Use in Agriculture) Regulations 111989. SI 1989 No.1263.

Appanna, V.D. and T. Viswanathan. 1986. Effects of some substrate analogues on aerobactin synthetase from *Aerobacter aerogens* 62-1. *FEBS Lett.* **202**:107-110.

Appanna, V.D. and C.M. Preston. 1981. Manganese elicits the synthesis of a novel exopolysaccharide in arctic *Rhizobium*. *FEBS Lett.* **215**:79-82.

Blake, A. 1991. Terraplan saves grower "serious" money. *Farmers Weekly*, 13 Dec.1991, p 43.

Brooks, P.C. and S.P. McGrath. 1984. Effects of metal toxicity on the size of the soil microbial biomass. *J.Soil Sci.* **35**:341-346.

Brooks, P.C. and S.P. McGrath. 1987. Adenylate energy charge in metal contaminated soil. *Soil Biol.Biochem.* **19**:219-220.

Brooks P.C., S.P. McGrath, D.A. Klein and E.T. Elliot. 1984. Effects of heavy metals on microbial activity and biomass in field soils treated with sewage sludge. p 574-583. In *Environmental Contamination*. CEP, Edinburgh.

Brooks P.C., S.P. McGrath, and C. Heijnen. 1986a. Metal residues in soils previously treated with sewage sludge and their effects as growth and nitrogen fixation by blue-green algae. *Soil Biol.Biochem.* **18**:345-353..

Brooks P.C., C. Heijnen, S.P. McGrath and E.D. Vance. 1986b. Soil microbial biomass estimates in soils contaminated with metals. *Soil Biol. Biochem.* **18**:383-388.

Chambers, B.C., M.A. Shepherd and J.H. Spink. 1994. Effect of heavy metas in sewage sludge on the growth of legumes. Paper presented at the *European Conference on Sludges and Organic Wastes*, Wakefield. 12-15 April 1994.

Chambers, B.C. 1995. Effects of heavy metals from sewage sludge on the growth and yield of legumes. *Review of MAFF R&D on Soil Protection*-24 July 1995. Environment Policy Group Science Report (unpublished).

Chumbley, C.G. 1971. Permissible levels of toxic metals in sewage used on agricultural land. *ADAS Advisory Paper No.10*, MAFF (Publications), Pinner, Middlesex.

Clapp, C.E., W.E. Larson and R.H. Dowdy. 1994 (eds). *Sewage Sludge: Land Utilization and the Environment*. American Society of Agronomy, Inc., Crop Scince Society of America, Inc., Soil Science Society of America, Inc., Madison, WI, USA, 258 pp.

Dao, C., R.J. Ooten and D. Cook. 1994. Evaluation of chlorination replacement chemicals: County sanitation districts of Orange county CA. p 1-7. Paper presented at *Water Envirnment Federation Conference*, Chicago, USA.

Davies, G.R. 1980. Pot experiments testing zinc, copper and nickel salts on the growth and composition of crops. p 191-204. *In* MAFF Refernce Book 326, *Inorganic Pollution and Agriculture*. HMSO, London.

DOE 1981. *Report of the Sub-Committee on the Disposal of Sewage Sludge to Land*. DOE/NWC Report No. 20, NWC London.

El-Aziz, R., J.S. Angle and R.L. Chaney. 1991. Metal tolerance of *Rhizobium mellilotii* isolated from heavy metal contaminated soils. *Soil Biol. Biochem.* **23**:795-798.

Giller, K.F., S.P. McGrath and P.R. Hirsch. 1989. Absence of nitrogen fixation in clover grown on soil subject to long term contamination with heavy metals is due to survival of only ineffective *Rhizobium*. *Soil Biol. Biochem.* **21**:841-848.

Lindsay, D.G. 1980. Evaluation of the impact of soil pollution on consumers. p 1-10. *In* MAFF Reference Book 326, *Inorganic Pollution and Agriculture*. HMSO London.

McGrath, S.P. and P.C. Brooks. 1988. Effects of potentially toxic metals in soils derived from past applications of sewage sludge on nitrogen fixation by *Trifolium repens*. L. *Soil Biol.Biochem.* **20**:415-424.

McGrath, S.P. and P.R. Hirsch. 1989. Effects of pollutants on diversity of soil microbes. p *77-78. Institute of Arable Crops Research Annual Report for 1989.*

MAFF 1986. *The Analysis of Agricultural Materials*. Reference Book 427. HMSO, London.

Marks, M.J., J.H. Williams and C.G. Chumbley. 1980. Field experiments testing the effects of metal contaminated sewage sludge on some vegetable crops. p 235-251. *In* MAFF Reference Book 326, *Inorganic Pollution and Agriculture*. HMSO, London.

Mathews, P.J. 1980. Discussion after papers 12-17, Trace element content of soils and effects of metals on crops. p 219. *In* MAFF Reference Book 326, *Inorganic Pollution and Agriculture*. HMSO, London.

Mills, C.F., J.K. Campbell, I. Bremner and J. Quarterman. 1980. The influence of dietary composition on the toxicity of cadmium, copper, zinc and lead to animals. p 11-21. *In* MAFF Reference Book 326, *Inorganic Pollution and Agriculture*. HMSO, London.

Misslebrook T.H., M.A. Shepherd and B.F. Pain. 1996. Sewage sludge application to grassland: influence of sludge type, time and method of application on nitrate leaching and herbage yield. *J. Agric. Sci.* **126**:343-352.

Obbard, J.P. and K.C. Jones. 1993a. The effect of heavy metals on dinitrogen fixation by rhizobium-white clover in a range of long-term sewage sludge amended and metal-contaminated soils. *Environmental Pollution* **79**:105-112.

Obbard, J.P. and K.C. Jones. 1993b. The use of cotton-strip assay to assess cellulose decomposition in heavy metal contaminated sewage sludge-amended soils. *Environmental Pollution* **81**:173-178.

Richardson, S.J. 1980. Composition of soils and crops following treatment with sewage sludge. p 252-278. *In* MAFF Reference Book 326, *Inorganic Pollution and Agriculture*. HMSO, London.

Richardson, S.J. and C.G. Chumbley. 1981. Yield and metal content of crops grown on sewage sludge amended soil - effects of double digging, peat, chalk and sulphur additions. Internal Report, ADAS Reading (unpublished).

Rother, J.A., J.W. Millbank and I. Thornton. 1982. Seasonal fluctuations in nitrogen fixation (acetylene reduction) by free-living bacteria contaminated with cadmium, lead and zinc. *J.Soil Sci.* **33**:101-114.

Rother, J.A., J.W. Millbank and I. Thornton. 1983. Nitrogen fixation by white clover (*Trifolium repens.* L) in grasslands on soil contaminated with cadmium, lead, and zinc. *J.Soil Sci.* **34**:127-136.

Royle, S.M., N.C. Chandrasekhar and R.J. Unwin. 1989. The effect of zinc, copper and nickel applied to soil in sewage sludge on the growth of white clover. Poster paper to *WRc Symposium*, York, WRc Report CP596.

Smith, S.R. 1991. The legacy of Woburn. Internal Report, WRc, Medmenham, Marlow (unpublished).

Vaidyanathan, L.V. 1976a. Relative differences in yields of red beet and celery to different levels of zinc, copper and nickel in digsted sewage sludges added to soil. p 51-52. *ADAS Experiments and Development in the Eastern Region 1975*. MAFF, Cambridge.

Vaidyanathan, L.V. 1976b. Residual effects of metal contaminated sewage sludge additions to soil on grass-clover ley. p 53-55. *ADAS Experiments and Development in the Eastern Region 1975*. MAFF, Cambridge.

Vaidyanathan, L.V., M.A. Shepherd and B.J. Chambers. 1991. Mineral nitrogen arising from soil organic matter and organic manures related to winter wheat production. p 315-327. *In* W.S.Wilson (ed), *Advances in Soil Organic Matter Research*. The Royal Society of Chemistry, Cambridge.

Webber, J. 1972. Effect of toxic metals on crops. *J.Water Pollution Control* **71**:404-413.

Webber, J. 1980. Effects of zinc and cadmium added in different proportions on the growth and composition of lettuce. p 205-210. *In* MAFF Reference Book 326, *Inorganic Pollution and Agriculture*. HMSO, London.

Williams, J.H. 1975. Use of sewage sludge on agricultural land and the effects of metals on crops. *J.Water Pollution Control* **74**:635-644.

Williams, J.H. 1980. Effect of soil pH on the toxicity of zinc and nickel to vegetable crops. p. 211-218. In MAFF Reference Book 326, *Inorganic Pollution and Agriculture*. HMSO, London.

Influence of Lime Stabilized Sewage Sludge Cake on Heavy Metals and Dissolved Organic Substances in the Soil Solution

Yongming Luo[1] and Peter Christie[1,2]

[1]DEPARTMENT OF AGRICULTURAL AND ENVIRONMENTAL SCIENCE, THE QUEEN'S UNIVERSITY OF BELFAST, NEWFORGE LANE, BELFAST BT9 5PX, NORTHERN IRELAND
[2]AGRICULTURAL AND ENVIRONMENTAL SCIENCE DIVISION, THE DEPARTMENT OF AGRICULTURE FOR NORTHERN IRELAND, NEWFORGE LANE, BELFAST BT9 5PX

Abstract

A randomised block glasshouse experiment was carried out in which barley plants (*Hordeum vulgare* L.) were grown for 40 days after seed germination in two contrasting arable soils following incorporation of a lime stabilized sewage sludge cake at rates equivalent to 0, 30, 90 and 120 t fresh product ha^{-1}. Some short term effects of the sludge product on dissolved organic substances (DOS) and heavy metal concentrations in the soil solution were studied. Soil solution (passed through a < 0.20 μm pore size filter) was extracted from soil sub samples (500 g from each pot) by a centrifugation and filtration method. The solution was analyzed immediately for DOS (predicted by absorbance at 360 nm) and for four heavy metals (Cu, Zn, Ni, and Cr). Concentrations of Cd and Pb were too low to be detected.

The concentrations of DOS increased significantly in both soils with increasing application rate of the sludge material. Increasing the application rate also led to an increase in the concentrations of Cu, Ni, and Cr in the soil solution, although the proportions of the metals applied that were found in the soil solution were small. The sludge product had no overall effect on the concentrations of Zn in the soil solution.

The effect on soil solution concentrations of DOS and Cu, Ni, and Cr was related to soil type. Significant positive correlations between each of these three metals and DOS were also observed. The findings indicate that DOS derived from the sewage sludge may have affected the concentrations of Cu, Ni, and Cr in the soil solution and that these metals may be mobile, largely as metal-organic complexes in the soils.

1 Introduction

Nearly half of the one million tonnes (dry solids) of sewage sludge generated annually in the UK is currently applied to less than 1% of the agricultural land (Department of the

Environment, 1993). The recycling of sewage sludge by application to agricultural land is likely to increase in the future as larger quantities of sludge are produced and the dumping of sludge at sea must cease in the UK from 1998 (Smith, 1996). Sewage sludge is a potentially valuable fertilizer for many crops, and it may improve the physical properties of agricultural and forest soils since it contains significant quantities of macronutrients, micronutrients, and organic matter. Therefore, beneficial recycling of sewage sludge to agricultural and forest land is usually the most economical disposal outlet, especially for inland sewage treatment works. However, there are constraints on the recycling of sewage sludge to land, mainly because of the presence of heavy metals which may either cause phytotoxicity, or put animal and human health at risk through the food chain. There has been increasing concern recently about soil and water protection after land application of sewage sludge (Ministry of Agriculture, Fisheries and Food, 1991, 1993; Bahri, 1994; McBride, 1995; Beck *et al.*, 1995; Harris *et al.*, 1995).

The Water Executive in Northern Ireland produces a lime stabilized sewage sludge cake ('Agri-Soil') by intimately mixing sewage sludge cake with cement kiln dust and subsequent short term composting (Love, 1990). Addition of liming agents for the composting of sewage sludge is also being considered by many municipal authorities in other parts of the world (Logan and Harrison, 1995; Wong *et al.*, 1995). The agricultural value of 'Agri-Soil' and that of a similar product ('N-Viro Soil') has been reported recently (Logan and Harrison, 1995; Luo and Christie, 1995a). The bioavailability of Zn, Cd, and Pb in an alluvial soil amended with 'N-Viro Soil', and of Cu and Zn in two arable soils amended with 'Agri-Soil' was reported by Pierzynski and Schwab (1993) and by Luo and Christie (1995b), respectively. The agronomic value, and the environmental importance over a six year period of applications of 'lime cake' municipal sewage sludge have also been reported (Horman *et al.*, 1994). The chemical speciation of Cu and Zn from 'Agri-Soil' associated with the solid phase of bulk soils (Luo and Christie, 1995b) and in the fine particles of aqueous extracts from a granite soil (Luo and Christie, 1995c) have also been studied. There here have been no published reports on the effect of this type of sludge product on heavy metals and dissolved organic substances (DOS) in the soil solution.

The importance of the soil solution chemistry of native and anthropogenic heavy metals in agricultural and environmental studies has been widely recognised (Sposito and Bingham, 1981; Kinniburgh and Miles, 1983; Campbell and Beckett, 1988; Leita and De Nobili, 1991; Jopony and Young, 1994; Lorenz *et al.*, 1994; Bierman *et al.*, 1995). Soil solution data can provide important information on the dynamics of heavy metals in the soil profile and their mobility and availability to plants and micro-organisms. Soil solution studies can also provide valuable data on the magnitude and rate of movement of heavy metals to surface waters. Thus there is a need to understand the chemistry of heavy metals in the soil solution following application of lime stabilized solid sludge products. This information is essential for the protection of soil and water quality, and eventually for the sustainable use of sewage sludge solids on agricultural land.

The effects of liquid or dried digested sewage sludges and their incinerator ash on heavy metals in the soil solution have been intensively investigated during last fifteen years. Numerous investigators have shown that sewage sludge markedly increased the concentrations of some heavy metals (e.g. Cd, Zn, Cu, and Ni) in the soil solution (Emmerich *et al.*, 1982; Behel *et al.*, 1983; Mullins and Sommers, 1986; Campbell and Beckett, 1988; Lamy *et al.*, 1993; Bierman *et al.*, 1995). However, a few studies showed no significant elevation of other heavy metals in the same solution (Behel *et al.*, 1983; Bierman *et al.*, 1995). Recently Harris-Pierce *et al.* (1995) reported increased Cu and Ni in runoff water with increasing rates of application of sludge in a semiarid grassland. In a study on arable land, Dowdy *et al.* (1994) found that the concentrations of Cd, Ni, and Pb in runoff were not affected by 10 years of application of liquid digested sludge, but the sludge did increase the concentration of Cu, and slightly elevated the concentration of Zn in the runoff water. When sludge has been incorporated into the soil, the concentrations of heavy metals in the soil solution may be influenced by many factors, including the composition and metal loading of the sewage sludge, and various properties of the soil and metal contaminants.

Sewage sludges from different sources may vary greatly in the quantities and chemical speciation of heavy metals (Largewerff *et al.*, 1976; Sidle and Kardos, 1977; Rudd *et al.*, 1988; Leita and De Nobili, 1991; McGrath and Cegarra, 1992), and this may affect the solubility and release of heavy metals from the sludge to the soil solution. The metal loading, the pH value, the SOM content, the soil cation exchange capacity, and the time after incorporation of the sludges can be important in determining the chemical forms and the concentrations of heavy metals in the soil solution (Emmerich *et al.*, 1982; Behel *et al.*, 1983; Mullins and Sommers, 1986; Campbell and Beckett, 1988; Sanders *et al.*, 1987; Lamy *et al.*, 1993; Bierman *et al.*, 1995; Harris-Pierce *et al.*, 1995). The different sequences of affinity of the heavy metals for oxides and organic matter in soil due to the properties of divalent metals (McBride, 1989; Alloway, 1995) may also alter the amounts of the metals in the soil solution. Applications of fertilizers (Lorenz *et al.*, 1994) and biological activity (Linehan *et al.*, 1985; Linehan *et al.*, 1989; Treeby *et al.*, 1989; Zhang *et al.*, 1991; Holm *et al.*, 1995) can change the concentrations of heavy metals in the soil solution. In addition, even soil solution extraction procedures can affect solute chemistry (Gillman and Bell, 1978; Litaor, 1988; Ross and Bartlett, 1990; Dahldren, 1993; Jones and Edwards, 1993; Jopony and Young, 1994; Lorenz *et al.*, 1994). Thus the quantities of the metals in the solution of sludge-amended soils are likely to be affected by numerous factors, and the soil solution data must be interpreted with caution.

Organic carbon is frequently the most abundant solute in the soil solution (Kinniburgh and Miles, 1983). Sewage sludge contains substantial quantities of biodegradable organic materials with a wide range of molecular weights. Application of sludge to soils increased the quantities of soluble organic substances (SOS) in the soils (Behel *et al.*, 1983; Campbell and Beckett, 1988; Lamy *et al.*, 1993) and thus induced the perturbations in the soil. The sludge-derived soluble substances could be important active organic components in the soil solution, or organic pollutants in the aquatic environment. These could complex native and applied heavy metals in the soil solution, and hence influence the reactivity, mobility, and bioavailability of the metals in the soils (Baham and Sposito, 1994; Lamy *et al.*, 1993; Bierman *et al.*, 1995) as low- and high- molecular weight organic compounds have high affinities for heavy metals (Daum and Newland, 1982; McBride, 1989; Harter and Naidu, 1995). Moreover, sewage sludge-borne soluble organic substances may also influence the fate (including transport) of other organic chemicals such as hydrocarbons, halogenated compounds, and pesticides in sludge-amended soils (Beck *et al.*, 1995) and in waters (Rebhun *et al.*, 1992; Rebhun and Rav-Acha, 1992). In this way organic substances and metals could move down the soil profile, or pass into surface waters and contribute to the chemical budgets of metals and carbon. It is therefore important to consider dissolved organic substances together with heavy metals when studying the environmental effects of sewage sludge on agricultural ecosystems. Unfortunately, there is little information published on the behaviour and fate of dissolved organic substances derived from sewage sludge in the soil solution.

The objectives of this work were:

(i) to investigate the impact of lime stabilized sewage sludge cake on heavy metals and dissolved organic substances in the soil solution of two contrasting arable soils;

(ii) to study the relationships between the metals and dissolved organic substances; and

(iii) to obtain information about the mobility and bioavailability of the sludge-derived metals in terms of the chemistry of the soil solution.

2 Materials and Methods

Two contrasting soils (0-20 cm) representing a large proportion of the agricultural area of Northern Ireland were used. One was a sandy loam derived from Silurian shale and sandstone, and the other was a clay loam derived from basalt. Some properties of the two soils are listed in Table 1. Lime stabilized sewage sludge cake ('Agri-Soil') was provided by the Water Executive of the Department of the Environment for Northern Ireland. This product is a partially composted mixture of de-watered sewage sludge cake and cement kiln dust. The sludge product had a relatively high content of organic carbon and had a high pH value, but the concentrations of heavy metals were relatively low because the material was made using rural sludges (Table 2).

Soil was collected from the top 20 cm of the profile. One kg (oven dry basis) of each soil was mixed with the sludge product and placed in a plant pot to give a mixture equivalent to application rates of fresh sludge product (50% dry matter) of 0, 30, 90 and 120 t ha^{-1}. Twelve barley plants (*Hordeum vulgare* cv Forrester) were grown in each pot in the glasshouse. There were four replicates of each treatment in a randomized block design. The soil moisture content in each pot was maintained (using distilled water) within the range of 60-70% of field capacity. The plants were harvested 40 days after seed germination. The shoots were harvested, weighed, dried at 80 °C for 48 hours after washing with deionized water, and then re-weighed. The soil remaining in the pots was stored at 4 °C prior to the extraction of the soil solution. The oven dried shoots were ground and digested in a mixture of nitric and perchloric acids (Ministry of Agriculture, Fisheries and Food, 1986).

The soil solution was extracted by low speed centrifugation and filtration. Randomly selected sub samples (500 g, oven dry weight, including roots) of fresh soil were transferred to acid-washed plastic centrifuge bottles. Distilled water was added to each bottle to adjust the soil subsamples to 70% of field capacity in order to extract an adequate volume of soil solution for chemical analysis. After shaking for one h and centrifuging at 3000 rpm for 30 min, the supernatant was immediately filtered through Whatman No. 41 filter paper into a polycarbonate jar, and the filtrate was then passed through, under negative pressure (1 bar), a 0.20 μm cellulose acetate filter. The filtrates were stored at 4 °C before analysis.

The filtered soil solutions were analysed for heavy metals (Cu, Zn, Ni, and Cr), and absorbance was measured at a wavelength of 360 nm (A_{360}). The heavy metals were determined using a graphite furnace atomic absorption spectrometer (AAS; Perkin Elmer model 5000). Cd and Pb could not be determined because the concentrations in all the soil solutions were below the detection limit (0.001 mg L^{-1}). Absorbance at 360 nm (A_{360})

Table 1 *Some physical and chemical properties of the soils*

Soil Parent Material	pH (in water)	Clay (%)	CEC mmol kg^{-1}	Organic Carbon (%)	Total* Cu mg kg^{-1}	Total Zn mg kg^{-1}	Total Ni mg kg^{-1}	Total Cr mg kg^{-1}	Total Cd mg kg^{-1}	Total Pb mg kg^{-1}
Silurian shale and Triassic sandstone	6.4	13.2	147	2.4	31	57	57	200	2	23
Basalt	6.1	29.2	475	4.7	132	194	205	725	7	40

* Total metal determined by X - ray Fluorescence Diffractometry, except in the case of Cd which was determined in an aqua regia digest by AAS

Table 2 *The pH of the sludge product and its concentrations of organic carbon and selected heavy metals*

Organic carbon	14%
pH	about 8
Total metal (mg kg^{-1})*	
Cu	79
Zn	314
Ni	77
Cr	58
Pb	251
Cd	4

* Total metal determined by X-ray Fluorescence Diffractometry, except for Cd which was determined in an aqua regia digest by AAS.

was measured using an SP6-550 UV/VIS spectrophotometer and 1-cm silica cells. The use of this simple optical method for estimation of DOS in the soil solution was based on the work of Lewis and Tyburezy (1974), Lewis and Canfield (1977), Grieve (1985b), and Rebhun and Rav-Acha (1992). The contents of DOS in the soil solutions were calculated from the measurements of A_{360}, using the equation of Grieve (1985b) after correcting for the different cell path length used. Results for DOS were converted to dissolved organic carbon (DOC), assuming a 50% carbon content for dissolved organic matter (Larson, 1978).

The results reported are the mean values from four blocks of each treatment. The contents of sludge-borne metals found in the soil solution are expressed as a percentage of the total amounts applied in the sludge product. The data were tested by analysis of variance, and the mean values within each variable were compared by least significant difference (L.S.D., $\alpha = 0.05$). The concentrations of Cu, Zn, Ni, and Cr in the soil solutions were compared with the concentrations of DOS by linear correlation and regression analysis.

3 Results

The influence of lime stabilized sewage sludge cake on the concentrations of heavy metals in the soil solution is shown in Table 3. In general, the concentrations of Cu, Ni, and Cr increased significantly with increasing sludge application rate, but Zn showed little change forty days after application of the sludge product. At the intermediate (90 t ha^{-1}) and higher (120 t ha^{-1}) application rates of the sludge product the concentrations of Cu, Ni, and Cr in the basaltic clay loam soil solutions were significantly lower than those in the sandy loam soil solutions.

The percentages subsequently found in the soil solutions of Cu, Ni, and Cr applied in the sludge product were calculated by subtracting the concentrations in the control soil solution from those in the amended soil solutions, then dividing by the amounts applied, and multiplying by 100. The results are given in Table 4. The percentage content of each

Table 3 *Influence of lime stabilized sewage sludge cake on concentrations of heavy metals in the soil solution*

Soil Type	Sludge Application Rate (t ha^{-1})	Cu (mg L^{-1})	Zn (mg L^{-1})	Ni (mg L^{-1})	Cr (mg L^{-1})
Silurian	0	42e	6b	18e	6f
sandy	30	42e	4c	32d	11e
loam	90	136b	5bc	88a	24b
	120	169a	5bc	93a	31a
Basaltic	0	21f	6b	19e	6f
clay	30	42e	6b	28d	8f
loam	90	90d	6b	54c	16d
	120	111c	9a	73b	19c
LSD ($\alpha = 0.05$)*		12	2	9	3

*LSD (a = 0.05), least significant difference at 5% protection level. For each metal, means followed by the same letter are not significantly different by LSD

metal was small, although it was higher at sludge application rates of 90 and 120 t ha^{-1} than it was at the 30 t ha^{-1} rate. On average, the percentages of Cu, Ni, and Cr were 3.1%, 1.7%, and 0.7%, respectively. Again, at the moderate and higher application rates of the sludge product, the percentages of Cu, Ni, and Cr in the basaltic soil were much lower than those in the Silurian soil.

Figure 1 shows the changes in the concentrations of dissolved organic substances in the soil solutions after application of the lime stabilized sewage sludge cake. The amounts of dissolved organic substances in both soils increased significantly with increasing rate of sludge application. The contents of DOS in the basaltic clay loam were significantly lower than those in the Silurian sandy loam. Linear correlation analyses showed a significant

Table 4 *The percentage* of Cu, Ni, and Cr applied in the sludge material that was subsequently found in the soil solution*

Soil Type	Sludge Application Rate (t ha^{-1})	Cu (%)	Ni (mg L^{-1})	Cr (mg L^{-1})
Silurian	30	0.0	1.6	0.8
sandy	90	3.5	3.6	0.9
loam	120	7.5	2.1	0.9
Basaltic	30	2.3	1.0	0.3
clay	90	2.6	1.3	0.5
loam	120	2.5	1.5	0.5
Mean		3.1	1.7	0.7

*[(amended soil solution concentration - control soil solution concentration)/amount applied] x 100

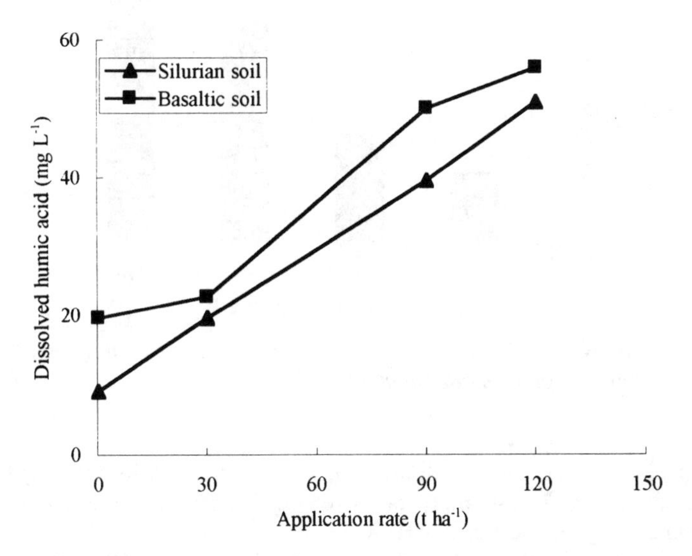

Figure 1 *Plot of dissolved humic acid versus application rate of sludge product in the Silurian and Basaltic soils*

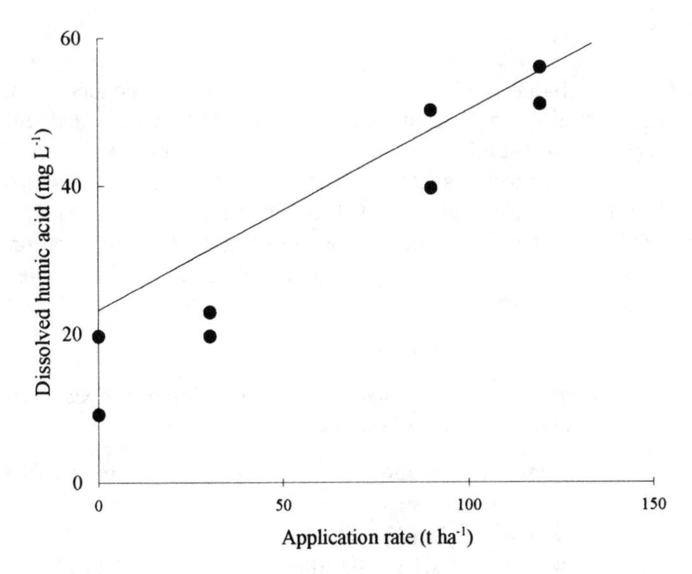

Figure 2 *Linear relationship between dissolved humic acid and sludge application rate;* $y = 13.338 + 0.337x$; $r = 0.967^{**}$

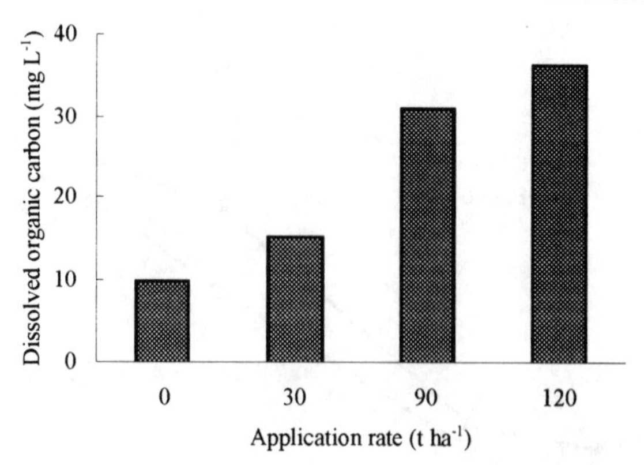

Figure 3 *Mean dissolved organic carbon in the soil solution assuming that the dissolved organic matter contains 50% carbon*

relationship between the concentration of dissolved organic substances and the sludge application rate (Figure 2). Similarly, dissolved organic carbon in the solutions increased with increasing sludge application rate (Figure 3).

Table 5 shows the relationships between heavy metals and DOS in the soil solution. Linear correlation and regression demonstrated that in the soil solution the concentrations of Cu, Ni, and Cr were positively correlated with the concentrations of DOS. In contrast, Zn showed no such relationship.

4 Discussion

The lime stabilized sludge product did not always affect the concentrations of heavy metals in the soil solutions in the same way. The soil solution was obtained using a low speed centrifugation and filtration method. A similar filtration method for extraction of soil solution was reported to be satisfactory for the measurement of trace metals (Jopony and Young, 1994). The concentrations of Cd and Pb in our soil solutions were so low that these could not be detected. Although Zn was measurable, its concentrations in the soil solutions were also low, even at the highest application rate of the sludge product. In general, the lime-stabilized sludge material had no overall effect on the concentration of Zn

Table 5 *Relationship between heavy metals (mg L^{-1}) and dissolved humic acid (DHA) in the sludge-contaminated soil solution*

Metal	Regression Equation (y = bx + a)	Correlation coefficient (r)
Cu	y = 0.0029[DHA] - 0.064	0.968*
Ni	y = 0.0017[DHA] - 0.061	0.981*
Cr	y = 0.0005[DHA] - 0.009	0.944*
Zn	y = -0.0029[DHA] + 0.0307	-0.528NS

**P < 0.01; NS = not significant

in the soil solution. In contrast, the concentrations of Cu, Ni, and Cr increased significantly with increasing rate of sludge application in both soils but remained within normal range (Campbell and Beckett, 1988). The results are partly in agreement with those of Emmerich *et al.* (1982), Campbell and Beckett (1988), and Harris-Pierce *et al.* (1995). The results do not support the findings of Behel *et al.* (1983), Mullins and Sommers (1986), and Lamy *et al.* (1993) who reported that digested sewage sludge elevated the concentration of Cd and/or Zn in the soil solution. Our results also do not support those of Bierman *et al.* (1995) who reported that sewage sludge incinerator ash increased Cd and Zn, but had no significant effect on the concentrations of Cu, Ni, and Pb in solution. These discrepancies probably result mainly from the different sources of sewage sludge used by the various researchers.

Although the sludge material markedly increased the soil solution concentrations of Cu, Ni, and Cr in both soils, only a small amount of the metals applied in the sludge was present in the soil liquid phase. The major part (over 96%) of the metals was present in association with the soil solid phase. This is in accordance with our previous work on the chemical distribution of sludge-derived heavy metals in the soil solid phase (Luo and Christie, 1995a). The sludge-derived metals in the soil solution can also exist in different chemical species (Emmerich *et al.*, 1982; Christensen and Lun, 1989; Holm *et al.*, 1995). Despite that, the amounts of the metals applied in the sludge product which were subsequently found in the solutions may be highly mobile and bioavailable.

It was not surprising that DOS in the soil solutions (that passed through the < 0.20 μm pore size filters) increased markedly in our experiment with increasing application rate of the solid sewage sludge material. The substantial increase observed accorded with the observation that the solutions had a yellowish brown colour, and became much darker with increasing application rate of the sewage sludge. The contribution of the sludge product to organic substances in the soil solution was due to the fact that the amounts of organic carbon present as water soluble and as solid forms in the sewage sludge (Table 2) were ca two to five times greater than that found in the mineral soils (Table 1). The soluble organic substances in the sludge product might enter the soil solution directly. Also, the added solid organics might undergo microbial transformations, and thus contribute indirectly to the carbon applied to the solution even though the sludge product had been lime stabilized and partially composted. Schaumberg *et al.* (1980) found that at moderate levels of liquid sludge amendment, much of the added organic carbon underwent microbial transformations near the soil surface and was degraded within 4 to 5 weeks. Hsieh *et al.* (1981) reported that after about two months of incubation, the mineralization rates of an activated and a digested sewage sludge in soil approached a square root function of time. Our preliminary experiment also showed that the absorption decreased monotonically with increasing wavelength for the soil solution (results not shown). The featureless absorption spectra were similar to those for dark river water (Lewis and Canfield, 1977) and moorland streamwater (Grieve, 1985a), and also to those for humic and fulvic acids extracted from soils (Ghosh and Schnitzer, 1979). Humic acids and fulvic acids are polyelectrolyte macromolecules and their UV and visible absorption spectra are featureless (Ghosh and Schnitzer, 1979). It can thus be inferred that dissolved organic substances from the solid sewage sludge were present at least partly as soluble humic substances in the solutions of sludge-amended soil. Holtzclaw *et al.* (1976) successfully extracted and purified the fulvic acid fraction of sludge-soil mixtures.

The increase in DOS in the soil solution following addition of the sludge material may be responsible for increasing the concentrations of Cu, Ni, and Cr in the soil solutions.

There were strong positive linear correlations between the concentrations of the metals measured and the DOS in the soil solutions. No such relationship was observed for Zn. Zn^{2+} has less affinity for organic matter than does Cu^{2+} and Ni^{2+} (McBride, 1989). Zn^{2+} may be present as a higher proportion of free ionic species in the soil solution (Jeffery and Uren, 1983; Sanders, 1983; Mullins and Sommers, 1986; Holm *et al.*, 1995). Therefore, when the soil pH increased with increasing rates of application of the sludge product (Luo and Christie, 1995a), the Zn^{2+} in the soil solution might be more tightly complexed by the soil solid components (Emmerich *et al.*, 1982; McBride, 1989, Bierman *et al.*, 1995). In contrast, the other three metals in the soil solution might form more stable metal-organic complexes with tendencies to remain in solution. Sanders (1982), and Jeffery and Uren (1983) found that copper concentrations in the soil solutions decreased only slightly as the solution pH increased, and most of the copper in all solutions was present as complexed organic species. James and Bartlett (1983) reported that Cr was usually present as a trivalent ion in soils under normal conditions, and water soluble organic matter from air dried soil kept Cr^{3+} in solution above pH 5.5, and prevented its immediate removal by soil. Moreover, the differences between the two soils in the concentrations of Cu, Ni, and Cr and DOS in the soil solutions can be ascribed to the differences in the contents of organic matter and clay, and in the cation exchange capacity (CEC) (Table 1). The interactions between heavy metals, organic matter, and clay minerals in soil has long been recognised (Huang, 1980; McBride, 1989; Baham and Sposito, 1994; Harter and Naidu, 1995).

The linear relationships mentioned suggest that Cu, Ni, and Cr in the solution of 'Agri-Soil'-amended soil are substantially associated with DOS, probably mainly as humic-metal complexes. These relationships also indicate that the metals may rely heavily on organic associations for mobilization. The organic ligand-metal complexes may move lower down the soil profile to groundwater and/or flow out from the soil body into surface waters. The implications of the soluble complexes for catchment budgets could be considerable (Pettersson *et al.*, 1993). In addition to its significance for contaminant transport, the metal-organic complexation may also affect soil genesis, fertility, and metal toxicity (Harter and Naidu, 1995). Therefore the biogeochemistry of metal-organic-mineral complexes in the aqueous phase of sewage sludge-amended soils deserves further study.

5 Conclusions

Application of lime stabilized sewage sludge cake led to a significant increase in dissolved organic (humic) substances or DOC in the soil solution. The increase was significantly related to sludge application rate and to the soil type.

Soil solution concentrations of Cu, Ni, and Cr also generally increased after relatively large application rates of the sludge material were applied to soil. However, the soil conditioner had no overall effect on the concentration of Zn in the soil solution, and Cd and Pb were undetectable. Although the percentage of the metals applied in the sludge product that was subsequently found in the soil solution was small, the mobility and availability of Cu, Ni, and Cr in the soils may have been influenced by the application of the sludge product.

The concentrations of Cu, Ni, and Cr in the soil solution were significantly correlated with the concentrations of DOS. In contrast, no such relationship was found for Zn. The concentrations of Cu, Ni, and Cr in the solution of sludge-amended soil may be buffered by DOS and the three metals might be mobile, largely as metal-organic complexes. However, the association of the metals with dissolved organics may affect the chemical behaviour and transport of the metals in the soils.

Acknowledgement

We thank the Water Executive of the Department of the Environment for Northern Ireland for financial support and for supplying the lime stabilized sewage sludge cake.

References

Alloway, B.J. 1995. Soil processes and the behaviour of heavy metals. p.11-37. *In* B.J. Alloway (ed.), *Heavy Metals in Soils*. Second edition. Blackie Academic & Professional, London.

Baham, J. and G. Sposito. 1994. Adsorption of dissolved organic carbon extracted from sewage sludge on montmorillonite and kaolinite in the presence of metal ions. *J. Env. Qual.* **23**:147-153.

Bahri, A. 1994. Impacts of sewage sludge application on drainage water quantity and quality. Proc. 15th Congr. Intern. Soc. Soil Sci. (Acapulco), **3a**:483-493.

Beck, A.J., R.E Alcock, S.C. Wilson, M.J. Wang, S.R. Wild, A.P. Sewart, and K.C. Jones. 1995. Long-term persistence of organic chemicals in sewage sludge-amended agricultural land: a soil quality perspective. *Adv. in Agronomy* **55**:345-391.

Behel, D., J.R. Darrell, W. Nelson, and L.E. Sommers. 1983. Assessment of heavy metal equilibria in sewage sludge-treated soil. *J. Env. Qual.* **12**:181-186.

Bierman, P.M., C.J.Rosen, P.R. Bloom, and E.A. Nater. 1995. Soil solution chemistry of sewage-sludge incinerator ash and phosphate fertilizer amended soil. *J. Env. Qual.* **24**:279-285.

Campbell, D.J. and P.H.T. Beckett. 1988. The soil solution in a soil treated with digested sewage sludge. *J. Soil Sci.* **39**:283-298.

Christensen, T.H. and X.Z. Lun. 1989. A method for determination of cadmium species in solid waste leachates. *Water Res.* **23**:73-80.

Dahldren, R.A. 1993. Comparison of soil solution extraction procedures: effect on solute chemistry. *Communications in Soil Science and Plant Analysis* **24**:1783-1794.

Daum K.A.and L.W. Newland. 1982. Complexing effects on behaviour of some metals. p. 129-139. *In* O. Hutzinger (ed.), the *Handbook of Environmental Chemistry. Volume 2, Part B: Reactions and Processes.* Springer-Verlag, Berlin.

Dowdy, R.H., C.E. Clapp, D.R. Linden, W.E. Larson, T.R. Halbach, and R.C. Polta. 1994. Twenty years of trace metal partitioning on the Rosemount sewage sludge watershed. p. 149-155. In: C.E. Clapp, W.E. Larson, and R.H. Dowdy (eds), *Sewage Sludge: Land Utilization and the Environment.* American Society of Agronomy, Crop Science Society of America, Soil Science Society of America, Madison, Wisconsin.

Department of the Environment. 1993. *UK Sewage Sludge Survey Final Report*, February 1993. Consultants in Environmental Sciences Ltd.

Emmerich, L.J., A.L. Lund, A.L. Page, and A.C. Chang. 1982. Predicted solution phase forms of heavy metals in sewage sludge-treated soils. *J. Env. Qual.* **11**:182-186.

Ghosh, K. and M. Schnitzer. 1979. UV and visible absorption spectroscopic investigations in relation to macromolecular characteristics of humic substances. *J. Soil Sci.* **30**:735-745.

Gillman, G.P. and L.C. Bell. 1978. Soil solution studies on weathered soils from tropical north Queensland. *Aus. J. Soil Res.* **16**:67-77.

Grieve, I.C. 1985a. Annual losses of iron from moorland soils and their relation to free iron contents. *J. Soil Sci.* **36**:307-312.

Grieve, I.C. 1985b. Determination of dissolved organic mater in streamwater using visible spectrophotometry. *Earth Surface Processes and Landforms* **10**:75-78.

Harris-Pierce, R.L., E.F. Reente, and K.A. Barbarick. 1995. Sewage sludge application effects on runoff water quality in a semiarid grassland. *J. Env. Qual.* **24**:112-115.

Harris, B.L., T.L. Nipp, D.K. Waggoner, and A. Weber. 1995. Agricultural water quality program policy considerations. *J. Env. Qual.* **24**:405-411.

Harter, R.D. and R. Naidu, 1995. Role of metal-organic complexation in metal sorption by soils. *Adv. in Agron.* **55**:219-263.

Holm, P.R.E.,.T.H. Christensen, J.C. Tjill, and S.P. McGrath. 1995. Speciation of cadmium and zinc with application to soil solutions. *J. Env. Qual.* **24**:183-190.

Holtzclaw, K.M., G. Sposito, and G.R. Bradford. 1976. Analytical properties of the soluble, metal-complexing fractions in sludge-soil mixtures: I. Extraction and purification of fulvic acid. *Soil Sci. Soc. Am. J.* **40**:254-258.

Hormann, C.M., C.E. Clapp, R.H. Dowdy, W.E. Larson, D.R. Duncomb, T.R. Halbach, and R.C. Polta. 1994. Effect of lime-cake municipal sewage on corn yield, nutrient uptake, and soil analyses. p. 173-183. *In* C.E. Clapp, W.E. Larson, and R.H. Dowdy (eds), *Sewage Sludge: Land Utilization and the Environment.* American Society of Agronomy, Crop Science Society of America, Soil Science Society of America, Madison, Wisconsin.

Hsieh, Y.P., L.A. Douglas, and H.L. Motto. 1981. Modeling sewage sludge decomposition in soil: I. Organic carbon transformation. *J. Env. Qual.* **10**:54-59.

Huang, P.M. 1980. Adsorption processes in soil. p. 45-59. *In* O. Hutzinger (ed.), *The Handbook of Environmental Chemistry, Volume 2, Part A, Reactions and Processes.* Springer-Verlag, Berlin.

James, B.R. and R.J. Bartlett. 1983. Behaviour of chromium in soils: V. Fate of organically complexed Cr^{3+} added soil. *J. Env. Qual.* **12:**169-172.

Jeffery, J.J. and U.C. Uren. 1983. Copper and zinc species in the soil solution and the effects of soil pH. *Aus. J. Soil Res.* **21**:479-488.

Jones, D.L. and A.C. Edwards. 1993. Effect of moisture content and preparation technique on the composition of soil solution obtained by centrifugation. *Communications in Soil Science and Plant Analysis* **24**:171-186.

Jopony, M. and S.D. Young. 1994. The solid-solution equilibria of lead and cadmium in polluted soils. *European J. Soil Sci.* **45**:59-70.

Kinniburgh, D.G. and D.L. Miles. 1983. Extraction and chemical analysis of interstitial water from soils and rocks. *Environmental and Scientific Technology* **17**:362-368.

Lamy, I., S. Bourgeois, and A. Bermond. 1993. Soil cadmium mobility as a consequence of sewage sludge disposal. *J. Env. Qual.* **22**:731-737.

Lorenz, S.E., R.E. Hamon, and S.P. McGrath. 1994. Differences between soil solutions obtained from rhizosphere and non-rhizosphere soils by water displacement and soil centrifugation. *European J. Soil Sci.* **45**:431-438.

Largewerff, J.V., G.T. Biersdorf, and D.L Brouwer. 1976. Retention of metals in sewage sludge. I. Constituent heavy metals. *J. Env. Qual.* **5**:19-23.

Larson, R.A. 1978. Dissolved organic matter of a low-colour stream. *Freshwater Biology* **8**:91-104.

Leita, L. and M. De Nobili. 1991. Water-soluble fractions of heavy metals during composting of municipal solid waste. *J. Env. Qual.* **20**:73-78.

Lewis, W.M. and D. Canfield, 1977. Dissolved organic carbon in some dark Venezuelan waters and a revised equation for spetrophotometric determination of dissolved organic carbon. *Archiv für Hydrobiologie* **79**:441-445.

Lewis, W.M. and J.A. Tyburczy, 1974. Amounts and spectral properties of dissolved organic compounds from some freshwaters of southeastern U.S. *Archiv. für Hydrobiologie* **74**:8-17.

Linehan, D.J., H. Sinclair, and M.C. Mitchell. 1985. Mobilisation of Cu, Mn and Zn in the soil solutions of barley rhizospheres. *Plant and Soil* **86**:147-149.

Linehan, D.J., H. Sinclair, and M.C. Mitchell. 1989. Seasonal changes in Cu, Mn, Zn and Co concentrations in soil in the root-zone of barley (*Hordeum vulgare* L.). *J. Soil Sci.* **40**:103-115.

Litaor, M.I. 1988. Review of soil solution samplers. *Water Resources Research* **24**:727-733.

Logan, T.J. and B.J. Harrison. 1995. Physical characteristics of alkaline stabilised sewage sludge (N-Viro Soil) and their effects on soil physical properties. *J. Env. Qual.* **24**:153-164.

Lorenz, R.E., R.E. Hamon, S.P. McGrath, P.E. Holm, and T.H. Christensen. 1994. Applications of fertilizer cations affect cadmium and zinc concentrations in soil solutions and uptake by plants. *European J. Soil Sci.* **45**:150-165.

Love, S.C.P. 1990. Developments in sludge treatment. Paper presented to the Institution of Water and Environmental Management in Dublin. Department of the Environment for Northern Ireland.

Luo, Y. M. and P. Christie. 1995a. Some short-term effects of a lime stabilised sewage sludge cake applied to arable and forest soils. *Irish J. Agric. and Food Res.* **34**:78.

Luo, Y. M. and P. Christie. 1995b. Chemical forms and plant uptake of Cu and Zn in soils amended with a lime stabilized sewage sludge cake. *Third International Conference on the Biogeochemistry of Trace Elements, Paris, France. Theme A1* (Abstract).

Luo, Y. M. and P. Christie. 1995c. Chemical fractions of copper and zinc in organic-rich particles from aqueous extracts of a metal-contaminated granite soil. *International Conference on Organic-Mineral Interactions in Soil and Sediments, Newcastle-upon-Tyne, UK.* (Abstract).

McGrath, S.P. and J. Cegarra. 1992. Chemical extractability of heavy metals during and after long-term applications of sewage sludge to soil. *J. Soil Sci.* **43**:313-321.

Ministry of Agriculture, Fisheries and Food. 1993. *Review of the Rules for Sewage Sludge Application to Agricultural Land: Soil Fertility Aspects of Potentially Toxic Elements.* MAFF Publication No. PB1561. HMSO London.

Ministry of Agriculture, Fisheries and Food. 1986. *The Analysis of Agricultural Materials.* MAFF/ADAS Reference book 427. HMSO London.

Ministry of Agriculture, Fisheries and Food. 1991. *Code of Good Agricultural Practice for the Protection of Water.* MAFF Publication No. PB0587. HMSO London.

McBride, M.B. 1989. Reactions controlling heavy metal solubility in soils. *Advances in Soil Science* **10**:1-57.

McBride, M. B. 1995. Toxic metal accumulation from agricultural use of sludge: are USEPA regulations protective? *J. Env. Qual.* **24**:5-18.

Mullins, G.L. and L.E. Sommers. 1986. Characterization of cadmium and zinc in four soils treated with sewage sludge. *J. Env. Qual.* **15**:382-387.

Pettersson, C., K. Hakansson, S. Karsson, and B. Allard. 1993. Metal speciation in a humic surface water system polluted by acidic leachates from a mine deposit in Sweden. *Water Res.* **27**:863-871.

Pierzynski, G.M. and A.P. Schwab. 1993. Bioavailability of zinc, cadmium, and lead in a metal-contaminated alluvial soil. *J. Env. Qual.* **22**:247-254.

Rebhun, M., R. Kalabo, L. Grossman, J. Manka, and C.H. Rav-Acha. 1992. Sorption of organics on clay and synthetic humic-clay complexes simulation aquifer processes. *Water Res.* **26**:79-84.

Rebhun, M. and C.H. Rav-Acha. 1992. Binding of organic solutes to dissolved humic substances and its effects on adsorption and transport in the aquatic environment. *Water Res.* **26**:1645-1654.

Ross, D.S. and R.J. Bartlett. 1990. Effects of extraction methods and sample storage on properties of solutions obtained from forested spodosols. *J. Env. Qual.* **19**:108-113.

Rudd, T., J. A. Campbell, and J. N. Lester. 1988. The use of model compounds to elucidate metal forms in sewage sludge. *Environmental Pollution* **50**:225-242.

Sanders, J.R. 1982. The effect of pH upon the copper and cupric ion concentrations in soil solutions. *J. Soil Sci.* **33**:679-689.

Sanders, J.R. 1983. The effect of pH on the total and free ionic concentrations of manganese, zinc and cobalt in soil solutions. *J. Soil Science* **34**:315-323.

Sanders, J.R., S.P. McGrath, and T McM. Adams. 1987. Zinc, copper and nickel concentrations in soil extracts and crops grown on four soils treated with metal-loaded sewage sludges. *Environmental Pollution* **44**:193-210.

Schaumberg, G.D., C.S. Levesque-Madroe, G. Sposito, and L.J. Lund. 1980. Infrared spectroscopic study of the water-soluble fraction of sewage sludge-soil mixtures during incubations. *J. Env. Qual.* **6**:47-52.

Sidle, R.C. and L.T. Kardos. 1977. Aqueous release of heavy metals from two sewage sludges. *Water, Air, and Soil Pollution* **8**:453-459.

Smith, S.R. 1996. Chapter One: Introduction. p. 1-7. *In* S.R. Smith (ed), *Agricultural Recycling of Sewage Sudge and the Environment.* CAB International, Biddles Ltd, Guildford.

Sposito, G. and F.T. Bingham. 1981. Computer modeling of trace metal speciation in soil solutions: correlation with trace metal uptake in higher plants. *Journal of Plant Nutrition* **3**:35-49.

Treeby, M., H. Marschner, and V. Römheld. 1989. Mobilisation of iron and other micronutrient cations from a calcareous soil by plant-borne, microbial, and synthetic metal chelators. *Plant and Soil* **114**:217-226.

Wong, J.W.C., S.W.Y. Li, and M.H. Wong. 1995. Coal fly ash as a composting material for sewage sludge: effects on microbial activities. *Environmental Technology* **16**:527-537.

Zhang, F., V. Römheld, and H. Marschner. 1991. Release of zinc mobilizing root exudates in different plant species as affected by zinc nutritional status. *J. Plant Nutrition* **14**:675-686.

The Effect of Long-Term Annual Sewage Sludge Applications on the Heavy Metal Content of Soils and Plants

M.N. Aitken and D.I. Cummins

SCOTTISH AGRICULTURAL COLLEGE, ENVIRONMENTAL SCIENCES DEPARTMENT, AUCHINCRUIVE, AYR KA6 5HW, SCOTLAND

Abstract

Three experimental sites were set up on grassland farms in Lanarkshire to study the effects of liquid sewage sludge applications on heavy metal behaviour, agricultural productivity, and long term soil fertility. Over the period 1985-1993, sewage sludge was applied at 0, 67, 135 and 270 m^3 ha^{-1} y^{-1} to grassland soils maintained at pH 5.5 or 6.5. After four years, the high rate of metal sludge applications increased soil concentrations of Cu, Zn, Pb, Ni and Cr to a depth of 100-150 mm. The largest accumulation was in the top 0-25 mm. Approximately 33% of each metal applied was recovered in this top layer. Soil pH did not affect the distribution of metals in the soil profile.

Acetic acid extracted similar amounts of Zn, but significantly lower amounts of Cu and Pb compared to the levels obtained using EDTA. This indicates that relatively high amounts of Cu and Pb were bound to the soil organic matter. Total soil analysis for Cu and Pb revealed a correlation with the respective EDTA extractable component, but the exact relationship varied between the sites.

Plant Cu and Zn concentrations were increased, but those of Cd, Ni, Cr, and Pb were unaffected by sludge application. Increasing the soil pH from 5.5 to 6.5 decreased plant uptake of Zn and Ni. A positive relationship between soil total Cu and herbage Cu was found at two sites, but this relationship was absent from the third. Although soil Cu, Zn, and Pb concentrations were close to or exceeded UK Regulations at one location, the herbage still did not contain toxic concentrations of these elements.

Sewage sludge appears to represent a sustainable soil conditioner, although long term use on grassland may lead to a build up of metals in the top 25 mm of the soil profile giving rise to a potential risk for grazing animals through soil ingestion, especially during winter.

1 Introduction

In Scotland, an application of 10 000 t of dry solid (DS) sewage sludge was made annually to agricultural land in 1990/91 (DoE, 1993). This represents about 13% of the total (78 900 t DS) produced annually. The UK Department of the Environment (DoE, 1993) predict that Scottish production of sludge will increase from 78 900 to around 185 000 t

DS y^{-1} by the year 2010 as a result of the Urban Waste Water Treatment Directive (91/271/EEC). Sea disposal currently accounts for approximately 75% of sludge disposal in Scotland, but will have to cease in 1998 to comply with the Directive requirements. This will lead to a substantial increase in the amount of sludge applied to agricultural land in Scotland (Aitken, 1996).

In the past sewage sludges contained appreciable amounts of heavy metals, but concentrations have decreased through improved treatment and effluent quality control imposed by the water industry. Domestic sources of heavy metals, such as Cu plumbing systems, Zn-based pharmaceutical products, and run-off from roads now provide the most important inputs of metals to sludges (Davis, 1989). However, there is continuing concern about the long-term effects of metals which can persist in the topsoil layer (0 - 200 mm) long after sludge additions have ceased (McGrath, 1987). Legislative permitted levels of metal concentrations in soil and maximum annual rates for additions of metals from sewage sludge are given in the Sludge (Use in Agriculture) Regulations, Statutory Instrument No 1263 (1989).

The spreading of suitably treated sewage sludge on agricultural land provides a convenient outlet for sludge disposal and a free source of nutrients for crop production. It can thus be considered as a sustainable soil management strategy, provided that certain accompanying criteria are met. These include:

(i) a reduction in use of fertilizers;
(ii) an overall improvement in soil quality; and
(iii) no adverse effects occur within or outwith the soil ecosystem.

It is the third, and perhaps most critical, factor that has provided most discussion amongst farming and scientific communities when the issue of widespread sludge application to land has been raised. This paper summarises some of the results obtained from three long-term field experiments carried out across Scotland, with the aim of clarifying some of the concerns and issues raised.

2 Materials and Methods

The experiments were conducted at three *Lolium perenne* dominated grassland sites over the period 1985-1993 (site details are given in Table 1). At each site four rates of anaerobically digested liquid sewage sludge (0, 67, 135 and 270 m^3 ha^{-1} y^{-1}) were applied every spring. The sludge contained metal concentrations close to the median levels of sludges used in UK farmland (DoE, 1993). The average annual metal loadings for the maximum 270 m^3 ha^{-1} y^{-1} sludge application were as follows: Cu 3.8 kg ha^{-1} y^{-1}; Zn 6.89 kg ha^{-1} y^{-1}; Pb 3.16 kg ha^{-1} y^{-1}; and Cd 0.027 kg ha^{-1} y^{-1}. The range of application rates tested encompassed the whole practicable range for more extensive use of sludge in the agricultural community. Results obtained in the field tests are thus directly interpretable for 'real life' scenarios.

Each plot size was 4 x 3 m and was split into two, with the subplot treatment of maintaining soil pH values at 5.5 or 6.5 [using either $Ca(OH)_2$ or sulphur] randomly allocated. Optimum doses of inorganic nitrogen, phosphorus, and potassium were applied to the control plots in March each year whilst the experimental sites received the prescribed doses of liquid sewage sludge using a watering can fitted with a splash plate to achieve uniform distribution.

Table 1 *Site details*

Site	Lower Carbarns (LC)	Watsonfoot (WF)	Garrionhaugh (GH)
Topsoil texture	Light clay loam	Clay loam	Sandy loam
Subsoil texture	Heavy clay loam	Clay	Sandy clay loam
Drainage	Imperfect	Poor	Free
Previous Sludge Use	Sludged for ~50 years	Rare sludge use	No applications previously

Herbage sampling was carried out each May, July, and September using hand shears to avoid soil contamination. Samples were analysed for total metal content via HCl digestion using HPLC. Whole plots were also harvested to assess fresh and dry matter yields.

The soil was sampled every four years by taking 10 replicate cores from each plot to 400 mm depth. Cores were subsequently divided by depth into 0-25, 25-50, 50-100, 100-150, 150-200 and 200-400 mm sub-samples. All sub-samples were treated with 2.5% acetic acid, 0.05M EDTA (buffered to pH = 7) solution and aqua regia, and analysed for Cd, Cr, Cu, Ni, Pb and Zn using standard atomic absorption spectrophotometry. The yield, soil and plant metal data were analysed as a split-plot randomised complete block experiment using the ANOVA directive in Genstat 5 (Lawes Agricultural Trust, 1993).

The environmental significance of the accumulation of sludge metals in soils depends on how much of the metals are available for plant or microbe uptake. Determination of total soil metal concentration indicates the extent of contamination, but not necessarily its biological significance, due to the likelihood that some of the metal 'load' may be present in 'non-available' form. Metals may be held in soil by a number of processes. These include ion exchange, chemisorption and chelation. Good reviews of soil sorption processes are given by McBride (1989) and by Swift and McLaren (1991).

Various soil extraction techniques have been developed to portion heavy metals between different soil fractions. The two most commonly utilised reagents are acetic acid and ethylene diamine tetracetic acid (EDTA). Dilute (0.44M) acetic acid, buffered with (calcium) acetate salt solution, has been used as an extracting reagent for the mobile (= 'acid-soluble') reserves of heavy metals resident within soils (Miller *et al.*, 1986a). More often, 2.5% acetic acid (pH 2.5) is used, and this is believed to extract 'specifically-bound' trace metals (Iyengar *et al.*, 1981). In contrast, EDTA is usually buffered at neutral pH (7), but due to its strong complexing potential it is able to remove those metals complexed with any organic matter present, as well as part of that component sorbed by sesquioxides (Miller *et al.*, 1986b). Use of aqua regia reagent (3:1, HCl:HNO_3) gives the 'total' amount of metals present in soil, although there is some doubt whether 100% metal extraction is possible using this combination (Beckett, 1989).

3 Results and Discussion

Soils

Total soil metal concentrations for the control (no applications of sludge), and for the 270 m^3 ha^{-1} y^{-1} treatment following four years of sludge applications are given in Tables 2 - 4. The historic use of sludge over 50 y at Site LC resulted in higher Zn, Cu, Cd and Pb

Table 2 *Site GH. Total soil concentrations (mg kg^{-1}) of Cu, Zn, Pb, and Cd in the soil profile after 4y of treatment with 0 and 270 m^3 sludge ha^{-1} y^{-1}*

Soil depth (mm)	Copper		Zinc		Lead		Cadmium	
	0	270	0	270	0	270	0	270
0-25	20	39	72	102	35	57	0.38	0.37
25-50	21	30	76	88	32	41	0.28	0.23
50-100	21	26	76	85	31	36	0.22	0.18
100-150	22	24	77	82	33	32	0.20	0.16
150-200	21	24	76	77	31	31	0.19	0.14
200-400	21	21	70	73	26	29	0.14	0.11
SED	1.370		5.132		3.413		0.033	

Table 3 *Site WF. Total soil concentrations (mg kg^{-1}) of Cu, Zn, Pb and Cd in the soil profile after 4 y of treatment with 0 and 270 m^3 sludge ha^{-1} y^{-1}*

Soil depth (mm)	Copper		Zinc		Lead		Cadmium	
	0	270	0	270	0	270	0	270
0-25	49	75	222	259	116	136	0.67	0.75
25-50	49	65	228	254	115	128	0.67	0.71
50-100	52	60	242	271	124	130	0.67	0.75
100-150	57	59	248	279	134	140	0.73	0.75
150-200	65	59	251	279	128	133	0.65	0.72
200-400	52	58	226	247	112	121	0.54	0.63
SED	3.738		16.1		7.417		0.061	

Table 4 *Site LC. Total soil concentrations (mg kg^{-1}) of Cu, Zn, Pb and Cd in the soil profile after 4 y of treatment with 0 and 270 m^3 sludge ha^{-1} y^{-1}*

Soil depth (mm)	Copper		Zinc		Lead		Cadmium	
	0	270	0	270	0	270	0	270
0-25	106	137	354	382	257	274	1.7	1.6
25-50	111	128	353	368	270	275	1.5	1.5
50-100	112	119	371	380	272	283	1.4	1.5
100-150	107	109	382	393	266	271	1.3	1.3
150-200	94	94	312	315	220	222	0.9	0.9
200-400	56	53	145	140	82	80	0.4	0.4
SED	7.268		39.81		36.65		0.118	

concentrations in the control plots compared to the other two sites. Site WF, which had received a small amount of sludge in the past, contained a higher concentration of these metals in the control plot compared to Site GH, which had never received sludge.

The concentrations of metals for the sludge treatments at all three sites clearly show downward movement through the soil profile, to a sampling depth of 100-150 mm. Downward mobility could be attained via either a dissolved liquid phase or in solid particulate form. Earthworm activity and preferential slippage of sludge down soil fissures

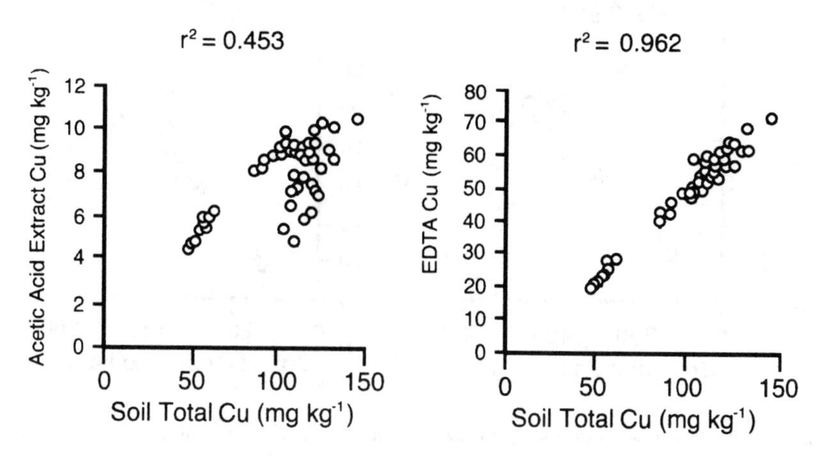

Figure 1 *Relationship between soil total Cu and the acetic acid and EDTA extractable Cu concentrations at site LC*

and channels were likely to be the main causes of the downward movement seen. Soil pH (5.5 or 6.5) had no significant effect on the distribution of the metals.

Results show that approximately 33% of each metal applied was recovered in the 0 - 25 mm layer of soil. This accumulation of added sludge metals in the surface layers of the soil has been noted by many (see Aitken, 1996). Barrow and Reaves (1985), for example, attributed this effect to the immobility of applied metals in soils and to the influence of organic matter (OM) which is known to bind several metals, especially Cu.

The significant metal accumulation at or near the soil surface of fields treated with sludge presents a risk of uptake of metals by grazing animals from soil ingestion when sward heights are low. Estimates of the amount of intake of soil by grazing livestock during the winter months, when grass growth is restricted, suggest that anything from 400 - 1400 g day^{-1} may be ingested by grazing cattle (Thornton, 1974). At the LC Site, a 'worst case' scenario would thus have cows assimilating approximately 5 g of Zn and 2 g of Cu within their body tissue if left to graze for a 10 day winter period.

Clearly, farmers need to be made aware of this soil ingestion risk if and when sewage application to land becomes more widespread. It would thus appear not sensible to overgraze livestock in winter on sward used extensively for sludge application. More work, however, needs to focus on this issue in the near future. In particular, further attention needs to be given towards establishing a better understanding between metal concentrations in surface soils, soil ingestion levels by livestock, and a corresponding build-up of metal in animal body tissue.

Graphical analysis of plots of total versus extractable metal concentration from the three study sites is informative. Significant differences are obvious when the results obtained using acetic acid as an extracting reagent are compared to their EDTA extractable counterparts. Cu and Pb in particular illustrate these differences well at the Lower Carbarns (LC) Site (Figures 1 - 3). Cu and Pb in the EDTA extracts are always higher than in their acetic acid counterparts. Moreover, correlations between extractable and total metal concentrations are much improved for the set of sub-samples treated with EDTA.

An insight into why correlation is less good for those samples treated with acetic acid

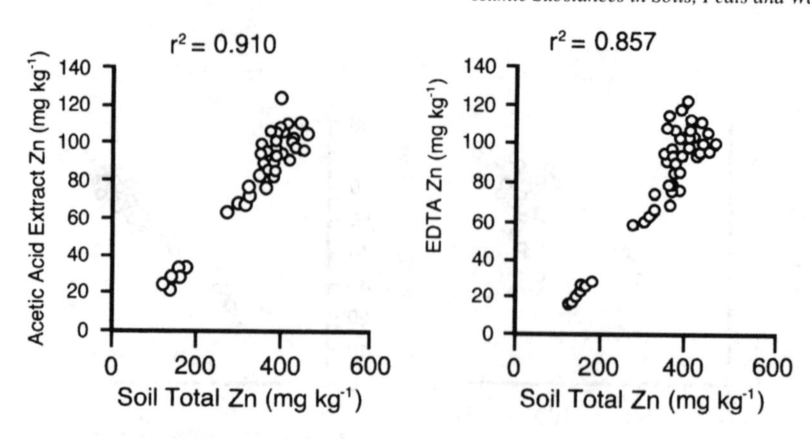

Figure 2. *Relationship between the soil total Zn and the acetic acid and EDTA extractable Zn concentrations at site LC.*

can be gained by separating the data given in Figures 1 and 3 into two depth ranges: a 'shallow' horizon extending down as far as 100 mm depth, and a 'deeper' band, lying in the 100 - 400 mm depth range (Figures 4 and 5). Surface soils at the LC site are very rich in OM, and these samples appear to show a lower than expected amount of acetic extractable (HAC) metal for a given total metal concentration. The binding action upon certain heavy metals (notably Cu and Pb) of soil OM, as noted previously (e.g. Barrow and Reaves, 1985), would seem an obvious explanation for this trend. Samples from lower depths seem to be less prone to this process because of the significantly lower OM content. It would appear that, for deeper soil horizons, a near fixed proportion of these metals is extracted by the weak acid, regardless of the total contents of the metals.

Figures 6 and 7 show that there is no single unifying relationship between the total and the EDTA-extractable metal concentrations within Scottish soils. Some level of correlation between the two variables is noted for all three sites, but the exact relationship is obviously different in all three cases. It is evident that if any significant correlation can be formulated between EDTA-extractable and the total metal concentrations within soils treated with sludge, its exact nature will be very much dependent upon local site factors.

Herbage

Herbage concentrations of Cu, Zn, Ni, Cd, Cr, and Pb after 8 y of sludge treatments are given in Tables 5 - 7. Metal concentrations in the herbage in the control treatments were related to soil concentrations, and were further correlated to previous historic sludge use. Compared to the other two sites, Site LC had higher concentrations of Cu, Zn, and Ni in the plants.

Subsequent sludge treatments over eight years influenced herbage Cu and Zn concentrations. Application rates of 270 m^3 ha^{-1} y^{-1} of sludge (supplying approximately 4 kg ha^{-1} Cu over 9 y) resulted in an increase of 3.2 (range 3.0 - 3.3) mg kg^{-1} in the herbage. The liming treatment had no effect on plant Cu contents. Similarly, ca 74 kg ha^{-1} Zn was supplied over this period, leading to an average increase in contents in the herbage of 24

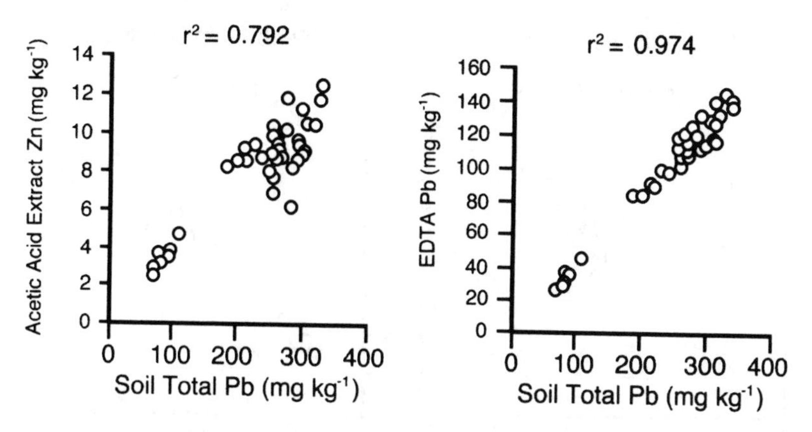

Figure 3 *Relationship between the soil total Pb and the acetic acid and EDTA extractable Pb concentrations at site LC*

(range 8 - 42) mg kg^{-1}. In this case, liming did decrease Zn contents by an average of 15 (range 4 - 30) mg kg^{-1}. This phenomenon is likely to be attributable to the accompanying increase in soil pH causing the formation of sparingly soluble metal hydroxides and carbonates which are less available for plant uptake.

Graphical plots of total and herbage Cu concentrations for all sites reveal a significant positive relationship for two of the three sites (Figure 8). No relationship was found at site LC. The correlations for sites WF and GH are different, and together with data for site LC data, highlight the limitations of using total soil Cu concentration as a predictor for

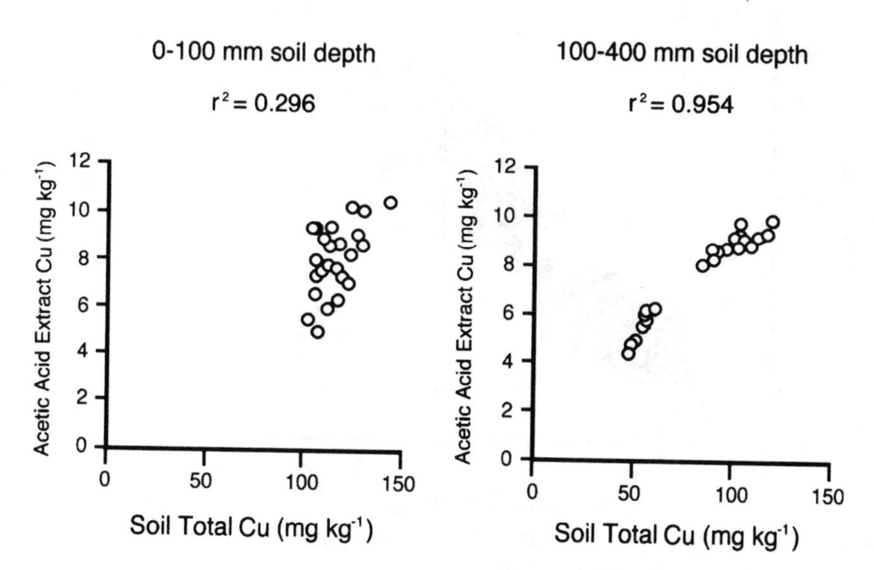

Figure 4 *Relationship between the soil total Cu and the acetic acid extractable Cu for the 0-100 mm and the 100 - 400 mm soil samples at site LC*

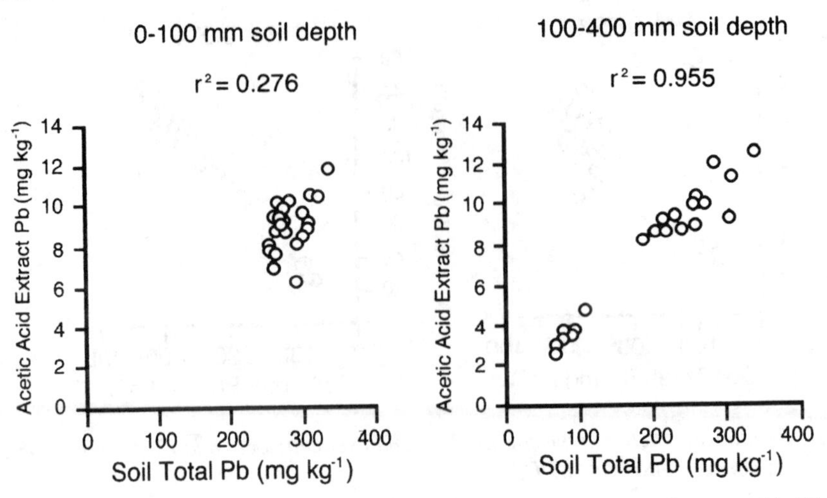

Figure 5 *Relationship between the soil total Pb and the acetic acid-extractable Pb for the 0 - 100 mm and the 100 - 400 mm soil samples at site LC*

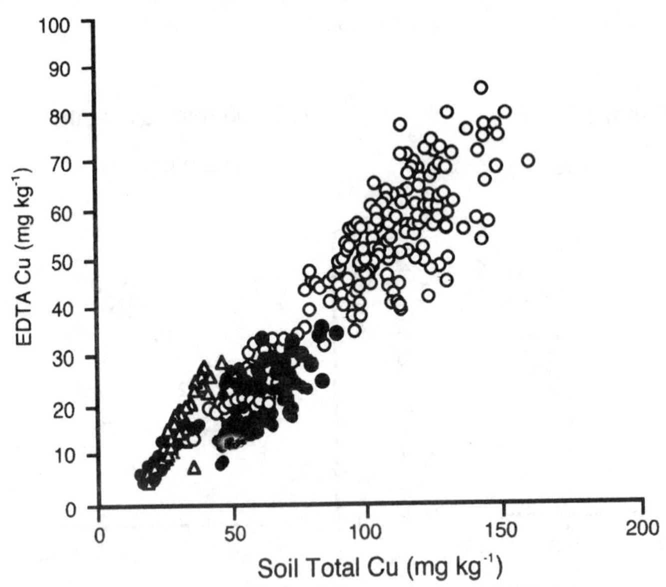

Figure 6 *Relationship between the soil total Cu and the EDTA extractable Cu conentrations for all three sites*

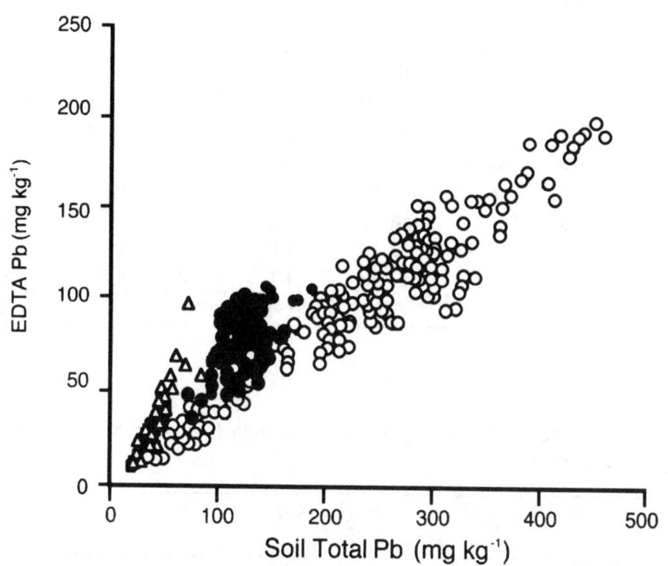

Figure 7 *Relationship between the soil total Cu and the EDTA extractable Pb concentrations for all three sites*

levels found within herbage. Local site factors clearly play an important role in determining the proportionate herbage uptake of Cu and other metals from the soil.

Plant Cr and Pb contents were not significantly affected by the sewage sludge treatments or by the liming sub-treatments. This indicates that despite these elements being supplied in high amounts by the sludge (9 kg ha^{-1} Cr and 33 kg ha^{-1} Pb), both remained in an unavailable form in the soil.

Despite the high levels in the soil, the heavy metal contents in the plants were well below those considered to be phytotoxic (MacNicol and Beckett, 1985) or zootoxic (Webber *et al.*, 1984). This result was achieved despite the fact that at Lower Carbarns, soil Cu and Zn contents were greater than the UK stipulated maximum concentration, and Pb was only slightly lower than the specified critical level (DoE, 1989). From the studies undertaken to date, it thus appears that the legislative limits on soil Cu, Pb, and Zn concentrations set out in the 1989 Sludge Regulations are set at levels which adequately protect the food chain from metal toxicity through grass uptake across a range of soil types (sandy loam to clay loam).

4 Conclusions

Annual use of liquid sewage sludge at three Scottish sites over a four year experimental period resulted in an increase in soil concentrations of Cu, Zn, Pb, Ni, and Cr down to a depth of 100 - 150 mm. The largest accumulation of metals occurred at a soil depth of

Table 5 *Site WF. Herbage contents (mg kg^{-1} DM) of metals at first cut after eight years of sludge treatment*

Sludge Rate m^3 ha^{-1} yr^{-1}	pH	Cu	Zn	Cd	Ni	Cr	Pb
0 5.5		8.65	38.55	0.10	1.35	0.48	1.00
0	6.5	7.25	27.90	0.11	0.54	0.22	0.44
67	5.5	6.74	32.60	0.13	1.13	0.21	0.63
67	6.5	6.33	23.15	0.06	0.47	0.20	0.41
135	5.5	9.16	41.50	0.11	1.07	0.24	0.60
135	6.5	9.56	38.73	0.13	0.94	0.32	0.72
270	5.5	11.75	47.10	0.11	1.19	0.27	0.72
270	6.5	11.22	42.68	0.10	0.82	0.28	0.57
SED		1.425	6.70	0.031	0.297	0.127	0.299

Table 6 *Site GH. Herbage contents (mg kg^{-1} DM) of metals at first cut after eight years of sludge treatment*

Sludge Rate m^3 ha^{-1} yr^{-1}	pH	Cu	Zn	Cd	Ni	Cr	Pb
0	5.5	8.14	31.73	0.06	1.88	0.30	0.57
0	6.5	7.89	27.95	0.06	1.13	0.29	0.47
67	5.5	8.16	32.83	0.08	1.97	0.30	0.66
67	6.5	8.14	28.00	0.06	1.00	0.29	0.54
135	5.5	8.29	39.38	0.06	2.10	0.30	0.57
135	6.5	8.16	30.22	0.06	0.85	0.30	0.50
270	5.5	11.58	52.95	0.06	2.50	0.35	0.85
270	6.5	11.85	41.25	0.06	1.50	0.34	0.69
SED		0.59	2.00	0.009	0.213	0.038	0.123

Table 7 *Site LC. Herbage contents (mg kg^{-1} DM) of metals at first cut after eight years of sludge treatment*

Sludge Rate m^3 ha^{-1} yr^{-1}	pH	Cu	Zn	Cd	Ni	Cr	Pb
0	5.5	10.48	56.07	0.13	4.47	0.29	0.91
0	6.5	10.17	46.83	0.11	2.16	0.33	0.81
67	5.5	11.97	61.00	0.13	3.97	0.44	0.88
67	6.5	11.30	51.89	0.10	2.06	0.63	1.91
135	5.5	11.92	71.33	0.13	4.06	0.32	1.13
135	6.5	10.13	47.50	0.14	2.47	0.55	2.97
270	5.5	13.52	98.05	0.13	5.57	0.42	1.13
270	6.5	12.58	67.68	0.13	3.06	0.56	2.16
SED		1.612	9.38	0.024	0.543	0.119	1.016

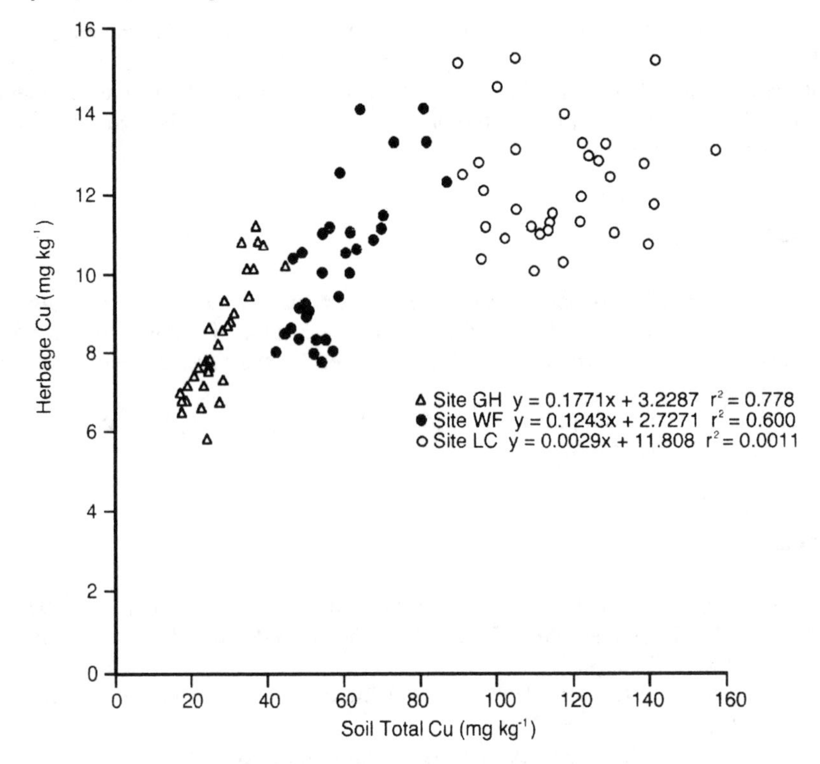

Figure 8 *Relationship between soil total Cu and herbage Cu contents for all three sites*

0-25 mm. Significantly lower amounts of the total soil Cu and Pb were extracted by acetic acid compared to EDTA. This indicates that relatively high amounts of Cu and Pb were bound to the SOM. Good correlations between total and EDTA-extractable Cu and Pb were found, but the exact relationship was different at all three sites.

The high rate of sludge use at one site (Lower Cabarns) prior to the commencement of this study has led to soil Cu and Zn levels at this site exceeding UK legislative levels. Elevated uptakes of heavy metals in livestock related to soil ingestion could arise if herbivores were allowed to graze upon this soil. Caution should thus be exercised when grazing livestock during winter months on grassland which has been receiving long term sludge applications. The risk of livestock ingestion of metals can be decreased by periodic ploughing of the fields.

Current legislative sampling requires soil samples to be taken within a 0 - 25 cm depth interval. Averaged metal concentrations within such a core could be significantly lower than those found within the top 25 mm soil depths. The present requirements do not then provide an adequate assessment of the risks to herbivores from ingestion of surface contaminated grassland soils. An additional sampling depth of 25 mm would provide a more useful risk assessment.

Zn and Cu metal levels within herbage were elevated in the plants grown on the sludge treatments. A significant positive correlation between soil and herbage Cu concentration existed at two out of the three sites. However, this relationship varied, demonstrating limited value in the use of total Cu as a predictor of herbage Cu content. Liming the ground to a pH of 6.5 was successful in decreasing plant uptake of Zn, but had little impact upon uptake of the other metals described here. At the Lower Carbans site, where soil metal contents were highest, the plant concentrations remained well below the toxic threshold levels for both plants and animals.

Acknowledgements

This work forms part of an ongoing research project and we acknowledge the valuable input by previous project leaders, R G Golightly and D A Klessa. Technical staff in SAC Environmental Sciences Department are warmly thanked for their assistance in field and laboratory studies.

SAC receives financial support from Scottish Office Agriculture, Environment and Fisheries Department (SOAEFD).

References

Aitken, M. N. 1996. Sustainable use of sewage sludge on agricultural land. p. 152-167. *In* A.G. Taylor, J.E. Gordon, and M.B. Usher (eds), *Soils, Sustainability and the Natural Heritage*. HMSO, Edinburgh.

Barrow, M. L. and G. A. Reaves. 1985. Extractable copper concentrations in Scottish soils. *Journal of Soil Science* **36**:31-43

Beckett, P. H. T. 1989. The use of extractants in studies on trace metals in soils, sewage sludges, and sludge-treated soils. p. 144-171. *In Advances in Soil Science, Vol 9.* Springer Verlag, New York.

Davis, R. D. 1989. Agricultural utilisation of sewage sludge- a review. *Journal of Institution of Water and Environmental Management* **3**:117-126.

DoE. 1989. UK Department of the Environment. *Code of Good Agricultural Use of Sewage Sludge*. HMSO, London.

DoE. 1993. Sludge Use in Agriculture 1990/91. *UK report to the commission under Directive 86/278/EEC Department of the Environment.* HMSO, London.

Iyengar, S. S., Martens, D. C. and Miller, W. P. 1981. Distribution and plant availability of soil zinc fractions. *Soil Sci. Soc. Am. J.* **45**:735-739.

Lawes Agricultural Trust. 1993. *Genstat 5 statistics software programme*. Copyright to, Statistics Department, IACR-Rothamsted Experimental Station, Harpenden, Hertfordshire.

McBride, M.B. 1989. Reactions controlling heavy metal solubility in soils. *Advances in Soil Science* **10**:1-56

McGrath, S. P. 1987. Long-term studies of metal transfer following application of sewage sludge. p.310-317. *In* P.J. Coughtrey, M.H. Martin, and M.H. Unsworth (eds), *Pollutants Transport and Fate in Ecosystems*. British Ecological Society Special Publication no. 6. Blackwell, Oxford.

MacNicol, R.D. and P.H.T. Beckett. 1985. Critical tissue concentrations of potentially toxic elements. *Plant and Soil* **85**:107-129.

Miller, W. P., D. C. Martens, and L. W. Zelazny. 1986a. Effect of sequence in extraction of trace metals from soils. *Soil Sci. Soc. Am. J.* **50**:559-560.

Miller, W. P., D. C. Martens, L. W. Zelazny, and E.T. Kornegay. 1986b. Forms of solid-phase copper in copper-enriched swine manure. *J. Env. Qual.* **15**:69-72.

Statutory Instrument. 1989. *UK Statutory Instrument No 1263. The Sludge (Use in Agriculture) Regulations 1989*. HMSO, London.

Swift, R.S. and R.G. McLaren. 1991. Micronutrient adsorption by soils and soil colloids. p.257-293. In: G.H. Bolt, M.F. DeBoodt, M.H.B. Hayes, and M.B. McBride (eds), *Interactions at the Soil Colloid-Soil Solution Interface*. Kluwer, Dordrecht.

Thornton, I. 1974. Biochemical and soil ingestion studies in relation to trace element nutrition of animals. p. 451-454. *In* W.H. Hoekstra, J.W. Suttie, H.E. Guathar, and W. Mertz (eds), *Trace Element Metabolism in Animals*. Proceedings of the 2nd International Symposium on Trace Elements. University Press, Baltimore.

Webber M.D., A. Kloke, and J. Tjell. 1984. A review of current sludge use guidelines for the control of heavy metal contamination in soil. p. 371-386. *In* P. L'Hermite and H. Ott (eds), *Processing and Use of Sewage Sludge*. Proceedings of the 3rd International Symposium. Elsevier, London.

Effect of Repeated Application of Sewage Sludge to Pasture on Metal Levels in Soil and Herbage

E.G. O'Riordan and D. McGrath[2]

TEAGASC. GRANGE RESEARCH CENTRE, DUNSANY, CO. MEATH, IRELAND
[2]TEAGASC. JOHNSTOWN CASTLE RESEARCH CENTRE, WEXFORD, IRELAND

ABSTRACT

The effects of repeated applications of sewage sludge on both soil and herbage metal (Cu, Pb and Zn) levels were studied at two sites. At one site, a metal-rich sludge was applied at rates up to a maximum of 75 m^3 ha^{-1} on three occasions each year for two years. This contributed, at maximum addition rates, 55 kg of Cu, 564 kg of Zn, and 9 kg of Pb ha^{-1}. At a second site, under the same application regime over a three year period, a low-metal sludge supplied a maximum of 4.3 kg of Cu, 8.0 kg of Zn and 1.9 kg of Pb ha^{-1}. Increased soil levels of both total and EDTA-extractable metals were observed for both applications. Increases were small or non-significant below a depth of 52 mm in soil for the metal-rich sludge treatment, and below a depth of 13 mm in soil for the low-metal treatment.

Effects on metal levels in herbage cut for silage were variable. A spring application of metal-rich sludge usually increased herbage metal concentrations, whereas later additions gave variable effects. In the year subsequent to the final sludge addition, Zn only was elevated in the herbage. Application of low-metal sludge had little or no effect on herbage metal levels. Results indicated that sludge applications, which conformed to Irish law giving effect to the European Union (EU) Directive on Sewage Sludge Use in Agriculture, were unlikely to cause environmentally undesirable increases in herbage metals.

1 Introduction

Within the EU, about 37% of the sewage sludge produced is used in agriculture (Newman *et al.*, 1989). In Ireland, the proportion is closer to zero. There is a number of reasons for this. Firstly, larger towns are on the coast and thus disposal to the marine has been an easy option. Secondly, landfill has to date had the capacity to accommodate sludges from existing treatment plants that service inland towns. Thirdly, Ireland is predominantly a grassland country and the sludge produced amounts to less than one per cent of the wastes from housed cattle and other animals. Not surprisingly, there is little demand from either the grassland sector or from the crop sector (which utilises less than 10% of the farmed-land area) for an additional organic waste.

By the year 1999, the marine disposal of sewage sludge must cease according to EU Directive 91/271 (EEC, 1991). In addition, sewage treatment facilities must be upgraded or built to cater essentially for all centres with populations in excess of 2000. It has been calculated (Watson, 1993) that sludge production in the Republic of Ireland will increase from the present level of 40 000 t dry matter (DM) to 130 000 t DM per annum by the year 2013, and with most of the increase occurring before 2005. Other disposal options, including use in agriculture, forestry, land-reclamation, composting and incineration, need to be developed. Of these, use in agriculture appears to be the one that could most easily be initiated and expanded to accommodate a large proportion of sludge requiring disposal.

Agronomic aspects relating to sludge are broadly similar in Ireland to those elsewhere. However, different practical considerations apply but because of local factors, including legal strictures on the quantity and quality of sludge that may be used, the grassland-based nature of the agricultural industry, and the vulnerability of the largely pristine environment. Irish legislation which gives effect to the EU Directive 86/278 (EEC, 1986) on sewage sludge use in agriculture stipulates that usage of sludge should not exceed 2 t DM ha^{-1} y^{-1}. On average, this amount of sludge contains phosphorus (O'Riordan *et al.*, 1986b) that is sufficient for many agricultural enterprises, including extensive grazing and cereal production (Gately, 1994). Because of variation in both the nitrogen content of sludges (O'Riordan *et al.*, 1986a) and in the availability of the nitrogen to the plant (O'Riordan *et al.*, 1987), it may be best to discount the nitrogen of sludge, except for crops where excess may diminish quality. Thus the major chemical/environmental questions relating to sludge use encompass the influence that may be exercised by the heavy metal components on soil and on the crop (mainly grass) quality. Irish leglislation has adopted the strictest provisions of the EU Directive (EEC, 1986) regarding acceptable heavy metal levels in both sludge and soil. Heavy metal composition of sludges has been established (Table 1) in a number of Irish studies (O'Riordan *et al.*, 1986; McGrath and Postma, 1996). This paper addresses briefly the polluting potential of the three heavy metals, copper, lead, and zinc that are present in greatest quantities in most sludges. One plot trial where a metal-rich sludge was used (O'Riordan *et al.*, 1994b; O'Riordan *et al.*, 1994c), and one where a low-metal sludge was used (O'Riordan *et al.*, 1994a) are considered with regard to soil enrichment by heavy metals, and to contamination/uptake of heavy metals by herbage. Both trials incorporated a considerable number of treatments, but only data for the controls and the high rates of sludge treatments are considered here.

2 Materials and Methods

Sludges. Two sludges were used. These included Sludge A from an anaerobic digestion treatment plant. This sludge was enriched in heavy metals. Sludge B ws from an extended aeration process and was known to have a low content of heavy metals.

Soils. Sludge A was used on a site at Johnstown Castle, Co Wexford, Ireland. The soil was a slowly-permeable, imperfectly-drained gley formed from an Irish Sea till parent material. The soil had been sown with a perennial ryegrass (*Lollium perenne* L)/white clover (*Trifolium repens* L) mixture three years before the start of the trial. Sludge B was used on a soil of the Rathowen soil series, a Grey Brown Podsolic soil, from Co. Westmeath, Ireland. The soil was a moderately well-drained clay loam, formed from limestone and shale drift parent material. It had been seeded with perennial ryegrass and clover one year prior to commencement of the trial.

Treatments. Plots, measuring 6 m x 12 m, received three separate applications of sludge per annum over two years for trial A, and over three years for trial B. Applications were made from a tanker wagon at rates of 0, 25, 50 and 75 m^3 ha^{-1}. Treatments were replicated six times in a randomised complete block design. Times of application were near the beginning of April, at the beginning of June, and at the end of July.

Harvesting and analysis. Grass was harvested 6 - 8 weeks after each application, and also at similar times in the year following the final application of sludge for Trial A. Soil samples were taken to depths of 0 - 13, 13 - 52, and 52 - 156 mm at the end of each year. Sludge and soil samples were dried at 40 °C. The sludge was ground in a hammer mill and sieved to pass a 1 mm mesh sieve. The soil was rolled to pass a 2 mm mesh sieve. Herbage was dried at 95 °C and ground to pass a 1mm mesh sieve. Sludge and soil analyses were carried out using an emission spectrograph, as described by O'Riordan *et al.* (1986). Heavy metals in herbage were also analysed spectrographically (O'Riordan *et al.*, 1994b; O'Riordan *et al.*, 1994c), except for Cu and Zn in Trial B which were analysed by flame AA following digestion with nitric-sulphuric-perchloric acid (7:2:1). Heavy metals in soils were also analysed by flame AA following extraction with 0.05M EDTA for 1 h on a peripheral shaker, as described by Byrne (1979).

3 Results and Discussion

Soil Cu, Pb, and Zn levels

Metal-rich sludge. Application of metal-rich sludge resulted in the addition of considerable amounts of heavy metals to the soil (Table 2) over a two year period. The amount added was equivalent to 92 years of application for sludges of mean Cu content, 69 years of application for sludge of mean Pb content, and 245 years application for sludge of mean Zn content (McGrath and Postma, 1996). Immediately after the final application, levels of Cu and Zn had increased by a significant ($P \leq 0.05$) amount in the 0 - 13 mm and 13 - 52 mm soil layers, but levels of Pb increased in the 0 - 13 mm soil layer only. Increases were not significant below 52 mm. One year later, levels in the treated 0 - 13 mm layer had decreased by 22% for Cu and Pb, and 45% for Zn (O'Riordan *et al.*, 1994a). Evidence, which was statistically significant for Zn only, was obtained for small increases of Cu and Zn in soil below 52 mm (O'Riordan *et al.*, 1994a). It was evident that most of the metal applied was retained in the topsoil. In other studies it has been shown to move very

Table 1 *Mean heavy metal compositions ($mg\ g^{-1}$) of Irish sewage sludges*

Heavy metal	O'Riordan (1978 - 1981)[1]	McGrath (1993)[1]	Legal limit
Cadmium	1	0.9	20
Chromium	88	24	1000
Copper	493	301	1000
Lead	269	64	750
Mercury	-	0.42	16
Nickel	83	17	300
Zinc	1537	1150	2500

[1]When the sludge was collected

Table 2 *Mean additions and levels in the soil where metal-rich sludge was applied*

Treatment		Cu	Pb	Zn
Applied (kg ha^{-1})		55.4	8.84	564
Soil level (mg g^{-1})				
At 0-13 mm	Control	9	26	120
	Treatment	102	45	1657
	[1]LSD	33	10	564
At 13-52 mm	Control	10	25	110
	Treatment	20	29	220
	LSD	8	7	68
At 52-156 mm	Control	10	24	88
	Treatment	12	30	115
	LSD	9	9	62

[1]LSD at $P \leq 0.05$

slowly (if at all) down the profile (McGrath and Lane, 1989; Dowdy *et al.*, 1991). Despite the massive doses of sludge applied, Cu and Pb levels in soil were estimated to be below the limits tolerated for soil in receipt of sludge. These are fixed at 50 μg g^{-1} for Cu and Pb in soil (0 - 60 mm sample). However, Zn in the treated soil exceeded considerably the limit which has been fixed under Irish leglisation at 150 μg g^{-1}.

Low-metal sludge. Annual rates of sludge application at 225 m^3 ha^{-1} were similar to those for the metal-rich sludge trial, but application was extended over three years compared to two years for the latter. However, the amounts of heavy metals added to soil

Table 3 *Metal additions and levels in the soil when sludge of low metal contents were applied*

Treatment		Cu	Pb	Zn
Applied (kg ha^{-1})		4.32	1.89	8.01
Soil level (mg g^{-1})				
At 0-13 mm	Control	14	20	160
	Treatment	26	27	152
	[1]LSD	10	4	74
At 13-52 mm	Control	13	19	132
	Treatment	24	24	154
	LSD	9	5	33
At 52-156 mm	Control	12	19	122
	Treatment	22	23	139
	LSD	8	5	29

[1]LSD at $P \leq 0.05$

Table 4 *Heavy metals extracted from soil, using 0.05M EDTA, after two y of sludge application*

Treatment		Cu	Zn
Metal Rich			
At 0-13 mm	Control	3	6
	Treatment	32	1271
	[1]LSD	18	394
At 13-52 mm	Control	2	3
	Treatment	13	107
	LSD	3	19
At 52-156 mm	Control	2	2
	Treatment	3	10
	LSD	1	7
Low Metal			
At 0-13 mm	Control	7.5	5.6
	Treatment	15.3	21.5
	[1]LSD	1.9	2.8
At 13-52 mm	Control	7.3	4.1
	Treatment	10.1	10.9
	LSD	1.7	2.7
At 52-156 mm	Control	7.2	4.5
	Treatment	6.2	3.4
	LSD	1.4	1.4

[1]LSD at $P \leq 0.05$

were much smaller (Table 3) and were not unlike those likely to be added under a responsible fertilization program conforming to Irish legislation. Annual additions of metal exceeded the quantities (0.60 kg of Cu, 0.128 kg of Pb, and 2.30 kg of Zn) that would accrue from the use of the average sludge applied in 1993 (Table 1), when applied at the rate of 2 t DM ha^{-1} but were considerably less than the maximum legally-allowable annual additions of 2.0, 1.5, and 5.0 kg of Cu, Pb, and Zn, respectively. Sludge addition, however, far exceeded the maximum tolerated 2t DM ha^{-1} y^{-1} which showed an agronomically unrealistic amount of P, in particular, was being applied (O'Riordan *et al.*, 1996). Increases in soil metal were small, and were significant mostly for Cu and Pb (Table 3).

Metals Extractable With 0.05M EDTA

The data indicated significant, if sometimes small, increases in Cu and Pb to a depth of 156 mm in soil for the metal-rich trial, and to 52 mm for the low-metal trial (Table 4). Pb was not examined. Added Zn was more completely extracted by EDTA solution than was native Zn, especially from the 0 - 13 mm soil layer. Also, added Zn was more completely extracted than was added Cu. This was the reverse of that found by Berrow and Burridge (1990) for two soils following massive addition of Cu- and Zn-enriched sludges.

Table 5 *Heavy metal levels of herbage (mg g^{-1}) over two y of applications of metal-rich sludge*

Treatment		Cu	Pb	Zn
Year 1				
Harvest 1	Control	9	0.8	34
	Treatment	28	1.6	127
	[1]LSD	10	2.0	57
Harvest 2	Control	14	1.2	46
	Treatment	16	1.6	66
	LSD	6	1.0	36
Harvest 3	Control	17	3.0	49
	Treatment	25	6.3	136
	LSD	4	2.0	40
Year 2				
Harvest 1	Control	9	0.7	19
	Treatment	17	1.7	82
	[1]LSD	4	0.4	20
Harvest 2	Control	7	0.5	22
	Treatment	7	0.5	53
	LSD	4	0.3	12
Harvest 3	Control	8	0.3	22
	Treatment	9	0.4	78
	LSD	2	0.3	20

[1]LSD at $P \leq 0.05$

Enhanced Levels of Metals in Herbage

Metal levels in herbage, as a consequence of sludge application, were often but not always increased when metal-rich sludge was used (Table 5). Increases were most striking for the first harvests. Largest increases were experienced for Zn, but increases in Cu and Pb were sometimes significant. Recorded elevations in Cu levels in herbage were considerably less than those experienced in trials which examined the disposal to pasture of Cu- and also Zn-enriched pig faecal slurry (Poole *et al.*, 1983). It is arguable that small increases of these

Table 6 *Mean heavy metal levels of herbage (mg g^{-1}) together with mean LSD values over nine harvests for applications of sludges of low metal contents*

Treatment	Cu	Pb	Zn
Control	9	1.1	24
Treatment	10	1.4	28
[1]LSD	2.3	0.6	4.6

[1]LSD at $P \leq 0.05$

Table 7 Heavy metal levels of herbage *(mg g^{-1}) in the year subsequent to sludge application when metal-rich sludges were applied*

	Treatment	Cu	Pb	Zn
Harvest 1	Control	4	1.8	11
	Treatment	4	1.5	46
	[1]LSD	2	0.5	14
Harvest 2	Control	6	0.4	23
	Treatment	6	0.5	45
	LSD	2	0.3	16
Harvest 3	Control	8	1.4	25
	Treatment	8	1.3	50
	LSD	3	1.5	26

[1]LSD at $P \leq 0.05$

metals serve to correct for deficiencies which are widespread in grazing animals (Poole, 1981; Hambridge *et al.*, 1986). Increased herbage metal levels probably depended on a number of variables, including rate of application of sludge, heighth of grass at application, herbage growth, and weather factors (Klessa and Desira-Buttigieg, 1992). Application of low-metal sludge had little effect on the heavy metal levels of herbage (Table 6); significant rises were mainly confined to Zn. When herbage was sampled in the year subsequent to the end of metal-rich sludge application, there was essentially no increase in Cu or Pb for any of the three harvests (Table 7). However, Zn levels were increased for all harvests. It is not known whether this increase was due to herbage contamination, or to uptake.

4 Conclusions

Significant increases in soil Cu, Pb, and Zn were demonstrated following the addition of a metal-rich sewage sludge to soil. Nearly all Cu and Zn remained in the top 52 mm of soil. Some movement of Cu and Zn from the 0 - 13 mm layer to lower layers was indicated especially using the more sensitive EDTA extractant. Contamination of herbage was variable in extent following the use of metal-rich sludge. Contamination in subsequent years was not significant, except in the case of Zn where the effect was considerable. Critically, addition of sludge metals at levels consistent with constraints imposed by Irish legislation caused no increase in herbage Cu and Pb in the three year treatment period. Significant but small increases were measured for Zn. However, these were considered to be nutritionally desirable rather than toxicologically undesirable (Hambridge *et al.*, 1986).

References

Berrow, M.L. and J.C. Burridge. 1990. Persistence of metal residue in sewage sludge treated soils over seventeen years. *Int. J. Environmental Anal. Chem.* **19**:173-177.

Byrne, E. 1979. *Chemical Analysis of Agricultural Materials.* An Foras Taluntais, Dublin.

Dowdy, R.H. J.J. Latterell, T.D. Hinesly, R.B. Grossman, and D.L. Sullivan. 1991. Trace metal movement in an Aeric Ochraqualf following 14 years of annual sludge applications. *J. Environmental Qual.* **20**:119-123.

EEC. 1986. Council Directive on the protection of the environment, and in particular of the soil, when sewage sludge is used in agriculture. *Official Journal L 181/6*, 12 June.

EEC 1991. Council directive concerning urban waste water treatment. *Official Journal L 135 40,* 21 May.

Gately, T.F. (ed.). 1994. *Soil Analysis and Fertiliser, Lime, Animal Manure and Trace Element Recommendations. 1994.* Teagasc, Johnstown Castle, Wexford, Ireland.

Hambridge, K.M., C.E.Casey, and N.F. Krebs 1986. Zinc. p. 1-138. *In* W.Mertz (ed.), *Trace Elements in Human and Animal Nutrition,* **Vol. 2,** Academic Press.

Klessa, D.A. and A. Desira-Buttigieg. 1992. The adhesion to leaf surfaces of heavy metals from sewage sludge applied to grassland. *Soil Use and Management* **8**:115-121.

McGrath, S.P. and P.W. Lane. 1989. An explanation for the apparent losses of metals in a long-term field experiment with sewage sludge. *Environmental Pollution* **60**:235-256.

McGrath, D and L. Postma. 1996. A note on the levels of trace contaminants in Irish sewage sludges. *Biology and Environment* (submitted). Proc. Royal Irish Academy.

Newman, P.J., A.V. Bowden, and A.M. Bruce. 1989. Production, treatment and handling of sewage sludge. p. 11-38. *In* A.H. Dirkzwager (ed) *Sewage Sludge Treatment and Use: New Developments, Technological Aspects and Environmental Effects*. Elsevier.

O'Riordan, E.G., Dodd, V.A., Fleming, G.A. 1994a. Spreading a low-metal sludge on grassland: Effects on soil and heavy metal concentrations. *Irish J. Agric. Res.* **33**:61-70.

O'Riordan, E.G., V.A. Dodd, G.A. Fleming, and H. Tunney, 1994b Repeated application of a metal-rich sewage sludge to grassland. 1. Effects on metal levels in soil. *Irish J. Agric. Res.* **33**:41-52.

O'Riordan, E.G., Dodd, V.A., Fleming, G.A. and Tunney, H. 1994c. Repeated application of a metal-rich sewage sludge to grassland. 2. Effects on herbage metal levels. *Irish J. Agric. Res.* **33**:53-60.

O'Riordan, E.G., V.A. Dodd, H. Tunney, and G.A. Fleming. 1986a. The chemical composition of Irish sewage sludges. 1. Nitrogen content. *Irish J. Agric. Res.* **25**:223-229.

O'Riordan, E.G., V.A. Dodd, H. Tunney, and G.A. Fleming. 1986b. The chemical composition of Irish sewage sludges. 2. Phosphorus, potassium, magnesium, calcium and sodium contents. *Irish J. Agric. Res.* **25**:231-237.

O'Riordan, V.A. Dodd, H. Tunney, and G.A. Fleming. 1986c. The chemical composition of Irish sewage sludges. 3. Trace element content. *Irish J. Agric. Res.* **25**:239-249.

O'Riordan, E.G, V.A. Dodd, H. Tunney, and G.A. Fleming. 1987. Fertiliser value of sewage sludge. 1. Nitrogen. *Irish J. Agric. Res.* **26**:45-51.

Poole, D.B.R., D. McGrath, G.A. Fleming, and J. Sinnott. 1983. Effects of applying copper-rich pig slurry to grassland. 3. Grazing trials: stocking rate and slurry treatment. *Irish Journal of Agricultural Research* **22**:1-10.

Poole, D.B.R. 1981. Implications of applying copper rich pig slurry to grassland - effects on health of grazing sheep. p. 273-282. P. L'Hermite and J. Dehandtschutter (eds). EEC Workshop, Bordeaux (1980).

Weston, F.T.A. Ltd. 1993. *Strategy study on Options for the Treatment and Disposal of Sewage Sludge in Ireland.* Vol. 1. Executive Summary and Recommendations.

Comparative Effects of Organic Manures on Spring Herbage Production of a Grass/White Clover (c.v. Grasslands Huia) Sward

J. Humphreys, T. Jansen, N. Culleton, and F. MacNaeidhe

TEAGASC, JOHNSTOWN CASTLE, WEXFORD, IRELAND

Abstract

A relatively low level of herbage production in the spring is a major disadvantage of grass/white clover [*Trifolium repens* (L.)] swards. The objective of this experiment was to investigate the effect of cattle slurry and farmyard manure (FYM) on spring herbage production from grass/white clover (c.v. Grasslands Huia) swards in temperate grassland. Four treatments were used. Cattle slurry [3000 L ha^{-1}; approximately 6% dry matter (DM)] was applied to plots in mid-August 1993 and 1994 (summer application), or mid-March 1994 and 1995 (spring application). Twenty t ha^{-1} of FYM (approximately 20% DM) was applied in mid-March 1994 and 1995. No manure was applied for the control. Treatments were arranged in a randomised complete block design and replicated six times. Herbage DM yields were assessed in May 1994 and 1995. In May 1994 treatments yielded an average of 3.42 t DM ha^{-1} and there was no significant ($P < 0.05$) difference between treatments. In May 1995 there was no significant ($P < 0.05$) difference in herbage DM yields between the FYM (1.77 t ha^{-1}) and the spring-applied slurry (1.71 t ha^{-1}) treatments. Both of these treatments had significantly ($P < 0.05$) higher yields than the summer-applied slurry (1.02 t ha^{-1}) and the control (0.98 t ha^{-1}) which were not significantly ($P < 0.05$) different from each other. Response to the cattle slurry and FYM applied in March was primarily in terms of increased grass production which accounted for ca 90% of spring herbage yields. The three manure treatments resulted in a decrease in the extent of white clover stolon development and in the white clover content of herbage compared to the control. However, the spring-applied slurry favoured stolon survival during winter, and the clover content of herbage in the spring; hence it favoured the persistency of the clover in the sward relative to the other manure treatments. The three manure treatments maintained soil phosphorus, potassium, and magnesium concentrations in line with the requirements of optimum production from grass/white clover swards.

1 Introduction

In Ireland 90% of the agricultural land is devoted to grassland farming. Moderate temperatures and relatively high levels of rainfall make it a favoured area for grassland production (Murphy, 1984). White clover (*Trifolium repens* L.) has traditionally played an important role in Irish grassland systems (Murphy *et al.*, 1986), and it is considered to be the most important herbage legume in temperate regions of the world (McDonald *et al.*, 1988). It is grown universally with a companion grass, most commonly perennial ryegrass (*Lolium perenne*) (Haggar, 1989).

In a moderately intensive grassland system only relatively small amounts of nutrients end up in food products and are sold off the farm (Hopkins, 1993). Nutrient elements such as phosphorus and potassium are strongly held in the soil substrate (Jakobsen, 1993) and are almost completely recycled within grass based systems (Frissel, 1978). Major losses from the system, however, involves unavoidable losses in nitrogen via:

(i) denitrification of nitrate-N by soil micro-organisms;
(ii) runoff and leaching of soil nitrate; and
(iii) volatilization of ammonia to the atmosphere (Sherwood, 1990).

Nevertheless, the most important natural addition of nutrients to agro-ecosystems is through symbiotic fixation of atmospheric nitrogen by *Rhizobium* bacteria in association with leguminous crops. Although fixation is an expensive source of nitrogen for agricultural production compared to artificial products, it is fuelled by the products of photosynthesis of the host legume, and therefore it is a sustainable source of nitrogen which can maintain soil nitrogen supplies for agricultural production (Frissel 1978).

The extent of nitrogen fixation by the white clover in grassland ranges between 50 and 300 kg ha^{-1} y^{-1} (Masterson and Murphy, 1983; Ledgard and Steele, 1992). During the summer and early autumn white clover can contribute up to 50% of the herbage dry matter (DM) yields and can increase the herbage production from grass/white clover swards considerably (Ledgard and Steele, 1992). However, a problem with grass/white clover swards is that white clover makes only a minor contribution to herbage production in the spring (Davies and Evans, 1990; Rhodes *et al.*, 1994). This is attributed to the higher temperature requirements of white clover (9 °C) compared to temperate grasses (6 °C) for the commencement of growth in the spring (Murphy, 1985; Roberts *et al.*, 1989).

Haggar (1989) suggested that it may be possible to increase spring herbage production from grass/white clover swards by the strategic application of animal manures. However, the nitrogen content of manure generally has a negative effect on the clover component of the sward. It has been widely reported that application of fertilizer nitrogen decreases the extent of nitrogen fixation and the white clover content of the sward (Davies and Evans, 1990; Nesheim *et al.*, 1990; Harris, 1994).

Nesheim *et al.* (1990) applied mineral nitrogen at 75 kg ha^{-1} or cattle slurry at 50 000 L ha^{-1} to perennial ryegrass/white clover swards in the spring. These treatments resulted in an increase in herbage yields; however, the yield of fixed nitrogen was decreased from 51.2 kg ha^{-1} to 17.2 kg ha^{-1} in the mineral nitrogen treatment and to 24.9 kg ha^{-1} in the cattle slurry treatment. Furthermore, nitrogen applied in the manure can lead to a reduction in the white clover content of swards (Chapman and Heath, 1988; Boller *et al.*, 1992). For example, Chapman and Heath (1988) applied cattle slurry at 30 000 and 60 000 L ha^{-1} (equivalent to 30 and 60 kg available nitrogen ha^{-1}, respectively) to a grass/white clover sward in the spring. Both slurry treatments increased first cut herbage yields, but the white clover content of the herbage was reduced by the higher slurry application. Boller *et al.* (1992) applied 25 000 and 50 000 L ha^{-1} of diluted cattle slurry to grass/white clover swards. Both treatments resulted in a decrease in the clover component of the sward, with the greatest decrease occurring where the higher slurry application rate was employed.

Although the nitrogen in cattle slurry may adversely affect white clover content and yield, other nutrients contained in the slurry may have a positive effect in situations where available soil phosphorus, potassium, and other nutrients are inadequate for sward growth (Christie, 1987; Chapman and Heath, 1988). The objective of this experiment was to examine the possibility of increasing spring herbage production from grass/white clover swards by applying cattle slurry either in the previous summer or in the early spring, or by applying farmyard manure in the early spring.

2 Materials and Methods

General

The experiment was carried out on the Teagasc Experimental Research Farm at Johnstown Castle in Wexford, Ireland. The soil type was classified as a loam with 30% silt and 16.2% clay. Soil analysis (following extraction, using Morgan's solution, of a volume of soil previously dried at 40 °C for 16 hours) at the beginning of the experimental period (March 7, 1993) gave the following results: pH 6.5, phosphorus 6 mg L^{-1}, potassium 105 mg L^{-1} and magnesium 345 mg L^{-1}. This site had been under permanent pasture for 15 y prior to this experiment.

The area was ploughed on March 9 and cultivated on April 21, 1993. A suitable seedbed was obtained with two passes of a power harrow. Perennial ryegrass (cultivar, Magella) was sown immediately following cultivation at a rate of 15 kg ha^{-1} using a Fiona drill. This cultivar is a persistent perennial ryegrass with an erect growth habit and is suitable for both silage and pasture production. It is high yielding particularly in mid- to late- summer and is considered to be a suitable companion grass for white clover (Gilliland, 1988). On the April 23, white clover seed (cultivar, Grasslands Huia) was broadcast onto the soil surface at a rate of 5 kg ha^{-1} in plots 10 m long and 5 m wide.

The Organic Manure Treatments

There were four organic manure treatments in this experiment:

1, the control treatment where no manure was applied;
2, cattle slurry (CS) applied on August 15 1993 and on August 15 1994;
3, CS applied on March 16 1994 and on March 7 1995; and
4, farmyard manure (FYM) applied on March 15 1994 and on March 6 1995.

The CS treatments were injected to a depth of approximately 5 cm at a rate of approximately 30 000 L ha^{-1} using a 'Greentrak Slurry Injector' fitted with a specially designed electronic weighing device. The objective was to apply the cattle slurry at a rate which supplied approximately 100 kg total nitrogen ha^{-1}. Nutrient contents of the cattle slurry treatments used in this experiment are presented in Table 1.

The FYM was applied to the sward surface at a rate equivalent to 20 t ha^{-1}. The FYM was removed from straw-bedded cattle houses in the autumn of each year. It was passed through a composting machine once during the winter before application in the spring. Nutrient contents in the farmyard manures are presented in Table 1.

Table 1 *Nutrient contents (kg t^{-1}) in cattle slurry and farmyard manure*

Nutrient:	Nitrogen	Phosphorus	Potassium	Magnesium
Cattle slurry				
August, 1993	4.0	0.65	9.4	0.6
August, 1994	3.5	0.85	3.7	0.7
March, 1994	4.3	0.65	5.2	0.6
March, 1995	3.2	0.65	4.3	0.6
Farmyard manure				
March, 1994	3.4	0.75	5.6	0.6
March, 1995	3.8	1.00	6.0	0.5

Herbage Yields

During 1994 all treatments were harvested on May 17, August 4, September 13, and October 26. In 1995 all treatments were harvested on May 16.

Herbage was harvested using a Haldrup herbage harvester to a cutting height of approximately 5 cm above ground level. After harvest a 'core sample' of herbage was taken from each treatment and dried at 100 °C for 16 h. Herbage DM yields per ha were calculated. A second 'grab sample' of herbage consisting of five handfuls were taken at random from each treatment. White clover herbage was separated from the total herbage. Both samples were dried at 100 °C for 16 h and the white clover content of the total herbage on a dry weight basis was calculated. White clover and grass DM yields per hectare were determined. Following each harvest herbage was grazed by dairy cows.

Morphological Measurement of White Clover

The length and dry weight of white clover stolons were measured on 10 x 10 x 10 cm sod samples. Sod samples were taken from the control treatments on November 1, 1993, and from all treatments on May 23, 1994, November 1, 1994, and on May 18, 1995.

Five sods were taken at random from treatments, placed in a glasshouse and allowed to dry for approximately one week at air temperature. The white clover stolon material was then carefully extracted from each sod. Roots and expanded leaves (laminae) were removed from the stolon material using scissors. The total length of all the stolon material was then measured. This included stolon from the first expanded leaf behind the terminal bud. Each sample of stolon was dried at 100 °C for 16 hours in order to determine weight of stolon m^{-2}. Stolon dry weight per metre length of stolon (stolon size) was ascertained.

Soil Mineral Concentrations and Soil pH

On August 11, 1994, soil samples from each treatment, following extraction using Morgan's solution. were analysed for soil P, K and Mg contents, and for soil pH.

Statistical Analysis

This experiment was laid out in a randomised complete block design and replicated six times. The data were analysed using analysis of variance procedures to evaluate the significance of treatment differences. Pair comparison between treatment means at $P < 0.05$ level of significance are presented using the least significant difference test (Gomez and Gomez, 1984). Correlation analysis was employed to examine the relationships between grass DM yields and herbage DM yields in May 1994 and May 1995.

3 Results

Herbage Dry Matter Yields

Table 2 gives herbage, grass, and white clover DM yields in May 1994 and May 1995.

In May 1994, treatments yielded an average of 3.42 t DM ha^{-1}, and there was no significant ($P < 0.05$) difference between treatments. Likewise there was no significant ($P < 0.05$) difference in white clover DM yields. However, grass DM yields were significantly ($P < 0.05$) lower in the control treatment compared to the other treatments which were not significantly ($P < 0.05$) different from each other. There was a significant ($P < 0.001$) positive correlation between grass DM yields and herbage DM yields in May 1994: $r = +0.954$ (Fig. 1a). No significant relationship existed between herbage DM yields and white clover DM yields or the white clover content of herbage in May 1994.

Total herbage, grass and white clover DM yields during 1994 are presented in Table 3. The FYM treatment had significantly ($P < 0.05$) higher herbage DM yields than the other three treatments. These were not significantly ($P < 0.05$) different from each other. The FYM treatment had significantly ($P < 0.05$) higher total grass DM yields than the control treatment, but not the two slurry treatments. Whereas the spring-applied slurry treatment had significantly ($P < 0.05$) higher total grass DM yields than the control, the summer-applied cattle slurry treatment did not. The control had significantly ($P < 0.05$) higher total white clover DM yields than the two slurry treatments but not the FYM treatment. There was no significant ($P < 0.05$) difference in total clover DM yields between these three latter treatments.

In May 1995 there was no significant ($P < 0.05$) difference in herbage and grass DM yields between the farmyard manure and the spring-applied slurry treatments (Table 2). Both of these treatments had significantly ($P < 0.05$) higher herbage and grass DM yields than the summer-applied slurry and control treatments, which were not significantly ($P < 0.05$) different from each other. There was no significant ($P < 0.05$) difference in white clover DM yields between the control and the spring-applied slurry treatments. Both of these treatments had significantly ($P < 0.05$) higher clover DM yields than the summer-applied slurry and farmyard manure treatments. These latter treatments were not significantly ($P < 0.05$) different from each other. As in May 1994, there was a significant ($P < 0.001$) positive correlation between grass DM yields and herbage DM yields in May, 1995: $r = +0.988$ (Figure 1b). No significant relationship existed between herbage DM yields and white clover DM yields or the white clover content of herbage in May, 1995.

White Clover Content of Herbage

Table 2 shows the white clover content of herbage DM in May, 1994 and May, 1995.

In May 1994 the white clover content of herbage was significantly ($P < 0.05$) higher in the control treatment compared to the other treatments. These latter treatments were not

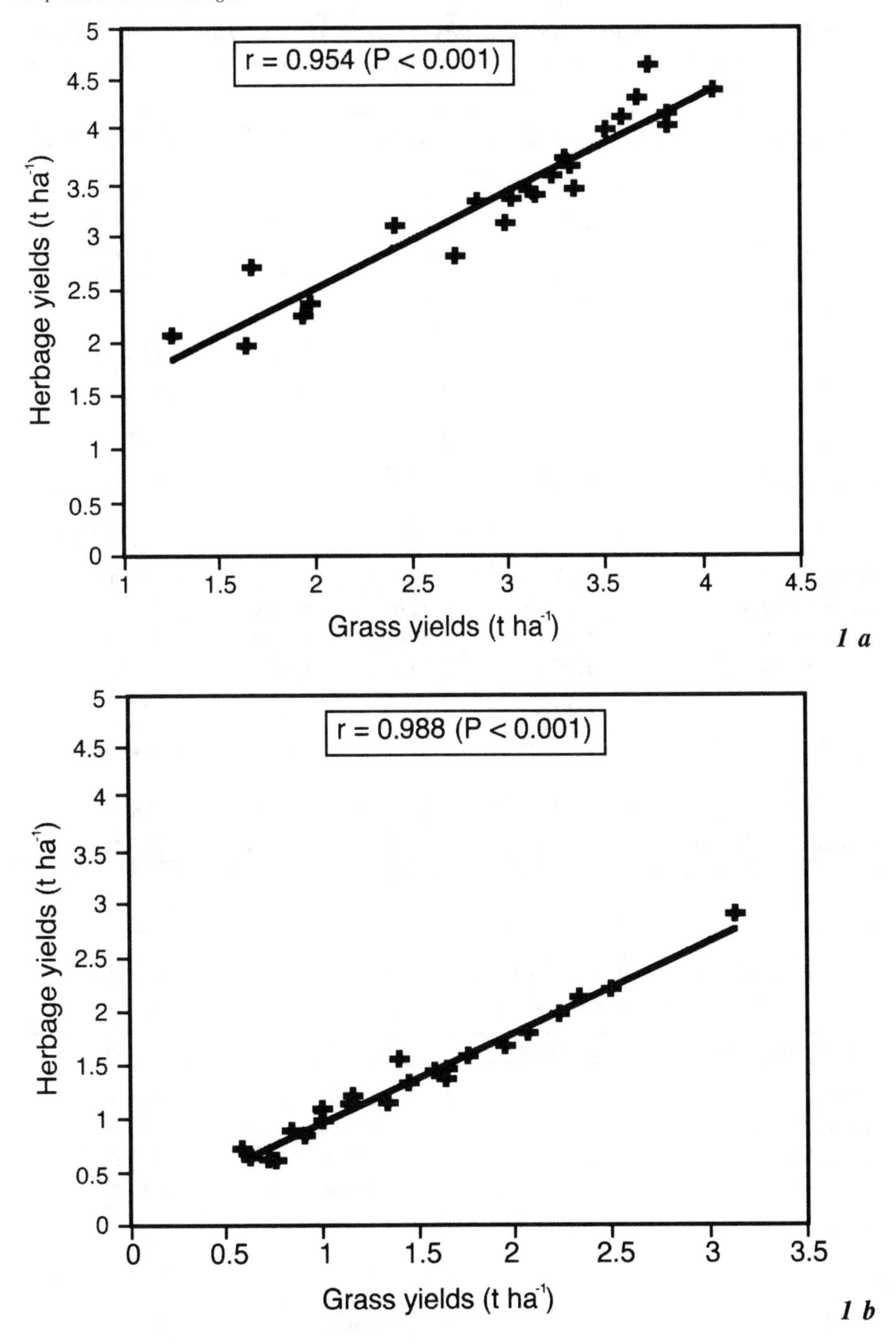

Figure 1 *The relationship between herbage and grass dry matter yields in May 1994 (a) and in May 1995 (b)*

Table 2 *Herbage, grass and white clover dry matter yields (t ha^{-1}) and the white clover content of herbage dry matter (%) in May, 1994 and in May, 1995 (mean of six replicates)*

Manure Treatment	Herbage	Grass	Clover	Clover content
		May 1994		
Control	2.96	2.36 b	0.61	22.54 a
Summer Slurry	3.43	3.09 a	0.35	10.31 b
Spring Slurry	3.78	3.37 a	0.41	10.59 b
Spring FYM	3.52	3.17 a	0.36	10.04 b
Mean	3.42	2.99	0.43	13.37
F test	n.s.	*	n.s.	*
LSD (5%):	___	0.70	___	8.97
Variation coefficient	16.6%	18.9%	49.1%	54.5%
		May 1995		
Control	0.98 b	0.76 b	0.22 a	22.8 a
Summer Slurry	1.02 b	0.98 b	0.04 b	3.9 c
Spring Slurry	1.71 a	1.55 a	0.16 a	10.2 b
Spring FYM	1.77 a	1.70 a	0.07 b	4.1 c
Mean	1.37	1.25	0.12	10.2
F test	**	**	***	***
LSD (5%:	0.50	0.48	0.067	4.5
Variation coefficient	29.7%	31.4%	44.8%	35.7%

Data followed by the same letter are not significantly different from each other (LSD test)

Table 3 *Total herbage, grass and white clover dry matter yields (t ha^{-1}) during 1994 (mean of six replicates)*

Manure Treatment	Herbage	Grass	Clover
Control	8.28 b	4.85 b	3.43 a
Summer Slurry	8.51 b	6.12 ab	2.39 b
Spring Slurry	8.87 b	6.68 a	2.19 b
Spring FYM	10.22 a	7.39 a	2.83 ab
Mean	8.97	6.26	2.71
F test	*	*	*
LSD (5%)	1.30	1.46	0.89
Variation coefficient:	11.8%	19.0%	26.7%

Data followed by the same letter are not significantly different from each other (LSD test)

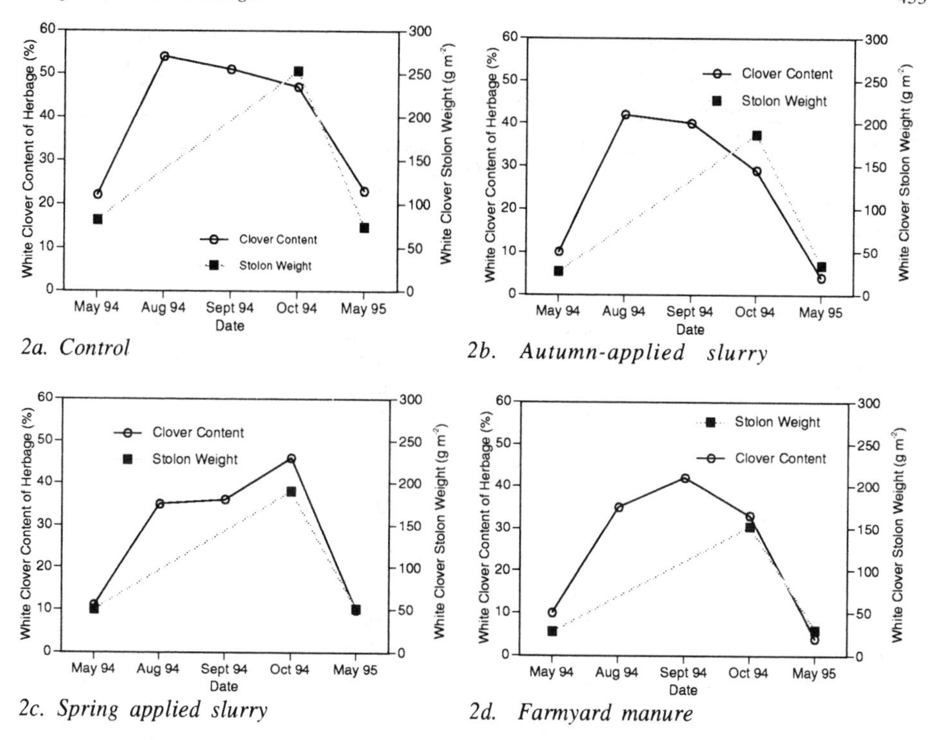

Figure 2 *The white clover content of herbage dry matter and stolon weight per unit area in each of the manure treatments between May, 1994 and May, 1995*

significantly ($P < 0.05$) different from each other.

The white clover content of herbage between May 1994 and May 1995 is presented in Figure 2. The clover content of herbage accounted for at least 40% of herbage DM yields in all treatments during the summer of 1994, but declined to relatively low levels during the winter of 1994/1995. The extent of increase during the summer and decline during the winter was dependent on manure treatment. The control treatment had relatively high contents throughout the experiment. The summer-applied slurry treatment had a relatively high content in August, but this declined markedly between September and October 1994. In contrast, there was a marked increase in the clover content of herbage in the spring-applied slurry treatment towards the end of the 1994 growing season. The greatest contribution of white clover to herbage DM yields in the farmyard manure treatment occurred during September. Nevertheless, the clover content of herbage in this treatment tended to be relatively low throughout the experiment.

In May 1995 the herbage in the control treatment had a significantly ($P < 0.05$) higher clover content than the other treatments (Table 2), and the spring-applied CS had a significantly ($P < 0.05$) higher content than the summer-applied CS and FYM treatments. There was no significant ($P < 0.05$) difference between the CS and FYM treatments. While the clover contents in the herbage of the control and spring-applied CS treatments in May 1995 were similar to those for the treatments of May 1994 (ca 22.5% and 10.5%, respectively), there was a decrease in the clover content from ca 10% in May 1995 in both the summer-applied slurry and the FYM treatments.

Table 4 *White clover stolon dry matter weight (g m^{-2}), stolon length (m m^{-2}) and stolon dry matter weight per metre length of stolon (stolon size, g m^{-1}) throughout the experiment (mean of six replications)*

Manure Treatment	Weight	Length	Size
	November 1993		
Control	55.7	39.4	1.5
Standard Error	6.7	5.9	0.1
	May 1994		
Control	81.3 a	95.3 a	0.85
Summer Slurry	28.8 c	36.8 c	0.79
Spring Slurry	53.3 b	65.3 b	0.82
Spring FYM	29.6 c	38.1 c	0.81
Mean	48.2	58.9	0.82
F test	***	***	n.s.
LSD (5%):	20.8	25.1	
Variation coefficient	35.1%	34.7%	15.3%
	November 1994		
Control	255.8	141.5	1.89
Summer Slurry	185.4	110.2	1.71
Spring Slurry	190.4	108.4	1.75
Spring FYM	155.0	90.3	1.67
Mean	196.7	112.6	1.75
F test	n.s.	n.s.	n.s.
LSD (5%):	-----	----	----
Variation coefficient	30.3%	32.2%	16.8%
	May 1995		
Control	73.8 a	81.9 a	0.90
Summer Slurry	33.3 b	39.1 b	0.85
Spring Slurry	50.4 ab	48.2 ab	1.04
Spring FYM	33.3 b	34.6 b	0.96
Mean	47.7	51.0	0.94
F test	*	*	n.s.
LSD (5%)	31.4	35.5	----
Variation coefficient	53.5%	56.6%	16.8%

Data followed by the same letter are not significantly different from each other (LSD test)

White Clover Morphological Measurements

White clover stolon DM weight m^{-2}, length m^{-2} and DM weight per metre length of stolon (stolon size) throughout the experiment are presented in Table 4. In the control treatment in November, 1993 there was 55.7 g of stolon DM m^{-2}, which consisted of 39.4 m of stolon m^{-2}. This stolon material weighed 1.5 g DM m^{-1} length of stolon.

In May 1994 the control treatment had significantly ($P < 0.05$) higher stolon weights m^{-2} and stolon lengths m^{-2} than the spring-applied slurry treatment. The spring-applied slurry treatment had significantly ($P < 0.05$) higher stolon weights m^{-2} and stolon lengths

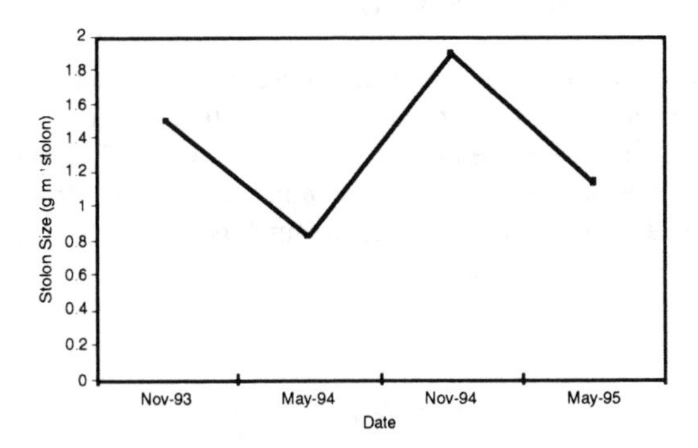

Figure 3 *Changes in stolon size (weight of stolon m^{-1} stolon) in the control treatment between November 1993 and May 1995*

m^{-2} than the summer-applied slurry and the FYM treatments. These two latter treatments were not significantly ($P < 0.05$) different from each other.

In November 1994 there were no significant differences in stolon weights m^{-2} or stolon lengths m^{-2} between the different treatments in this experiment.

In May, 1995 the control treatment had significantly ($P < 0.05$) higher stolon weights m^{-2} and stolon lengths m^{-2} than the summer-applied slurry and farmyard manure treatments, but was similar to the spring-applied slurry treatment. Furthermore, there was no significant ($P < 0.05$) difference in stolon weights m^{-2} and stolon lengths m^{-2} between these three latter treatments.

Stolon weights m^{-2} increased between May and November, 1994 and declined between November, 1994 and May, 1995 (Figure 2).

There was no significant difference in stolon size between treatments throughout this experiment. Stolon size fell between November, 1993 and May, 1994, increased between May, 1994 and November, 1994, and decreased between November, 1994 and May, 1995 (Figure 3).

Table 5 *Soil phosphorus, potassium, and magnesium concentrations (mg L^{-1}) and soil pH in August, 1994 (mean of six replicates)*

Manure Treatment	Phosphorus	Potassium	Magnesium	pH
Control	4.58	87.0	393.5	6.6
Summer Slurry	3.67	119.0	372.8	6.5
Spring Slurry	4.83	113.5	402.2	6.5
Spring FYM	4.33	118.3	380.2	6.5
Mean	4.35	109.5	387.2	6.5
F test	n.s.	n.s.	n.s.	n.s.
LSD (5%)	----	----	----	----
Variation coefficient	26.0%	23.2%	9.3%	2.0%

Soil Mineral Concentrations and Soil pH

Soil phosphorus, potassium, and magnesium concentrations, and soil pH measured on samples taken in August 1994 are presented in Table 5. There were no significant differences in either the soil mineral concentrations measured, or in soil pH between treatments in this experiment, although the potassium concentration in the control treatment tended to be lower than in the other treatments.

4 Discussion

Herbage Yields

In 1995 the application of cattle slurry and FYM in March increased herbage yields in May over those in the control treatment. A similar response to the application of spring-applied cattle slurry and FYM did not occur in May 1994. Furthermore herbage yields were lower in May 1995 compared to May 1994. These distinctions may have been due to differences in soil nitrogen availability and/or to differences in growing conditions between the two years.

It is likely that the increased availability of nitrogen associated with the spring-applied slurry and FYM treatments resulted in an increase in herbage yields in May 1995 similar to that recorded by Chapman and Heath (1988), Nesheim *et al.* (1990) and Boller *et al.* (1992). The increase in herbage yields was mainly attributable to increased grass yields. This accounts for the strong positive correlation between spring herbage DM yields and grass DM yields in both years. This response is somewhat similar to that in grass/white clover swards following fertilizer nitrogen applications (Castle *et al.*, 1965; Davies and Evans, 1990; Sfegaard, 1990).

The summer-applied cattle slurry treatment did not increase spring herbage yields compared to the control in May 1995 and therefore it is likely that nitrogen supplied by this treatment was probably lost by various processes such as volatilization, leaching, or denitrification during the winter (Sherwood, 1990; Smith and Chambers, 1993). The efficiency of utilization of nitrogen in cattle slurry appears to have been highest following the spring applications. Smith and Chambers (1993) and Vertès and Simon (1993) stated that nitrogen recovery from applied cattle slurry is greater in spring (because of greater uptake by grass) than later in the growing season.

Nitrogen uptake by grass is greater in the spring primarily because higher rates of grass growth occur during the spring. Therefore, the optimum time to apply cattle slurry in order to obtain maximum response from its nitrogen content is in early March once soil conditions are suitable for application. This will prevent loss of valuable nutrients and minimise pollution of the environment (Tunney, 1980; Carton *et al.*, 1993).

White Clover Stolon Development

In this experiment there was a build up in the amount of stolon present at the soil surface in all treatments throughout the growing season (Figure 2). Maximum amount of stolon generally occurs towards the end of the growing season in the months of September and October (Simon *et al.*, 1989). Stolon mortality and decomposition occurred throughout the winter months and, as in the present experiment, Woledge *et al.* (1988) and Marriott and

Smith (1992) found that only approximately one-third of the original stolon present in the autumn remained alive and healthy in the spring.

Collins and Rhodes (1995) stated that abundant stolon development is important to maintain a high white clover content of the herbage in the spring. This plays an important role in promoting the long-term survival of the white clover component of the sward. There was a clear relationship between the extent of stolon development and the white clover content of herbage throughout the experiment. In May 1995 the control treatment had the highest white clover stolon weight m^{-2} and clover content of herbage compared to the other treatments (Figure 2a). In contrast, the summer-applied slurry and the farmyard manure treatments had relatively low stolon weights m^{-2} and low clover contents (Figure 2b and 2d, respectively). In this respect, the spring-applied slurry had a clear advantage compared to the summer-applied slurry and farmyard manure treatments. The spring-applied slurry supported relatively high white clover stolon development and clover content of the sward in May 1995.

Much of the increase of stolon weight per unit area during the summer, and the subsequent decrease during the winter can be accounted for in terms of increase or decrease in length of the stolon. However, stolon thickness also tended to increase during the summer and to decrease during the winter (Figure 3). This appears to be associated with the seasonal dynamics of the white clover stolon (Hay *et al.*, 1987) and was not greatly influenced by the manure treatments in this experiment.

The White Clover Content of Herbage

The contribution of white clover to spring herbage yields in both years was relatively low ranging from 10.0% to 22.5% in May 1994 and from 3.9% to 22.5% in May 1995. This was due in part to the later commencement of growth of this species in the spring compared to grass (Murphy, 1985; Roberts *et al.*, 1989). Moreover, the white clover content of the herbage was far higher in the control treatment compared to the other treatments in the spring of both years. This indicates that the increased nitrogen availability associated with the applications of organic manure resulted in a depression of the white clover content of the sward in the spring of both years. Similar reductions have been reported by a number of authorities (Nesheim *et al.*, 1990; Davies and Evans, 1990; Harris 1994).

The nitrogen in cattle slurry consists of approximately 50% ammonium nitrogen and 50% organic nitrogen (Kiely, 1987; Sherwood, 1990). The organic nitrogen fraction is not immediately available for uptake by plants. The organic fraction must be broken down by soil micro-organisms (mineralised) before the nitrogen becomes available to the plant. The rate at which this organic fraction is mineralised is slow and variable (Tunney, 1980; Carton *et al.*, 1993). In contrast, the ammonium fraction of cattle slurry is often considered to be as available as fertilizer nitrogen to a crop (Paul and Beauchamp, 1993; Smith and Chambers, 1993), and only this fraction should be considered in terms of a direct crop response to the applied nitrogen in slurry (Carton *et al.*, 1993). The supplementary nitrogen supplied by the spring-applied slurry and FYM treatments increased spring herbage yields, particularly that of the grass component of the sward but lowered the white clover content of the sward. This reduction was probably mainly due to competition between the grass and clover components of the sward for light, soil moisture, various soil minerals, and other edaphic factors.

In addition to the animal excreta contained in cattle slurry, FYM also contains bedding material, usually straw, in varying quantities. The resultant manure may have undergone varying degrees of decomposition in the process of being converted into microbial material and humic substances. These influence the biological stability of nitrogen complexes and hence nitrogen availability (Allison, 1973). Farmyard manure generally contains less soluble nitrogen and tends to make nitrogen available to plants over a longer time period compared to cattle slurry (Jakobsen, 1995). The depression of the white clover content of the sward was greater in the FYM treatment than in the spring-applied slurry treatment. This is probably due to the differences in the manner in which nitrogen was made available to plants from these two manure treatments.

In the current experiment it appears that the spring-applied cattle slurry treatment tended to depress the white clover content of the herbage between May and September 1994, but consequently the white clover content of herbage increased markedly between September and October 1993 (Figure 2c). This favoured the profusion of stolon prior to the winter, the winter survival of the white clover and the clover content of herbage in May 1995. The FYM treatment produced higher total herbage yields than the other treatments during 1994, but the white clover content of herbage was decreased in this treatment (which tended to give relatively low clover contents throughout the growing season). The FYM treatment therefore decreased the accumulation of white clover stolon prior to the winter and consequently the clover content of herbage in May 1995.

The value of FYM is due as much to its organic matter content as to the fertilizer elements it contains (Allison, 1973). It enriches the soil with organic matter which improves soil properties, including water retention capacity, structure, drainage, aeration and workability (Dressel *et al.*, 1993; Jakobsen, 1995). However, in this experiment the FYM was detrimental to the persistency of white clover in the grass/white clover sward. The spring-applied slurry gave an increase in spring herbage yields similar to that of the FYM treatment, but the former treatment had the added advantage that it had a relatively high clover content of the sward in the spring.

The summer-applied slurry treatment produced a relatively low white clover content (3.9%) compared to the control treatment (22.5%) and the spring-applied slurry treatment (10.2%) in May 1995. The application of cattle slurry in August tended to depress the white clover content of herbage between September and October 1994 (Figure 2b). This depression had a negative effect on the white clover content of herbage during the autumn and on stolon survival during the winter. This accounts for the low white clover content of herbage in this treatment in May 1995. Therefore, the application of cattle slurry in August did not increase spring herbage yields, it was likely to have caused substantial nitrogen losses from the system, and also it had a damaging effect on the white clover persistency in the sward.

Soil Mineral Concentrations and Soil pH

The three organic manure applications maintained similar soil phosphorus, potassium, and magnesium concentrations and soil pH levels in August 1994. These levels were probably adequate to maintain herbage yields in May 1995 (Spedding and Diekmahns, 1972; Gately, 1993).

The soil phosphorus concentrations were 6 mg L^{-1} at the beginning of the experiment. It is likely that the ploughing in 1993 may have buried much of the phosphorus which was previously concentrated at the soil surface. This accounts for the decrease in soil phosphorus concentrations between March 1993 and August 1994. Soil phosphorus concentrations do not generally show a rapid decline under grassland (Tunney, 1990). The phosphorus in fresh FYM and cattle slurry is present mainly in insoluble organic forms and will not result in immediate uptake by crops following application. However, slow decomposition of manure releases soluble phosphorus which contributes to soil concentrations over a period of years (Dressel *et al.*, 1993).

The potassium in cattle slurry and in FYM is mostly in a readily available form, and is a valuable source of readily available potassium for plant growth (Kiely, 1987; Christie, 1987; Smith and Chambers, 1993). In this experiment, the soil potassium concentrations were approximately 87 mg L^{-1} in the control treatment, and approximately 115 mg L^{-1} in each of the manure treatments in August. This indicates that soil potassium concentrations tended to decline where no manure was applied, and that soil potassium concentrations were maintained where organic manures were applied.

5 Conclusions

The application of cattle slurry and FYM in the March increased spring herbage yields. This response was primarily in terms of increased grass production which accounted for approximately 90% of spring herbage yields. The application of cattle slurry in August did not increase spring herbage yields.

The three manure treatments resulted in a decrease in the extent of stolon development and the clover content of herbage compared to the control. However, the spring-applied slurry treatment favoured stolon survival during the winter, the clover content of herbage in the spring, and hence the persistency of the clover in the sward relative to the two other manure treatments.

The summer-applied slurry treatment was likely to have caused substantial nitrogen losses from the system by various processes, such as volatilization, leaching, or denitrification during the winter.

The three manure treatments maintained soil phosphorus, potassium, and magnesium concentrations in line with the requirements of optimum production from grass/white clover swards.

Acknowledgements

This study was supported by research contract No. 3C 92-0776 under the European Commission AIR Programme.

References

Allison, F.E. 1973. Soil organic matter and its role in crop production. *Developments in Soil Science* **3**:178-398. Elsevier, Amsterdam.

Boller, B., Nesheim, L., Lehmann , J. and Walther, U. 1992. Effect of slurry and mineral N fertilizer on nitrogen fixation by white clover. *Landwirtschaft Schweiz* **5**:149-151.

Carton, O.T., O'Bric, C., Cuddihy, A. and Magette, L.W. 1993. p. 1-11.Total and available nutrients in animal wastes. *Teagasc, Grassland Production Seminar*, Wexford,

Castle, M.E., Reid, D. and Heddle, R.G. 1965. The effect of varying the date of application of fertilizer nitrogen on the yield and seasonal productivity of grassland. *Journal of Agricultural Science, Cambridge* **64**:177-184.

Chapman, R. and Heath, S.B. 1988. The effect of cattle slurry on the clover content of grass/clover swards. *In* Papers presented at a Research Meeting at the Welsh Agricultural College, Aberystwyth, September 1988, Session IV, Paper 4.

Christie, P. 1987. Some long term effects of slurry on grassland. *Journal of Agricultural Science, Cambridge* **108**:529-541.

Collins, R.P. and Rhodes, I. 1995. Stolon characteristics related to winter survival in white clover. *Journal of Agricultural Science, Cambridge* **124**:11-16.

Davies, A. and Evans, E.M. 1990. Effects of spring defoliation and fertilizer nitrogen on the growth of white clover in ryegrass/clover swards. *Grass and Forage Science* **45**: 345-356.

Dressel, J., Weigelt, W. and Mockel, D. 1993. Changes in the nutrient potential of a soil according to long-term dressing of farmyard manure and mineral fertilizer. Agrobiological Research - *Zeitschrift Fur Agribiologie Agrikulturchemie Okologie* **46**: 321-330.

Frissel, M.J. (ed.) 1978. *Cycling of Mineral Nutrients in Agricultural Ecosystems.* Elselever Scientific Publishing Co. Developments in Agricultural and Managed-forest Ecology, **3**:356.

Gately, T.F. (ed.) 1993. *Soil analysis & fertilizer, lime, animal manure and trace element recommendations.* Teagasc, Johnstown Castle, Wexford.

Gilliland, T. 1988. Grasses and clovers: 1988/1989 guide to recommended varieties. *Agriculture in Northern Ireland,* May, 1988, **No. 8**:14-15.

Gomez, K.A. and Gomez, A.A. 1984. *Statistical Procedures for Agricultural Research.* An International Rice Research Book. A Wiley-Interscience publication.

Haggar, R.J. 1989. Agronomic limitations to production of forage legumes. In: P. Plancquaert and R. Haggar (eds), *Legumes in Farming Systems,* ECSCEEC EAEC, Brussels and Luxembourg. pp. 64-69.

Harris, S.L. 1994. Nitrogen and clover. p. 22-28. *Proc. 46th Ruakura Farmers Conference.*

Hay, M.J.M., Chapman, D.F., Hay, R.J., Pennel, C.G.L., Woods, P.W. and Fletcher, R.H. 1987. Seasonal variation in the vertical distribution of white clover stolons in grazed swards. *New Zealand Journal of Agricultural Research* **30**:1-8.

Hopkins, A. 1993. Rye and forage herbs as companions for improving nutrient uptake in clover/grass swards. *In* J. Frame, (ed.), *White Clover in Europe, State of the Art.* REUR Technical Series-FAO Regional Office for Europe (1993) **29**:113-115.

Jakobsen, S.T. 1995. Aerobic decomposition of organic wastes .2. Value of compost as a fertilizer. *Resources Conservation and Recycling* **13**: 57 - 71.

Jakobsen, S.T. 1993. Interaction between plant nutrients, part III, Antagonism between potassium, magnesium and calcium. *Plant and Soil Science* **43**:1-5.

Kiely, P.V. 1987. Slurry spreading, trying alternative methods. Teagasc, *Farm and Food Research* **18**:7-9.

Ledgard, S.F. and Steele, K.W. 1992. Biological nitrogen fixation in mixed legume/grass pastures. *Plant and Soil* **141**:137-153.

Marriott, C. and Smith, M. 1992. Senescence and decomposition of white clover stolons in grazed upland grass/clover swards. *Plant and Soil* **139**:219-227.

Masterson, C. and Murphy, P. 1983. Legumes and nitrogen fixation. *An Foras Talúntais, Grass Production Seminar*, Wexford. Paper **8**:116-148.

McDonald, P., Edwards, R.A. and Greenhalgh, J.F.D. 1988. *Animal Nutrition*. Longman, Essex.

Murphy, P.M., Turner, S. and Murphy, M. 1986. Effect of spring applied urea and calcium ammonium nitrate on white clover (*Trifolium repens*) performance in a grazed ryegrass-clover pasture. *Irish Journal of Agricultural Research* **25**:251-259.

Murphy, P.M. 1985. Legumes and nitrogen fixation. p. 97-115. *An Foras Talúntais, Grassland Production Seminar*, Wexford,

Murphy, W.E. 1984. Nutrient balances in farming systems - Intensive grassland in Northwestern Europe. p. 63-70. *18th Colloquium of the International Potash Institute, Bern*.

Nesheim, L., Boller, B.C., Lehmann, J. and Walter, U. 1990. The effect of nitrogen in cattle slurry and mineral fertilizer on nitrogen fixation by white clover. *Grass and Forage Science* **45**:91-97.

Paul, J.W. and Beauchamp, E.G. 1993. Nitrogen availability for corn in soils amended with urea, cattle slurry, and solid and composted manures. *Canadian J. Soil Sci.* **73**:253-266.

Rhodes, I., Collins, R.P. and Evans, D.R. 1994. Breeding white clover for tolerance to low temperature and grazing stress. *Euphytica* **77**:239-242.

Roberts, D.J., Frame, J. and Leaver, J.D. 1989. A comparison of a grass/ white clover sward with a grass sward plus fertilizer nitrogen under a three-cut silage regime. *Research and Development in Agriculture* **6**:147-150.

Sherwood, M. 1990. Management of wastes to prevent leaching of nitrates. p. 87-101. *Seminar on the Environmental Impact of Landspreading of Wastes*, Teagasc, Johnstown Castle, Wexford.

Simon, J.C., Gastal, F. and Lemaire, G. 1989. White clover morphologenesis and light competition: leaves and growing points emergence. *Agronomie* **9**:383-389.

Smith, K.A. and Chambers, B.J. 1993. Utilizing the nitrogen content of organic manures on farms - problems and suitable solutions. *Soil Use and Management* **9**:105-112.

Sfegaard, K. 1990. Strategic nitrogen application to white clover/grass sward. *Tydsskrift for Planteavl* **94**:457-464.

Spedding, C.R.W. and Diekmahns, E.C. (eds) 1972. Bulletin 49. *Grasses and legumes in British Agriculture*. Commonwealth Bureau of Pastures and Field Crops.

Tunney, H. 1990. A note on a balance sheet approach to estimating the phosphorus fertilizer needs of agriculture. *Irish Journal of Agricultural Research* **29**:149-154.

Tunney, H. 1980. Fertilizer value of animal manures. Teagasc, *Farm and Food Research* **11**:78-79.

Vertès, F. and Simon, J.C. 1993. Nitrate leaching measurements in grazed grasslands. *In* J. Frame (ed.), *White Clover in Europe, State of the Art.* REUR Technical Series-FAO Regional Office for Europe (1993) **29**:116-118.

Woledge, J., Tewson, V., and Davidson, I.A. 1988. The physiology of seasonal changes in the growth of grass/clover mixtures. p. 2. *In* Papers presented at a Research Meeting at the Welsh Agricultural College, Aberystwyth, September 1988, Session III, Paper 5.

Cheese Whey Applications to Two Contrasting Soil Types: Effects on Grass Growth, Soil Properties, and Drainage Water Composition

M.V. Ross and G. J Mullen

DEPARTMENT OF LIFE SCIENCES, UNIVERSITY OF LIMERICK, NATIONAL TECHNOLOGICAL PARK, LIMERICK, IRELAND

Abstract

Cheese whey, when used as a fertilizer and applied in moderate amounts has been shown to contain sufficient quantities of plant available nutrients to replace conventional fertilizer effectively in pasture and tillage crop production. However, there are no detailed recommendations in either Ireland or Britain on whey usage practices designed to prevent water pollution. The objective of this study was to assess the value of whey as a fertilizer alternative on grassland and to determine its potential for environmental pollution. Cheese whey was applied out-of-doors to grass on 160 soil columns at rates equivalent to 0, 40, 80 and 160 $m^3 ha^{-1} y^{-1}$ Two soils of contrasting agricultural significance were compared, and there were 20 replications of each soil/whey combination. Five of the twenty replications (i.e. 40 soil columns) were dismantled at 6 monthly intervals and a range of physical and chemical soil parameters were measured at six depths. Drainage water was collected once per month from columns and analysed for nitrite nitrogen, nitrate nitrogen, ammonium nitrogen, and orthophosphate. Grass was cut on all columns on three occasions per growing season, and the herbage dry matter production was determined.

Data from drainage water analyses give cause for concern, and suggest that repeated applications of whey at high rates may lead to groundwater contamination. The application of whey supported low but acceptable levels of herbage production.

1 Introduction

The current level of cheese production in Ireland, 105 000 tonnes per annum, represents an increase of 58 per cent on the quantity produced in 1985 (Anon., 1995). Much of this increase has been the result of a dramatic development in farmhouse cheese production, as farmers diversify from the traditional dairy products of milk and butter. Large cheese producers utilise whey to produce saleable products, such as whey powder, lactose, lactic acid, and alcohol, and these are used for human and animal nutrition and for industrial purposes. Small production units, on the other hand, cannot afford the machinery needed to manufacture such products and must dispose of whey in some other manner. Robinson

(1986) has pointed out that, for each tonne of cheese produced, an outlet must be found for 8 tonnes of whey. Due to the large amounts of whey produced, and because of its potential as a pollutant if discharged into natural waterways (Kosikowski. 1982), disposal can be a problem in cases where it cannot be fed locally to farm livestock. Transport costs are considerable, even over short distances (Modler, 1987), and Ghaly and Singh (1985) have stated that whey could only be fed to pigs economically when transport distances are less than 30 - 40 km. In this context, land spreading of whey becomes an obvious alternative means of disposal.

Berry (1923), in Scotland, was the first researcher to report on the merits and demerits of spreading cheese whey on land. Since then studies by Sharratt *et al.* (1959), Peterson *et al.* (1979), Morris *et al.* (1985), Radford *et al.* (1986), Robbins and Lehrsch (1992), Radford (1992), and Ross and Mullen (1993), amongst others, have researched various aspects of this practice. Cheese whey has been classified as an "allowable supplementary fertiliser" under EU Regulation 2092/91. In view of recent developments in the EU leading to the encouragement of extensification in agriculture and the development of organic farming (Lampkin 1990), there appears to be a need for further detailed research into the spreading of whey on land.

Fertilizer Value of Whey

When used as a fertilizer, and applied in moderate amounts, whey has been shown to contain sufficient quantities of plant available nutrients to replace effectively conventional fertilizer in pasture (Morris *et al.*, 1985; Radford *et al.*,1986; Ross and Mullen, 1993) and in tillage crop production (Cain, 1956; Peterson *et al.*, 1979; Radford *et al.*, 1987; Jones *et al.*, 1993a). Most of the nitrogen in whey is initially in the organic form, as protein. The rate of release of plant-available nitrogen in the first year is estimated to be in the range of 30-60 per cent. The remaining nitrogen, although still bound as protein, is slowly released to give a continuing benefit (Sharratt *et al.*, 1959; Kelling and Peterson 1981, Radford *et al.*, 1986; Ross and Mullen 1993). Peterson *et al.* (1979) reported that whey applications on maize (*Zea mays*) in Wisconsin gave growth responses for 4 years after application on a silt loam soil. Phosphorus is mostly present in the inorganic form. Lactic and sulphuric casein wheys applied at 40 000 L ha^{-1} provide approximately 25 - 26 kg P ha^{-1}, whereas cheese whey provides 16 kg P ha^{-1} (Radford *et al.*, 1986). Peterson *et al.* (1979) found that annual applications of 8.4 cm of sweet whey over an 18-year period to a Michigan soil increased the surface soil phosphorus to abundant levels and did not inhibit the production of hay crops.

Potassium, also, is mostly present in whey in the inorganic form, and 59-60 kg K ha^{-1} are provided when whey is applied at 40 000 L ha^{-1} (Radford *et al.*, 1986). Peterson *et al.* (1979) showed that the build-up of potassium in soil was directly proportional to the amount of whey applied. They found that potassium remained in the upper horizon in soil, and their data indicated that very little downward movement took place, except when excessive applications of whey were made. Cheese whey applied at 45 000 L ha^{-1} supplied 16 kg ha^{-1} sulphur, 27 kg ha^{-1} calcium, 20 kg ha^{-1} sodium, and 4 kg ha^{-1} magnesium (Morris *et al.*, 1985). Peterson *et al.* (1979) analysed whey for sulphur, zinc, copper, boron, manganese, and chromium, and reported low concentrations of these elements, which is typical for whey. They found that these minor elements had no significant impact on on silt loam soils in Wisconsin but, with the exception of zinc and manganese, were all present in corn leaf tissue.

Whey and Soil pH

The effect of whey on soil pH depends on the pH of the whey applied and on the initial pH of the soil. The pH of whey ranges from 4.0 to 6.6 (Kosikowski 1982). Sharratt *et al.* (1959) studied the effects of adding a composite of three cheese wheys with a pH of 4.0 to two silt loam soils having pH values of 6.8 and 5.2. The decrease in soil pH, and the length of time that the decrease persisted, appeared to depend on the quantity of whey added. There was a gradual increase in soil pH in the neutral soil, and the pH value eventually exceeded that of the control. However, the authors concluded that in the strongly acid soil the addition of whey might temporarily lower the pH to a point injurious to the plant metabolism. In a survey undertaken by the New Zealand Dairy Research Institute on 22 farms, Morris *et al.* (1985) showed that whey gave a definite increase in soil nutrient and in pH levels. The trial showed that whey has a definite 'liming' effect as the soil pH increased by 0.15 and 0.13 pH units from applications of lactic acid and sulphuric acid casein wheys, respectively. This was thought to result not only from the calcium present but also from the increased microbial action in the soil following whey applications. Robbins and Lehrsch (1992) and Jones *et al.* (1993a) showed that acid wheys with low pH values decreased soil solution pH and increased calcium solubility. The lowered pH caused most micronutrient cations to become available to plants grown on reclaimed sodic soils.

Whey and the Reclamation of Sodic Soils

Lehrsch *et al.* (1994) have stated that the most difficult impediment to the reclamation of sodic soils is slow infiltration into the dispersed, often puddled surface of the soil. Acidic cottage cheese whey is effective as a chemical amendment for sodic soils (Robbins and Lehrsch, 1992; Jones *et al.*, 1993a, 1993b), as it lowers the soil pH, the sodium adsorption ratio, and the exchangeable sodium percentage to a depth of 30 cm while increasing aggregate stability to a depth of 15 cm. These changes speed up the leaching of exchangeable sodium from sodic soil profiles when sufficient water is passed through the soil. The lowered soil pH makes most micronutrient cations more available for plant growth on the reclaimed site (Lehrsch *et al.*, 1994).

Robbins and Lehrsch (1992) have pointed out that cottage cheese whey would not be expected to help reclaim saline (low sodium, high soluble salts) soils, but would be likely to increase the salinity problems.

Effects of Whey on Soil Physical Properties

Addition of acid whey to sodic soils has been shown to increase the infiltration rate, to decrease clay dispersion (Jones *et al.*, 1993a), and to increase aggregate stability (Lehrsch *et al.*, 1994). In non-sodic soils, the percentage of soil aggregates greater than 0.25 mm diameter was increased in the spring by all of a number of whey application rates on a silt loam soil cropped to maize (*Zea mays*) or alfalfa (lucerne, *Medicago sativa*) in Wisconsin (Sharratt *et al.*, 1959, 1962). Kelling and Peterson (1981), also in Wisconsin, found that applications of 50 to 600 mm of whey all led to improved soil aggregation. In contrast, Ross and Mullen (1993) failed to observe any change in the per cent aggregate stability 19 months after moderate applications of sweet whey were made to permanent grassland on a sandy loam soil in Co. Cork, Ireland. Watson *et al.* (1977) recorded large increases in infiltration rates into a fallow silt loam soil three months after application of cheese whey.

Effects of Whey on Crop Yields

In a greenhouse experiment, using an infertile sandy soil, Berry (1923) found in Scotland that whey had a beneficial effect on the growth of oats. When the same author applied whey for several years on pasture at the rate of 2.25 tons per acre per day from June until September, he observed a greater growth of the coarser grasses. Berry concluded that whey was beneficial to the soil but that transport costs would probably be too high to make its use profitable in most cases. In a greenhouse study in Wisconsin, using a silt loam soil, Cain (1956) found that oats which received a single 100 tons per acre application of whey at the time of planting were initially stunted, but recovered and gave a higher yield than those which did not receive whey. The same author applied whey on grassland in a field experiment and concluded that an increased yield could be seen for each addition of whey. Sharratt *et al.* (1962) applied whey at five rates, from 0 to 12 acre-inches in April/May, on field plots in Wisconsin before planting maize. They reported increased yields in the first and second year after application of whey, although they noted that the full benefits of the higher application rates were not observed until the second year. This was attributed to a combination of high salt levels, and possibly to reducing conditions in the soil in the first growing season which inhibited crop growth. These conditions had disappeared by the second growing season. Their findings indicated that the optimal yearly application rate was of the order of 4 inches (100 mm).

Peterson *et al.* (1979) came up with broadly similar results, clearly showing that excessive applications caused decreased yields of maize, but they identified 200 mm as the optimal yearly application rate. These rates are well in excess of the maximum recommended yearly application rate of 25 - 38 mm suggested by Kelling and Peterson (1981) in order to safeguard against groundwater contamination by the nitrogen not utilized by the crop. Morris *et al.* (1985) and Radford *et al.* (1986) have indicated that the typical recommended amounts of whey applied to New Zealand dairy pastures are of the order of 40 - 45 000 L (4.0 - 4.5 mm) ha^{-1} y^{-1}. Radford *et al.* (1987) state that these amounts supply quantities of nutrients comparable with those supplied by typical application rates of conventional fertilizers in New Zealand. It is obvious then, as pointed out by Radford (1992), that the rates of whey applied in the work of Sharrat *et al.* (1962), Peterson *et al.* (1979), and Kelling and Peterson (1981) were much greater than were necessary for the adequate supply of plant nutrients. In a glasshouse study, Radford *et al.* (1987) reported that grass dry matter yield increased linearly for whey application at rates of 0, 25 000, 50 000, and 70 000 L ha^{-1}. They also reported positive responses to whey application for lettuce, maize grain, kiwi fruit and rhododendron. In a field experiment in New Zealand, Radford (1992) applied whey and fertilizer (potassic superphosphate with urea) at four paired rates (which supplied equivalent amounts of plant nutrients) to old permanent pasture on a soil of low fertility. The whey application rates were 0, 40 000, 80 000 and 160 000 L (0, 4, 8 and 16 mm) ha^{-1} y^{-1}. When compared with yields on control plots, each rate of application gave significantly ($P < 0.012$) increased yields of pasture, but there were no significant differences between comparable rates of application. Ross and Mullen (1993), in a similar experiment on permanent pasture on a sandy loam soil in Ireland, compared equivalent applications of sweet whey and fertilizer (0:10:20 compound with calcium ammonium nitrate). Where plots were cut once in mid season, the total herbage yield was greater on the plots which received the fertilizer spread. A similar result was observed when the plots were cut three times. When cut six times, however, no differences were observed between whey and fertilizer treatments.

Environmental Impact of Whey Use

Whey has a biological oxygen demand (BOD) value in the range 4000 - 4800 mg L^{-1} (Scott 1986), and so it has obvious potential to cause pollution. However, very little of the research carried out on the spreading of whey on land has placed emphasis on water quality and pollution. Much of the information presented is based on inference. The adverse effects of applications to crops of very high levels of whey have been attributed to the possibility of toxic conditions caused by increased biological activity combined with salt accumulation (Sharratt *et al.*, 1962), "excess salts and/or organic overloading" (Jones *et al.*, 1993b), and were not commented on by Peterson *et al.* (1979). Kelling and Peterson (1981) have pointed out that groundwater contamination by non-utilized nitrogen can take place at excessive application levels of whey, and they suggested an upper limit of application of 25 to 38 mm y^{-1} to avoid this problem. McAuliffe *et al.* (1982) reported that a single application of 35 mm of simulated whey led to a 50% decrease in saturated hydraulic conductivity (K) within 2 days. Repetitive applications in some soils caused decreases of more than 99 per cent. Biologically induced recovery occurred to varying extents within 1 to 3 weeks. The authors recommended the avoiding of overloading to prevent ponding, and allowing intervals between applications to allow K recovery.

Modler (1987) recommended that no more than 10 to 20 mm should be applied at a time in order to avoid run off. Run off is dependent on soil type and on rainfall. He also listed areas with flooding, excessive drainage, high water tables, and shallow depth to rock as high risk areas for pollution. Modler also drew attention to odour pollution from stored whey, and recommended the spreading of the whey in smaller amounts, its injecting into the soil, or tilling after irrigation. He pointed out that odours from storage reservoirs can be detected at up to 2 km from source, and those from spreading operations may be detected at 4 km. Peterson *et al.* (1979) reported that whey when diluted 1:8 in irrigation waters did not produce observable effects on the underlying groundwaters. They did not provide information on rates of application of whey, on soil type, or on depth to groundwater table. In a laboratory study, Ghaly and Singh (1985) applied 32 mm of cheddar cheese whey to soil columns and leached these repeatedly with 98.5 mm of water on days 0, 8, 16, 24, and 32. Analysis of water samples collected every 4 days showed that the organic nitrogen, ammonium nitrogen, and nitrite nitrogen concentrations in the leachate were very low. The concentration of nitrate nitrogen was high (4.9-7.5 mg L^{-1}) on day 4, and it gradually declined thereafter. It was concluded that continuous application of cheese whey at rates higher than used might result in groundwater contamination.

In Ireland, the Local Government Water Pollution Act (1977) and the Water Pollution Amendment Act (1990) state that "a person shall not cause or permit any polluting matter to enter waters". Legislation outlines the penalties for water pollution, but does not recommend application rates. The British Code of Practice (Hobson and Robertson, 1977; Ball and Bell 1994) provides some useful guidelines on the disposal of organic wastes. In so far as the authors are aware, no detailed recommendations are available in Britain or Ireland with regard to whey disposal methods that would avoid soil and water pollution.

The objectives of the present work were to evaluate:

1, the merit of cheese whey as an organic amendment on grassland; and
2, the effects, in the context of environmental pollution, of repeated applications of

whey on the physical and chemical properties of soils, and on the composition of drainage waters.

2 Materials and Methods

Location and Soils

The experiment was set up in 1993 in the Life Sciences Department of the University of Limerick. Two contrasting soils, an Elton gravelly loam and a Howardstown clay loam were selected. These soils are extensively found in Munster and South Leinster (Gardiner and Radford 1980), and are of contrasting agricultural significance. The Elton series is a well drained, eluviated brown earth of wide use range and is associated with arable cropping and grazing. The Howardstown series is a poorly drained gley of limited use range, and is associated with pasture (Finch and Ryan 1966). Soils were collected from a former Economic Test Farm of An Foras Taluntais (The Irish Agricultural Research Institute). With the aid of a detailed soil map of the farm (Finch 1964), three sampling locations, which were considered to be fully representative of the soil type, were chosen for each soil. Using the available information (Finch 1964) as a guide, the soils were selected and carefully removed, using spades and shovels, and placed, horizon by horizon, in heavy duty plastic bags. These were sealed to prevent moisture loss during transport to the laboratory. Before leaving each collection site, undisturbed soil samples were collected from each horizon for determination of bulk density. These were placed in plastic bags and sealed to prevent moisture loss during transport to the laboratory.

Preparation of Experimental Units (Soil Columns)

The objective was to construct 75 cm long columns of soil having properties closely resembling those in the field, but capable of being dismantled and sub-divided into layers after a period of time. Split-pipe/outer jacket units were constructed in accordance with Mulchrone *et al.* (1980). The split-pipes were carefully packed with 6 mm sieved soil to achieve the bulk densities of the undisturbed soil transported from the field. Split-pipes were packed horizon by horizon; each horizon was packed in three equal sized sub-units to achieve uniform bulk density throughout. Columns were insulated with 4 cm of polystyrene (commercial pipe lagging) covered with black plastic to make the insulation waterproof. Insulated columns were then installed in a vertical position (using wooden stakes and cable ties) sitting in 16 cm plastic saucers designed to collect leachate. A square plastic sheet was fitted tightly to the base of each column to prevent rainwater entering the saucer and to minimise evaporation in hot weather. Soil columns were carefully monitored over a 3-month period to ensure that all possessed a satisfactory degree of permeability and were then sown with perennial ryegrass (*Lollium perenne*, variety Magella). When the grass was successfully established, each column was thinned to 10 plants.

Design of the Experiment

The design was a fully randomized 2 x 4 x 20 factorial, using two soil variables, four whey application variables, and twenty replications. Whey applications were 0, 40 000, 80 000, and 160 000 L ha^{-1} y^{-1}, applied at the various times shown in Table 1.

Table 1 *Cheese whey applications*

Month	Whey application (L ha^{-1})			
	W1	W2	W3	W4
March		13 330	13 330	17 770
April	No			17 770
May	whey		13 330	17 770
June	applied	13 330	13 330	17 770
July	(Control)			17 770
August			13 330	17 770
September		13 330	13 330	17 770
October				17 770
November			13 330	17 770

The application rate W2 represented moderate use, as recommended in New Zealand (Morris *et al.*, 1985; Radford *et al.*, 1986). W3 represented a more intensive application; it exceeds the two-monthly rate recommended by the Code of Practice for spreading animal excreta in the UK (Hobson and Robertson 1977), but it does not exceed the recommended yearly application rate. W4 represented the very high application rate and was chosen to approximate to dumping. It exceeds all recommended applications (Hobson and Robertson 1977; Anon., 1987; Anon., 1989) for animal excreta. Distilled water was applied to all soil columns on all application dates in amounts designed to ensure that all columns received the same volume of liquid on each date and, therefore, per annum as a consequence of whey application. The whey used was collected from a farmhouse cheddar producer in Glenroe, Kilmallock, Co Limerick. It was placed in plastic bags and stored in a freezer at -10 °C. The cheese whey was removed from the freezer prior to being applied to the soil and it was allowed to thaw at room temperature for 24 h.

Experimental Procedure

The experiment was located out-of-doors. Whey was applied from November 1993 to September 1995. Leachate samples were collected once per month over this period and either analysed immediately, or frozen and stored for subsequent analyses. Five of the twenty soil column replications for each soil (a total of 40 soil columns) were dismantled at 6 monthly intervals and the soil was carefully subdivided into its constituent horizons. A range of parameters was measured on all horizons on each occasion.

Measurements recorded. Saucers were emptied of leachate once per month and the volume recorded. The leachate was mixed and 100 mL aliquots were removed for analyses. Nitrate nitrogen and nitrite nitrogen were measured using the methods described by Bremner (1965).

Ammonium nitrogen and orthosphosphate phosphorus were determined by the methods described by APHA (1985). During each dismantling of soil columns, undisturbed samples were carefully excavated from each horizon for determination of the bulk density. This was determined by a modification of the clod method (Blake, 1965). Paraffin wax was used to coat the clods. Values were corrected for mass and for volume of stones and gravel greater than 2 mm in size. At the same time, approximately 100 g of soil was collected from each horizon and weighed immediately for determination of moisture content (by gravimetry and oven drying; Gardner 1965). The rest of the soil from each horizon was sieved (2 mm sieve), air dried, and used for the remaining soil determinations.

Aggregate stability was determined using a modification of the wet sieving method, as described by Kemper (1965). Air dry aggregates in the 1-2 mm size range were used. Aggregates were soaked for 30 minutes and wet sieved on a mechanical sieving machine for 3 minutes. A correction was made for sand particles larger than 0.25 mm diameter.

pH. A 2:1 water:soil mixture was was stirred twice over a period of 30 min, then allowed to stand for one h before measuring the pH of the suspension (Byrne, 1979).

Electrical conductivity was determined using a conductivity meter (Bower and Wilcox, 1965). A 1:5 soil:water mixture was shaken mechanically for one h. One drop of sodium hexametaphosphate solution (0.01%) was added to prevent precipitation of calcium carbonate, and samples were allowed to stand for one hour.

Water extractable nitrate nitrogen and water extractable nitrite nitrogen were determined by the method of Bremner (1965) after extraction with distilled water.

Available phosphorus, available potassium, available calcium and available magnesium were determined after soil extraction with Morgan's solution (Byrne 1979).

***Organic carbon** (%)* was determined by the Walkley and Black method (Allison, 1965). Excess dichromate was titrated with ferrous ammonium sulphate, using phosphoric acid and diphenylamine indicator solution. A recovery factor of 75 per cent was used.

Herbage dry matter production was determined on all soil columns three times per year. The grass was cut uniformly to a height of 1 cm, dried in a Gallenkamp fan oven at 105 °C to constant weight.

Statistical analysis of results. Analysis of variance was carried out on an IBM compatible personal computer using the SOLO package (BMDP Statistical Software Inc. 1989). Reference was made to Steel and Torrie (1960), Wardlaw (1985) and Hintze (1989) in the interpretation of results.

3 Results

Leachate

Whey application affected ($P < 0.05$) the concentration in leachate of all nitrogen measurements on all dates from November 13, 1993 to June 7, 1994, inclusive, and on most dates from June 6, 1995 to October 8, 1995. On these dates the concentrations, in general, showed the following trend: W4 > W3 > W2 > W1. This trend was observed for both soils. Temporal fluctuations in concentration appeared to be related to weather fluctuations and to the moisture contents in the soil columns. Orthophosphate concentration was affected ($P < 0.05$) by whey application on eight occasions only. Again, no differences were observed that were attributable to soils. The highest concentrations (mg L^{-1}) recorded for the parameters were: nitrite nitrogen 6.8; nitrate nitrogen 5.3; ammonium nitrogen 16.4; orthophosphate 2.5. Table 2 gives the dilution of the concentration of measured ions, averaged over all soil and whey application treatments.

Soil Columns

The effects of soil, whey application, and depth (layer/horizon) on the measured soil

Table 2 *Dilution of measured parameters with time in drainage water*

Concentration in whey (mg L^{-1})	Sampling date	Concentration in leachate (mg L^{-1})			
		Nitrite nitrogen	Nitrate nitrogen	Ammonium nitrogen	Orthophosphate
Nitrite N 3.4	13.11.93*	0.11	2.79	0.33	1.09
Nitrate N 8.8	10.12.93*	0.08	1.20	0.29	0.26
Ammonium 41.0	18.12.93	0.06	1.67	0.25	0.59
Orthophosphate 287.0	05.01.94	0.03	0.86	0.23	0.20
	11.02.94	0.16	0.40	0.15	0.25
	08.03.94*	0.05	0.19	0.19	0.17
	05.04.94*	0.15	0.71	0.68	0.26
	06.05.94*	0.15	0.43	0.54	0.30
	07.06.94*	0.09	0.36	0.64	0.31
	13.07.94*	0.00	0.04	0.09	0.21
	05.08.94*	0.02	0.00	0.26	0.30
	01.09.94*	0.08	0.15	0.31	0.19
	09.10.94*	0.00	2.02	0.14	0.37
	08.11.94*	0.00	0.23	1.40	0.00
	08.12.94	0.00	0.12	1.13	0.04
	06.01.95	0.00	0.15	1.31	0.06
	04.02.95	1.70	0.13	2.02	0.01
	12.03.95*	1.62	0.12	2.45	0.01
	12.04.95*	0.04	0.02	0.00	0.00
	06.05.95*	0.00	0.08	0.00	0.00
	06.06.95*	0.16	0.31	0.09	0.00
	18.07.95*	0.07	0.96	0.31	0.00
	10.08.95*	0.00	0.00	0.00	0.00
	08.09.95*	1.86	0.13	6.55	0.02
	08.10.95	3.81	0.41	9.83	0.00

* = dates when applications of whey *were made*

parameters are summarised in Table 3. The effects of soil are to be expected, in view of the initial differences between the soils used. The same reasoning applies to the effects of depth observed. The effects of whey application and the interaction effects have not yet been fully studied and are, therefore, not commented on here.

Herbage Dry Matter Production

Whey application did not have any effect on herbage dry matter production in the first

Table 3 *Significance of the effects of soil, whey application (Appl), and depth (layer) on the physical and chemical soil parameters measured after 6, 12, 18 and 24 months (*** = very highly significant ($P < 0.001$); ** = highly significant ($P < 0.01$); *: = significant ($P < 0.05$); NS = not significant*

Parameters Measured	6 Months			12 Months			18 Months			24 Months		
	Soil	Whey Appl	Depth	Soil	Whey Appl	Depth	Soil	Whey Appl	Depth	Soil	Whey Appl	Depth
pH	NS	NS	NS	***	NS	***	***	NS	***	***	NS	***
Electrical conductivity	NS	NS	NS	***	NS	***	*	NS	NS	**	NS	***
Magnesium	***	***	***	**	NS	***	NS	***	NS	NS	NS	NS
Calcium	***	***	***	**	NS	NS	NS	***	***	***	***	***
Potassium	**	NS	***	***	***	***	*	***	***	***	***	***
Phosphate	***	NS	***	***	NS	***	NS	***	***	*	NS	NS
Water extractable nitrite N	***	NS	***	NS	NS	NS	NS	NS	NS	NS	**	***
Water extractable nitirate N	**	NS	NS	*	**	**	NS	***	NS	NS	***	NS
Aggregate stability	NS	NS	***	NS	NS	***	NS	NS	***	NS	*	***
Bulk density	***	***	***	***	NS	***	***	NS	***	***	NS	***
Organic carbon	*	NS	***	*	NS	***	*	NS	***	*	NS	***
Moisture content	***	NS	***	NS	NS	***	***	NS	***	***	NS	***

year of the study. However, it increased ($P < 0.01$) yield in all cuts in the second year. Dry matter production levels were low (mean: 5.5 tonnes ha^{-1}).

4 Discussion and Conclusions

It is clear from the data presented here that whey application gave rise to levels of nitrogenous and phosphorus compounds in drainage water which can cause some concern. Much of the impact of these levels on water quality depends on their subsequent dilution or concentration in field situations (Mason 1996; Sherwood 1993). It is likely, however, that continuous application of whey at high rates to grassland may lead to ground water contamination, at least at a local level, under Irish climatic conditions. The increased herbage dry matter yield observed in the second year suggests that organic nitrogen in the whey was mineralized slowly, as suggested by Sharratt *et al.* (1959), Kelling and Peterson (1981), Radford *et al.* (1986) and Mullen and Ross (1993). The fact that no yield increase attributable to whey application was observed in year one may be due to the masking effect of residual plant nutrients from artificial fertilisers.

Acknowledgements

The authors wish to thank Wavin (Ireland) Ltd. and Southern Chemicals Ltd. for financial support. The assistance of Mr Denis Collins, Herberstown, Co Limerick who provided the soils is greatly appreciated, as is the advice on analytical methods given by the laboratory technical staff of the Department of Life Sciences, University of Limerick.

References

Allison, L.E. 1965. Organic carbon. p. 1367-1378. *In* C.A. Black (ed.), *Methods of Soil Analysis. Part 2. Chemical and microbiological properties* American Society of Agronomy, Madison.

Anon. 1987. *Farmyard wastes and pollution.* An Foras Faluntais, Dublin

Anon. 1989. *Intensive farming and the impact on the environment and the rural economy of restrictions on the use of chemical and animal fertiliser.* Commission of the European Communities, Luxembourg.

Anon. 1995. *Annual Report;* Bord Bainne, Dublin

APHA. 1985. *Standard Methods for the Examination of Water and Wastewater.* 16th Edition. American Public Health Association, New York.

Ball, S. and B. Bell. 1994. *Environmental Law.* Blackstone Press, Ltd., London

Berry, R.A. 1923. The production, composition and utilisation of whey. *J. Agric. Sci.* **13**:192 - 239.

Blake, G.R. 1965. Bulk density. p. 374-390. *In* C.A. Black (ed.), *Methods of Soil Analysis. Part 1. Physical properties* American Society of Agronomy, Madison.

Bower, C.A. and L.V. Wilcox. 1965. Soluble salts. p. 933-951. In C.A. Black (ed.), *Methods of Soil Analysis. Part 2. Chemical and microbiological properties.* American Society of Agronomy, Madison.

Bremner, J.M. 1965. Inorganic forms of nitrogen. p. 1179-1237. *In* C.A. Black (ed.), *Methods of Soil Analysis. Part 2. Chemical and microbiological properties*. American Society of Agronomy, Madison.

Byrne, E. 1979. *Chemical analysis of agricultural materials.* An Foras Taluntais, Dublin.

Cain, J.M. 1956. Utilisation of whey as a source of plant nutrients. M.S. Thesis. Library, University of Wisconsin, Madison.

Finch, T.F. 1964. *Economic test farm Herbertstown, Co Limerick. Soil Survey Bulletin No 7.* An Foras Taluntais, Dublin

Finch, T.F. and P. Ryan. 1966. *Soils of Co Limerick, Soil Survey Bulletin No 16.* An Foras Taluntais, Dublin

Gardiner, M.J. and T. Radford. 1980. *Soil associations of Ireland and their land use potential.* An Foras Taluntais, Dublin

Gardner, W.H. 1965. Water content. p. 82-127. *In* C.A. Black, (ed.), *Methods of Soil Analysis. Part 1. Physical properties.* American Society of Agronomy, Madison.

Ghaly, A.E. and R.K. Singh. 1985. Land application of cheese whey. p. 546-553. *Proceedings of the 5th international symposium on agricultural waste utilisation and management* American Society of Agricultural Engineers, Chicago.

Hintze, J.L. 1989. *SOLO statistical system - version 3.* BMDP Statistical Hardware Inc., LosAngeles.

Hobson, P.N. and A.M. Robertson. 1977. *Waste Treatment in Agriculture.* Elsevier Applied Science Publications Ltd., London.

Jones, S.B., C.W. Robbins, and C.L. Hansen. 1993a. Sodic soil reclamation using cottage cheese(acid) whey. *Arid Soil Research and Rehabilitation* 7:51-61.

Jones, S.B. C.L. Hansen, and C.W. Robbins. 1993b. Chemical oxygen demand fate from cottage cheese (acid) whey applied to a sodic soil. *Arid Soil Research and Rehabilitation* 7:71-78.

Kelling, K.A. and A.E. Peterson. 1981. *Using whey on agricultural land - a disposal alternative.* Serial No. A3098, Cooperative Extension Programme, University of Wisconsin, Madison.

Kemper, W.D. 1965. Aggregate stability. p. 511-519. *In* C.A. Black (ed.), *Methods of Soil Analysis. Part 1. Physical properties* American Society of Agronomy, Madison.

Kosikowski, F.J. 1982. *Cheese and Fermented Milk Foods.* F.W. Kosikowski and Associates, Brooktondale, New York.

Lampkin, N. 1990. *Organic Farming.* Farming Press, Ipswich.

Lehrsch, G.A., C.W. Robbins, and C.L. Hansen. 1994. Cottage cheese (acid) whey effects on sodic soil aggregate stability. *Arid Soil Research and Rehabilitation* **8**:19-31.

Mason, C.F. 1996. *Biology of Freshwater Pollution.* 3rd Edition. Longman, London.

McAuliffe, K.W., D.R. Scotter, A.N. MacGregor, and K.D. Earl. 1982. Casein whey wastewater effects on soil permeability. *J. Env. Qual.* **11**:31-34.

Modler, H.W. 1987. *The use of whey as animal feed and fertiliser. Bulletin No. 212.* International Dairy Federation, Brussels.

Morris, S., J. Nixon, and R. Kilgour. 1985. Whey: feed or fertiliser. *Proceedings of the Ruakuru Farmer's Conference* **37**:113-116.

Mulchrone, S., M.A. Morgan, and G.J. Mullen. 1980. A method for obtaining undisturbed samples from soil columns. *Agronomy J.* **72**:170-172.

Peterson, A.E.., W.G. Walker, and K.S. Watson. 1979. Effect of whey applications on chemical properties of soils and crops. *J.Agric. Fd. Chem.* **27**:654-658.

Radford, J.B. 1992. Utilization of whey and its derivatives as a fertiliser replacement for pasture. *In* P.E.H. Gregg and L.D. Currie (ed.), *The use of wastes and by-products as fertilisers and soil amendments.* Occasional Reports No. 6, Fertiliser and Lime Research Centre, Massey University, Palmerston North.

Radford, J.B., D.B. Galpin, and M.F. Parkin. 1986. Utilisation of whey as a fertiliser replacement for dairy pasture. *New Zealand Journal of Dairy Science and Technology* **21**:65-72.

Radford, J.B., D.B. Galpin, and M.F. Parkin. 1987. Utilisation of whey as a fertiliser replacement for pasture and crops. In *New approaches to effluent disposal.* The Australian and New Zealand Association for the Advancement of Science, Palmerston North.

Robbins, C.W. and G.A. Lehrsch. 1992. Effects of acidic cottage cheese whey on chemical and physical properties of a sodic soil. *Arid Soil Research and Rehabilitation* **6:**127-134.

Robinson, R.K. 1986. *Modern Dairy Technology. Volume 1.* Elsevier Applied Science Publications, Ltd., London.

Ross, M.V. and G.J. Mullen. 1993. A comparison of the effects of cheese whey and conventionalfertiliser application on herbage dry matter production, soil organic carbon content and soil aggregate stability. *Irish Journal of Agricultural and Food Research* **32**:98 (Abstract).

Scott, R. 1986 *Cheesemaking Practice.* Elsevier Applied Science Publications Ltd., London

Sharratt, W.J., A.E. Peterson, and H.E. Calbert. 1959. Whey as a source of plant nutrients and its effect on the soil. *Journal of Dairy Science* **42**:1126-1131.

Sharratt, W.J., A.E. Peterson, and H.E. Calbert. 1962. Effect of whey on soil and plant growth. *Agronomy J.* **54**:359-361.

Sherwood, M. 1993. Impact of agriculture on surface water and groundwater quality. p. 108-118. *In* C. Mollan (ed.), *Water of Life.* Royal Dublin Society, Dublin.

Steel, R.G. and J.M. Torrie. 1986. *Principles and Procedures of Statistics with Special Reference to Biological Sciences.* Edward Arnold Publishers, Ltd. London

Wardlaw, A.C. 1985 *Practical Statistics for Experimental Biologists.* Wiley, Chichester.

Watson, K.S., A.E. Peterson, and R.D. Powell. 1977. Benefits of spreading whey on agricultural land. *Journal of Water Pollution Control Federation* **49:**24-35.

The Use of Peat in Treating Landfill Leachate

H.J. Lyons and T.J. Reidy

REGIONAL TECHNICAL COLLEGE, TRALEE, Co. KERRY, IRELAND

Abstract

The proposal for a Council directive on the landfill of waste by the Commission of the European Communities in May 1991, coupled with the establishment of the Environmental Protection Agency in Ireland in March 1993 has ensured that sites for the disposal of municipal and industrial waste are subjected to more rigorous scrutiny than before. This study has investigated the potential of unmodified virgin peats from high moor and low moor sources for the treatment of leachates from landfill having both domestic and industrial wastes. A steady feed rate of leachate was applied to peat columns overlying drainage media, and the raw leachates and filtrates were analysed periodically [for biological oxygen demand (BOD), chemical oxygen demand (COD), phosphorus (P), ammonia (NH_3), nitrate (NO_3^-), copper, iron, manganese, lead, chromium, arsenic, zinc, mercury, nickel, cobalt, and phenols] until saturation of the peat (when contaminants in the filtrate were greater than those of the raw leachate) had occurred.

The percentage decreases of concentrations of pollutants were computed, and the period to peat saturation for each parameter was calculated. Periodic flow tests checked the potential flow rates through the filter bed. Substantial decreases took place for all parameters, with the exception of manganese. The combination of parameters showed that even though peat can have a substantial affinity for an individual species, its behaviour is different when presented with a mixture of components, because each element competes for the available adsorption spaces on the surface of the peat. The decrease ranged from 53% for mercury to 86% for lead and phosphorus, and the majority of the decreases were in the 70% and 80% ranges. The results are encouraging, and would justify the setting up of a pilot project to test if similar results could be achieved when peat is used as a treatment method in a landfill. Virgin peat would be suitable as a biological filter, as distinct from a chemical filter, should the problem of peat clogging be overcome.

1 Introduction

Historical

The problem of environmental pollution arising from the disposal of metal bearing industrial effluents is a cause of serious concern, and there have been extensive studies of applications of peat for trapping such metals (Chaney and Hundermann, 1978; Dissanayake and Weerasooriya, 1976; Eger *et al.*, 1980; Kadlec and Tilton, 1979; Loxham, 1980; Loxham and Burghardt, 1983; Tinh von Quach *et al.*, 1971). These studies have variously used peat to extract elements such as cadmium, chromium, lead, vanadium, nickel, or copper from industrial waste waters, and the abilities of peat to sorb substantial quantities of these elements have been clearly demonstrated. Peat filtration is an effective method of domestic wastewater treatment (Coupal and Lolancette, 1976; Ekman, 1976; Farnham and Brown, 1972, 1976; Osborne, 1975; Rock *et al.*, 1982, 1984; Sridhar and Coupal, 1973; Stanlick, 1985; Tinh von Quach *et al.*, 1971). These reports indicate that peat successfully decreases BOD and nitrate levels in septic tank effluents, but it is effective in decreasing phosphate levels only when it contains elements such as calcium, aluminium, and iron (Nichols, 1980).

Milled peat filter beds can be effective for the removal of colour in some industrial effluents (Dufford and Ruel, 1972; Poots and McKay, 1980). Studies in Finland, Canada and the U.S.A. have used undisturbed peatland to attenuate industrial waste water (Eger *et al.*, 1980; Farnham, 1974; Guntenspergen *et al.*, 1980; Loxham and Burghardt, 1983; Nichols, 1980; Surakka and Kamppi, 1971; Kadlec and Tilton, 1979; Tyler *et al.*, 1978). The data indicate that BOD can be lowered, and nitrates, phosphates and metals removed from effluent in this manner. However, because the metals are sorbed to the peat to varying depths, migration of such pollutants to groundwater systems cannot be ruled out.

The Use of Peat to Attenuate Contaminants in Landfill Leachate

Landfill leachate contains all the common eutrifiers, such as phosphorus and nitrate, as well as transition elements and heavy metals. The attenuating abilities of high moor and low moor virgin peats for these substances have been investigated using both a single pass and a two pass system Reidy (1994). The aims of that research were:

1, to achieve a substantial lowering of contaminants in the filtrate;
2, to observe the interactions between the various elements in the sorbent mixtures as competition took place for sorption sites on the peat; and
3, to investigate the flow rates of the leachate through the filter beds.

The Choice of Peat

The composition of peat varies with location and depth, even within a given bog. The geological factor of greatest importance in peat genesis is the relationship of the water in the peat deposit to the main groundwater system of the adjacent mineral soils. If the bog water system is continuous with the mineral and groundwater system, the peatland is said to be low moor peat. Peat deposits whose water system is significantly above the groundwater for adjacent mineral soils are referred to as high moor peats (Fuchsman, 1980). The relation of the peat and the groundwater system controls the availability of dissolved inorganic components, and determines whether the peat is low or high moor. In high moor bogs the available water comes directly from rain or snow, and the mineral content of the water is correspondingly low. The botanical assessment of peat involves identifying the plants that grow most vigorously, and these are likely to be the same as the plants whose decomposed remnants form the peat immediately at the surface. High moor peats are characterized principally by mosses (*Sphagnum* spp.), cotton grasses (*Eriophorum* spp.), and heath plants (*Ericaceae*). Low moor peats are characterized by frondiferous mosses (*Hypnum* spp.), reeds (*Phragmites*), sedges (*Carex* spp.), and by woody plants such as willow (*Salix*), birch (*Betula*), and alder (*Alnus*). Plants such as the rush and bulrushes are common to high and to low moor peats. This general classification was further developed to give a total of 38 classes.

The Von Post (Fuchsman, 1980) humification scale relates extents of decomposition and transformations to field conditions, and allows distinction to be made between the two broad classes of peats (Kaila, 1956). This is a ten point scale, in which H1 corresponds to slightly humified peat and H10 corresponds to highly humified peat. The high moor sphagnum peats range from H1 to H7, and the low moor peats range from H1 to H10.

Ash contents, pH, and bulk density values are used in the laboratory to classify peats into their two main groups. The percentage ash (measured on a dry weight basis) for low moor peat ranges from 3 to 14 per cent, the pH ranges from 4 to 7.5 (Fuchsman, 1980), and the bulk density for peat from *Carex* ranges from 0.075 to 0.16 g cm^{-3} (Paivanen, 1969). For high moor peats, the pH ranges from 3 to 5, ash is usually less than 3 per cent, and the bulk density is in the range 0.065 to 0.104 g cm^{-3}.

Leachate Constituents

The quality of a leachate is determined by the type of waste stored in the landfill, and by the leachability of the compounds. Putrescible organic materials in landfill are readily degraded biologically, and volatile fatty acids are formed initially (the acidification phase). The leachate is characterized during this stage by a relatively low pH (5.5 to 6.5), and a high organic pollution, and the chemical oxygen demand (COD) can be as high as 40 000 mg L^{-1}. The acidity can cause many heavy metals to leach out. In the time span between one half and two years the acidification phase will change slowly to the methanogenic phase in which methane is produced, and the pH increases to 7-8.

Objectives of the Present Study

The objectives of the present study were to determine the efficiencies by which columns of high moor and of low moor peats decreased contaminating parameters as the effluents from a municipal landfill site were passed through the columns.

2 Materials and Methods

Choice of Leachate Parameters

Approximately 400 constituents, for which a plausible case could be made for inclusion in a monitoring programme for landfill leachate, are listed in Appendix II of the Federal Register on Solid Waste Disposal (E.P.A. 1991). A two year monitoring programme was carried out on the municipal landfill site in Tralee, Co. Kerry, Ireland, and the averages and the ranges of various parameters in the leachates during that time are given in Table 1. The parameters chosen give indications of the total polluting capacity of the leachate.

Materials.

Based on the geological data available, and data from laboratory analyses, a high moor peat from the Reamore/Lyreacrompane area, which form part of the Stacks Mountain Range in

Table 1 *The concentrations, and their range of selected component parameters in the leachate from the Tralee municipal landfill*

Parameter/units	Average Concentration	Range	
BOD, mg L^{-1} O_2	561.5	120	1587
COD, mg L^{-1} O_2	928.5	70	2100
pH	7.4	6.9	
	8.2		
Copper, mg L^{-1}	0.15	0.01	0.5
Iron, mg L^{-1}	10.7	3	24
Manganese, mg L^{-1}	4.5	1.07	13
Phosphorus, mg L^{-1}	48.7	3.5	141
Ammonia, mg L^{-1}	60	9	173
Nitrates, mg L^{-1}	57	0	210
Aluminium, mg L^{-1}	0.36	0.2	0.85
Lead, mg L^{-1}	0.32	0	0.97
Chromium, mg L^{-1}	0.22	0	0.6
Zinc, mg L^{-1}	1.4	0.1	16
Nickel, mg L^{-1}	0.24	0	0.56
Cobalt, mg L^{-1}	0.26	0	0.8
Phenols, mg L^{-1}	1.13	0.1	5
Tin, mg L^{-1}	15.84	0	113
Total solids, mg L^{-1}	14000	600	2860

North Kerry, and a low moor peat from Raecaol bog Kilcummin near Killarney were used in the present study.

Acrylic plastic cylinders or columns, 2 m x 115 mm (i.d.) were mounted in a vertical position on a concrete structure. An automatic feed system provided a steady feed rate of leachate to the cylinders. Columns A and B contained low moor peat and high moor peat, respectively, in a single pass system (i.e the leachate passed through the peat layer only once). The peat in Column A was taken at depths between 0.5 m and 1.0 m, and was thoroughly mixed. It had a pH value of 5.73, and an ash content of ca 26%. That in Column B was taken from the 0.4 m depth, had a pH of 3.34, and an ash content of 3.3%. Columns C and D contained high moor peat and low moor peat, respectively, in a two pass system (where alternate layers of peat and drainage media were used). The peat in column C was taken from depths between 0.9 m and 1.3 m, and was thoroughly mixed. It had a pH of the order of 4, and the ash content was low (ca 1.1%). The peat in Column D was taken from the 0.5 to 1.0 m depth, and was thoroughly mixed.

Vertical columns containing peat, sand, and gravel were set up (Figure 1). Column A contained 830 mm of loosely packed low moor peat, 245 mm of washed sand (80% with a particle size < 0.3 mm, and ca 45% was < 0.2 mm), and 280 mm of pit gravel (70% with a particle size between 6.3 and 14 mm, and the remainder was less than 6 mm). The sand and gravel were included to provide suction. The top layer (100 mm) of gravel enabled a uniform distribution of the leachate over the entire surface of the peat. Raw leachate was added at a rate of 3 L day^{-1} (i.e. ca 17 cm day^{-1}), and the filtrate was collected for analysis. Analyses were continued on the raw leachate and on the filtrate until the peat had reached

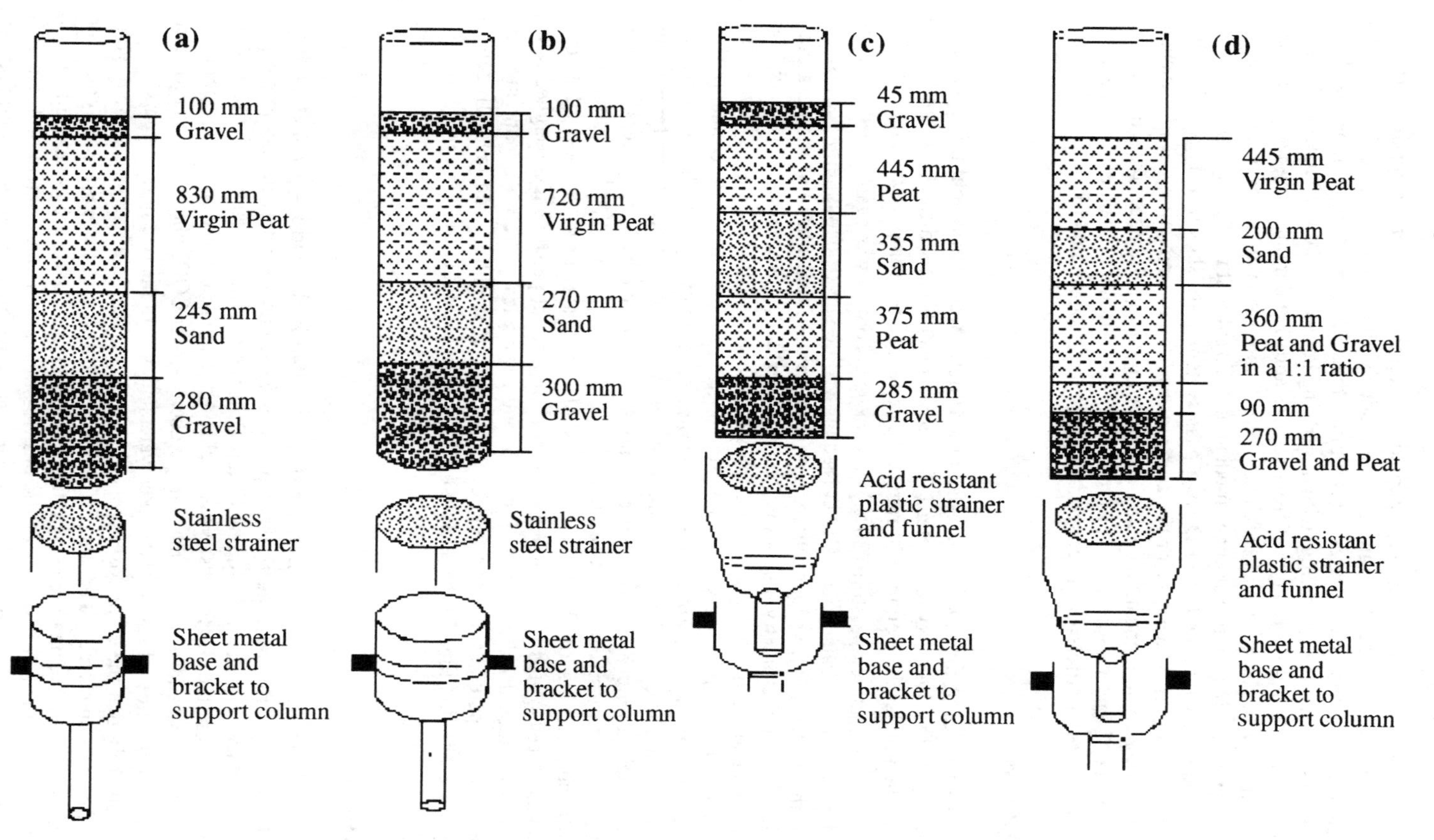

Figure 1 *The arrangements of peat, sand, and gravel layers in Columns A and B (a, b, single pass system) and in Columns C and D (c, d, double pass system). See the text for details*

saturation (i.e when all the negatively charged sites on the peat were saturated with the cations in the leachate). The flow rate was continuously monitored.

Column B had a top layer of gravel (100 mm), then 720 mm of high moor peat from the 0.4 m depth in the Reanmore bog, with underlying layers of sand (270 mm) and gravel (300 mm). A flow rate of 2 L day^{-1} of leachate was added and monitoring was continued until the capacity of the peat was reached. Direct comparisons can be made between columns A and B, using similar drainage media but different peat types.

The set up for Columns C and D (two pass systems) are shown in Figure 1. This allowed investigations of whether or not a greater hydraulic conductivity of the leachate and a greater degree of attenuation could be achieved by using alternate layers of peat and drainage media. Leachate was added (3 L day^{-1}) till saturation was reached. Attenuations were calculated from differences in concentrations in the leachates and filtrates.

Methods Used

Analytical analyses of the raw leachates and the filtrates were carried out using standard procedures. A dissolved oxygen meter (WTW/oximeter) was used to check the oxygen content in the raw leachate and the filtrate on day one, and the samples were incubated in darkness at a constant temperature (20°C). The oxygen content was again checked on day five and the BOD was calculated. Direct reading, visible spectrophotometry was used to measure the COD, reactive phosphate, ammonia, nitrate and phenols. Flame atomic absorption spectrophotometry, using the Perkin-Elmer 280, was used to measure copper, iron, manganese, lead, chromium, zinc, nickel, cobalt and tin. Cold vapour atomic absorption spectrophotometry was employed to measure mercury concentration. Electrothermal atomic absorption spectrophotometry, employing the Thermo-Jarel Ash 11E instrument, was used to measure aluminium concentration. Acetylene flame, electrothermal atomic absorption spectrophotometry, using hydride generation, was employed to measure arsenic concentration.

For measurements of the bulk densities of the peats, peat samples of known dimensions were isolated from the bogs, wrapped in airtight plastic bags, weighed in the the laboratory, then dried at 105 °C for 24 h, and reweighed. The water content of the samples was computed as a percentage of the volume at sampling, and the bulk density was calculated by dividing the dry weight by the constant volume.

3 Results

A comparison is given in Table 2 of the decreases (%) in the various parameters of the landfill leachates after passage through peats in Columns A and B of the single pass system and through peats in Columns C and D of the two pass system. The data show that the peats from the single pass system achieved greater decreases in BOD, ammonia, nitrate, manganese and lead, and that bigger decreases in COD, phosphorus, chromium, total suspended solids, phenols, and nickel were achieved in the two pass systems.

It is unlikely that the extra cost involved in layering peat in a two pass system would be justified in commerce. The data in Table 2 indicate that there were no consistent differences between the treatments. On that basis, both methods were grouped together in order to evaluate the effectiveness of the peat to attenuate the parameters in Table 2. It was found that, in the cases of some of the parameters in some instances, saturation was achieved when the analyses were carried out. The data in such instances were not considered and the results in Table 3 are slightly higher than the averages given in Table 2.

Table 2 *Comparison of decreases (%) in the compositions of various parameters after passing landfill leachates through peats in columns A and B (single pass system), and in columns C and D (double pass system)*

Parameter mg L^{-1}	Decrease (%) Column A	Decr. (%) Column B	Decr. (%) Column C	Decr (%) Column D
BOD	75	86	78	78
COD	75	80	85	75
Phosphorus	65	75	93	84
Ammonia	90	90	78	72
Nitrate	93	78	74	83
Manganese	69	84	15 (-)	23
Chromium	89	80	94	75
Total suspended solids	56	86	77	73
Phenols	76	61	81	89
Aluminium	68	52	62	71
Lead	97	92	71	95
Nickel	30	75	73	69

4 Discussion

Landfill leachate contains all the common eutrifiers, such as phosphorus and nitrate, as well as many transition elements and heavy metals. The small uptake of manganese (see the negative value in Table 2) reflects its relative inability to compete with the other species in the mixture for adsorption sites on the peat. Peat will attenuate phosphorus only when elements such as iron, calcium, and aluminium are contained in the medium. On average the leachate from the Tralee municipal landfill contained 10.7 mg L^{-1} of iron and 0.37 mg L^{-1} of aluminium (Table 1); these interacted with and precipitated the phosphate.

It is important to be aware of the abilities of peat to bind metals when combinations of these are in the in leachates. The data in Tables 2 and 3 indicate that, with the exception of managanase in the the two pass systems (Columns C and D), peat brought about substantial decreases in the various parameters listed. Increasing levels of manganese were found in the filtrate during the first eight days of passage through the peat, and afterwards, until the experiment was completed 18 days later, the manganese content in the eluate exceeded that in the leachate. The pH of the eluate increased during the treatment process, but the fact that other metals did not follow the trend for manganese would suggest that a selective ion exchange process displaced the weakly bound manganese.

The layering of the peat and sand provided very good attenuations, and the results would indicate that this combination can provide a useful method in order to achieve high percentage decreases of pollutants of low concentrations in the landfill leachates. The washed pit sand used had little effect in creating greater suction and improving the hydraulic conductivity of the peat. A high ash content of the peat in Column A (26%) indicates that mineral soil was contained in the peat, and that might suggest that clay in the peat would lengthen the time to saturation of the column, and give improved hydraulic

Table 3 *Overall average decreases, using peat filters, in polluting materials in landfill leachate*

Parameter	Average	Change (%)
BOD (raw)	601.9	81.22
BOD (filtrate)	113	
COD (raw)	968.6	83.91
COD (filtrate)	155.9	
pH (raw)	7.5	7.19
pH (filtrate)	6.9	
Copper (raw) (mg L^{-1})	0.17	81.68
Copper (filtrate) (mg L^{-1})	0.03	
iron (raw) (mg L^{-1})	12.8	79.81
iron (filtrate) (mg L^{-1})	2.6	
Manganese raw) (mg L^{-1})	5.12	40.28
Manganese (filtrate) (mg L^{-1})	3.05	
Phosphorus (raw) (mg L^{-1})	47.9	86.76
Phosphorus (filtrate) (mg L^{-1})	6.3	
Ammonia (raw) (mg L^{-1})	75.9	75.89
Ammonia (filtrate) (mg L^{-1})	18.31	
Nitrate (raw) (mg L^{-1})	74.4	83.63
Nitrate (filtrate) (mg L^{-1})	12.2	
Aluminium (raw) (mg L^{-1})	0.4	62.49
Aluminium (filtrate) (mg L^{-1})	0.14	
lead (raw) (mg L^{-1})	0.34	86.65
Lead (filtrate) (mg L^{-1})	0.05	
Chromium (raw) (mg L^{-1})	0.25	84.29
Chromium (filtrate) (mg L^{-1})	0.04	
Arsenic (raw) (mg L^{-1})	0.00	None present
Arsenic (filtrate) (mg L^{-1})	0.00	
Zinc (raw) (mg L^{-1})	1.60	82.30
Zinc (filtrate) (mg L^{-1})	0.23	
Mercury (raw) (mg L^{-1})	0.21	53.44
Mercury (filtrate) (mg L^{-1})	0.09	
Nickel (raw) (mg L^{-1})	0.24	55.48
Nickel (filtrate) (mg L^{-1})	0.10	
Cobalt (raw) (mg L^{-1})	0.33	79.57
Cobalt (filtrate) (mg L^{-1})	0.06	
Phenols (raw) (mg L^{-1})	1.27	80.29
Phenols (filtrate) (mg L^{-1})	0.25	
Tin (raw) (mg L^{-1})	21.10	82.88
Tin (filtrate) (mg L^{-1})	3.61	
T.S.S (raw) (g L^{-1})	3.19	74.55
T.S.S (filtrate) (g L^{-1})	0.81	

conductivity. Local Authorities have shown reluctance to use peat as cover material at landfill sites because of the danger that thin layers of peat would be ignited during periods of prolonged dry weather, and because of the suggestion that peat has preservative properties and might inhibit the biological degradation of the organic materials in the landfill. Our studies would suggest that layers of peat might be used in combination with the preferred inert covering material, and the peat would provide an inexpensive and efficient means of treating the leachate prior to the final treatment processes. Modern landfills are designed in cellular formation, and provided the cell is filled and capped in the recommended time interval, the extent of metal release should not present a problem.

The findings from this research suggest possibilities for the use of virgin peat as a biological filter, as distinct from a chemical filter. However, although the biological processes continue to be very effective, as time progresses clogging of the peat takes place, and the suspended solids in the leachate cause severe decreases in the flow rate through the peat bed. Further research is required to search for other organic fibrous media which, in combination with peat, could be used to overcome this problem. Our data clearly show the abilities of peat to attenuate the concentrations of pollutants in the leachates. The data in Table 3, for example, outline the efficiency with which iron was removed. Iron decreased almost linearly in the leachate, from 24 mg L^{-1} to 5 mg L^{-1} during passage through the high moor peat in Column C. Peat could thus have uses as iron filters.

It is realised that economic factors will determine the choice of treatment for landfill leachate. Conventional treatments include tertiary processes where adequate provision is made for the treatment of transition and heavy metals, along with the common eutrifiers reflected in BOD, COD, nitrate and phosphate, and the use of reverse osmosis. These costly treatments should be considered only when landfill sites are close to population centres where these facilities are already in use for the treatment of sewage effluent.

The difficulties in procuring land for landfills adjacent to centres of population is causing these to be sited in bogs and wooded areas, and away from cities and towns. It is in this context that peat must be considered as a means of leachate treatment.

The research described in this paper is based on the use of unmodified peat which eliminates the costs associated with the milling and drying processes for other alternatives. The extraction and the fine chopping of the virgin peat can readily be achieved using existing machinery. The removal of the extruder and sod former from peat cutting machines should provide the conveying from the auger to the treatment cell peat of suitable particle size and composition. It is envisaged that leachate should pass through a number of these cells in order to achieve a filtrate that would comply with E.U. guidelines. It would be appropriate to spray the leachate onto the filter media to enable oxidation processes to take place and thereby to assist the treatment process. Because clogging of the media presents problems, further research into the use of fibrous media, such as tree bark, wood chips, sawdust, finely chopped rushes, etc., in combination with the peat might be carried out. An appropriate combination with such media would help to achieve more comprehensive treatment, and provide the additional desirable features of biological filtration, especially in instances where heavy metals do not present a problem. Research along these lines is ongoing at the Regional Technical College, Tralee .

References

Chaney, R. L. and P.T. Hundermann. 1978. Use of peat moss columns to remove cadmium from waste water. *J. WPCF* **51**:17-21.

Coupal B and J.M. Lolancette. 1976. The treatment of waste waters with Peat Moss. *Water Research* **10**:1017-1076.

Dissanayake, C. B. and S.V.R. Weerasooriya. 1976. Peat as a metal trapping material in the purification of industrial effluents. *Intern. J. Environmental Studies* **17**:233-238.

Dufford, J. and M. Ruel. 1972. Peat moss a s an adsorbing agent for the removal of colouring matter. *Proc. 4th. Internat. Peat Congress* (Helsinki) **4**:299-231..

Eger P., K. Lapakko, and P. Otterson. 1980. Trace metal uptake by peat: interaction between a white cedar bog and mining stockpile leachate. p. 542-547. *Proc. 6th Intern. Peat Congress* (Duluth Minnesota).

Ekman E. 1976. The possibility of using peat for the natural environment protection. p. 79-87. *Proc. 5th International Peat Congress* (Poland).

E.P.A. 1991 (Oct.). p. 1 - 141. Federal Register E.P.A. Solid waste disposal criteria. E.P.A. Washington.

Farnham R. S. 1974. Use of organic soils for waste water filtration. p.111-118. *Soil Sci. Society of America, Special Publication No. 6.* Soil Sci. Soc. Am., Madison.

Farnham, R.S. and J.L. Brown. 1972. Use of peat and peat sand filtration media. *Proc. 4th. Internat. Peat Congress* (Helsinki) **4**:271-286.

Farnham, R.S. and J.L. Brown. 1976. Use of peat for wastewater filtration - principles and methods. p. 349-357. In *Proc. 5th. Internat. Peat Congress* (Poland).

Fuchsman, C.H. 1980. *Industrial Chemistry and Technology*. Academic Press, NY.

Guntenspergen G., W. Kappel, and F. Stearns. 1980. Response of a bog to application of Lagoon sewage: The Drummond Project. An operational trial. p. 559-562. *Proc. 6th Intern. Peat Congress* (Duluth).

Kadlec R. H. and D.L. Tilton. 1979. The use of fresh water wetlands as a tertiary waste water treatment alternative. p. 185-212. *CRC. Crit. Rev. Environ. Contr.*

Kaila A. 1956. Determination of the degree of humification in peat samples. *Sci. Agr. Soc. Finland* **28**:18-35.

Loxham M. 1980. Theoretical considerations of transport of pollutants in peats. p. 600-606. *Proc. 6th International Peat Congress* (Duluth).

Loxham M. and W. Burghardt. 1983. Peat as a barrier to the spread of micro-contaminants to the ground water. p. 337-349. *Proc. Int. Symp. on Peat Utilisation* (Minnesota).

Nichols D.S. 1980. Nutrient removal from wastewater by wetlands. p. 638-642. *Proc. 6th Intern. Peat Symp.* (Duluth).

Osborne, J.M. 1975. Tertiary treatment of camp ground waste using a native Minnesota peat. *J. Soil and Water Conservation* **30**:235-236.

Paivanen, J. 1969. The bulk density of peat and its determination. *Silva Fennica* **3**:1-19.

Poots, V.J.P. and G. McKay. 1980. Flow characteristics and parameters relating to the use of peat and wood as cheap adsorbent materials for wastewater purification. *Royal Dublin Soc. Series A* **6**:408-441.

Reidy T.J. 1994. *The use of peat in attenuating landfill leachate*. M. Sc. Thesis. Regional Technical College, Tralee, Ireland.

Rock, C.A., J.L. Brooks, S.A. Bradeen, and J. Struchtemeyer. 1984. Use of peat for on site waste water treatment. *J. Environ. Qual.* **13**:518-529.

Rock, C.A., J.L. Brooks, S.A. Bradeen, and F.E. Woodward. 1982. Treatment of septic tank effluent in a peat bed. p. 116-123. *Proc. 3rd National Symp. on land and Small Community Sewage Treatment.* Soc. Agric. Eng.

Sridhar H. and B. Coupal. 1973. Continuous production of modified peat for pollution control applications. p.1-20. *Proc. 4th Joint A.I.C.H.E - C.S.C.H.E., Chemical Eng. Conf.* (Vancouver).

Stanlick H. T. 1985. Treatment of septic tank effluent using underground plant filters. p.1-28. U.S. Forest Service, 633 W Wisconsin Ave., Wisconsin U.S.A.

Surakka S. and A. Kamppi. 1971. Infiltration of waste water into peat soil. (Finnish - English Summary). *Suo.* **22**:51-58.

Tinh von Quach, R. Leblanc, and L.M. Janssens. 1971. Peat mosas - A natural adsorbing agent for the treatment of polluted waters. p. 99-105. *CIM Bulletin*.

Tyler E .J , R. Laak, E. McCoy, and S.S. Sandhu. 1978. The soil as a treatment system. p. 22-37. *Second National Home Sewage Treatment Symposium.* ASAE Chicago.

Subject Index

—A—

Acid deposition, 278, 287
Acid hydrolysis of soils, 126
Acidification of lake water, 299, 309
Acidification of watershed, 299, 309
Acidity and sorption, 215
Adsorbent IUPAC definition, 190
Adsorption, 26
  IUPAC definition of, 189
Adsorption by soil of
  2,4-D, 35
  atrazine, 35
  CIPC, 36
  lindane, 35
  linuron, 35
Adsorptive IUPAC definition, 190
Afforestation of peatland, 262-271
  ecological impacts of, 263
Agglomeration tests, 238, 239
  impact of H-TS on, 244
Aggrigate stability
  measurement of, 468
'Agri-Soil', 411
Alachlor
  solubilities of, 211, 212
  structure of, 210
Aleization, 259
Alkylammonium ions, 220, 221
Altex valve system, 194
Aluminium hydroxide agglomeration, 244
  impact of H-TS on, 244
Aluminium hydroxide seed, 239
Aluminium pillard clays, 222, 223
  sorption of fulvic acids by, 224
Aluminium smelters, 239
Amine cations, bicyclic, 220, 221
Amino acid contents,
  of FAs, 19
  of HAs, 19
  of XAD-4 acids, 19
Amino acids
  effect of management on, 115
  in marine humic substances, 321, 323
  order of abundences, 115, 145, 146
Ampholite, 12
Anomeric carbon, 18
Anomeric carbon signal, 19
Antarctic diatoms
  extrudates from, 323
Anthropogenic, 5
Anthropogenic organic chemicals (AOC), 5, 26
  biological activity of, 200
  measurement of solubilities, 210, 211
  transportation by dissolved humics, 208, 217
  transportation of, 208
Aquatic humic acid
  pyrene fluorescence and KBr, 43
  pyrene fluorescence in, 43
Aromaticity, in humic substances, 16, 18, 19, 25
Aromaticity and sorption, 215
Arsenal
  herbicide, 199
  structure of, 200
Assert
  herbicide, 199
  structure of, 200
Atrazine adsorption by soil, 36

—B—

Bacillus spp, 361
Bark composts, 363
Base cation ratios
  relationships, 288, 297
Batch slurry
  measurement of sorption, 201
Bayer liquor, 238
  agglomeration tests for, 238, 239, 240
  foaming of, 238, 240, 243
  oxalate breakpoint in, 238
Bayer liquor foaming
  impact of H-TS on, 241
Bayer process, 237
Bayer synthetic liquor, 242
  H-TS in, 241
Benzene polycarboxylic acids, 21
Benzoquinones
  model experiments with, 351
Beringite, a calcined schist, 227, 228
  fractionation of fulvic acids by, 231, 232
  sorption of fulvic acids by, 230, 231
BET surface area, 221
Biochemical oxygen demand, 480
  in waters, 326
Biogeochemical weathering, 281
Biological oxygen demand, 482
  in waters, 327
Biomass, 4
Blanket peatland, 262
Blanket peats
  coniferous trees in, 263
  forestry impact on, 263
BOD, 482
Boreal zone, 266

—C—

CIPC
  adsorption by soil, 36
C3 plants
  decomposition of, 311, 317
C4 plants
  decomposition of, 311, 317
Ca^{2+}-exchanged humic acids
  sorption of imidazolinones by, 204, 206
Caffeic acid, 353
Calluna moorland podzols
  pH of, 282
Carbohydrates
  analyses, 301, 302
Carbohydrates in soils, 299, 300, 301
Carbohydrates in waters, 299, 300, 301
Carbohydrates TFA hydrolysis of, 300, 301
Carbon cycling, 266
Carbon dioxide flux
  measurement of, 311, 312
Carbon nitrogen ratio, 313
Cation exchange capacity
  by isotope dilutiuon, 178
  determination of, 177, 181
  values of, 182, 183
Cations in precipitation and in waters, 291, 297
CEC, 14
Cellular protection against, 340
Cheese whey
  and reclamation of sodic soils, 464
  and soil pH, 464
  and soil physical properties, 464
  applications to soil, 462, 472
  effect on soil physical and chemical properties, 471
  effects on crop yields, 465
  environmental impact of, 466
  fertilizer value of, 463
Chemical oxygen demand, 480, 482
  in waters, 326, 327
Chemical shift, 18
Chlorinated organics, 219
Chlorinated water
  and cancers, 227
Citrate buffer, 192
Clay-derived materials, 226
Clays, 219
Clay-size fraction of soil
  isolation of, 179
COD, 482
Coir fibre
  as soil amender, 373, 374
Compost in control of
  Fusarium, 364
  Phytophthora root, 364
  Pythium, 364
  Rhizoctonia solani, 364
Compost steepages
  and disease control, 363, 364
Composts, 359, 364
  and control of plant diseases, 359, 364
  bark, 363
  mineralization rates, 363
Composts in control of
  Flavobacterium, 360
  Gliocladium virens, 360
  Phytophthora root, 360, 363
  Pythium, 363
  Streptomyces, 360
  Trichoderma, 360
Continuous flow
  measurement of sorption, 201
Continuous flow procedures
  for measuring sorption, 190, 195
Copper sorption, 189, 197
CPMAS ^{13}C NMR, 15, 18. *See* NMR
  for predicting disease suppression of compost, 364
Cuprizone
  development of colour with Cu, 194
Cuprizone in FIA, 193

—D—

D_2O exchange in NMR, 53
DDT
  solubilities of, 211, 212
  structure of, 210
Decarboxylation of humics, 20
Decomposition of soil organic matter
  influences of clays on, 123
Degradation (chemical)
  digest products of, 148, 157
Degradation digests, 22
Degradation (chemical) products of, 152, 153, 154
Delta ^{13}C
  analysis, 161
  of clay fractions, 166
  of coals, 168
  of fulvic acids, 163, 165
  of humic acids, 163, 165
  of peat fulvic acids, 167
  of peat humic acids, 167, 168
  of plant materials, 158, 159
  of soil fractions, 166
  of water humics, 169, 170, 171, 172
  of XAD-4 acids, 163
  values for whole soils, 163, 164
Denitrification, 447
Diafiltration, 193
Dicarboxylic acids, 21
  decanedioic, 21
  ethanedioic, 21
Differential thermal analysis (DTA), 180
Dissolved humic acids
  influence on solubility of atrazine, 209
  influence on solubility of DDT, 209
  influence on solubility of terbuthylazine, 209
Dissolved organic carbon, 10
  in oceans, 320

in soil solution, 418
Dissolved organic matter (DOM)
downward transport of, 320
Dissolved organic nitrogen
in marine humic substances, 322
DMSO, 6
in identification of clays, 180
DMSO/HCl
fractionation of HS, 6
DNA adduct formation, 341, 342
DNA damage
by MX, 340, 341
DOC
increase from afforestation, 270
DOM cycling in polar regions, 320
Dystrophic lake, 300

—E—

EDTA
extractable metals, 34
extractable zinc, 37
Electron probe X-ray micro analysis, 93
spectra of humic acids, 95, 96, 98
Electron microscopy, 93
sample preparation for, 94
Electron spin resonance (ESR), 15
computer simulated spectra, 70
first derivative spectra of, 65, 66, 67
flavonoid A-ring, 70
flavonoid A-ring, oxidation of, 70
flavonoid B-ring, 70
flavonoid B-ring, oxidation of, 70
second derivative spectra of, 65, 66, 67
spectra, 15, 63, 71
Elevated CO_2 levels
effects on soil carbon, 311, 312
Enhanced chemiluminescence (ECL), 326, 332
1,3,5-benzenetricarboxylic acid, 330
1,4-benzoquinone, 330
3,4-dimethoxybenzaldehyde, 330
3-hydroxybenzoic acid, 330
4-hydroxybenzoic acid, 330
4-hydroxybenzophenone, 330
IHSS standards, 330
assay, 327, 328
catechin, 330
catechol, 330
catechol and, 328
gentisic acid, 330
lignin, 330
phenol, 330
quinol, 330
resorcinol, 330
salicylic acid, 330
sinapinic acid, 330
tannic acid, 328, 330
vanillic acid, 330
Epidemiological studies, 226
and water chlorination, 337, 338
Ether functionalities, 25
Exchangable aluminium, 182

—F—

Farmyard manure
as nutrient source, 369, 370, 371
crop response to, 446, 456
mineral nutrients in, 446
as soil conditioner, 369
Ferric stagnopodzol, 124
Flow injection analysis (FIA), 192, 193, 194
Fluorescence quenching, 39, 40
Fractionation, 12
XAD-8/XAD-4 in tandem, 7
Fractionation of HS
DMSO/HCl, 6
Freundlich K values, 206, 207
Freundlich plots, 203, 206
FTIR, 24
Fulbent
sorption of fulvic acid by, 231, 232, 233, 234
Fulvic Acids, 26
^{1}H NMR spectra of, 88, 89, 90
amide protons in, 53
aromatic protons in, 53
cation exchange capacity, 214
COSY spectra of, 53, 54, 57
CPMAS ^{13}C NMR spectra, 15
definition of, 6
enhanced chemiluminescence assay of, 327, 328, 329
estimates of, 33
exchangable protons in, 53
fluorescence spectra of, 83, 86
fractionation of with XAD-8, 6
HMQC spectra of, 57, 58, 59
K_{oc} values for, 42
NOESY spectra, 59
PAH interactions with, 42
reference, 9
ROESY spectra, 59
solubility of AOC, 212
standard, 9
TOCSY spectra of, 55, 56, 57
Fulvic acids and
solubility of AOC, 213
Fulvic acids from drainage waters
abundances of sugars in, 303, 306
neutral sugar contents of, 303, 306
neutral sugar ratios of, 303, 306
Fulvic acids in lake waters
neutral sugar abundances in, 307, 308
Fusarium, 359
Fused aromatic structures, 21

—G—

Gallic acid
dimerisation of, 348
transformations of, 347, 348
GCMS, 22, 23
Gel chromatography, 11
Gel permeation chromatography, 11

Genetic toxicity
and water chlorination, 337, 343
Genotoxic products, 338
Global carbon cycle, 3
role of peats in, 266
Global carbon cycle models, 124
Global warming, 5
Glycine
in marine humic substances, 322
GPC, *see* Gel permeation chromatography
Grassland, 15
Grey brown earth, 124

—H—

$[H^+]$ v $[Ca^{2+}]$-deposition
relationship with soil pH_{CaCl2}, 285
$[H^+]$ v$[Ca^{2+}]$-depoisition
relationship with soil pH_{CaCl2}, 286
H^+-humic acids
isotherms for sorption of imidazolinones, 203, 204, 205
Healing effects oh humic substances
biochemistry of, 346, 355
Heavy metals
acetic acid extractable, 427, 429
and dissolved humic acid, 418
aqua regia extractable, 427
domestic sources of, 426
EDTA extractable, 427, 429, 430, 438, 442
in sludge treated soil, 418
Heavy metals from sludges, 430, 435
concentration in herbage, 430, 433
Herbage dry matter production, 470
Herbicide phytotoxicity, 36
High altitude podzols, 253, 254
High moor peats, 476
characteristics of, 476
plants contributing to, 476, 477
Hippocrotes
and the quality of drinking water, 226
Horseradish peroxidase, 327
Humic, 5, 18, 19
Humic acid degradations
degradation products, 154
methylation of, 150
with alkaline permanganate, 150, 152, 153, 154
with sodium amalgam, 149, 151, 152, 153, 154
with sodium amalgam, 152, 153, 154
with sodium in liquid ammonium, 150, 152, 153, 154
Humic acid fraction, 6
Humic acid radicals, 63
ascorbic reduction and, 63, 71
ESR spectra of humic acids, 63, 71
Humic acids, 18, 19, 25
^{13}C NMR studies of, 15, 130, 132, 133
^{1}H NMR, 19
^{1}H NMR spectra of, 88, 89, 90
^{31}P NMR specrum of, 104, 105
activated carboxyls in, 206
and sludge applications, 417
Ca^{2+}-exchange of, 200
cation exchange capacity, 206, 214
definition of, 6
enhanced chemiluminescence assay of, 327, 328, 329
ESR spectra in alkali, 71
estimates of, 33
extraction of, 8, 9, 10, 191
fluorescence spectra of, 83, 85
fractionation of, 6, 11, 12, 191
fractionation of with DMSO, 6
fractionation of with XAD-8, 6
HA/FA ratio, 33
heavy metals in, 418
isolation of, 64, 126, 149, 200
neutral sugars of, 18
peptide structures in, 18
properties of, 182
pyrene interactions with, 41
reference, 9
semiquinone moieties in, 71
solubilities of AOC in, 211, 212
sorption of Cu by, 196, 197
standard, 9
Humic acids from drainage waters
abundances of sugars in, 302, 306
neutral sugar contents of, 302, 306
neutral sugar ratios of, 302, 306
Humic acids in lake waters
neutral sugar abundances in, 307, 308
Humic Substances, 5, 10, 12, 18, 20, 23
1-D NMR, 19
^{1}H NMR spectroscopy of, 84, 88, 89, 90
2-D COSY, 20
2-D HMQC, 20
2-D NMR, 19
2-D TOCSY, 20
adsorption to clays, 26
adsorption to hydroxides, 26
amino acid contents of, 19, 25, 115, 145, 146, 321, 323
amino acids in, 20
aromaticity of, 18
as surfactant, 11
'backbone' structures of, 20, 148
building blocks of, 20
cation bridging, 8
cation bridging of, 7
'core' structures of, 148
'core-structures', 47
as micellar structures, 39
cage-like structure, 42, 43, 44
carbohydrates bound to, 300, 301
chemical degradations of, 47
compositions of, 14
concepts of structures, 24
CPMAS ^{13}C NMR spectra of, 15, 329
critical micelle concentration, 11, 39

decarboxlation of, 20
definitions of, 5
degradations of, 148, 157
elemental analyses, 47
enols in, 26
enzymatic attack, 25
ESR of, 14
ether linkages, 25
extraction with DMF, 9
extraction with DMSO, 8
extraction with DMSO/HCl, 9
extraction with sodium hydroxide, 8
extraction with sodium pyrophosphate, 8
FABS of, 23
Fatty acids in, 25
fluorescence of, 84
fractionation of, 11, 84
fractionation of by charge density difference, 8
fractionation of by dialysis, 12
fractionation of by electrophoresis, 12
fractionation of by gel chromatography, 11
fractionation of by isoelectric focusing, 12
fractionation of by organic acids, 11
fractionation of by ultrafiltration, 12
fractionation of by resins, 327
FTIR of, 24
fused aromatic structures from, 23
gel filtration, 39, 40
good organic solvents for, 9
HCl/HF treatment of, 9
healing effects of, 346, 355
hydrocarbon structures in, 25
hydrogen bonding in, 7, 12
hydrolysis of, 148
hydrophilic neutral, 7
hydrophobic cage-like structures, 40
hydrophobic neutrals, 7
hydrophobic properties, 25
hydrophobic sites in, 24
in polar regions, 319
in polar waters, 323
in tropical soils, 176, 184
infrared spectroscopy of, 84
isolation from water, 7-11, 209, 301
isolation from water with reverse osmosis, 10
isolation from water with XAD resins, 10
isolation fron soil, 7
isolation of with XAD-8/XAD-4, 7
LDMS, 23
lignin sources, 25
MALDI of, 23
mechanisms of degradation, 23, 24, 149-154
micelle like aggragates, 11
mutagenic compounds from, 226
neutral sugars in, 20, 75, 77-81, 109-114, 144-145, 303-306, 308
NMR of, 14
NMR applications to, 18-20, 46
NMR resonance assignments, 18
olefines in, 25
oxidative degradation of, 22, 25
partition-like sorption, 43
polydispersity of, 26
purines in, 20
pyrene fluorescence quenching, 40
pyrimidine in, 20
pyrolysis of, 23
Raman spectroscopy of, 24
range of acid strengths in, 25
reductive degradation of, 22, 25, 149-154
resistance to degradation of, 4
roles in soils and water, 5
salting out of, 13
sample preparation for isolation from soil, 94
selective precipitation of, 13
sequential degradation procedures, 23
sequential extraction of, 8, 50
sequential/exhaustive extraction, 9
solvent properties for, 7
solid/gel conformations, 96, 97, 98
solution conformations of, 8
spectroscopic studies, 47
sugar contents of, 25
titration data of, 24, 25
two radical species in, 71
XAD-4 acids, 7
Humic substances chemical degradation by
alkaline cupric oxide, 22
alkaline permanganate, 20, 150, 152, 153, 154
BF_3, 20
phenol, 22
sodium amalgam, 20, 149, 151, 152, 153, 154
sodium in liquid ammonia, 150, 152, 153, 154
sodium sulphide, 22
zinc dust distillation, 22
zinc dust fusion, 22
Humic substances compositions of
from elemental analyses, 14
Humic substances compositions of from
spectroscopy
ESR, 15
IR, 14
NMR, 15
Raman, 14
Ultra violet-visible, 15
Humic substances degradation products by
GCMS, 22
Humic substances isolation
solvent properties for, 7
Humic substances from graves
amino acids in, 144
cation exchange capacities of, 143
cation exchange capacity of, 138, 141
DMSO treatment of, 143
E4/E6 ratios, 139, 140, 142
elemental analyses, 142
fulvic acid fraction, 138
isoaltion in DMSO, 138
isolation of, 137, 138
potentiometric titration of, 138
sugar analyses, 139
yields of, 140, 141

Humic substances from podzols, 74
E4/E6 ratios, 74, 77
elemental analyses, 75, 76
isolation and fractionation, 74
neutral sugars in, 75, 79, 80, 81
UV/VIS spectroscopy, 74, 78
Humic substances in aqueous solution
bromide suppression of pyrene fluorescence, 42
pyrene fluorescence in, 42
Humic substances in drainage waters
amino acids in, 110, 114, 115
CPMAS ^{13}C NMR of, 110, 116, 117
δ^{13}C values of, 109, 111, 112
δ^{15}N values of, 109, 111, 112
effect of management on, 113
elemental analyses of, 109, 111
fulvic acids in, 109, 111, 112
humic acids in, 109, 111, 112
infrared spectroscopy of, 110, 116, 117
isolation of, 109
neutral sugars in, 109, 110, 113, 114
neutral sugars order of abundences, 113
recovery from, 109, 111
XAD-4 acids in, 109, 111, 112
Humic-type substances
impact on Bayer process, 237
isolation from the Bayer liquor, 239, 240
Humification, 18
Humified human body, 137
Humified soil organic matter
isolation of, 178
Humin, 6
definition of, 6
isolation of in DMSO, 6
Humin in tropical soils, 181
DMSO extraction of, 181
Humus, 4
Hydrophilic,
acids, 10
bases, 10
neutrals, 10
Hydrophobic
acids, 10
bases, 10
neutrals, 10
Hydrophobic bonding, 11
Hydrophobic fulvic acids
isolation of, 126
XAD resins in, 127
Hydroquinones and amino acids
healing effects of interactions, 355
Hydroquinones and proteins
healing effects of interactions, 355
2-Hydroxy-1,4-naphthoquinone, 350
3-Hydroxythymoquinone, 350, 351
6-Hydroxythymoquinone, 350, 351
Hydroxyquinones
additions of glycine, 355

—I—

Identification of clays, 180
Identifications of digest products, 152, 156
Iodine retention in soil, 35
4-Iodophenol
as chemiluminescence enhancer, 327
Imidazolinone herbicides, 199, 207
Arsenal, 199
Assert, 199
mechanisms of sorption, 206
Scepter, 199
Imidazolone structures
pK_a and protonation of, 205
Inferences from digest products, 152, 156
Inorganic carbon, 182
Interment sites, 136, 157
humic substances from, 136, 157
Irish podzols, 249, 260
Iron oxides
amounts in soils, 182
determination of, 179
Iron pillard clays, 221
fulvic acid sorption by, 221, 222
humic acid sorption by, 222
Iron/aluminium pillard clays, 223, 224
Isolation of humic substances
solvent properties for, 7
Isoproturon
solubilities of, 211, 212
structure of, 210

—K—

Keggin ion, 231, 233

—L—

Labile plant materials, 4
decomposition of, 4
Lake Skjervatjern, 300, 309
Landfill
putrescible organics in, 477
Landfill leachate, 478, 480, 481
metals in, 478, 480, 481
Landfill leachates, 482
oxygen demands and metal contents, 482
Lignin, 16, 23, 27
^{13}C NMR chemical shift assignments, 16
^{13}C NMR of, 15
Lime stabilized sludge
metal contents from in soil solution, 416
Lindane adsorption by soil, 36
Linuron adsorption by soil, 36
Lipophilic organics, 35
Liquid scintillation counting
for measuring desorption, 201, 203
for measuring sorption, 201
Low altitude podzol, 255, 256
Low metal sludges, 441, 442
Low moor peats 476

characteristics of, 476, 477
plants contributing to, 476, 477
Luminol, 327

—M—

MALDI, 23, 24
Marine humic substances
and phythoplankton, 320
hydrophilic components, 323
hydrophobic components, 323
isolation of, 321
of algal origins, 323
origins of, 320
sorption on resins, 323, 324
UV radiation and, 320
XAD isolates, 321
Mean residence time, 4
Melamine resins
for electron microscopy, 94
Metal rich sludges, 440, 442
Metals in landfill leachate, 478, 480, 481
2-Methoxy-1,4-benzoquinone
transformations of, 347
Methylation of humic substances, 150
Montmorillonite, 219
Moorland podzols, 279
Moss
1-D NMR of moss humic substances, 51
humic substances from, 51, 52, 53
MX, 338
absorption of, 339
as risk to human health, 342
experiments with in vitro, 339
experiments with in vivo, 339
MX and mutagenicity, 227
MX chlorination product, 227

—N—

Natural abundance of ^{13}C
in aquatic humic substances, 159
in atmospheric CO_2, 159
in C3 plants, 158, 159
in C4 plants, 158, 159
Natural organic matter (NOM)
hydrophobic interactions with, 41
PAH interactions with, 41
Net primary production, 3
NPP, 4
Neutral sugars,
of humic substances, 8-20, 75, 79-81, 109-114, 144-145, 303-306, 308
Nitrogen fixation
by white clover, 447
NMR, 15, 16, 18, 20, 24, 53, 84, 88-90, 104-105
anomeric carbon, 18
chemical shift assignments, 15, 19
CPMAS ^{13}C assignments, 15
of humic substances, 19
1-D NMR, 19, 49
2-D COSY, 20, 53, 54, 57
2-D HMQC, 20, 57-59
2-D NMR, 19, 55-59
2-D TOCSY, 20, 55-59
^{1}H NMR of humic substances, 19
^{31}P NMR chemical shifts of
choline, 101
inorganic orthophosphate, 101
orthophosphate esters, 101
pyrophosphate, 101
^{31}P NMR of humic substances
chemical shift data, 103
preparation of samples, 102
^{31}P NMR spectroscopy
of humic acids, 100, 105
NMR and humic substances chemical shift data
heteronuclear experiments, 49
homonuclear experiments, 49
information from, 48, 49, 50
sample preparation for, 50

—O—

Oakwood podzols, 256, 257, 258, 259
Ombotrophic peats, 279
Organic macromolecules, 4
non-humic, 4
Organic manures
effects on crop production, 446, 447, 449, 450
effects on white clover content, 451, 456
effects on white clover stolon, 456, 457
Organic manuring, 375, 378
crop yields from, 375, 378
Organic matter
amounts present, 181
cation exchange capacity of, 177
estimates by Walkey black method, 32
extraction of, 181
light fraction of, 177
Organic matter fractions
radiocarbon dating of, 129, 131
Organic matter modelling
Century model, 123
Rothamsted model, 123
Oxalate breakpoint
determination of, 240, 241
impact of H-TS on, 240, 242
Oxidative deamination
mechanisms of, 354, 355
Oxisol, 124

—P—

Particulate amino acids
and chlorophyll a, 323
Particulate organic matter (POM)
in oceans, 320
Peat, 475
a resource, 367
in build up of soil organic matter, 378, 379
as CO_2 sink, 368

as soil amender, 371
as soil conditioner, 379
in treating landfill leachate, 475
Peat chemistry
relationship to pollutant levels, 289, 290
Peatlands
drainage of, 267
organic acidity in, 269
phytomass production in, 267
rehabilitation of, 271
sea salts, effects on, 269
sulphate and nitrate deposition in, 269
surface runoff from, 268
surface water quality from, 268
Peats
and climate regulation, 266
and global carbon cycle, 266
effects of cultivation on, 265
erosion of, 264
Japanese birch on, 265
Lodge pine on, 265
methane from, 266
mineralization of, 264
oxidation of, 264
phosphorus in, 264
physical and chemical properties of, 263
Scots pine on, 265
Sitka spruce on, 265
subsidence in, 264
Pesticides
K_d values, 35
pH and solubilities, 212
pH Soil, 279
Phenol, 22, 23
Phenolic, 21
Phenolic hydroxyl, 22
Phenolic moities
autooxidation of, 71
Phragmites peat, 265
Phthalic acid, 22
Phytophthora root, 359, 361
Phytoplankton
as source of organic material, 320
Phytotoxicity of herbicides, 36
Pillard clays, 219, 228
Mössbauer spectra of, 222
iron acetate, 221
pillaring process, 229
preparation of, 220
sorption of fulvic acids by, 230
surface area of, 221, 222
X-ray diffraction of, 229, 231
Plant litter
analysis of, 313, 314, 315
Plant remains
resistance to degradation, 4
Podzol
formation of, 249
hill, 251
horizons of, 250
low altitude, 256
mountain and hill, 252
raw humus in, 249, 250
rolling lowland, 252
vegetation, 249, 257
Podzol ecosystem relationships, 255, 256
Podzols
A_h horizon, 73
B_h horizon, 50, 73, 74
high altitude, 253, 254
horizons of, 257
low altitude, 255
Podzols from granite sites, 282
Podzols under oak, 83
Polynuclearhydroxy metals, 220, 221
Polysaccharide, 5
POM cycling in polar regions, 320
Proteinaceous compounds, 125
Protocatechuic acid, 64
Protocatechuic acid
transformations of, 347, 355
Protocatechuic acid semiquinone radical
effects of ascorbate on, 67, 68
Pulp mills
MX from, 227
Purine, 20
Purpurogallin-8-carboxylic acid, 348
Putrescible organics
in landfill, 477
Pyrazone adsorption by soil, 36
Pyrene fluorescence, 40
bromide quenching of, 40
Pyrene fluorescence quenching, 44
Pyrolysis, 23
Fisher Curie point pyrolysis, 23
Pyrophosphate, 15
Pythium, 361

—Q—

Quercertin, 64
reduction of semiquinone radical by ascorbate, 69
Quinones and peptide adducts, 354
Quinonoid structures, 350
Quinonylbenzene
as model for formation of humic substances, 350

—R—

Radiocarbon dating, 127
Raman Spectroscopy
Surface-Enhanced Raman Spectroscopy, 15
Refactory DOM, 323
Resin, 18
XAD-8, 6
Reverse osmosis, 10
Rhizoctonia solani, 361, 362

—S—

Salmonella typhimurium
in Ames test, 226

Salt
  influence of solubilisation of AOC, 212
Sample preparation
  from soils, 161
  from waters, 161
Sanskrit lore
  and impure water, 226
Scanning electron microscopy (SEM), 93
Scanning tunnelling electron microscopy
  (STEM), 93
Scepter
  herbicide, 199
  structure of, 200
Schists, 226
Semiquinone radicals, 65, 68
Sequential degradation procedures, 23
Settlement lake
  Bayer cycle, 239, 240
  isolation of humic type substances, 240
Sewage sludge
  and biodegradable organics, 412
  and crop growth, 384, 407
  and crop yields, 388, 389, 390, 391, 396,
    398, 400, 402
  and heavy metal contents of soils, 425, 434
  and metal movement, 412
  disposal at sea, 439
  effect on barley, 413
  effects on beans, 399
  effects on clover, 393, 394, 398, 402, 403,
    439, 440
  enhanced metals in herbage from, 443, 444
  EU directives on, 438
  heavy metals in, 438, 440
  husbandry methods with, 392, 393, 395
  metal bonding of, 412
  metals in, 385, 386, 387, 389, 390, 391, 396,
    398, 400
  phosphorus in, 439
  properties, 415
Sewage sludge and heavy metals
  concentration in soil, 428
  permitted levels of heavy metals, 426
Simazine
  solubilities of, 211, 212
  structure of, 210
Smectite clays, 219
Snape burial site, 136, 157
Sodium alginate
  as soil conditioner, 372, 373
Sodium amalgam, 21, 23
Sodium ascorbate, 64
Sodium oxalate cake
  extraction of humic type substances from, 240
  in Bayer process, 239
Sodium pyrophosphate, 15
Soil, 4
  CO_2 flux in, 4
Soil acidification, 279
Soil aggregates
  influence of fungi on, 122
Soil analyses
  of fractions, 280
  pH measurement, 280
  X-ray diffraction, 279
Soil carbohydrates, 34
Soil carbon loss
  rate of, 5
Soil decalcification, 126
Soil drainage waters
  neutral sugar contents of, 302, 307
Soil fractionation, 178
Soil inorganic components
  differential thermal analysis (DTA) of, 181
  X-ray diffraction of, 180
Soil organic carbon
  as supplier of nutrients, 31
  determination of, 178
  global estimates of, 3
  influence on soil applied pesticides, 36
  mean residence time, 4
  removal of, 179
Soil organic matter (SOM)
  and root extrudates, 123
  and soil structure, 122, 123, 369
  carboxylic groups of, 122
  cultivation and, 5, 122
  depletion of, 5
  effects of peat on, 380
  fractionation of, 125, 126
  gums in, 122
  in clay rich soils, 107
  isolation of, 125, 126
  labile plant components, 4
  organic nitrogen in, 122
  polysaccharides in, 122
  radiocarbon dating of, 123
  refractory fraction of, 5
  uses of XAD resins in, 125
Soil respiratory activity
  data for, 313, 314
  measurement of, 312
  rates of, 316, 317
Solvent
  IUPAC definition, 190
Sorption
  IUPAC definition of, 189
Sorptive
  IUPAC definition of, 190
Sphagnum peat, 362
Stagnohumic gley, 124
Stream water chemistry, 289, 297
Sugar ratios, 80, 81
  in drainage waters, 302, 306
Sutton Hoo burial site, 136, 157
Synthetic humic acids
  health effects of, 353, 354, 355

—T—

Tannin
  ^{13}C NMR of, 15
Terrestrial biomass, 4
Thymoquinones, 348

o-, *p*-, tautomerism in, 349
Titration data
of HS, 25
Trichoderma, 362
Trifluralin
solubilities of, 211, 212
structure of, 210
Trihalomethanes in drinking water, 227
Tris metal chelate, 221, 223
Tropical soils, 128
radiocarbon analysis of, 128, 129
Tryptophan
transformation of indoleacetic acid, 355
Typhoid and drinking water, 226

—U—

Ultrafiltration, 10, 12
Ultrafiltration cell, 193
Upland soils, 124
Uronic acids, 34
UV, 15

—V—

Vanillic acid
transformations of, 346, 348
Vertisol soil, 124
Void volume calculation, 195

—W—

'Wash-in', curves, 191, 195, 196
'Wash-out', curves, 191
Water
colour removal from, 227
Water chlorination, 227, 337, 343
Water disinfection
chlorination products, 337, 338
Wet Ca^{2+}-depoisition
relationship with soil pH_{CaCl2}, 285, 286
Wet H^{+}-depoisition
relationship with soil pH_{CaCl2}, 282, 283, 285, 286

—X—

XAD-4 acids
^{1}H NMR spectra of, 88, 89, 90
cation exchange capacity, 214
CPMAS ^{13}C NMR spectra, 15
enhanced chemiluminescence assay of, 327, 329
fluorescence spectra of, 83, 86, 87
XAD-4 acids and
influences on solubilies, 213
XAD-4 acids from drainage waters
neutral sugar contents of, 303, 306
XAD-4 acids in lake waters
neutral sugar abundances in, 308
XAD-8 and XAD-4 resins in tandem, 9

—Z—

Zinc equivalent, 385
Z-MX, 338